10th International School-Conference on Materials for Extreme Environment (MEEDPA10)

Development, Production and Application

IOP Conference Series: Materials Science and Engineering
Volume 130

Moscow, Russia
19 – 23 October 2015

Editors:

Boris A. Kalin
Gennady N. Elmanov
Pavel S. Dzhumaev

ISBN: 978-1-5108-2415-7
ISSN: 1757-8981

Printed from e-media with permission by:

Curran Associates, Inc.
57 Morehouse Lane
Red Hook, NY 12571

Some format issues inherent in the e-media version may also appear in this print version.

Printed by Curran Associates, Inc. (2016)

For permission requests, please contact the Institute of Physics
at the address below.

Institute of Physics
Dirac House, Temple Back
Bristol BS1 6BE UK

Phone: 44 1 17 929 7481
Fax: 44 1 17 920 0979

techtracking@iop.org

Additional copies of this publication are available from:

Curran Associates, Inc.
57 Morehouse Lane
Red Hook, NY 12571 USA
Phone: 845-758-0400
Fax: 845-758-2633
Email: curran@proceedings.com
Web: www.proceedings.com

Table of contents

Volume 130

10th International School-Conference on Materials for Extreme Environment: Development, Production and Application (MEEDPA10)
19–23 October 2015, Moscow, Russia

Accepted papers received: 13 April 2016
Published online: 18 May 2016

Preface

Papers

Structural materials: steels, Zr- and V- alloys, refractory alloys

Technology: brazing, welding, sintering, ion-plasma treatment

Methods of characterization and modeling of materials

IOP Conf. Series: Materials Science and Engineering **130** (2016) 011001 doi:10.1088/1757-899X/130/1/011001

10th International School-Conference «Materials for Extreme Environment: development, production and application»

Moscow, Russia, 19-23 October 2015

Organizers:

National Research Nuclear University «MEPhI»

Center of Nuclear Systems and Materials

Scientific and Technical Material Scientists Society

Preface

These proceedings are a collection of the selected articles presented at the 10th International School-Conference for young scientists and specialists "New Materials", meeting "Materials for Extreme Environment: development, production and application" (MEEDPA10), held in Moscow, Russian Federation on 19-23 October 2015. This conference is regularly organized by the *Center of Nuclear Systems and Materials* (NSM) of *National Research Nuclear University MEPhI* (*Moscow Engineering Physics Institute*), as well as the *Scientific and Technical Material Scientists Society* (MSS), Russian Federation.

The School-Conference was sponsored by the Competitiveness Growth Program of MEPhI ("5–100" Russian Academic Excellence Project - http://5top100.ru).

One of the main goals of this School-Conference was to train young scientists in matters of interdisciplinary scientific communications. Meetings were held in the form of review lectures of the leading scientists with free discussion and personal oral and poster presentations of the participants.

The general scope of the School-Conference covered the materials and nuclear fuel for PWR/BWR and fast breeder reactors.

The conference was attended by 143 participants from 8 countries (Russia, Belarus, Ukraine, Kazakhstan, France, UK and USA). During the four days of the conference 14 invited lectures, 25 oral and 47 poster presentations were delivered. However, only some papers were qualified for being published in the Open Access IOP Conference Series: Materials Science and Engineering (MSE) journal.

Based on the peer-review process, of the reports presented at the school-conference, 66 selected papers were included in this volume. These papers present new research in the various fields of Materials Science and Technology for Nuclear Engineering. We sincerely hope that the reports presented in this volume will contribute to the advancement of knowledge not only in these fields, but also others.

All papers in the issue are grouped into four sections, as well as in the conference program: (1) Structural materials: steels, Zr- and V- alloys, refractory alloys; (2) Functional materials and nuclear fuel; (3) Technology: brazing, welding, sintering, ion and plasma treatment; (4) Methods of research and modeling of materials.

The organizing committee of the conference thanks all the participants for their fruitful work and personal contribution to the development of these conference proceedings.

The Editors

Dr. Boris A. Kalin
Dr. Gennady N. Elmanov
Dr. Pavel S. Dzhumaev

LISTS OF COMMITTEES

International advisory committee

Chair: Bannikh Oleg A. - IMMS RAS (Russia)

Alekseev Pavel N. - NRC «Kurchatov Institute» (Russia)
Vladimirov Pavel - KIT (Germany)
Ganchenkova Maria G. - NRNU «MEPhI» (Russia)
Francis A. Garner - REC (USA)
Dub Aleksey V. - ROSATOM (Russia)
Kalin Boris A. - NRNU «MEPhI» (Russia)
Maksimkin Oleg P. - Institute of Nuclear Physics (Kazakhstan)
Lavrenuk Petr I. - JSC «TVEL» (Russia)
Salikhov Sergei V. - Minobrnauki of Russia
Troyanov Vladimir - ROSATOM (Russia)
Tuzov Alexander A. - JSC «SSC RIAR» (Russia)
Hannu Hänninen - Aalto University (Finland)
Lin Shao - Texas A&M University (USA)
Michael Short - MIT (USA)
Shtrombah Yaroslav I. - NRC «Kurchatov Institute» (Russia)
Smagul Karazhanov - IFE (Norway)

Technical program committee

Chair: Kalin Boris A.- NRNU «MEPhI» (Russia)

Ageev Valeriy S. -OJSC "VNIINM" (Russia)
Baranov Vitaly G. - NRNU «MEPhI» (Russia)
Chernov Ivan I.- NRNU «MEPhI» (Russia)
Dolgov Alexei B. - OJSC "TVEL" (Russia)
Grigoryev Evgeniy G. - NRNU «MEPhI» (Russia)
Novikov Vladimir V.- OJSC "VNIINM" (Russia)

Risovanyi Vladimir D. - State Corporation "Rosatom" (Russia)

IOP Conf. Series: Materials Science and Engineering **130** (2016) 011001 doi:10.1088/1757-899X/130/1/011001

Peer review statement

All papers published in this volume of *IOP Conference Series: Materials Science and Engineering* have been peer reviewed through processes administered by the proceedings Editors. Reviews were conducted by expert referees to the professional and scientific standards expected of a proceedings journal published by IOP Publishing.

High-strength economically alloyed corrosion-resistant steels with the structure of nitrogen martensite

O Bannykh, V Blinov and E Lukin

Institute of Metallurgy and Material Science RAS, Leninskiy av. 49, Moscow, 119991, Russia

Email: bannykh@imet.ac.ru

Abstract. The use of nitrogen as the main alloying element allowing one both to increase the corrosion resistance and mechanical properties of steels and to improve their processability is a new trend in physical metallurgy of high-strength corrosion resistant steels. The principles of alloying, which are developed for high-nitrogen steel in IMET RAS, ensure the formation of the structure, which contains predetermined amounts of martensite (70-80%) and austenite (20-30%) and is free from δ-ferrite, σ-phase, and $Cr_{23}C_6$ carbide. These principles were used as the base for the creation of new high-strength corrosion-resistant weldable and deformable 0Kh16AN5B, 06Kh16AN4FD, 08Kh14AN4MDB, 09Kh16AN3MF, 27Kh15AN3MD2, 40Kh13AN3M2, and 19Kh14AMB steels, which are operative at temperatures ranging from -70 to 400°C. The developed nitrogen-containing steels compared with similar carbon steels are characterized by a higher resistance to pitting and crevice corrosion and are resistant to stress corrosion cracking. The new steels successfully passed trial tests as heavy duty articles.

1. Introduction

New and very wide possibilities of the application in different industrial branches are opened for corrosion-resistant steel with the structure of nitrogen martensite. The urgency of such steels, which can reliably operate under the joint actions of high static, cyclic, and dynamic loads as well as cavitation and corrosion-active media, is determined by the fact that the reserves of improvement in the level and combination of the above properties of conventional stainless steels are almost exhausted. The most widely used stainless steel with the structure of nitrogen martensite is the 1Kh15N5AM3 steel (the RussianVNS-5 grade and the international AM-350 and AM-355 grades) [1, 2]. However this carbon containing steel in large sections tends to the precipitation of grain-boundary chromium carbides, which abruptly decreases the crack resistance at low temperatures (CRT = 5 J/cm^2) and the resistance to stress corrosion cracking in salt spray chambers (σ_{cc} = 300 MPa). Furthermore, this steel contains expensive molybdenum.

2. Results

As a result of studying the effects of alloying elements as well as heat treatment and thermomechanical treatment regimes on the structure, physicomechanical, chemical, and technological properties, our Institute developed new high-strength corrosion-resistant weldable and deformable 0Kh16AN5B, 06Kh16AN4FD, 08Kh14AN4MDB, 09Kh16AN3MF, 27Kh15AN3MD2, 40Kh13AN3M2 and 19Kh14AMB steels with the structure of nitrogen martensite (Table 1) [3, 4].

IOP Conf. Series: Materials Science and Engineering **130** (2016) 012001 doi:10.1088/1757-899X/130/1/012001

Table 1. Mechanical properties of the new high-strength corrosion-resistant Cr-Ni-N steels.

Steel	Treatment*	Mechanical properties			
		σ_B (MPa)	$\sigma_{0.2}$ (MPa)	δ (%)	Ψ (%)
05Kh16AN5B(0.13% N)	H.R. + Q. 1000°C + T. 400°C	1530	1320	23	68
	C.R. ε = 25%	1850	1760	10	46
08Kh14AN4MDB(0.19% N)	H.R. + Q. 1050°C + T. 500°C	1664	1466	19	61
	Q. 1050°C + C.R. ε = 25%	1920	1877	14	53
09Kh13AMB (0.21% N)	H.R. + Q. 950°C – air	1928	1625	7	16
19Kh14AMB (0.15% N)	H.R. + Q. 950°C - air	1949	1548	8.5	30
40Kh13AN3M2 (0.13% N)	F + C.T.-196°C + T. 200°C	2083	1454	12	10
09Kh16AN3MF (0.19% N)	H.R. + Q. 1090°C + C.T. -70°C + +T. 450°C	1780	1546	16	49
	H.R. + Q. 1090°C + T. 200°C + +T. 400°C	1785	1500	17	59
27Kh15AN3MD2 (0.13% N)	H.R. + C.T. -196°C + T. 400°C	1880	1740	12	43

*H.R. – hot rolling, F. – forging, C.R. – cold rolling, Q. – quenching, T. – tempering, C.T. – cold treatment

The Kh15N5-type steel additionally alloyed with nitrogen, molybdenum, and copper and microalloyed with vanadium and rare-earth metals [3] was selected as the basis for the development of high-strength economically alloyed stainless steel.

Formation of the martensitic-austenitic structure providing a combination of high strength, corrosion resistance, and workability is achieved under the following conditions:

$12 < Cr_{eqv} < 17;$
$Cr = 12.0 \div 17.0;$
$Mo = 0 \div 2.0;$
$V = 0 \div 0.1;$
$Nb = 0 \div 0.1;$

$9 < Ni_{eqv} < 14;$
$C = 0.05 \div 0.25;$
$N = 0.10 \div 0.25;$
$Ni = 0 \div 5.0;$
$Cu = 0 \div 2.0;$
$0.15 < C + N < 0.50;$

Such correlations of alloying elements in the steel provide the martensitic-austenitic structure (M = 70 ÷ 80%, A = 20 ÷ 30%) and the absence of δ-ferrite, σ-phase, and $Me_{23}C_6$ type carbides in the structure, while the thermoplastic processing provides a grain size of 5 - 15 microns. The formation of such structures provides improved corrosion resistance, strength, ductility, and workability in the conditions of manufacturing articles of complex shapes and various sizes.

The indicated carbon and nitrogen concentrations in the steel are necessary for the attainment of high strength. As the carbon and nitrogen contents exceed 0.30 and 0.20%, respectively, it is difficult to reach satisfactory levels of plasticity and impact toughness and to obtain a high-quality metal free from porosity caused by a limited solubility of nitrogen in steel.

The aim of the 12.0-17.0% Cr addition is to provide the desired corrosion resistance and to increase nitrogen solubility. As the chromium and nickel concentrations exceed 17.0% and 2.5%, respectively, the steel is low-ductile, especially at low temperatures, because of the appearance of δ-ferrite, which elevates the ductile-brittle transition temperature. As the nickel content exceeds 5%, the nitrogen solubility in the steel decreases.

The aim of the addition of 0.5-1.5% manganese is to increase the nitrogen solubility and, of course, the steel deoxidation. As the manganese concentration exceeds 1.5%, the content of retained austenite increases, and this deteriorates the strength characteristics of the steel.

The addition of 0.1% vanadium provides the formation of fine-grained structure. At higher vanadium contents, the strength of the steel decreases because the solid solution becomes depleted of nitrogen because of the formation of vanadium nitrides.

Alloying with 0.5-2.0% molybdenum improves the corrosion resistance, increases the solubility of nitrogen, and inhibits the formation of carbonitrides at grain boundaries and thereby improves the impact toughness of the steel.

Alloying with 1.0-2.0% copper allows one to eliminate the delta ferrite in the microstructure and to improve the corrosion resistance and strength as a result of aging with the precipitation of fine particles of the copper-rich phase.

The presence of 0.005-0.030% cerium and 0.005-0.030% calcium helps to reduce the content of impurities at grain boundaries, thereby altering the kinetics of particle formation at the boundaries and reducing the degree of embrittlement. Alloying with lanthanum and yttrium promotes deoxidation and grain refinement of the steel.

According to the Scheffler diagram, the chromium equivalent (Cr_{eqv}) should range from 12 to 17, and the nickel equivalent (Ni_{eqv}) should range from 9 to 14. This correlation provides the absence of δ-ferrite, σ-phase, and $Cr_{23}C_6$ chromium carbides in the structure (Figure 1).

The solubility limit of nitrogen in the steel was calculated using the interaction parameters obtained by Feichtinger [4]. The calculation showed that the solubility of nitrogen in such steels at maximum content of elements raising and lowering the solubility varies from 0.20 to 0.12%, respectively.

Mechanical and technological properties of such steels substantially depend on their phase composition.

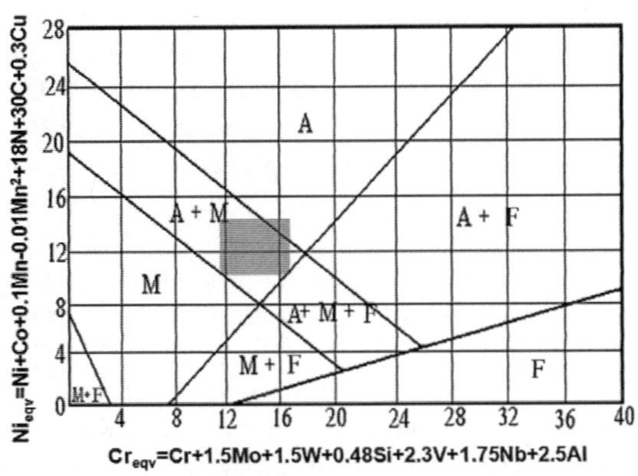

Figure 1. A. Scheffler phase diagram of corrosion-resistant steels.

The change in the phase composition and lattice parameters of the α- and γ-phases directly upon heating from 20 to 1000°C is shown (Figure 2) by the example of the 08Kh14AN4MDB steel [6]. At room temperature, the contents of α- and γ-phases are 78 and 22%, respectively. Upon heating to a temperature of 600°C, the phase composition remains virtually unchanged. However, starting from 300°C, the temperature dependence of the α-phase lattice parameter (Figure 2a) deviates from a straight line, and this shows that the nitrogen concentration in this phase reduces because of the precipitation of fine chromium nitride particles.

The temperature dependence of the γ-phase lattice parameter (Figure 2b) shows positive deviation from a straight line, especially at temperatures above 400°C. This shows that the nitrogen content in austenite increases. The α→γ transformation begins at temperatures above 600°C (Figure 2a), and the austenite content reaches 50% at 700°C and then increases to 80% at 800°C and to 100% at 1000°C.

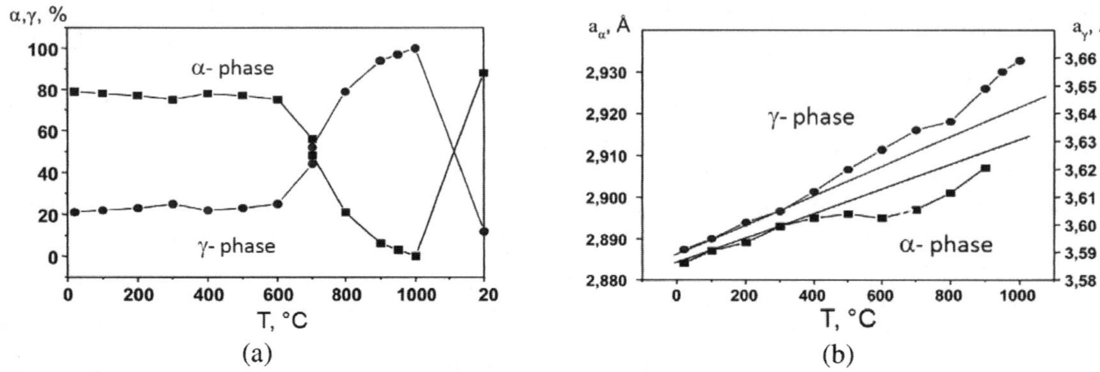

Figure 2. Changes in the phase composition and lattice parameters of the 08Kh14AN4MDB steel upon heating in a range of 20 - 1000°C [6].

In a temperature range of 600-800°C, the temperature dependence of the γ-phase lattice parameter (Figure 2b) is parallel to the line corresponding to the effect of linear expansion, and this indicates that the nitrogen content in the γ-phase is constant. Above 800°C, the slope of the temperature dependence of the lattice parameter substantially grows, suggesting that the nitrogen content in austenite in these temperature conditions increases.

The above results [6] can be interpreted as follows. The initial sample quenched from 1050°C contains α-phase (~ 80%), austenite, and fine particles of chromium nitride precipitated from austenite upon its decomposition in the process of quenching. Upon heating, the α phase at 300-600°C decomposes with the precipitation of nitride and, possibly, carbide phases. Thus, the nitrogen content decreases in martensite and increases in austenite. At 600-800°C, the α phase transforms to austenite, and this is accompanied by dissociation of nitrides. At temperatures of 800-1000°C, the transformation of the α-phase to austenite is completed, and the nitrogen content in it continuously increases. The dissolution of the nitride phase in austenite in this temperature range is intensified and is completed at 1000°C, as is evidenced by a sharp increase in austenite grain size [6].

The hardness, grain size, and amounts of the α and γ phases of the steels depend on heat treatment conditions (Figure 3). As the temperature of heating for water quenching increases from 950 to 1150°C, the hardness of the 08Kh14AN4MDB steel is reduced from 43 to 36 HRC, and the grain size is increased from 15 to 110 microns (Figure 3a) [6, 7]. According to the XRD data (Figure 3b), the decrease in the steel hardness with increasing quenching temperature is caused by the reduction in martensite content and the increase in the content of retained austenite.

The characteristic structure of the steels after quenching consists of packet martensite with interlayers of retained austenite (Figures 4a and 4b) [7].

According to the established [7] specific features of the quenched 08Kh14AN4MDB steel structure, the retained austenite is present in several morphological versions: in the form of thin layers at the boundaries of α-crystals inside the packets (Figures 4c and 4d), in the form of sufficiently broad, extended regions between α-crystals in the packets and at the junctions of packets (Figure 4), and in the form of sufficiently long randomly shaped regions near the boundaries of former austenite grains (Figure 4d).

The high level of the steel strength is reached after tempering as a result of the precipitation of fine carbonitride particles [7-10]. A maximum strength level of 497-507 HV for the 08Kh14AN4MDB steel quenched from 1050 - 1100 °C is achieved after tempering at 500°C for 0.5-2 hours. This can be attributed to the formation of fine carbonitride particles upon tempering. Tempering at 600 and 700°C leads to the carbonitride coagulation resulting in softening of the steels.

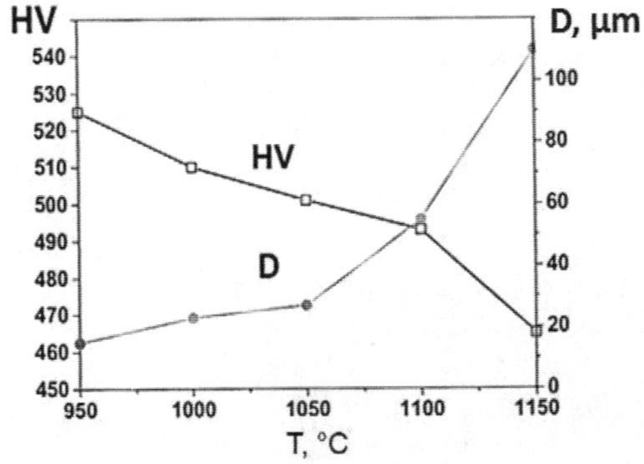

(a)

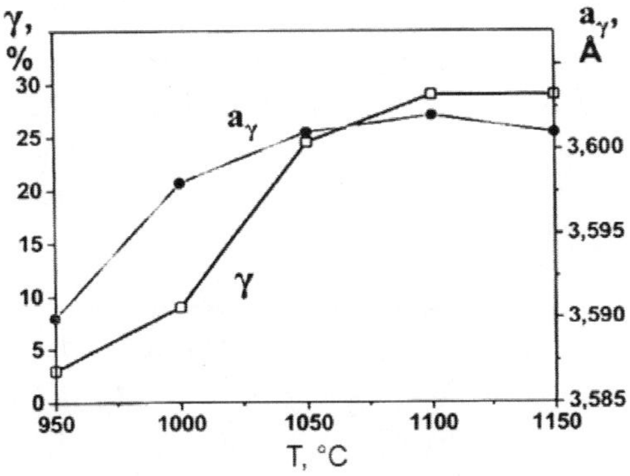

(b)

Figure 3. Effect of quenching temperature on the structure (a) and properties (b) of the 08Kh14AN4MDB steel [7].

(a)

(b)

(c)

(d)

Figure 4. Microstructure of the 08Kh14AN4MDB steel after quenching from 1050°C [7].

Fine precipitates (Figures 5a and 5b) were observed after tempering in the structure of the 08Kh14AN4MDB steel. The coarsest precipitates reach 70 nm in length and 15 nm in width. Their

IOP Conf. Series: Materials Science and Engineering **130** (2016) 012001 doi:10.1088/1757-899X/130/1/012001

density is relatively low, and they are distributed relatively uniformly within the α-crystals. The particles are also precipitated at the α-crystal boundaries [7].

The tensile tests of the steels at elevated temperatures showed that the 05Kh16AN5 steel after quenching from 1000°C and tempering at 400°C and the 08Kh14AN4MDB steel after quenching from 1050°C, cold treatment at -70°C, and tempering at 500°C are insignificantly softened at temperatures up to 500°C (Table 2).

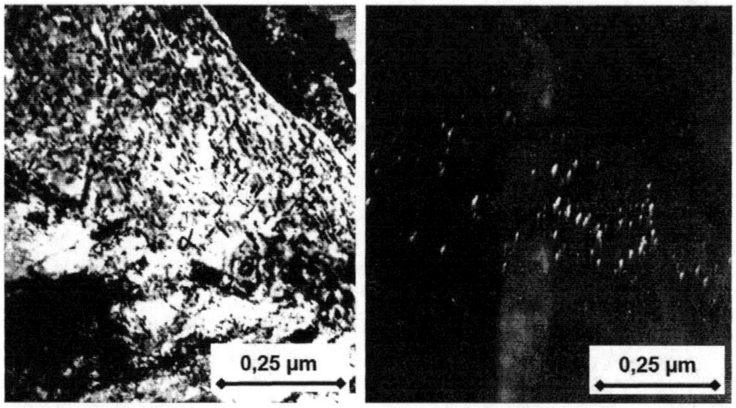

(a) (b)

Figure 5. Microstructure of the 08Kh14AN4MDB steel after quenching from 1050°C and tempering at 500°C: (a) bright-field image and (b) dark-field image [7].

Table 2. Mechanical properties at different test temperatures of the 05Kh16AN5B steel (after quenching from 1000°C and tempering at 400°C) and the 08Kh14AN4MDB steel (after quenching from 1050°C, cold treatment at-70°C, and tempering at 500°C) [7].

Steel	Test temperature (°C)	σ_B (MPa)	$\sigma_{0.2}$ (MPa)	δ (%)	Ψ (%)
	20	1450	1125	23	68
	300	1170	1000	19	49
05Kh16AN5B	400	1090	932	15	49
	500	980	821	10	56
	600	570	527	7	75
	20	1595	1070	14	45
	300	1200	1035	10	42
08Kh14AN4MDB	400	1170	990	9	42
	500	1000	885	7	45
	600	660	604	4	62

Thermomechanical treatment of the 14Kh15AN4M steel is accompanied by drastic changes not only in the phase composition, but also in the substructure of individual phases [11, 12]. Deformation leads to the austenite decomposition, and the resulting martensite inherits the austenitic-matrix deformation defects, which are added by the lattice defects caused by martensitic transformation upon cooling of the steel after rolling. The latter is related also to the austenite, which adds the transformation-induced strengthening upon the martensitic transformation to the substructure formed by rolling. These effects can be quantitatively estimated from the α- and γ-phase diffraction reflection widths (Figures 6 and 7), which are proportional to the accumulated deformation energy or the density of lattice defects.

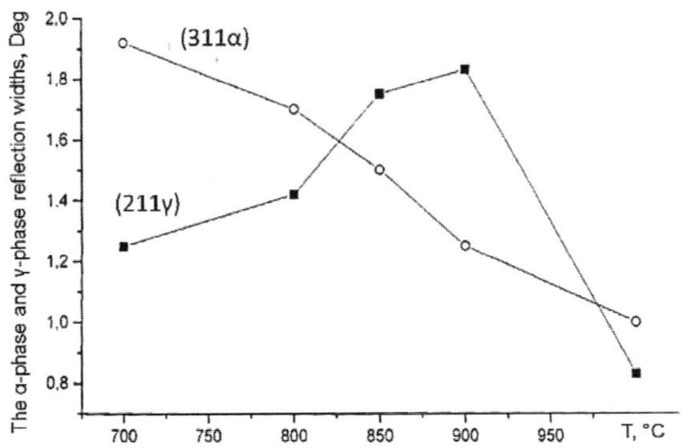

Figure 6. Widths of the (211) α-phase and (311) γ-phase reflections as a function of rolling temperature for the 14Kh15AN4M steel (a reduction of 70%) [11].

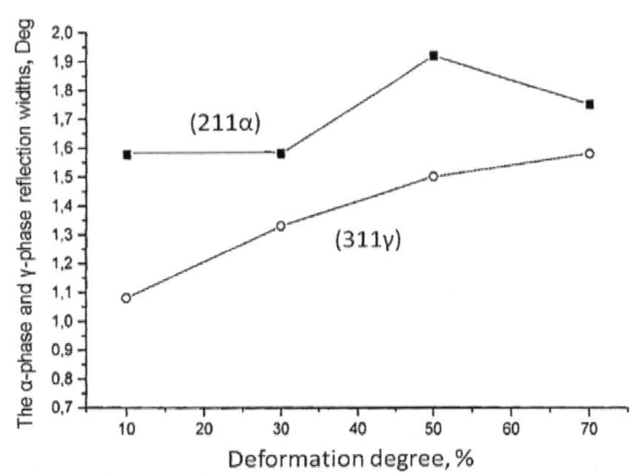

Figure 7. Widths of the (211) α-phase and (311) γ-phase reflections as a function of the degree of deformation after rolling of the 14Kh15AN4M steel at a temperature of 850°C [11].

Change the α-phase reflection width in Figure 6 is nonmonotonic. An increase in the defect density in the martensite crystal lattice with increasing rolling temperature from 700 to 900°C virtually coincides with the change in the austenite content, i.e., in fact, the defect density of martensite lattice is inversely proportional to the martensite content. This can be explained only by the fact that, unlike austenite, in which lattice defects are mainly caused by deformation and are uniformly distributed throughout the volume, martensite is characterized by the maximum defect density localized at the martensite/austenite interface. An abrupt reduction in the density of crystal defects at a temperature of 1000°C (Figure 6) is due, first, to decrease in the density of defects inherited from the matrix and, second, to decrease in fraction of the martensite/austenite interfaces relative to that observed after rolling at 900°C (Figure 6).

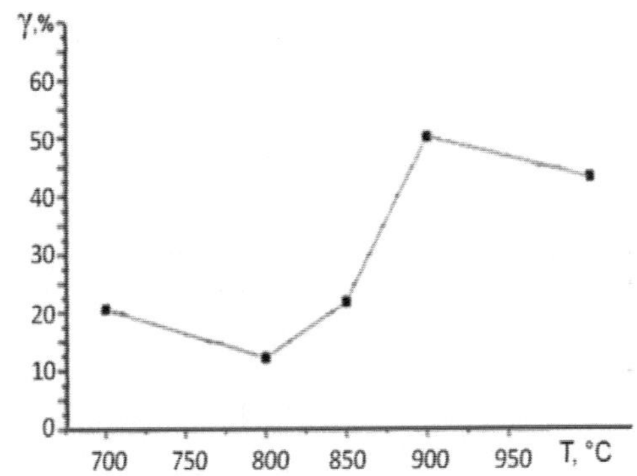

Figure 8. Effect of rolling temperature (reduction of 70%) on the austenite content in the 14Kh15AN4M steel [11].

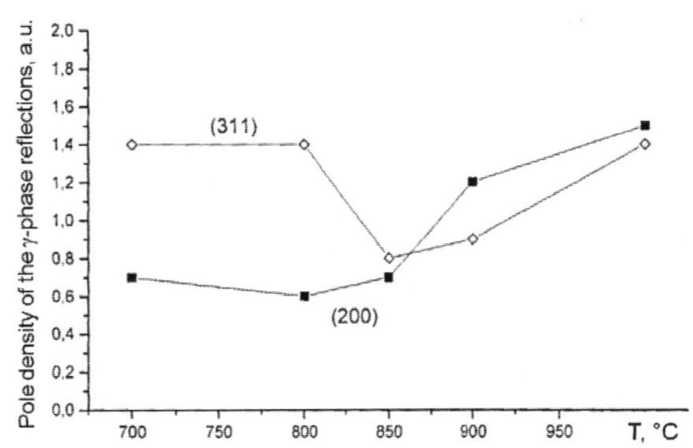

Figure 9. Pole density of the (200) and (311) γ-phase reflections as a function of rolling temperature (reduction of 70%) for the 14Kh15AN4M steel [11].

The monotonic increase in the defect density in the austenite with increasing degree of deformation at 850°C (Figure 7) is quite natural and is associated with increasing work hardening of the phase, in which deformation occurs. The defect density in martensite in much lesser extent varies with increasing degree of deformation. In this case, the effect of the degree of deformation on the reflection width is opposite to its effect on the austenite content (Figure 8), while the analogous dependences on temperature (Figures 6 and 8) are similar in character. This result requires a further study because it shows important features of the steel behavior upon thermomechanical processing.

Figures 9 and 10 show the dependences of the pole densities of various γ-phase reflections on rolling temperature and on the degree of deformation, respectively. They can be used to estimate the texture formation as a function of rolling conditions. We can conclude that no pronounced texture is formed under all deformation conditions, even in the austenite, which is directly subjected to deformation. This suggests the occurrence of dynamic recrystallization of austenite upon rolling, which is accompanied by the nucleation and growth of grains free from any preferred orientation. The martensite formed from the textureless matrix moreover should not have any preferred orientation, since the occurring multivariate transformation results in a textureless martensite state even in the case of a pronounced matrix texture [11].

A higher strength level of the steels is achieved by cold rolling. An increase in the total reduction to 75% upon cold rolling of the 08Kh14AN4MDB steel increases the martensite content from 79 to 96% (Figure 11), and this enhances the hardness from 40 to 55 HRC (Table 3). The accompanying growth of martensite tetragonality in the 08Kh14AN4MDB steel from 1.0033 to 1.0124 can be attributed to more substantial imperfection of its crystal lattice (Table 4). As the degree of reduction ranges between 10 and 25%, deformation of austenite occurs preferably without any martensite formation, and this leads to high strengthening at insignificant decrease in ductility (Figure 11). Warm rolling of the steels at 300, 400, 500 and 600°C with an increase in the total degree of reduction to 75% also results in increased hardness to 50-53, 51-52, 49-53, and 43-44 HRC, respectively (Table 3).

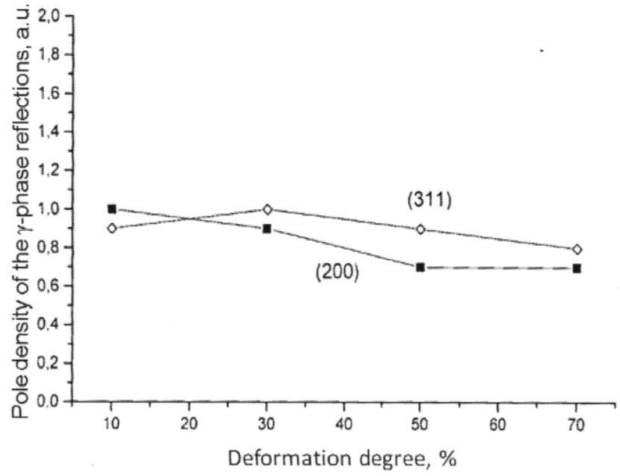

Figure 10. Pole density of the (200) and (311) γ-phase reflections of the 14Kh15AN4M steel as a function of the degree of deformation at 850°C [11].

Table 3. Hardness of the 08Kh14AN4MDB steel after quenching from 1050°C and rolling at various temperatures to various degrees of reduction [12].

Degree of reduction, %	Rolling temperature (°C)				
	20	300	400	500	600
0	40	41	41	42	38
25	48	44	44	44	40
30	50	46	46	47	41
45	52	49	49	51	42
60	55	50	50	52	43
75	55	53	52	53	44

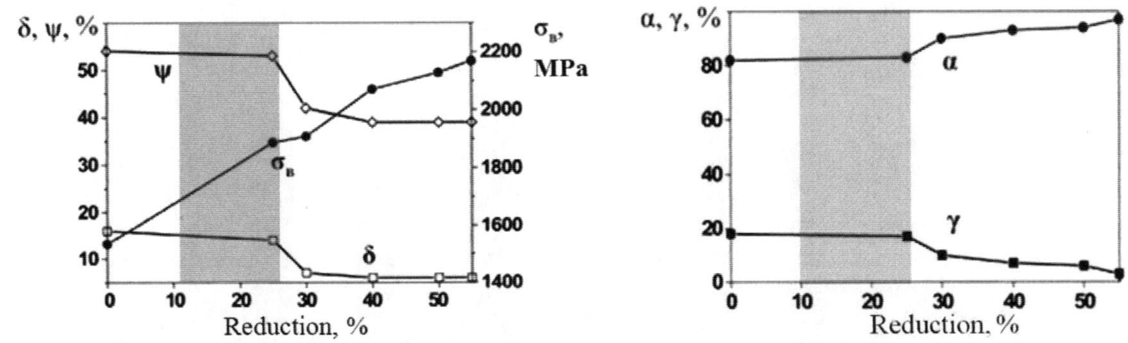

(a) (b)

Figure 11. Mechanical properties (a) and the α- and γ-phase contents (b) of the 08Kh14N4AMBD steel after quenching from 1050°C and subsequent cold rolling to different reductions [12]. The highlighted areas correspond to the optimum degrees of reduction upon rolling.

At a small degree of reduction (25%) upon cold rolling, the steels undergo mainly deformation of austenite without formation of martensite. This leads to high strengthening of austenite with a slight decrease in ductility (Figure 3). Rolling to a reduction of 30-55% leads to a higher strengthening due to the deformation of martensite formed upon quenching and the formation of stress-induced martensite from austenite (γ → α transformation). However, this is accompanied by a significant decrease in ductility of the steel. According to the X-ray diffraction data, an increase in the degree of reduction above 25% upon cold rolling increases the content and tetragonality (imperfection) of martensite (Table 4).

Table 4. Austenite lattice parameter and martensite tetragonality of the 08Kh14AN4MDB steel after quenching from 1050°C and cold rolling to various degrees of reduction [12].

Degree of reduction (%)	Austenite lattice parameter and martensite tetragonality	
	a_γ (Å)	c/a
0	3.601	1.0033
25	3.592	1.0034
30	3.592	1.0066
45	3.598	1.0071
50	3.596	1.0096
60	3.596	1.0109
65	3.594	1.0118
75	3.593	1.0124

The 05Kh16AN5 steel after quenching from 1000°C and tempering at 400°C for 2 h and the 08Kh14AN4MD steel after quenching from 1050°C and tempering at 500°C for 2 h have high fatigue strength (Figure 12) [9]. A higher fatigue limit based on loading 10^7 cycles is characteristic of the 05Kh16AN5 steel.

Corrosion resistance of the 05Kh16AN5B and 08Kh14AN4MDB steels can be considered as a function of the contents of chromium and nitrogen as well as heat treatment conditions [13]. The optimum heat treatment regimes for the experiments evaluating corrosion resistance are as follows: quenching from 1000-1050°C and tempering at 400 - 600°C.

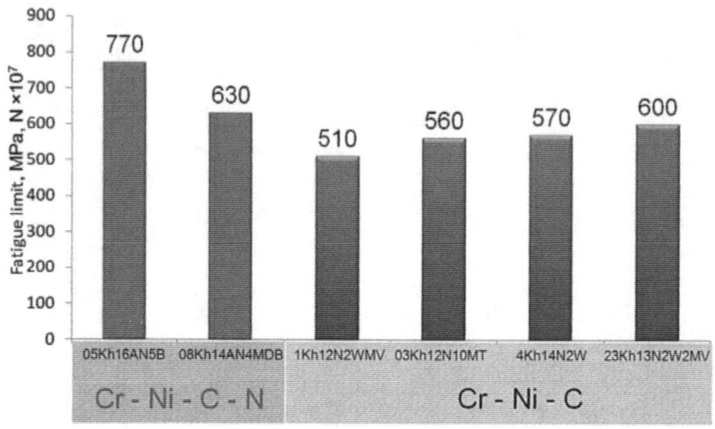

Figure 12. Fatigue strength of commonly used corrosion-resistant Cr-Ni-Cu martensitic steels and new Cr-Ni-N-C martensitic-austenitic steels [9].

The results of the study of general corrosion demonstrated that the corrosion rate of the martensitic-austenitic nitrogen-containing 05Kh16AN5 and 08Kh14AN4MDB steels as well as the austenitic 08Kh18N10T steel after all heat treatment conditions decreases with increasing duration of holding in 3.5% NaCl solution. A typical dependence of such attenuation of corrosion with increasing duration of the contact with a corrosive medium is shown in Figure 13 [13].

In the conditions under consideration, the worst corrosion resistance is exhibited by the 08Kh18N10T austenitic steel. The quenched steel in corrosion resistance ranks significantly below the 05Kh16AN5 08Kh14AN4MDB steels. However, the precipitation of carbonitride particles slightly increases corrosion rate of the latter steels.

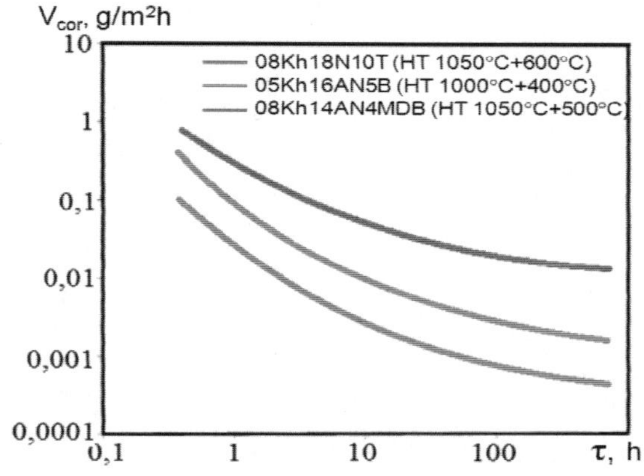

Figure 13. Corrosion resistance of the 08Kh14AN4MDB, 0Kh16AN5B, and 08Kh18N10T steels in 3.5% NaCl solution [13].

The results on the corrosion kinetics of the 08Kh14AN4MDB steel after tempering at 400 and 500°C (Figure 13) show that, regardless of the tempering temperature, the corrosion rate decreases by a factor of more than 25 as the holding time in the corrosive medium increases from 20 to 530 hours. The study of this steel revealed that the ratio between the martensite and austenite contents and the degree of their decomposition in tempered steel insignificantly affect their corrosion resistance in the NaCl solution.

Upon short-term holdings (20 hours), the 08Kh14AN4MDB steel exhibits a tendency to decreasing corrosion rate with elevating tempering temperature. Longer holdings lead to the growth of corrosion rate with elevating tempering temperature [12].

3. Summary

The technical effects of the creation of the new corrosion-resistant low-carbon steels with the structure of nitrogen martensite are as follows:

- the development of a new class of corrosion-resistant structural steels exceeding analogous steels in strength will provide a breakthrough in the field the design of advanced machines, instruments and constructions and improving the existing ones;
- an improvement in the reliability and service life of such products at a simultaneous reduction in the cost of their manufacture as a result of using cost-effective materials and processes, which do not require the development of new metallurgical equipment.

Conclusions

1. The chemical composition, which upon quenching and tempering of the steel provides the structure with predetermined contents of martensite (70-80%) and austenite (20-30%) and is free from δ-ferrite, σ-phase, and $Cr_{23}C_6$ type chromium carbides, has been selected and justified for the realization of improved characteristics of strength, ductility, and corrosion resistance of economically alloyed chromium-nickel-nitrogen steels.
2. The phase transformations occurring upon heating and cooling of chromium-nickel-nitrogen steels free from molybdenum and copper (05Kh16AN5B) and containing these elements (08Kh14AN4MDB) have been studied. The temperature dependences of the α- and γ-phase contents, the $\alpha \rightarrow \gamma$ transformation temperature range (600-800°C), and the temperatures (300°C) of the start of the precipitation and complete dissociation (1050°C) of chromium carbonitrides in the 08Kh14AN4MDB steel are determined. In the 05Kh16AN5 steel, the $\alpha \rightarrow \gamma$ transformation begins and ends at temperatures of 470 and 680°C, respectively.
3. The optimum quenching temperatures providing almost complete chromium carbonitride dissolution at the retention of fine-grained structure (with an average grain size of about 27-29 μm) have been established to be 1000°C for the 05Kh16AN5 steel and 1050°C for the 08Kh14AN4MDB steel.
4. The specific features of the structure of retained austenite, which is present in the quenched 08Kh14AN4MDB steel both in the form of thin layers between martensite crystals and in the form of massive regions with a high density of dislocations and twins, have been revealed by electron microscopy.
5. It is shown that the structure of packet martensite with thin interlayers of retained austenite free from fine precipitates is retained in the 08Kh14AN4MDB steel after quenching from 1050°C and tempering at 400°C. The strengthening fine carbonitride particles up to 70 nm long and 15 nm thick are present in the steel structure after quenching from 1050°C and tempering at 500°C.
6. The maximum strengthening of the 05Kh16AN5 steel ($\sigma_в$=1833 МПа) and the 08Kh14AN4MDB steel ($\sigma_в$=2168 MPa) is achieved by cold rolling to a reduction of 50-55%. These steels have the best combinations of mechanical properties after cold rolling to a reduction of 25% (the 05Kh16AN5 steel: $\sigma_в$=1680 МПа, $\sigma_{0,2}$=1410 MPa, δ=18%, ψ=60%; the 08Kh14AN4MDB steel: $\sigma_в$=1887 MPa, $\sigma_{0,2}$=1677 MPa, δ =14%, ψ =53%) and after warm rolling at temperatures of 300-500°C to a reduction of 60% ($\sigma_в$=1730-1790 MPa, $\sigma_{0,2}$=1300-1400 MPa, δ=18- 23%, ψ=50-63%).

References

[1] Bratuhin A G 1996 *Technological maintenance of high quality, reliability, service life of aviation equipment* (Moscow: Mashinostroenie) pp 524 (in Russian)
[2] Voznesenskaya N M, Izotov V I, Ulianova N V, Popova L S and Potak Ya M 1971 *Metal Science and Heat Treatment* **1** 32 (in Russian)

[3] Kostina M V, Bannykh O A and Blinov V M 2000 *Metal Science and Heat Treatment* **12** 3 (in Russian)
[4] Satir-Kolorz A H and Feichtinger H 1991 *Mettalkunde* **82** (9) 689
[5] Kostina M V, Bannykh O A, Blinov V M and Stepanov G A 2002 *Proc. Conf. on Strength mater. and structures at low temp. (Russia, Saint-Petersburg)* 26
[6] Bannykh O A, Betsofen S Ya, Blinov V M, Il'in A A, Kostina M V and Blinov E V 2006 *Metally* **5** 1 (in Russian)
[7] Bannykh O A, Blinov V M, Kostina M V, Afanasiev N A, Betsofen S Ya and Adyev S V 2004 *Metally* **6** 3(in Russian)
[8] Bannykh O A, Blinov V M, Kostina M V Lukin E I, Blinov E V and Rigina L G 2015 *Metally* **4** 72 (in Russian)
[9] Blinov V M, Bannykh O A, Kostina M V, Nemirovskiy Yu R and Khadyev M S 2000 *Metally* **3** 64
[10] Bannykh O A, Blinov V M, Shalkevich A B and Voznesenskaya N M 2005 *Metally* **3** 51
[11] Bannykh O A, Betsofen S Ya, Lukin E I, Blinov V M, Voznesenskaya N M, Tonysheva O A and Blinov E V 2015 *Metally* **1** 32
[12] Kostina M V, Bannykh O A and Blinov V M 2001 *Metal Science and Heat Treatment* **7** 3
[13] Blinov V M, Kalinin G Yu, Kostina M V, Mushnokova S Yu, Popova V I and Kharkov A A 2003 *Metally* **4** 84

Phase diffusionless $\gamma\leftrightarrow\alpha$ transformations and their effect on physical, mechanical and corrosion properties of austenitic stainless steels irradiated with neutrons and charged particles

O P Maksimkin

Institute of Nuclear Physics, 1 Ibragimov st., Almaty, Kazakhstan
National Research Nuclear University MEPhI (Moscow Engineering Physics Institute) Kashirskoe highway 31, Moscow, 115409, Russia

E-mail: maksimkin@inp.kz

Abstract. The work presents relationships of $\gamma\rightarrow\alpha'$ and $\alpha'\rightarrow\gamma$-transformations in reactor 12Cr18Ni10Ti and 08Cr16Ni11Mo3 austenitic stainless steels induced by cold work, irradiation and/or temperature. Energy and mechanical parameters of nucleation and development of deformation-induced martensitic α'-phase in the non-irradiated and irradiated steels are given. The mechanisms of localized static deformation were investigated and its effect on martensitic $\gamma\rightarrow\alpha'$ transformation is determined. It has been shown that irradiation of 12Cr18Ni10Ti steel with heavy Kr ions (1.56MeV/nucleon, fluence of $1\cdot10^{15}$ cm^{-2}) results in formation of α'-martensite in near-surface layer of the sample. Results of systematic research on reversed $\alpha'\rightarrow\gamma$-transformation in austenitic metastable stainless steels irradiated with slow (VVR-K) and fast (BN-350) neutrons are presented. The effect of annealing on strength and magnetic characteristics was determined. It was found that at the temperature of 400°C in the irradiated with neutrons samples (59 dpa) an increase of ferromagnetic α'-phase and microhardness was observed. The obtained results could be used during assessment of operational characteristics of highly irradiated austenitic steels during transportation and storage of Fuel Assemblies for fast nuclear reactors.

1. Introduction

In recent years, researchers, working in the area of radiation materials science, are interested in effects of temperature, deformation and irradiation on phase $\gamma\rightleftarrows\alpha$ transformations taking place in austenitic stainless steels. According to [1] this interest is related to the fact that after a long term operation of WWR, diffusionless transformations (a second order effects) start to play an important role in developing corrosion and mechanical properties of reactor materials. This research studies diffusionless martensitic transformations in irradiated structural materials of nuclear reactors. It is known that high-alloyed austenitic stainless steels and alloys in accordance with thermodynamics laws (figure 1) and phase diagrams (figure 2) in the combined effects of stress, strain, radiation and temperature exhibit the metastability, showing the whole number of phase-structural transformations, radically affecting operational characteristics [2-5]. In figure 1 T_0 is the equilibrium temperature of prior austenitic (γ) and martensitic (α') phases, $\Delta\sigma_{M_t}^{\gamma\rightarrow\alpha}$ is the critical value of the chemical driving force required to initiate $\gamma\rightarrow\alpha'$ - transformation under cooling with no load. Martensitic transformation may begin at T_1, $M_S < T_1 < T_0$ if the external load eases the critical nucleus formation, i.e. if it decreases at some value U required chemical driving force $\Delta\sigma^{\gamma\to\alpha}_{M_{\grave{o}1}}$ $\Delta\sigma^{\gamma\to\alpha}_{M_t}$.

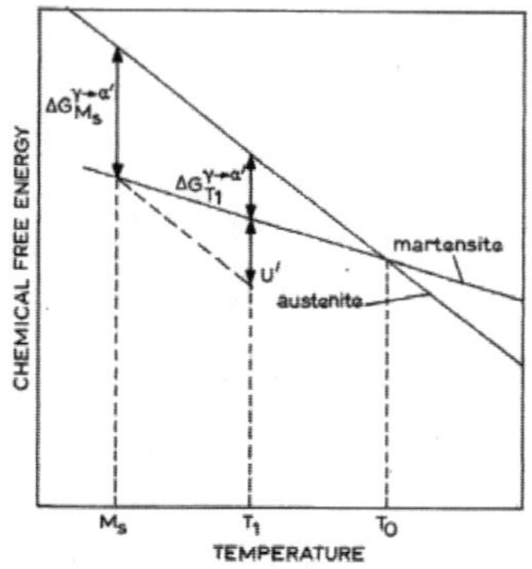

Figure 1. The free energy changing as a function of temperature [3].

2. The role of stacking fault energy (SFE) in $\gamma\to\alpha'$ transformation

One of the important characteristic that causes structural-phase transformations in austenitic steels is stacking fault energy (γ_{sfe}) of metal material, which largely depends on concentrations in the austenitic solution such element as C, N and Ni [6]. In general it is considered that in stainless steels SFE increases with Ni and C; whereas Cr, Mn and Si and N decrease SFE. The attempts to establish simple linear correlations between SFE and element concentrations in the stainless steels lead, in particular, to the following expressions:

$$\gamma = -53 + 6.2\%Ni + 0.7\%Cr + 3.2\%Mn + 9.3Mo\ [MJ/m^2]\quad [7];$$
$$\gamma = 1.2 + 1.4\%Ni + 0.6\%Cr + 17.7\%Mn - 44.7Si\ [MJ/m^2]\quad [8]; \qquad (1)$$
$$\gamma = 16.7 + 2.1\%Ni - 0.9\%Cr + 26\%C\ [MJ/m^2]\quad [9].$$

Martensitic $\gamma\to\alpha'$ - transformation results in the transition from the initial austenitic FCC lattice (γ - solid solution) to martensitic BCC lattice (α' - phase), what occurs diffusionless with the saving austenitic initial elementary composition in accordance with orientation ration of Kyrdyumov Zaks $(111)_\gamma \parallel (011)_\alpha$, $[\bar{1}01]_\gamma \parallel [\bar{1}\,\bar{1}\,1]_\alpha$ (figure 2, a). More details on martensitic transition starting temperature, the value of the applied stress (σ) and temperature dependence of the stable austenite yield strength (σ_T) are illustrated in figure 2, b. At the temperature M_S first martensite crystals arise without any external load. At temperature range $M_S<T<M^{\sigma}_{t}$ the martensite crystals formation is initiated by elastic stresses $\sigma<\sigma_T$.

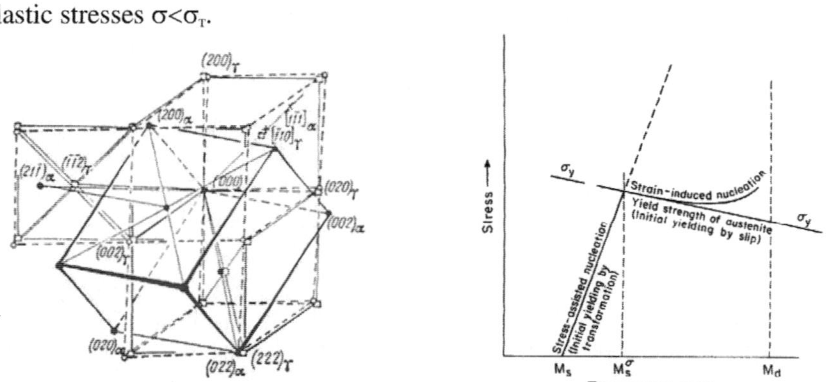

a b

Figure 2. Phase $\gamma\to\alpha'$-transition under the effect of stresses and temperature [4].

The obtained under these conditions α-martensite was named as the stress-induced martensite. At $M_t^{\sigma} < T < M_d$ plastic deformation precedes the martensite appearance, which was named as deformation-induced martensite.

During cold deformation austenitic steels are characterized by low stacking fault energy; α′-martensite arises at the intersection of the split dislocations and the phase transformation takes place at temperatures $M_s - M_d$ (figure 3).

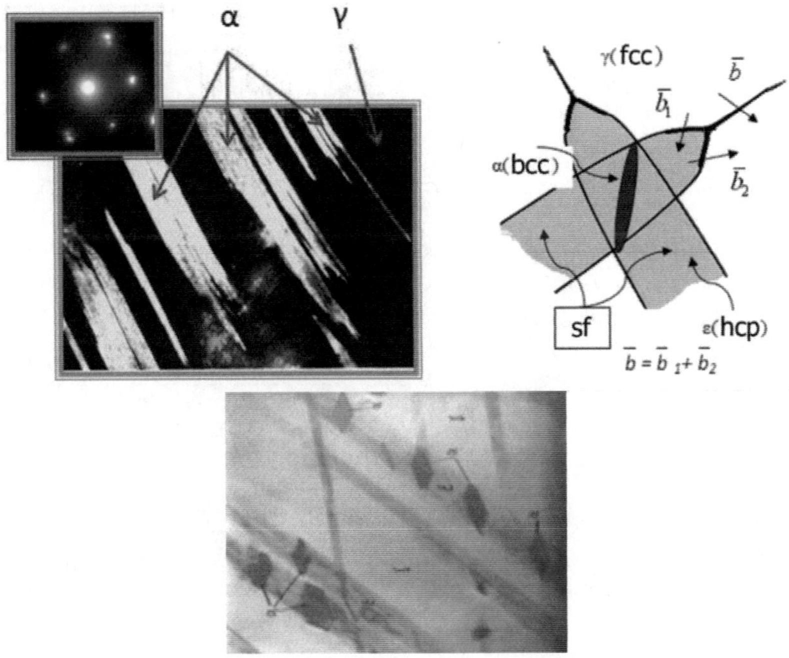

Figure 3. The incipience of martensitic α′-phase during deformation at the intersection of stacking faults.

Above the M_d temperature is not possible to induce the martensitic transformation by plastic deformation. Each stainless steel is characterized by M_d temperature setting value. The work [10] suggested the empirical temperature relationship M_d from composition of alloying elements for chromium-nickel steels:

$$M_d(°C) = 551 - 462(C+N) - 9.2Si - 8.1Mn - 13.7Cr - 29(Ni+Cu) - 18.5Mo - 68Nb \quad (2)$$

We have found that for 12Cr18Ni10Ti steel the M_d temperature above which deformation-induced α′-phase is not formed is 100°C both for unirradiated and irradiated with neutrons samples up to the fluence of $2\cdot10^{20}$ n/cm^2. Using the energy approach to the consideration of the α′-phase formation and destruction of metastable reactor steels the study proposed a new definition of M_d as the temperature at which the energy required for the starting γ→α′ transition becomes equal of mechanical energy which one needs to impart to the material to its failure. Thus at deformation temperatures more than M_d the sample fails before it has time to perform γ→α′-transition (figure 4).

3. Critical parameters of deformation-induced α′-phase formation

In addition, it was found [11] that martensitic α′-phase in the irradiated and deformed metastable steels is characterized by a low stacking fault energy values, and it forms only when the cellular dislocations start to dominate at a certain stage of the defect structure evolution. At the intersection of the cell walls, consisting of a number of split dislocations with ε-phase (hcp), increases the probability of α-phase formation (figure 5) at a reaction of γ→ε→α′ transformations.

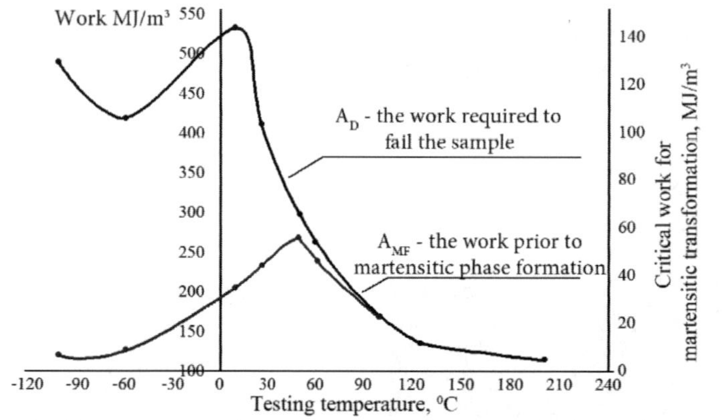

Figure 4. Temperature dependences of mechanical work, needed to originate martensitic γ→α′-transformation and to destruct unirradiated 12Cr18Ni10Ti steel samples at deformation.

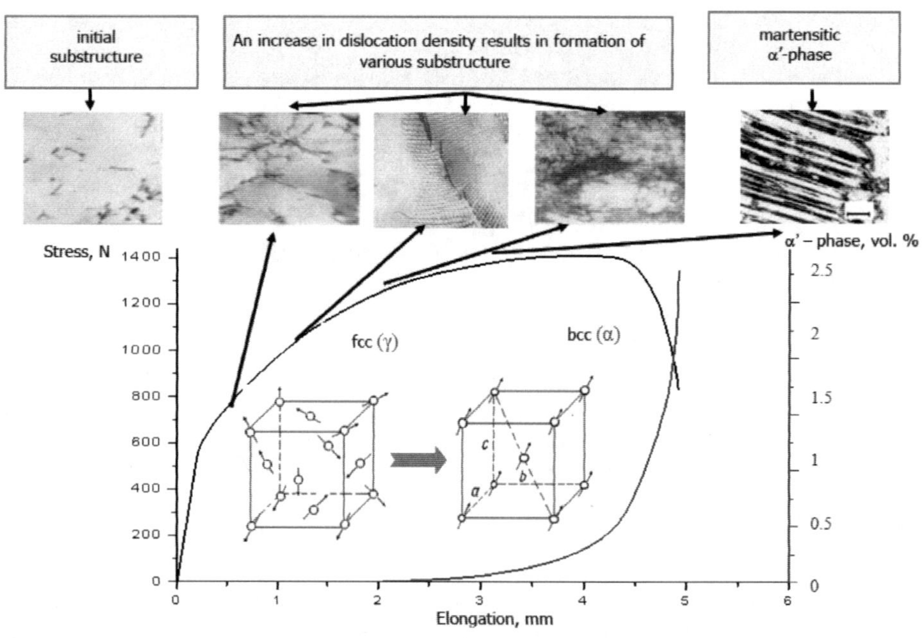

Figure 5. The evolution of defect structures and deformation-induced γ→α′- phase transformation in in metastable stainless steels at low temperatures.

Generally, at low temperature static tension α′-martensite of deformation forms at some critical deformation values δ_{cr}, stresses σ_{cr} or mechanical work density $A_{cr}^{\gamma\to\alpha}$, and with increasing doses of neutron irradiation values δ_{cr} and $A_{cr}^{\gamma\to\alpha}$ decrease and σ_{cr} remain almost unchanged what is shown in the experimental results (figure 6 and table 1).

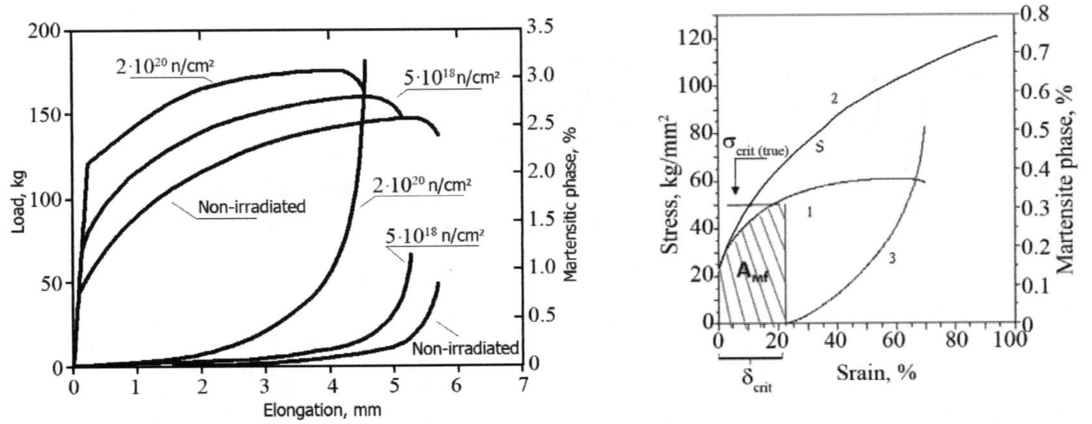

Figure 6. The formation and development of the deformation-induced martensitic α′-phase in 12Cr18Ni10Ti steel samples, non-irradiated and irradiated with slow neutrons (VVR-K).

Table 1. Critical nucleation α′-phase parameters at stretching non-irradiated and irradiated with neutrons 12Cr18Ni10Ti steel samples (T_{test} = 20°C, v = 0.1 mm/min).

Sample number	Fluence (n/cm^2)	σ_{cr} (true) (kg/mm^2)	δ_{cr} (%)	$A_{cr}^{\gamma\to\alpha}$ (MJ/m^3)
510	0	62	20.8	60
356	$2{\cdot}10^{18}$	62-63	10.2	55
369	$5{\cdot}10^{18}$	59	9.1	40
95	$2{\cdot}10^{20}$	61-65	4	30

4. Kinetics of martensitic α′-phase accumulation at deformation

Being formed, α'-phase continues to accumulate in the deformed sample and as it turned out, in all of randomly selected for observation sample areas (except neck), M_f accumulation occurs equally at the same power law. To describe the kinetics of deformation-induced γ→α′-transformation in the irradiated reactor steel the equation of Ludvigson-Berger [12] was chosen.

$$\frac{M_f}{1-M_f} = A\varepsilon^B ,$$

where the A parameter characterizes a tendency of deformed steel to martensitic transformation; B is the autocatalysis index. As a result, it was found that the experimentally determined B coefficient (B≈3.7) is similar for 12Cr18Ni10Ti steel samples both non-irradiated and irradiated with neutrons ($2{\cdot}10^{20}$ n /cm^2) [13, 14]. This is probably due to the fact that the neutron irradiation affects mainly positively on α-phase nucleation and facilitates it by creating defect areas. At the same time, originated, martensitic lens hardly develops in hardened by defects irradiated austenitic matrix that eventually leads to the formation of a large number of small plates of α'-phase.

5. γ→α′-transformation and localization of plastic deformation

It was shown that martensitic γ→α'-transformation plays a big role in localization processes of plastic deformation. The resulting α'-phase is stronger than the austenitic one and this induces localized inhomogeneous deformation in the form of strips, consistently and quickly moving along the working length of the sample (figure 7), and then transformed into the neck.

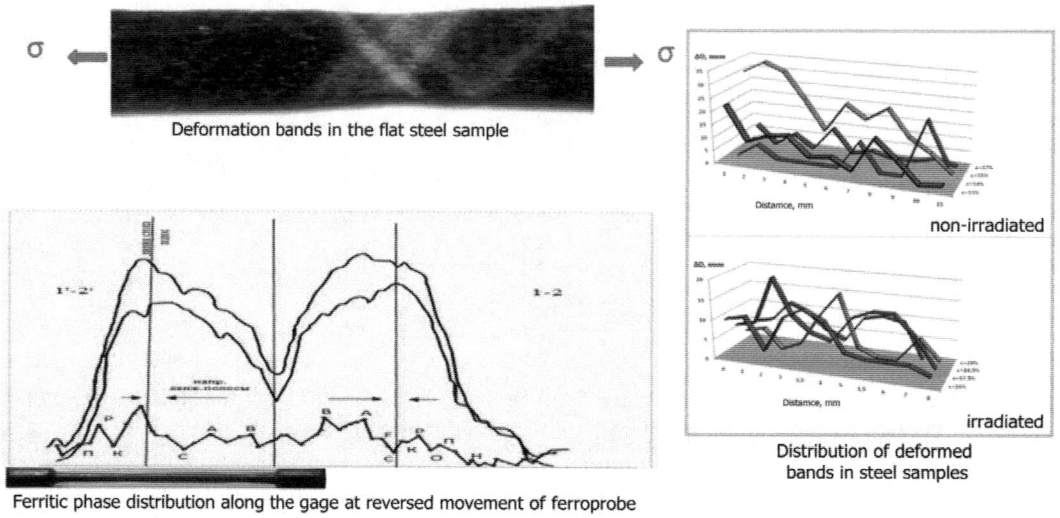

Figure 7. Strips of localized deformation in the 12Cr18Ni10Ti steel irradiated with neutrons.

It was also established that at the pre-destruction stage the maximum number of α'- ferrophase contained in that section of the working part of the sample, in which the destruction will happen. Thus, the precipitation kinetics of α'-phase for different cross-sections of the sample may help to monitor the process of localization and judge the speed of shape changes. The fact of the percolation possibility of $\gamma\rightarrow\alpha'$-transformations in metastable austenitic stainless steel during deformation can be used to detect the early signs and origin place of necking in a deformed sample.

Figure 8 shows that, given the localization of deformation kinetics of accumulation α'-martensite in the neck has an increasing character, predicted by various theories. Here the value of the parameter B for neutron-irradiated steel less than the non-irradiated. Thus, deformation-induced martensitic $\gamma\rightarrow\alpha'$ -transformation can have a significant effect on the local deformation in the reactor stainless steels. Estimates of the strips moving rates in unirradiated (V_m) and irradiated (V_{irr}) with neutrons ($2\cdot10^{20}$ n/cm^2) 12Cr18Ni10Ti steel samples deformed at 20°C at a strain rate of 0.5 mm/min, showed that at the deformations ε_{unirr} = 33% and ε_{irr}=18% the speed of strips in the irradiated sample is higher than in unirradiated sample and constitute V_m = 6.5 mm/min V_{irr} = 12.5 mm/min.

6. Deformation-induced wave

In some cases, such as when deformation (T = 20°C) highly irradiated (56 dpa) 12Cr18Ni10Ti steel sample, we observed the movement modeled on macrowaved deformation that led to the achievement of anomalously high plasticity brittle material after irradiation [15-18].

At the same time initial round markers, applying the paint to the surface of the sample, have evolved into ellipses, and at the diagram of stretching fixed plateau with the same value of the true flow stress (figure 9). It is found that in this case the rate of "wave" was ~0.04 mm/s at a grip rate of 0.008 mm/s.

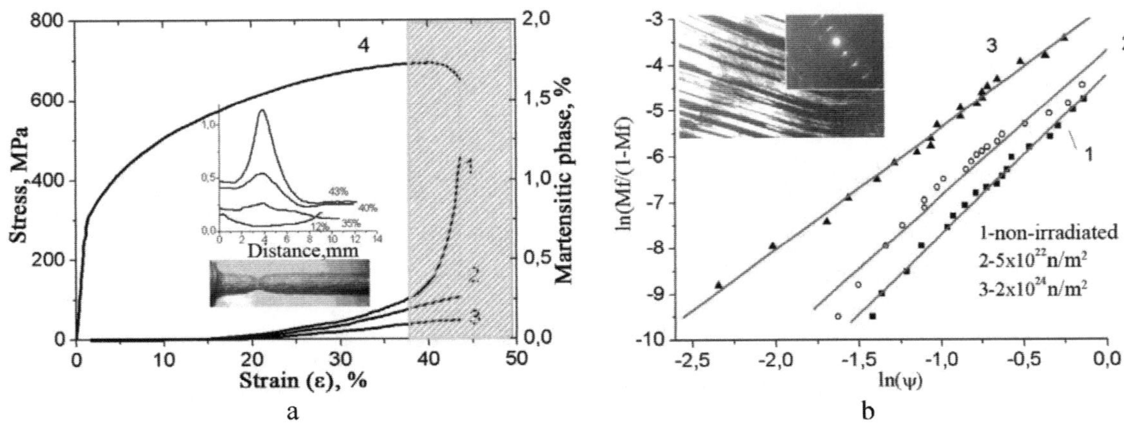

Figure 8. Kinetics of the formation and accumulation of deformation-induced α′–phase in the 12Cr18Ni10Ti steel: a) Engineering stress-strain diagram (4), kinetic curves of α′-phase (1)- in localization zone (Mf) and (2, 3) - near the neck of irradiated steel up to $5{\cdot}10^{22}$ n/m^2 (at inset martensitic phase distribution along the sample); b) the trend α′-phase curve in neck (inset TEM image and electron diffraction pattern α′-martensitic phase).

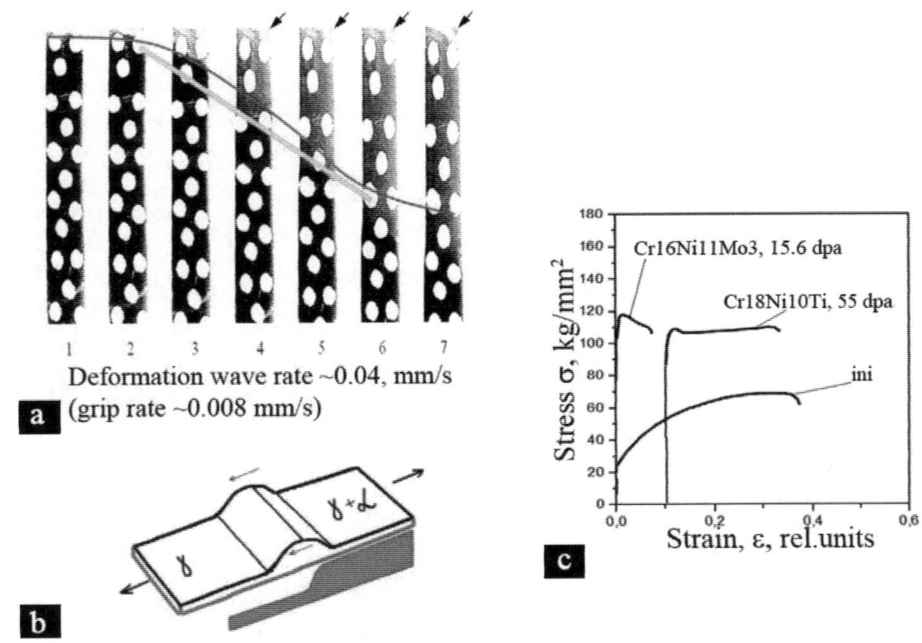

Figure 9. Deformation wave (a), its circuit (b) and the engineering stress-strain diagram of high irradiated samples of chromium-nickel steels (c) for cases of "no wave" (1.3) and "with a wave" (2).

The conducted systematic investigations of localized deformation found that a greater role of martensitic γ→α′-transformation plays during the formation and development of a stable neck in a deformed sample of metastable austenitic steel [19-21].

By analyzing the results of numerous experiments carried out to investigate the effect of neutron irradiation dose on the mechanical characteristics and power characteristics of reactor of austenitic steels, it was found that irradiation to some critical neutron fluence Ft=$3{\cdot}10^{22}$ n/cm^2, γ→α′-transformation in steels containing radiation defects may start spontaneously, without any additionally induced cold deformation (figure 10).

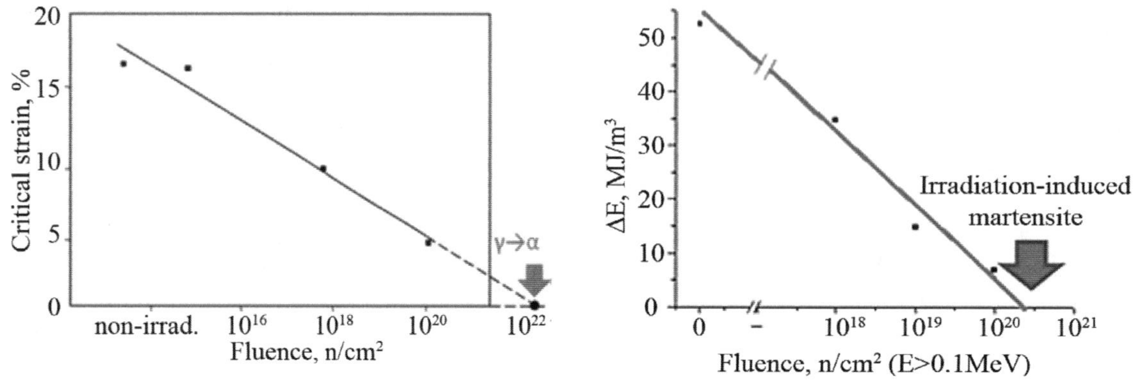

Figure 10. The effect of neutron fluence on the critical deformation (left) and energy density (right) required for $\gamma \rightarrow \alpha'$-transformation in 12Cr18Ni10Ti steel (E>0.1 MeV).

7. Irradiation-induced martensite

In the past we have repeatedly registered that the austenitic stainless steel samples were cut from different levels of the wall covers of spent fuel assemblies of the BN-350 nuclear reactor, detected the presence of α'-phase (figure 11).

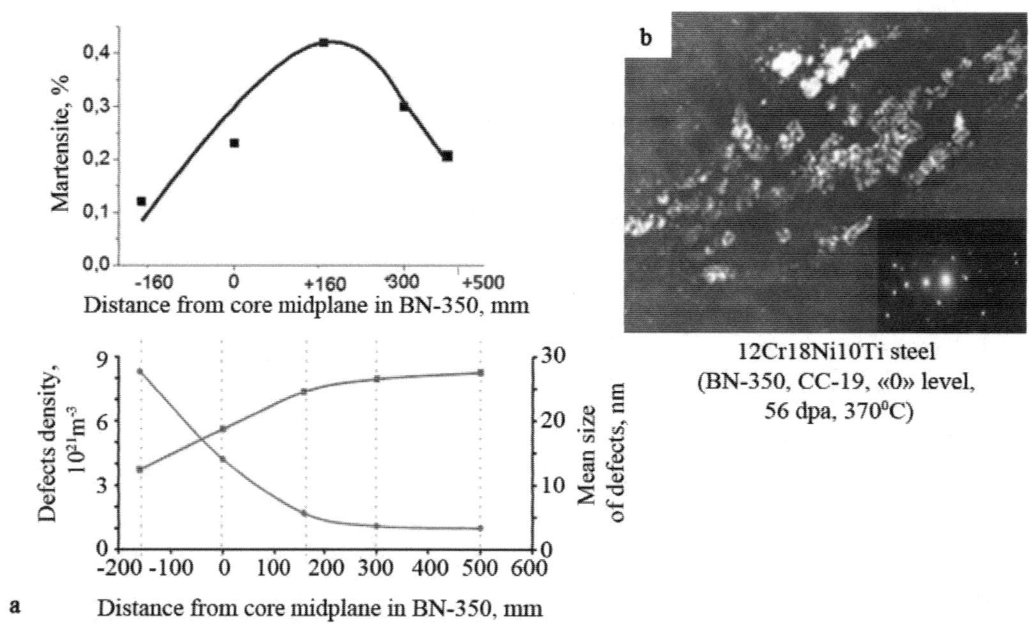

Figure 11. Formation α'- austenitic steels under irradiation with fast neutrons (12Cr18Ni10Ti steel, BN-350, CC-19, «0mm», 56 dpa, 370°C) (**a**); Frank loops, x50000 (**b**).

It was suggested that formation of martensitic phase is related to Frank loops. According to TEM studies these were the main radiation defects in irradiated material. This result, in particular, is confirmed by data obtained at ORNL (USA) under neutron irradiation of 304 steel (up to 12 dpa) with different silicon content in BOR-60 [24].

Based on the analysis of numerous experimental data it has been suggested that in 12Cr18Ni10Ti austenitic stainless steel ferromagnetic martensitic α-phase can be formed under neutron irradiation without introducing additional deformation. Apparently, in this effect a big role is assigned to the stored energy; upon reaching a critical value in the material $\gamma \rightarrow \alpha'$-transformation takes place. Using this approach α-phase in the austenite matrix should be formed not only in the case of neutron irradiation, but also by irradiation with charged particles. Indeed, as has been shown in [25] using EBSD method, that irradiation of 12Cr18Ni10Ti steel with heavy krypton ions with an energy of 1.56 MeV/nucleon at a fluence of 10^{16} particles/cm^2 leads to formation of α-phase (figure12). It was suggested to call this phase as irradiation-induced martensite by analogy with established concepts in the literature: "martensite of deformation", "stress" and "heat treatment" [26].

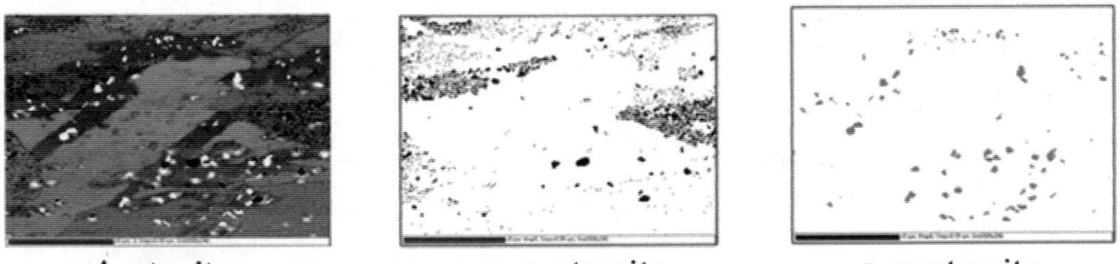

Figure 12. α-martensite irradiation in 12Cr18Ni10Ti steel after irradiation with heavy ions in the DC-60 accelerator (Kr 1.56 MeV/ nucleon) ~10^{16} particles/cm^2. Quantity α-martensite (BCC-lattice) – 9%; ε-martensite (HCP-lattice) – 2%.

It is worth noting that the formation of α-phase in the US austenitic steels was studied earlier in [27-30], and in the Russian steels this effect was discovered and described in [31, 32].

Taking into account the currently available information, we concluded that a phase transition $\gamma \rightarrow \varepsilon \rightarrow \alpha'$ can occur during irradiation when the defect structure evolves directly to the Frank loops at the by mechanisms proposed in [33]. The mechanisms impose restrictive conditions on the size of loops or, in other words, the stacking fault energy of steel. This idea found even greater confirmation when anomalously large stacking fault tetrahedrons were observed in the structure of steel samples irradiated with fast neutrons (figure 13). Both the stacking faults and Frank loops are related to the appearance of ferromagnetism. In fact, taking into account the generally accepted mechanism of formation of the tetrahedron, at every plane of which develops a defective dislocation loop and on the ribs can occur intersection of loops, i.e, $\varepsilon \rightarrow \alpha'$ transformation.

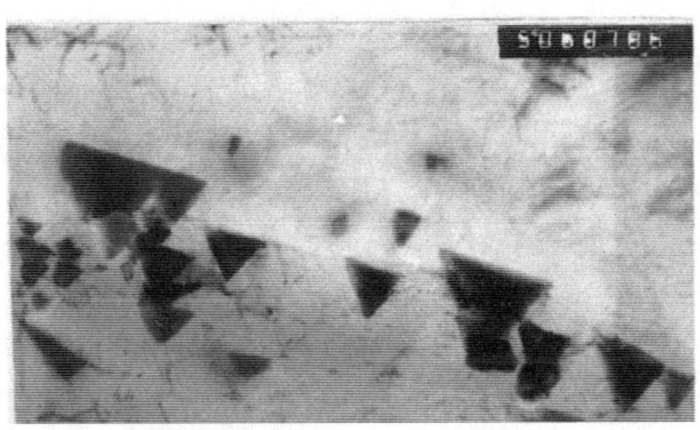

Figure 13. Tetrahedrons of stacking faults and α- phase in the microstructure of 08Cr16Ni11Mo3 austenitic stainless steel, irradiated with fast neutrons (BN-350, H-214(II), «0 mm», 15.6 dpa T_{irr}=337°C, $9 \cdot 10^{22}$ n/cm^2).

It has been suggested [26] the following formation scheme of α-phase in the metastable austenitic steel irradiated with neutrons at various temperatures to various damaging doses (figure 14). Irradiation-induced martensite is formed in the region 1 and is characterized by relatively low values of damaging dose (10-60 dpa) and irradiation temperature (200-400°C). In region 3 is also formed α-phase, but its nature is different – it is likely to be deformation (stress)-induced martensite. As the structure of α-phase is the bcc lattice, which itself is loose and therefore swells less than the fcc structure, so in the two temperature-dose areas 1 and 3, we can expect that swelling therein will be relatively less. Some experimental confirmation of this idea is obtained in [34].

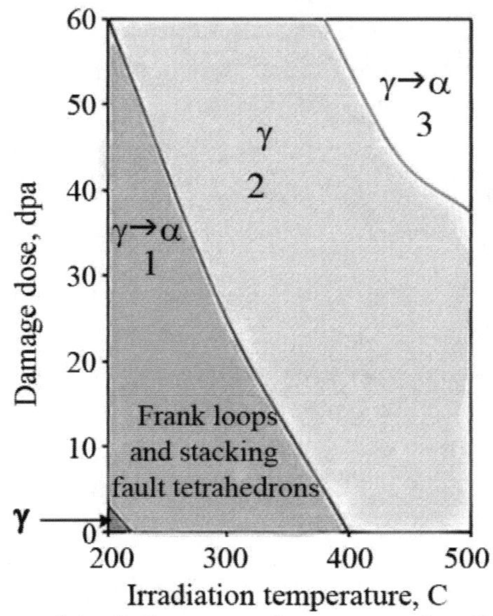

Figure 14. Influence of neutron irradiation parameters on the possibility of α'- phase formation in 12Cr18Ni10Ti stainless steel.

8. Reversed α′→γ-transformation

Stated above is attributed to α-martensite induced by deformation or irradiation of austenitic reactor steels. Being formed, α-phase can exist in the crystal as long as the material temperature is not increased to T = 350-400°C. At higher temperatures the amount of α'-phase will continuously decrease as temperature increases and upon reaching T = 700-800°C the microstructure will be completely annealed (figure 15a) with the formation of precipitation-hardened austenite [35]. Thus process of α′→γ-transformation is a multistage process. During magnetization occur 3 or 4 maximums (peaks), which position on the temperature scale depends on the radiation dose [36].

Analysis of the experimental data including results of Mössbauer, X-ray diffraction and TEM studies allows to assume the following sequence of changes in the structure, elemental and phase composition occurring during annealing of cold stainless steels in the identified stages. At 400°C the mobility of the chromium atoms increases leading to separation of the solid γ-solution with the formation of concentration inhomogeneities - spherical Guinier-Preston zones enriched with chromium. In the temperature range of 400-450°C α'-phase and Frank loops start to anneal and the swelling rate reaches 1% / dpa. It is worth noting that in some cases during the annealing of cold or irradiated steel samples around 400 °C it was observed an increase of α'-phase and this effect was also accompanied by an increase in microhardness and thermal effects (figure 16) [37]. In the range of 400-475°C occurs annealing of the newly formed α′-phase having a microscopic size. This annealing step is athermal. In the temperature range of 450-675°C an opposite α′→γ-transition occurs, which is carried out by a cooperative rearrangement of atoms by martensitic mechanism. This process is accompanied by a decrease in volume of the material. The α′→γ-transition ends by formation of netted dislocation structure in austenite.

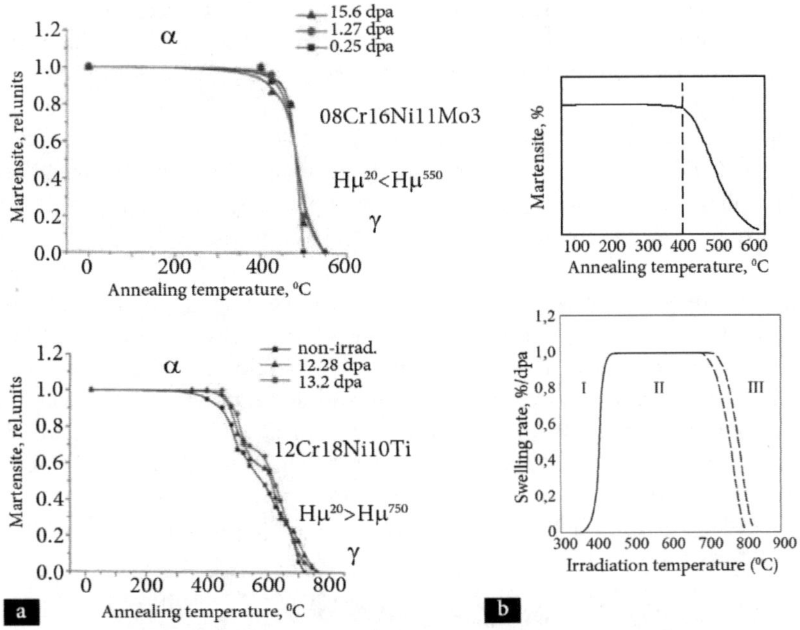

Figure 15. a) The inversed α'→γ martensitic transition in the stainless steels, irradiated with neutrons (BN-350); b) 400°C – The starting temperature of inversed α'→γ- transformation, annealing of Frank loops and swelling at a rate of 1%/dpa.

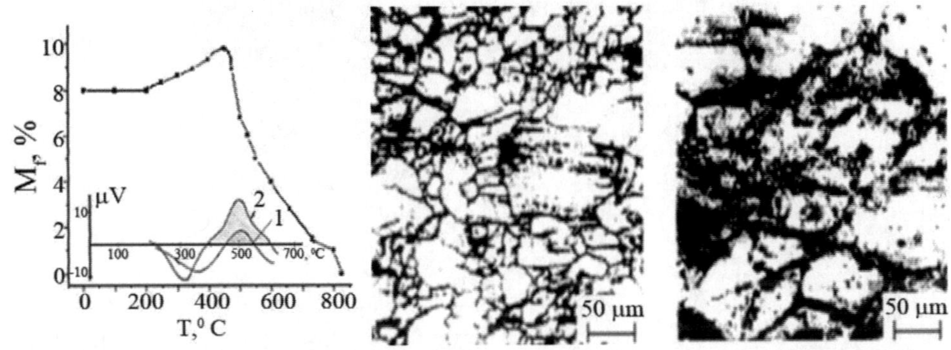

Figure 16. a) The observed effect of an increase in α'-phase during annealing of deformed 12Cr18Ni10Ti steel samples irradiated with neutrons to a damaging dose of 55.7 dpa; (in the insert – thermal effects accompanying α'-martensitic transition (200-400°C – heat absorbtion, 400-600°C – heat generation)); b, c) 08Cr16Ni11Mo3 steel microstructure (BN-350, FA H-214(II), «-1200mm», 0,25 dpa: б) before annealing; c) after annealing at 800°C (1 hour).

A decrease of annealing rate in the area of 550-700°C is associated with an increase in the volume fraction of α'-phase due to the fact that heating at temperatures 400-700°C is caused by ageing of steel, the formation of carbide and intermetallic phases and resulted in austenite destabilization and the formation of secondary α'-martensite due to depletion of solid solution in austenite – stable elements– C, N and Ni. In the range of 550-800°C magnetization of steel is decreasing due to the annealing of secondary α'- martensite. This third step of the martensitic transformation to austenite occurs by a mechanism controlled by diffusion.

9. The effect of α′-martensitic phase on corrosion in 12Cr18Ni10Ti stainless steel

The performed studies of the deformed and irradiated austenitic stainless steels have shown a decrease in the resistance of structural materials of nuclear reactors on fast neutrons to localized corrosion due to the formation and growth of induced deformation and irradiation α-martensite [37]. It was determined that samples of 12Cr18Ni10Ti steel irradiated with neutrons (10^{19} n/cm^2) in the VVR-K reactor and deformed at -60°C to 30% strain or more, apart from structural inhomogeneity of the phase are characterized by phase heterogeneity and the emergence of major corrosion pits, concentrated near the deformation strips and grain boundaries, where martensitic α'-phase is mainly accumulated.

In [38] gives data on corrosion resistance exposed to high damaging doses (~59 dpa) 12Cr18Ni10Ti steel samples cut from the walls of the covers of spent fuel assemblies from the operational life of the reactor BN-350. It was found that samples cut from different levels of from the center of the reactor core, contain different amounts of magnetic α'-phase, what correlated with the features of the defect structure (presence Frank loops and stacking fault tetrahedra). Established the intergranular nature of corrosion damage to the inner wall surface of the cover screen FA H-214 (1) at the level of "75 mm" from the center of the CAZ, where radiation damages are particularly noticeable. Figure 17 shows the deformation dependence of mass loss of steel samples 12Cr18Ni10Ti in aggressive solution (10% $FeCl_3$, 500 hours), which shows ambiguous changes in corrosion damage on the degree of cold deformation.

Figure 17. Effect of deformation martensite at resistance to pitting corrosion of cold deformed 12Cr18Ni10Ti steel (given in parentheses the volume content α'-martensitic phase in %).

10. Conclusions

There was carried out a complex research on materials science of structural materials for fast reactors and thermonuclear installations, and presented data analysis results. It is shown that in complex-alloyed austenitic stainless steel an intensive external influences (radiation, deformation, temperature) may result in phase-structural transformations, including direct (γ→α') and reversed (α'→γ) martensitic transformations, largely determining the operating characteristics of these materials. Considered some features of formation and development of deformation (stress)-induced and irradiation-induced martensite. Established the influence of γ→α'-transformation at localized deformation in the form of strips and stable necking. Obtained data allow to suggest that α'-phase, formed in the austenitic stainless steel, has a significant impact on corrosion properties and play an important role in long-term (>50 years) storage of spent fuel assemblies of fast reactors.

References

[1] Gussev M N, Busby J T and Garner F A 2015 *17-th International Conference on Environmental Degradation of Materials in Nuclear Power Systems – Water Reactors* (Ottawa, Canada)
[2] Filippov M A, Litvinov V S and Nemirovskyi Yu P 1988 (Moscow: Metallurgy) p 255
[3] Tamura I 1982 *Material Sciences* **16** 253
[4] Olson G B and Cehen M 1972 *Less-Common Metals* **28** 107
[5] Wayman C M 1985 *7-th Int. Conf. (Montreal 12-16 Aug.)* **3** 1779
[6] Maksimkin O P 2010 *Stacking faults, their energy and influence on the properties of irradiated metals and alloys* (Almaty) p 70
[7] Schvamm R E, Read R P 1975 *Metall Trans* A **6** 1345
[8] Rhodes C G and Thompson A W 1977 *Metall Trans* A **8** 1905
[9] Brofman P J and Ansell G S 1978 *Metall Trans* A **9** 879
[10] Kijohiko N, Yutaka O and Nabuo M 1977 *J. Iron and Steel Inst.* **63** 772
[11] Maksimkin O P and Aithozhin E S 2006 *Transactions of the National Academy of Science* **6** 77
[12] Ludwigson D C and Berger J D 1969 *J. Iron and Steel Inst.* **207** 63
[13] Maksimkin O P, Naltayev A and Rahashev B K 2009 *Transactions of the National Academy of Science* **1** 37
[14] Maksimkin O P, Yemelyanov A I, Naltayev A, Berdaliev D T and Rahashev B K 2008 *KazNU Bulletin – physical series* **1** 58
[15] Maksimkin O P, Gussev M N and Toktogulova D A 2008 *NNC RK Bulletin* **4** 27
[16] Maksimkin O P, Ruban S V, Gussev M N and Rybin S V 2010 *NNC RK Bulletin* **2** 25
[17] Gussev M N, Maksimkin O P, Osipov I S, Silniagina N S and Garner F A 2010 *Journal of ASTM international* **7** 210
[18] Gussev M N, Maksimkin O P and Garner F A 2009 *J. Nucl. Mater* **386-388** 273
[19] Maksimkin O P, Takieva A M, Naltayev A, Mazhenov N, Berdaliev D T and Rahashev B K 2013 *Bulletin of Karaganda University – physical series* **472** 34
[20] Maksimkin O P and Takieva A M 2014 *NNC RK Bulletin* **1** 45
[21] Tivanova O V 2008 *PhD thesis (Almaty)* p 139
[22] Maksimkin O P and Berdaliev D T 2009 *Transaction Ministry of Education and Science – physical and math series* **2** 19
[23] Maksimkin O P 2013 NNC RK Bulletin **3** 17
[24] Gussev M N, Busby J T, Tan L and Garner F A 2014 *J. Nucl. Mat.* **448** 294
[25] Alontseva D, Maksimkin O P and Russacova A 2014 *Materials science* **20** 15
[26] Maksimkin O P 2015 *NNC RK Bulletin* **1** 17
[27] Porter D L 1979 *J. Nucl. Mater.* **79** 406
[28] Stanley J T and Hendrickson L F 1979 *J. Nucl. Mat.* **80** 69
[29] Reynolds M B, Zow J B and Sullivan Z O 1955 *J. of Metals* **7** 555
[30] Iton M, Onose Sh and Yuhara Sh 1987 *J. Japan Inst. Metals* **51** 1101
[31] Voronin I M, Porollo S I 1989 *Atomic energy* **66** 33
[32] Chucalkin Yu G 1988 *Atomic energy* **65** 254
[33] Kashenko M P 2010 *Wave model of martensitic increase in* $\gamma - \alpha$ *transformation in iron-based alloys* (Moscow: Izhevsk) p 280
[34] Baron J G, Cadalbert R and Delapace J 1974 *Journal of Nucl. Mat.* **51** 266
[35] Malyshev K A, Sagaradze V V and Sorokin I P 1982 *Phase hardening of austenitic Fe-Ni alloys* (Moscow: Science) p 260
[36] Rahashev B K 2010 *PhD thesis (Almaty)* p 116
[37] Ruban S V, Maksimkin O P and Rybin S V 2011 *NNC RK Bulletin* **3** 98
[38] Maksimkin O P, Yarovchuk A V, Tsay K V and Ruban S V 2014 *NNC RK Bulletin* **1** 38
[39] Maksimkin O P and Kadyrzhanov K K 2007 *Safety Related Issues of Spent Nuclear Fuel Storage* (Springer) p 329

IOP Conf. Series: Materials Science and Engineering **130** (2016) 012003 doi:10.1088/1757-899X/130/1/012003

Issues of intergranular embrittlement of VVER-type nuclear reactors pressure vessel materials

O Zabusov [1, 2]

[1]National Research Centre "Kurchatov Institute", Kurchatov sq.1, Moscow, 123182, Russia
[2]National Research Nuclear University MEPhI (Moscow Engineering Physics Institute), Kashirskoe highway 31, Moscow, 115409, Russia

E-mail: Zabusov_oo@nrcki.ru

Abstract. In light of worldwide tendency to extension of service life of operating nuclear power plants – VVER-type in the first place – recently a special attention is concentrated on phenomena taking place in reactor pressure vessel materials that are able to lead to increased level of mechanical characteristics degradation (resistibility to brittle fracture) during long term of operation. Formerly the hardening mechanism of degradation (increase in the yield strength under influence of irradiation) mainly had been taken into consideration to assess pressure vessel service life limitations, but when extending the service life up to 60 years and more the non-hardening mechanism (intergranular embrittlement of the steels) must be taken into account as well. In this connection NRC "Kurchatov Institute" has initiated a number of works on investigations of this mechanism contribution to the total embrittlement of reactor pressure vessel steels. The main results of these investigations are described in this article. Results of grain boundary phosphorus concentration measurements in specimens made of first generation of VVER-type pressure vessels materials as well as VVER-1000 surveillance specimens are presented. An assessment of non-hardening mechanism contribution to the total ductile-to-brittle transition temperature shift is given.

1. Introduction

The VVER-type reactors pressure vessels (RPV) are produced from cylindrical shells joint by welding. The materials of RPV are low-carbon low-alloyed steels. The chemical content and manufacturing technology of these steels provide high strength properties and brittle fracture resistance ensuring safe operation of a nuclear power plant during its service life.

The influence of neutron irradiation and high temperature leads to an increase in yield strength and a decrease in brittle fracture resistance [1]. The ductile-to-brittle transition describable by critical transition temperature T_K is characteristic feature of bcc RPV steels. T_K is determined by impact tests of V-notched Charpy specimens. The tests allows taking into account two mechanisms of the mechanical properties degradation: the strengthening of ferrite matrix caused by formation of nanosized precipitations under irradiation enriched by Ni and Mn in the case of VVER-1000 RPV and generally by Cu in the case of VVER-440 RPV [2, 3] that leads to increase in yield strength and T_K and nonstrengthening mechanism caused by segregation of impurities (P in the first place) and alloying elements on prior austenite grain boundaries that leads to intergranular embrittlement [3].

Taking into account existing projects to extend service life of VVER-1000 up to 60 years and more NRC "Kurchatov Institute" has initiated a number of works on investigations of non-hardening mechanism contribution to the total embrittlement of reactor pressure vessel steels. The main results of these investigations are described in this article. Results of grain boundary phosphorus concentration measurements in specimens made of first generation of VVER-type pressure vessels materials as well as VVER-1000 surveillance specimens and assessment of non-hardening mechanism contribution to the total ductile-to-brittle transition temperature shift is given are presented.

2. First generation of VVER-type reactors

The intergranular embrittlement of the 1st generation of VVER-type RPV materials in the USSR was not considered as a key factor, limiting their service life, because the critical element of these RPV is the weld joint metal (WM). This phenomenon is not characteristic of these WM because of microstructure peculiarity [4]. But the fact itself of grain boundary (GB) segregation of phosphorus and alloying elements in the base metal (BM) of VVER-440 RPV was mentioned in literature [5] where the non-equilibrium irradiation-induced character of the segregations was noted.

For the first time a significant effect of intergranular embrittlement was revealed when investigating BM trepans cut out of a decommissioned reactor-prototipe pressure vessel wall [6]. The irradiation temperature was 275°C, fast neutron (E > 0.5 MeV) flux at inner surface of RPV wall was about $1 \cdot 10^{16}$ $m^{-2}s^{-1}$.

The chemical content of this BM is quite similar to this of VVER-1000 but impurities content is characteristic of VVER-440 BM and Cr content is increased: C – 0.24 wt. %, Si – 0.28 %, Mn – 0.49 %, S – 0.04 %, Cr – 3.3 %, Mo – 0.40 %, Ni – 1.07 %, P – 0.018 %, Cu – 0.01 %, Fe – bulk. It must be noted that such a combination of rather high content of P, Ni and Cr can promote GB segregation of P [3, 7].

Results of mechanical tests (impact and tensile tests) of this BM are presented in table 1 [6]. Few through-thickness (12 cm) trepans were cut out of RPV wall. The trepans were cut into layers, standard Charpy specimens (10x10x55 mm) and tensile specimens were manufactured of every layer. The results demonstrate a critical increase in T_K that implies an extremely high level of embrittlement. The character of fracture surfaces of specimens fractured at temperatures lower than T_K was almost fully brittle and predominantly intergranular (almost fully intergranular at the highest fluence). This denotes mainly intergranular embrittlement mechanism. Results of P content in GB measurements obtained by Auger electron spectroscopy (AES) are presented in figure 1. Measurement procedures are described in [6]. An increase in P content in GB with the fast neutron fluence is seen in the figure and, as will be shown below, these values at fluences above $4 \cdot 10^{23}$ m^{-2} are much higher than the values typical of VVER RPV BM.

Unfortunately the reference unirradiated BM of this RPV was absent but the data presented affirms that the non-hardening mechanism is able to make an important contribution to RPV embrittlement under neutron irradiation. A significant share of intergranular fracture in RPV upper cover specimens fracture surface must be noted as well. This material was subjected to the operation temperature influence only during the whole operation time. It can be an indirect evidence of intergranular embrittlement under the operational temperature, though the absence of the reference material does not allow to assert this.

Table 1. Results of mechanical tests of RPV BM (steel 25Cr3NiMo).

State of BM	T_K (°C)	ΔT_K (°C)	$Rp_{0,2}$ (MPa)
Inner layer of trepan, F=6,2·10^{23} m^{-2}	more than +260	more than 305	
1/4 of trepan thickness, F=4,2·10^{23} m^{-2}	+168	165	949
3/4 of trepan thickness, F=2,2·10^{23} m^{-2}	+107	152	
Outer layer of trepan, F=1,4·10^{23} m^{-2}	+120	165	
Annealed material (650°C/2 h)	-45*		
RPV upper cover	-22	23	612

*accepted as T_0.

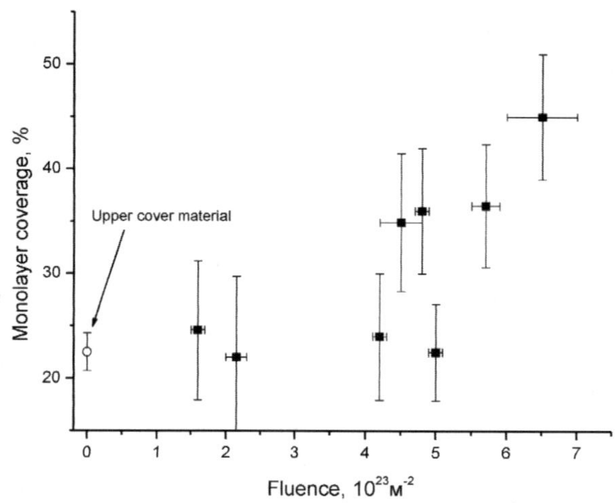

Figure 1. P content in GB of RPV BM (steel 25Cr3NiMo).

One more example of intergranular embrittlement of VVER-type reactors era results of investigation of two trepans of another decommissioned reactor [8]. The material was irradiated up to fast neutron fluence (E>0.5 MeV) of 1.05·10^{24} m^{-2} (inner layer of the RPV wall) at 290°C. Fast neutron flux at inner surface of RPV wall was equal to 3.4·10^{15} $m^{-2}s^{-1}$.

Table 2. The results of Charpy impact tests.

Location	T_{K0} (°C)	F (10^{22} m^{-2})	T_K (°C)	ΔT_K (°C)
Trepan 1, inner layer		105	+34	78
Trepan 2, outer layer	-44	11.8	+29	73
Trepan 1, re-irradiated inner layer		150	+46	90

The chemical content of the steel is: C– 0.14 wt. %, S – 0.26 %, Mn– 0.46 %, S – 0.024 %, Cr – 2.6 %, Mo – 0.59 %, Ni – 0.34 %, V – 0.26 %, P – 0.016 %, Cu – 0.096 %. The same steel was used to manufacture VVER-440 RPV shells but the irradiation temperature is higher than that in VVER-440 (270°C).

The results of Charpy impact tests are presented in table 2.

Though detailed SEM study of Charpy specimens fracture surfaces has not been carried out, brief survey showed the significant part of intergranular areas on the surfaces of specimens tested at DBTT (up to 80-90%). The specimens were cut out of different locations of the trepans and investigated using AES. Additionally a part of Trepan 1 inner layer was annealed at 475°C/100h and another part was re-irradiated with a fluence of $0.5 \cdot 10^{24}$ m-^{2} at a flux 3 10^{16} $m^{-2}s^{-1}$ and temperature of 270°C. The re-irradiated material was also investigated by AES. The results of the investigations are presented in figure 2.

Though the re-irradiation conditions differ from that at initial irradiarion, the lower temperature and the higher neutron flux during the re-irradiation could mitigate the GB segregation process rather than promote it. The reference un-irradiated material was absent too, this is why P content in the initial state in figure 2 was calculated according the equilibrium GB segregation model [9].

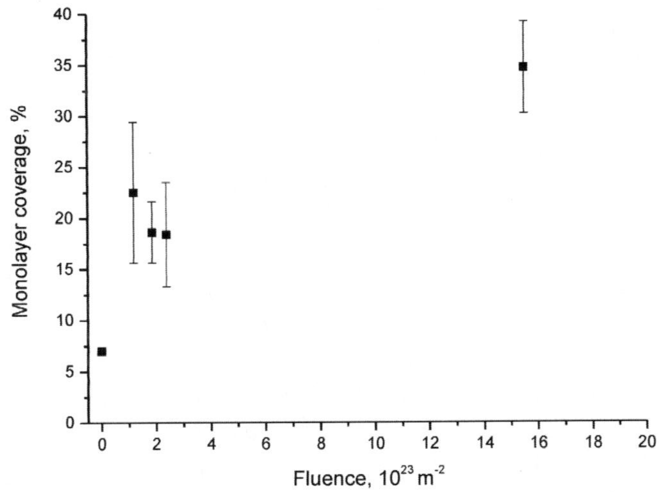

Figure 2. P content in GB of RPV BM (steel 15Cr2MoVA).

Even taking into account uncertainties mentioned above, an irradiation dose dependence of P content in the GB of this BM can be supposed. The changes in T_K were significant but lay under the normative limitations and P content in the GB was considerably lower than in the previous case. Nevertheless the prevailing intergranular fracture of irradiated specimens denotes an contribution of GB embrittlement.

Thus the above data shows that increased irradiation temperatures and Ni content can promote intergranular embrittlement of VVER-type RPV. The both factors are typical of VVER-1000 reactors.

3. VVER-1000 reactors

Though typical P content in VVER-1000 RPV materials is about two times lower than that in the 1st generation steels, the necessity of taking into account of the non-hardening mechanism contribution to the total embrittlement is conditioned by existing projects to extend service life of VVER-1000 up to 60 years and more. The results obtained in NRC "Kurchatov Institute" by AES [10, 11] in the frame of surveillance specimens investigations definitely reveal a significant increase in P content in GB of both base and weld metals (together with brittle intergranular fracture share [3]) at fluences typical to unextended service life. Some of the results are presented in figure 3. The AES procedures were improved by means of use of modern AES spectrometer, by increase of measurement number and proper selection of areas in fracture surfaces [10].

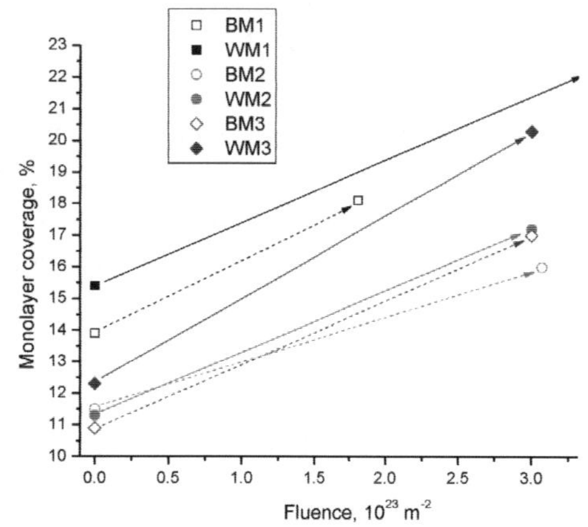

Figure 3. P content in GB of VVER-1000 RPV materials.

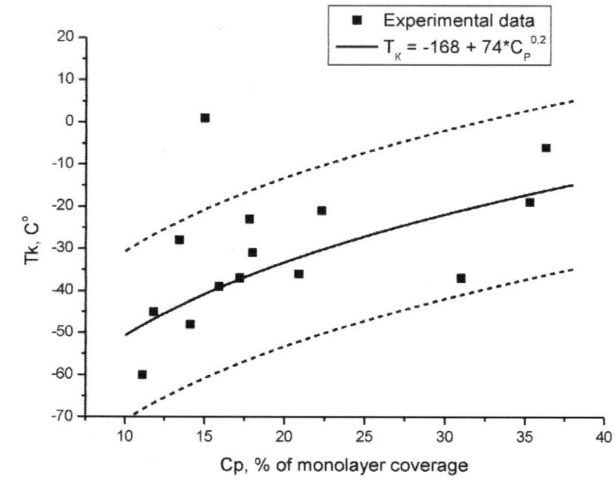

Figure 4. AES and impact test results data base.

Moreover the following tendencies were revealed:

- The increase in P content in GP during fast neutron irradiation is evident but significantly lower in comparison with the 1st generation materials due to lower P content;
- The effect of intergranular embrittlement is higher in the case of WM where Ni content is higher (up to 2 wt. % when compared with about 1% in BM);
- Segregation of P to GB takes place at the operational temperature 320°C in unirradiated BM of nozzle shells, although to a lesser degree;
- Irradiation in a research reactor at high fast neutron fluxes (up to 2 orders of magnitude higher than in the case of surveillance specimens) does not lead to a significant increase of P content in GB of these materials;
- An annealing at 565°C/100 h provides recovery of irradiated WM to the initial state including GB.

It is impossible to predict VVER-1000 RPV materials behavior under irradiation during extended service life basing on the surveillance specimens tests because of the absence of specimens irradiated up to 60 years. The only way to do this is the use of accelerated irradiations in research reactors. But, as was mentioned above, the irradiation at high flexes can lead to underestimation of GB embrittlement. On the basis of mechanical tests data base analysis and a model of thermally-stimulated P accumulation in GB a procedure of using the results of accelerated irradiation to expand the application range of the dependence based on surveillance specimens testing [12]. This dependence presents T_K shift as a sum of two items. The first is the T_K shift after accelerated irradiation and it takes into account hardening mechanism only (due to irradiation defects and irradiation-induced precipitation). The second item takes into account intergranular embrittlement using the model of P segregation.

To estimate possible changes in T_K due to increase in P content in GB a data base of AES results vs Charpy tests results was used. To avoid an influence of ferrite matrix hardening, only unirradiated VVER-1000 BM were selected with the similar yield stress values. The materials included unirradiated surveillance specimens, thermal sets of surveillance specimens and BM subjected to special heat treatments – see figure 4.

Using the least-squares method a dependence was derived: $T_K = -168 + 74*C_P^{0,2}$ (σ=10°C). The dependence is also presented in figure 4 with upper and lower envelope. It predicts up to 2°C increase in T_K with 1% increase of P coverage of GB. Therefore the total T_K shift during extended service life can reach tens of °C and must be controlled carefully.

4. Conclusions

The non-hardening mechanism conditioned by segregation of impurities (mainly P) in GB of VVER-type RPV materials under operating factors (fast neutron fluence, high temperature) can lead to a significant degradation of mechanical characteristics of the steels. The first generation RPV base metals demonstrate a propensity to intergranular embrittlement particularly at high Ni content that is typical of VVER-100 RPV.

Both VVER-1000 base and especially weld metals are liable to intergranular embrittlement but in the lesser degree than the 1st generation materials. The necessity of VVER-1000 service life extending up to 60 years forces to pay special attention to this phenomenon and take it into account when forecasting mechanical properties of RPV materials during extended service life using accelerated irradiation in research reactors. The total T_K shift during extended service life can reach tens of °C and must be controlled carefully. The annealing procedure can be used to recover the operation characteristics of the steels.

Acknowledgements
This work was performed within the framework of the Center of Nuclear Systems and Materials supported by MEPhI Academic Excellence Project (contract № 02.a03.21.0005, 27.08.2013).

References

[1] Alekseenko N, Amaev A, Gorynin I and Nikolaev V 1983 Radiation *Damage of Reac tor Pressure Vessel Steels* (Moscow: Energoatomizdat) (in Russian)
[2] Gurovich B, Kuleshova E, Erak D, Chernobaeva A and Zabusov O 2009 *J. Nucl. Mater.* **389** 490
[3] Gurovich B, Kuleshova E, Shtrombakh Y, Fedotova S, Zabusov O, Prikhodko K and Zhurko D 2013 *J. Nucl. Mat.* **434** 72
[4] Kuleshova E, Gurovich B, Shtrombakh Y, Erak D and Lavrenchuk O 2002 *J. Nucl. Mat.* **300** (2-3) 127
[5] Miller M, Jayaram R and Russell K 1995 *J. Nucl. Mater.* **225** 215
[6] Zabusov O, Dementjev A, Koroljov Y and Krasikov E 1995 *Proceedings of the Academy of Sciences: Metals* **3** 141 (in Russian)
[7] Seah M 1977 *Acta Metall.* **25** (3) 345
[8] Zabusov O, Krasikov E, Kozodaev M, Suvorov A, Pareige P and Radiguet B 2003 *VANT: Physics of Radiation Damages and Phenomena in Solids* **83** 66 (in Russian)
[9] Seah MP 1977 *Acta Metall.***25** (3) 345
[10] Zabusov O, Saltykov M, Gurovich B, Kuleshova E, Fedotova S and Zhurko D 2013 *VANT, Physics of Radiation Damages and Phenomena in Solids* **84** (2) 82 (in Russian)
[11] Zabusov O, Gurovich B, Kuleshova E, Saltykov M, Fedotova S and Dementjev A 2014 *Key Engineering Materials* **592-593** 573
[12] Erak D, Gurovich B, Zhurko D, Kuleshova E, Chernobaeva A, Papina V, Zabusov O and Hodan A 2015 *Proc. Int. Conf. SMiRT (Manchester)* Paper ID 383

IOP Conf. Series: Materials Science and Engineering **130** (2016) 012004 doi:10.1088/1757-899X/130/1/012004

Influence of structural-phase state of ferritic-martensitic steels on the helium porosity development

I I Chernov, M S Staltsov, B A Kalin, I A Bogachev, L Yu Guseva, P S Dzhumaev, O V Emelyanova, M V Drozhzhina, K V Manukovsky and I D Nikolaeva

National Research Nuclear University MEPhI (Moscow Engineering Physics Institute) Kashirskoe highway 31, Moscow, 115409, Russia

E-mail: i_chernov@mail.ru

Abstract. Transmission electron microscopy (TEM) has been used to study the effect of the initial structural-phase state (SPhS) of ferritic-martensitic steels EK-181, EP-450 and EP-450-ODS (with 0.5 wt.% nanoparticles of Y_2O_3) on the of helium porosity formation and gas swelling. Different SPhS of steel EK-181 was produced by water quenching, annealing, normalizing plus tempered, intensive plastic deformation by torsion (HPDT). Irradiation was carried out by He^+-40 keV ions at 923 K up to fluence of $5{\cdot}10^{20}$ He^+/m^2. It is shown that the water quenching causes the formation of uniformly distributed small bubbles ($\bar{d}$ ~ 2 nm) of the highest density (ρ~ 10^{25} m^{-3}). After normalization followed by tempering as well as after annealing bubbles distribution is highly non-uniform both by volume and in size. Very large faceted bubbles (pre-equilibrium gas-filled voids) are formed in ferrite grains resulting in high level of gas swelling of the irradiated layer with S = 4,9 ± 1,2 and 3.8 ± 0.9 % respectively. Nano- and microcrystalline structure created by HPDT completely degenerate at irradiation temperature and ion irradiation formed bubbles of the same parameters as in the annealed steel. Bubbles formed in EP-450-ODS steel are smaller in size and density, which led to a decrease of helium swelling by 4 times (S = 0.8 ± 0.2 %) as compared to the swelling of the matrix steel EP-450 (S = 3.1 ± 0.7 %).

1.Introduction

One reason for degradation of properties of fusion reactor structural materials and shortening of the useful life of its structural components can be helium, accumulating as a result of neutron bombardment and injection from plasma [1]. Many studies have shown that the chemical composition has a significant effect on radiation damage of structural materials under irradiation in reactor [1–3] as well as on the behavior of helium and the development of gas porosity in them [4–11]. At the same time, structural-phase state (SPhS) of materials can also affect the formation of helium porosity and gas swelling. However, this problem has not been studied.

The aim of this study was to determine the gas porosity formation peculiarities in the surface layer of the reactor ferritic-martensitic steels of different initial SPhS implanted by helium ions

2. Experimental procedure

Chemical composition of the investigated steels shown in Table 1. Various SPhS of steel EK-181 create by different thermo-mechanical treatment (TMT): 1) water quenching from 1373 K / 1 hours

(HT-1); 2) normalizing (1373 K / 40 min, air cooling) + tempering (993 K / 3 hours, air cooling) (HT-2); 3) hot rolling + annealing at 1083 K / 1 hours (HT-3); 4) intensive plastic deformation by torsion (p = 6 GPa, T = 673 K, number of revolutions is 10, rotation speed is 0,5 min^{-1}) (IPDT). Steels EP-450 and EP-450-ODS were after normalization at 1423 K / 1 hours, air cooling + tempering at 993 K / 2 hours, air cooling.

Table 1. Chemical composition of steels (wt. %).

Steel	C	Cr	Mo	V	W	Nb	Ta	Others
EK-181	0.1–0.2	10–12	< 0.01	0.2–1.0	1.5	< 0.01	0.17	0.03Ni; 0.0045B; 0.044N
EP-450	0.10–0.15	12–14	1.2–1.8	0.1–0.3	–	0.2–0.4	–	≤ 0.3Ni; 0.004B
EP-450-ODS	0.10–0.15	12–14	1.2–1.8	0.1–0.3	–	0.2–0.4	–	≤ 0.3Ni; 0.004B; $0.5Y_2O_3$

HT-1 used for achieving of supersaturated by alloying elements and impurities solid solution, although this kind of heat treatment is not applicable to ferritic-martensitic steel. HT-2 is the standard heat treatment for these type of steels, but HT-3 is sometimes considered for use to steel EK-181 also. Consideration of HPDT due to the fact that at one time in the scientific community conceived the idea of the creation of nano- and microcrystalline structure could effectively restrain radiation swelling of materials.

All electropolished steel samples were irradiated under identical conditions by 40 keV He^+ ions at 923 K up to fluence of $5\cdot10^{20}$ m^{-2}. Steel EK-181 in the HT-2 and HT-3 conditions was irradiated to the same fluence in the temperature range of 573–903 K.

Samples for TEM were prepared by two techniques: 1) jet electropolishing from unirradiated side in the device Tenupol-5 for the research front microstructure; 2) cutting out of thin foils perpendicular to the irradiated surface by a focused ion beam (FIB technology) for investigation of microstructure along the trajectory of the ion beam. The microstructure of the irradiated samples was studied in a TEM Libra-120 and FEI Tecnai G2 F20 at accelerating voltages of 120 and 200 kV respectively.

3. Results and Discussion

Figure 1 shows typical TEM images of the EK-181 steel irradiated after the HT-1 and HT-2. Parameters of bubbles are shown in Table 2.

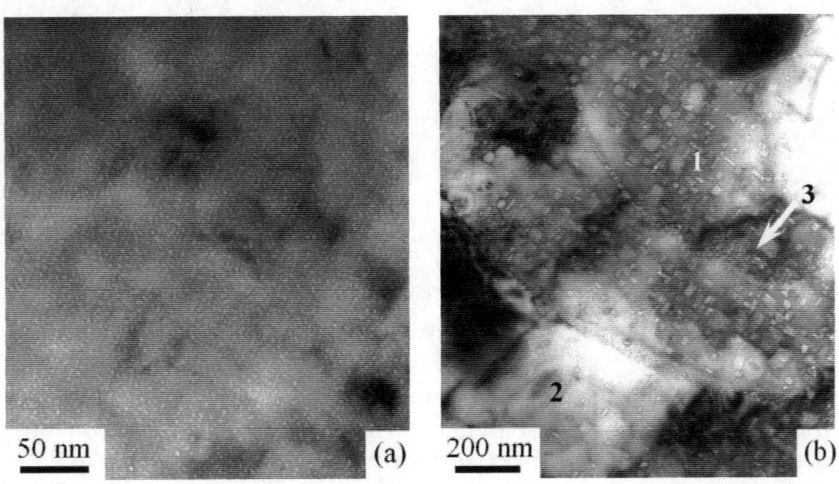

Figure 1. Helium bubbles in steel EK-181 irradiated by He^+ at 923 K: (a) HT-1, (b) HT-2 (Tenupol-5).

Table 2. Parameters of helium bubbles in irradiated steels: d_{max} and $\bar{d}$ are maximum and average sizes respectively, ρ is bulk density of bubbles, S is gas swelling.

Steel and its TMT	d_{max} (nm)	$\bar{d}$ (nm)	ρ (10^{22} m^{-3})	S (%)
EK-181 (HT-1)	~ 3	~ 2	~ 1000	< 0,03
EK-181 (HT-2)	~ 40	12.1	9 ± 3	4.9 ± 1.2
EK-181 (HT-3)	~ 41	11.4	2.2 ± 0.6	3.8 ± 0.9
EK-181 (IPDT)	~ 30	13.1	1.6 ± 0.4	3,8 ± 0,9
EP-450	~ 22	8.2	2.9 ± 0.7	3.1 ± 0.7
EP-450-ODS	~ 27	6.4	1.0 ± 0.3	0.8 ± 0.2

As can be seen in Figure 1(a) and in Table 2, uniformly distributed in the sample volume smallest (d ~ 2 nm) bubbles of highest density (ρ ~ 10^{25} m^{-3}) formed in water quenched steel. The standard heat treatment (HT-2) resulted in formation of bubbles extremely non-uniformly distributed both by volume and in size. Namely, large faceted bubbles (gas-filled pores) are in the ferrite grains (1), also faceted, but more rare bubbles are in sorbite (2) as well as interlayers of small bubbles (3) are in the grain boundary volumes (see Figure 1(b)).

Figure 2(a) shows the depth distribution of helium bubbles in steel EK-181 irradiated by He^+ ions in a HT-3 state. It can be seen that the largest bubbles develop in the layer of a high concentration of radiation vacancies instead of maximum implanted concentration of helium. There is a high density of small bubbles in some volumes of samples at great depths (more than 300 nm) where no radiation vacancies and should be virtually no helium. Probably, this is due to helium permeation through a ion mixing and/or channeling in certain crystallographic directions of the BCC lattice of steel.

Figure 2. Depth distribution of bubbles in EK-181 steel irradiated by He^+ ions at 923 K in a state of HT-3 (FIB); (b) is an enlarged view of bubbles shown in (a). The depth distribution of radiation vacancies and helium calculated by the program SRIM.

Character of the porosity development in the steel with HT-3 (Figure 2(b)) almost the same as the development of porosity in steel irradiated in a state of HT-2 (see Figure 1 (b)). Bubble parameters and helium swelling of annealed steel (HT-3) also differ slightly from the samples irradiated in the state of HT-2 (see Table 2).

Once there was the idea of creating materials with nano- and microcrystalline structure for suppression a radiation swelling. Figure 3 shows the character of bubbles development in the EK-181

IOP Conf. Series: Materials Science and Engineering **130** (2016) 012004 doi:10.1088/1757-899X/130/1/012004

steel irradiated by He^+ ions in the SPhS created using IPDT. As seen in Figure 3, the type of porosity does not differ from the nature of the porosity of the steel irradiated after HT-3 (see Figure 2(b)). As this takes place, the bubble parameters and level of gas swelling are practically identical (see Table 2). This is due to a complete degeneration of nano- and micro-crystalline structure created by IPDT even at an annealing temperature of 873 K (Figure 4 (c)), especially at more higher irradiation temperature of 923 K.

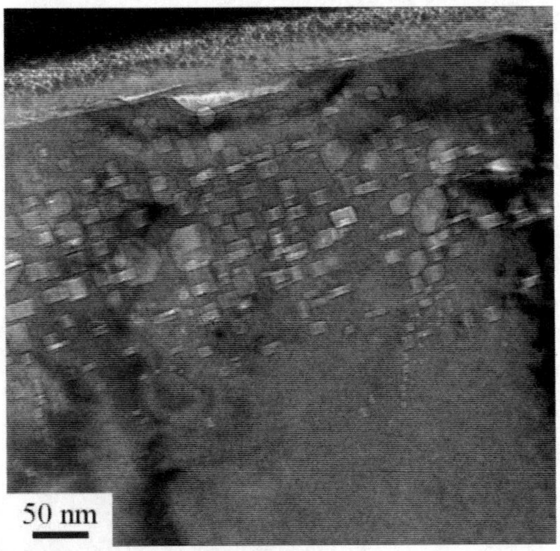

Figure 3. Helium bubbles in EK-181 steel irradiated by He^+ ions at 923 K in IPDT condition (FIB).

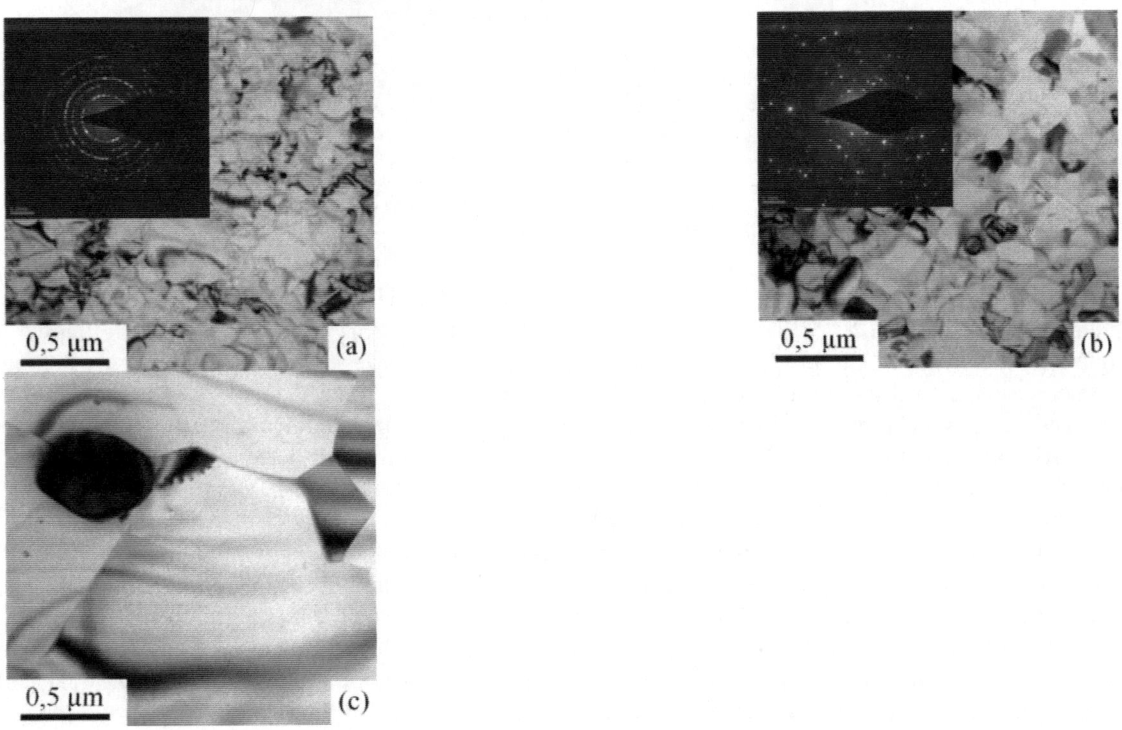

Figure 4. Change of structural-phase states of EK-181 steel under annealings: (a) after IPDT, mean grain size is d_g = 150 nm; (b) annealing at 723 K / 1 hours, mean grain size is d_g = 170 nm; (c) annealing at 873 K / 1 hours, mean grain size is d_g = 1,2 μm (SEM).

Figure 5 shows the temperature dependence of the helium swelling of EK-181 steel irradiated in two initial states. Visible in TEM helium bubbles appear already at a irradiation temperature of T_{irrad} = 573 K in annealed at 1083 K / 1 hours (HT-3) EK-181 steel. Bubble sizes and gas swelling are increasing continuously with increasing of T_{irrad} and swelling is about 0.6% at T_{irrad} = 903 K (see Figure 5). In contrast to the HT-3, the standard heat treatment of steel (HT-2) restrains the bubbles formation up to T_{irrad} = 773 K, but swelling increases significantly and reaches to 1,3 % at high irradiation temperature (T_{irrad} = 903 K) due to the formation in ferrite grains larger bubbles with d_{max} ~ 20 nm.

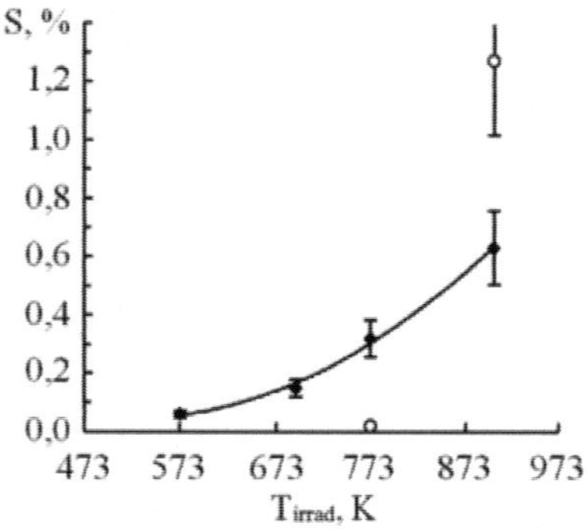

Figure 5. Temperature dependence of the gas swelling in steel EK-181 irradiated by He^+ ions in the two initial states: after the HT-2 (○) and HT-3 (●).

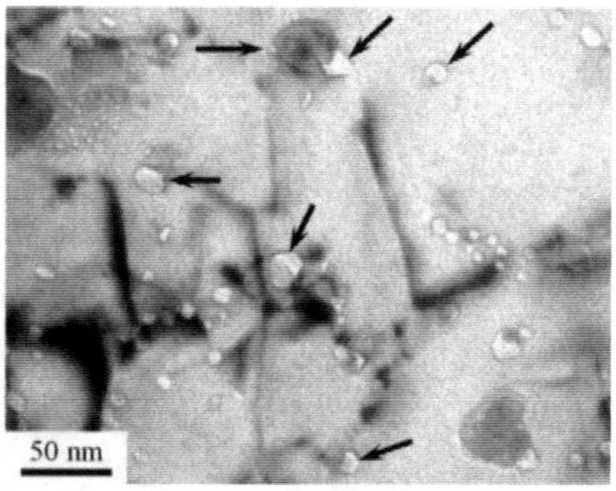

Figure 6. A typical view of helium bubbles in steel EP-450-ODS irradiated by He^+ ions at 923 K (Tenupol-5).

For all the investigated ferritic-martensitic steels irradiated by He^+ ions at a high temperature up to fluence of $5{\cdot}10^{20}$ m^{-2} [12–15], helium bubbles formed about the same character, similar to shown in Figure 1(b). Formed in EP-450-ODS steel bubbles are smaller in size and lower in density, which led to a decrease the swelling 4 times as compared with swelling of the matrix steel EP-450 (see Table 2). Features of gas bubbles development in the steel EP-450-ODS are: 1) bubbles formed at the boundaries of almost all dispersed particles of Y_2O_3, 2) large volumes of the sample does not contain bubbles (Figure 6). Having a high mobility in the metals, the atoms of helium accumulate to the incoherent or semicoherent interfaces of "strengthening particles – matrix", as the particles of Y_2O_3 can not be coherent fully with the matrix due to nonisomorphism of lattices of steel and yttrium oxide (ferrite has a BCC structure with lattice parameter a = 0.287 nm and Y_2O_3 has cubic structure of C-type with a = 1,059 nm). This development of bubbles in the ODS steel restrains the formation of large gas bubbles in the matrix and reduces a gas swelling. At the same time the bubbles on the boundaries of dispersed particles have a high binding energy to them causing the appearance of additional high temperature peaks of gas release in thermal desorption studies [15].

4. Conclusions

Based on the obtained experimental data about the influence of the initial SPhS of reactor ferritic-martensitic steels on the helium porosity formation pecuiliarities it can make the following conclusions.

1. It was shown that the state of a supersaturated solid solution of alloying elements and impurities created by quenching of steel EK-181, restrains the formation of helium porosity and gas swelling effectively.

2. It was established that under standart heat treatment (normalization + high temperature tempering) or annealing, high gas swelling observed because of large bubbles formation (gas-filled pores) in the ferrite grains of ferritic-martensitic steel EK-181.

3. It was found that nano- and microcrystalline structure created by IPDT degenerates fully with heating temperature of 873 K, and steel EK-181 swells in a similar manner as in the annealed condition under irradiation by He^+ ions at 923 K.

4. It was shown that in the steel EP-450-ODS bubbles formed smaller in size and in density, which resulted to a decrease in the helium swelling of the 4 times as compared with swelling of the matrix steel EP-450.

Acknowledgements
This work was performed within the framework of the Center of Nuclear Systems and Materials supported by MEPhI Academic Excellence Project (contract № 02.a03.21.0005, 27.08.2013) and was carried out under implementation of state assignment in the sphere of scientific activities №3.483.2014/K dated 10.06.2014.

References

[1] Kalin B A, Platonov P A, Tuzov Yu V, Chernov I I and Shtrombah Ya I 2012 *Physical Materials Science* (*Structural Materials for Nuclear Technology* vol 6) ed B A Kalin (Moscow: NRNU MEPhI) 736 pp (in Russian)
[2] Zelensky V F, Nekludov I M and Chernyaeva T P 1988 *Radiation Defects and Swelling of Metals* (Kiev: Naukova Dumka) 296 pp (in Russian)
[3] Zaluzhnij A G, Sokurskij Yu N and Tebus V V 1988 *Helium in Reactor Materials* (Moscow: Energoatomizdat) 224 pp (in Russian)
[4] Kalin B A, Chernov I I, Bogachev A G, Kalashnikov A N, Markin A V, Reutov I V and Korshunov S N 1992 *Mater. Sci. Forum* **97&99** 373
[5] Kalin B A, Chernov I I, Kalashnikov A N and Solovyev B G 1996 *Plasma Devices and Operations* **4** 313
[6] Kalin B A, Chernov I I, Kalashnikov A N and Timofeyev A A 1996 *J. Nucl. Mater.* **233-237** 1142–47
[7] Kalin B A, Chernov I I, Kalashnikov A N, Binyukova S Yu, Timofeyev A A and Dedyurin A I 2002 *Atomic Energy* **92** 50
[8] Kalashnikov AN, Chernov I I, Kalin B A and Binyukova S Yu 2002 *J. Nucl. Mater.* **307-311** 362
[9] Chernov I I, Staltsov M S, Kalin B A, Mezina O S, Kyi Zin Oo and Chernov V M 2011 *Atomic Energy* **109** 176–83
[10] Staltsov M S, Chernov I I, Aung Kyaw Zaw, Kalin B A and Kyi Zin Oo 2014 *Atomic Energy* **116** 35
[11] Staltsov M S, Chernov I I, Kalin B A, Kyi Zin Oo, Polyansky A A, Staltsova O S, Aung Kyaw Zaw, Chernov V M and Potapenko M M 2015 *J. Nucl. Mater.* **461** 56
[12] Chernov I I, Kalashnikov A N, Kalin B A and Binyukova S Yu 2003 *J. Nucl. Mater.* **323** 341
[13] Chernov I I, Binyukova S Yu, Kalin B A, Myo Htet Win, Than Swe, Chubarov S V, Kalashnikov A N, Yoltukhovsky A G and Leontyeva-Smirnova M V 2007 *J. Nucl. Mater.* **367-370** 468
[14] Binyukova S Yu, Chernov I I, Kalin B A and Than Swe 2007 *J. Nucl. Mater.* **367-370** 500
[15] Chernov I I, Kalin B A, Staltsov M S, Kyi Zin Oo, Binyukova S Yu, Staltsova O S, Polyansky A A, Ageev V S and Nikitina A A 2015 *J. Nucl. Mater.* **459** 259

The impact of the initial state on the kinetics of oxidation ion-modified fuel cladding alloy E110

B A Kalin, N V Volkov, R A Valikov, A S Yashin and T V Yakutkina

National Research Nuclear University MEPhI (Moscow Engineering Physics Institute) Kashirskoe highway 31, 115409 Moscow, Russia

E-mail: Romansbk@bk.ru

Abstract. The paper examines the impact of the initial state (the presence of impurities, surface preparation), and surface alloying on the kinetics of the oxidation of fuel cladding alloy E110. The studies concluded that the use of ionic polishing instead of traditional chemical polishing helps to reduce the rate of oxidation of zirconium alloys. Also studied the effect of alloying elements introduced in the surface layers of claddings by ion mixing on the kinetics of the oxidation of the alloy E110.

1. Introduction

Increasing resistance to corrosion of fuel cladding alloys of zirconium in the working conditions of pressurized water reactors is an important task of modern reactor materialscience. A significant role in the kinetics of oxidation has surface quality (the presence of impurities, roughness) [1]. Currently used mechanical polishingof the outer surface of the fuel rods can not achieve high levels of surface roughness and wherein the surface is contaminated with abrasive particles. The use of chemical polishing significantly pollute surface residues of acids used for etching. In this regard, it proposed to use the method of ion polishing at the final stage of manufacture of fuel elements for cleaning the outer surface from dirt and upgrading of surface roughness.

2. Experimental methods

The samples cladding diameter of 9.15 mm and a wall thickness of 0.65 mm from several batches on the basis of E110 alloy (Zr-1%Nb). In comparison with the original alloy E110 in the part of thebatches introduced impurity of nitrogen and carbon. Samples of each batch were divided into groups: the first group took only mechanical grinding, the second part of the samples was subjected to chemical etching and polishing in the electrolyte CH_3COOH + 10% $NClO_4$, and part of the samples was purified argon ion beam.

The doping of the surface layers of samples carried out by ion mixing of the deposited film by irradiation beams of Ar^+ ions with a broad energy spectrum on the "VOKAL" [2] (the average energy of the ions in the beam $\langle E \rangle$ = 10 keV, the ion current density j = 25 mA/cm^2, dose F= (1-5)$\times 10^{18}$ iom·cm^{-2}, the temperature of the samples under irradiation T≤200°C, residual gas pressure p = 1×10^{-6} Pa).

After preparation and doping samples were tested in the autoclave with water at temperature of 350 °C at a pressure of 17 MPa, which corresponds to the operating conditions of VVER reactors. At various stages of the autoclaving the samples were studied by scanning ion microscopy (SIM), which allows you to assess the condition and thickness of the oxide film.

IOP Conf. Series: Materials Science and Engineering **130** (2016) 012005 doi:10.1088/1757-899X/130/1/012005

3. Experimental results and discussions

It is known that oxidation processes are highly dependent on the surface topography. Developed relief significantly increases the area of interaction products with a corrosive environment, thereby increasing the rate of permeation of oxygen to the boundary of the "metal oxide". Measurement of surface roughness of the samples after preparation was conducted on the profilometer TR-200 (Japan). Value of roughness R_a of the samples after mechanical grinding was 1.7±0.2 μm (Figure 1), after the chemical polishing – 0.7±0.2 μm. In samples subjected to argon ion polishing value of roughness (R_a) reached 0.6±0.1 μm (Figure 2).

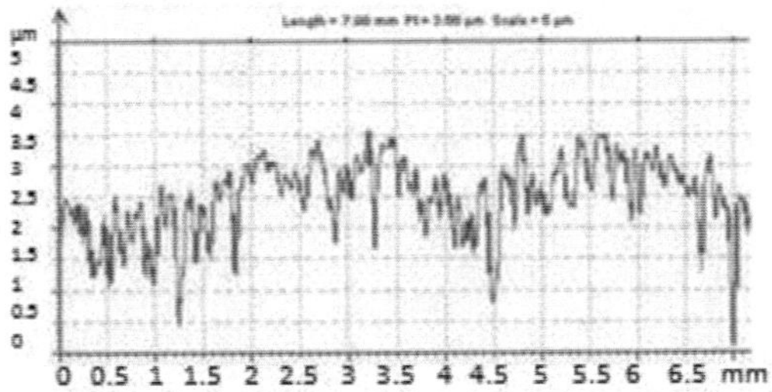

Figure 1. Profilogram sample surface after mechanical polishing.

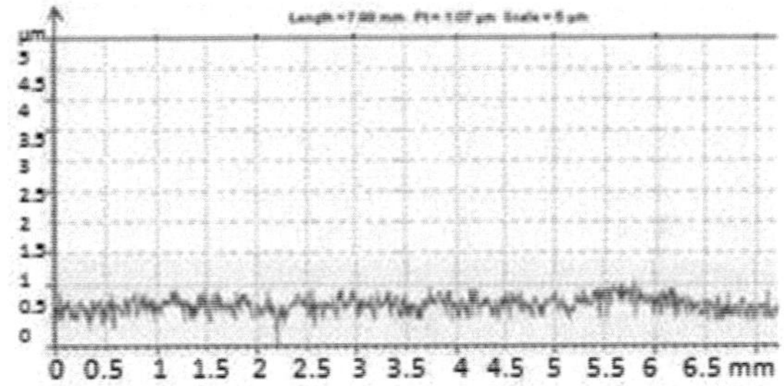

Figure 2. Profilogram sample surface after polishing ions Ar^+.

Previous research of the state of oxide films have shown that the structure of the films in the main repeating structural features of the substrate, in consequence of which there is a different film growth rate in individual grains [3, 4]. Therefore, the presence of impurities and contaminants on the surface of the sample promotes violatehomogeneity of the oxide film, an increase of internal stresses in the oxide,and as a result, poor adhesion of the oxide to the metal substrate. The presence of contaminants on the surface of the samples was determined by X-ray fluorescence analysis, which showed the presence of significant amounts of chlorine and carbon on the surface of samples after the chemical polishing. For samples that have passed the ion cleaning and polishing the presence of impurities on the surface is minimal.

Autoclaved test of the samples were carried out during 4320 hours. To study the influence of the initial state on the oxidation kinetics, oxide films of the samples were researched after 300, 720 and 2160 hours of oxidation. SIM method used to measure the thickness of the oxide film, was determined presence of pores, cracks and inclusions on the surface and in the bulk oxide (Figures 3, 4).

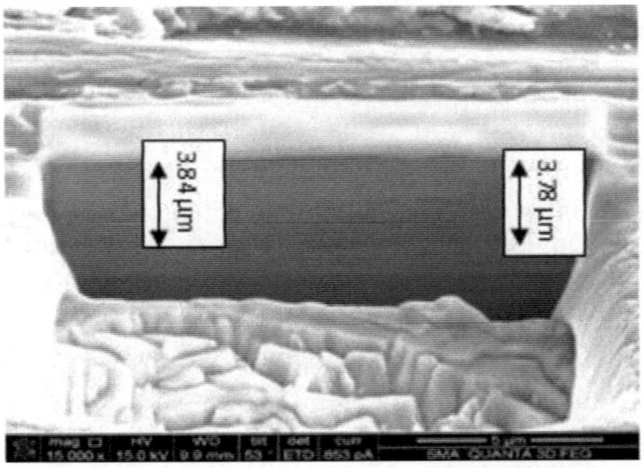

Figure 3. Photograph of a cross structure of the oxide film of the sample after the mechanical grinding and autoclave test for 4320 hours.

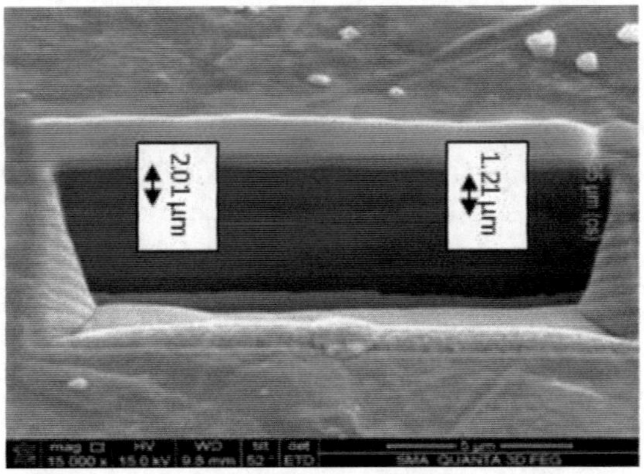

Figure 4. Photograph of a cross structure of the oxide film of the sample after the chemical polishing treatment + Ar^+ ion beam and autoclave test for 4320 hours.

As shown in Table 1 for the first 720 hours of oxidation on all samples of E110 alloy grows homogeneous solid oxide film regardless of the method of surface preparation. However, the use of ion treatment external surface slows down its rate of growth. After 2160 hours of autoclave tests in samples of untreated and polishing of the ion beam in the oxide film are formed pores and cracks that occur due to internal stresses in the oxide caused by microscopic irregularities of surface topography and the presence of contaminants and undesirable impurities.

Table 1. Results of the autoclave test.

	mechanical-polishing	chemical-polishing	mechanicalpolishing + ion polishing	chemicalpolishing + ion polishing
300 hours of autoclave test	Oxide thickness – 1.0±0.1 μm, No pores	Oxide thickness – 0.9±0.1 μm, No pores	Oxide thickness – 1.0±0.1 μm, No pores	Oxide thickness – 0.8±0.1 μm, No pores
720hours of autoclave test	Oxide thickness – 1.6±0.1 μm, No pores	Oxide thickness – 1.3±0.1 μm, No pores	Oxide thickness – 1.3±0.1 μm, No pores	Oxide thickness – 1.2±0.1 μm, No pores
2160hours of autoclave test	Oxide thickness – 2.1±0.1 μm, Pore density – 0.5%	Oxide thickness – 1.7±0.1 μm, Pore density – 0.5%	Oxide thickness – 1.7±0.1 μm, No pores	Oxide thickness – 1.7±0.1 μm, No pores
4320 hours of autoclave test	Oxide thickness – 3.8±0.1 μm, Pore density – 2.0%	Oxide thickness – 3.0±0.1 μm, Pore density – 2.5%	Oxide thickness – 2.3±0.1 μm, Pore density – 0.5%	Oxide thickness – 1.9±0.1 μm, No pores

The presence of pores and cracks in the bulk oxide facilitates the penetration atoms of oxygen to the boundary of the "metal oxide", which causes the acceleration of the kinetics of oxidation of the samples that passed the mechanical and chemical polishing of the surface, which significantly according to the thickness of oxide films of the samples after 4320 hours of autoclave tests. In addition, the presence of pores in the oxide causes an increase in internal stress, which leads to the further development and increase the number of voids in the oxide film.

Samples of special melting E110 alloy with impurities of carbon and nitrogen, depending showed similar growth rate of the oxide film, depending on the initial state of the surface, but have a lower corrosion resistance as compared with the reference samples of E110 alloy.

4. Conclusion

The experiments confirmed the considerable influence of the surface state and the ways of its preparation on the kinetics of oxidation zirconium alloys. It is shown that mechanical grinding cannot produce sufficiently high surface roughness in comparison with the chemical polishing and surface treatment of ionic claddings. Also, because a sufficiently high ductility alloy E110 during mechanical polishing are abrasive particles. As a result, the corrosion resistance of E110 alloy is reduced due to the state of the surface.

Chemical polishing allows to achieve high values of surface roughness, however, for improving the corrosion resistance needs to cleaned the surface from impurities remaining from the etchant, such as chlorine and carbon.

Surface claddings argon ion beam, as a final step of manufacturing the fuel elements can slow the growth rate of the oxide film on the outer surface of the fuel cladding of VVER reactors by 30-40% due to the removal of impurities and smoothing the microscopic irregularities of the relief, the remaining after mechanical polishing. Thus, application of the technology of ion polishing in addition to mechanical polishing or chemical polishing can significantly improve corrosion resistance of the alloy product E110.

Acknowledgements
This work was performed within the framework of the Center of Nuclear Systems and Materials supported by MEPhI Academic Excellence Project (contract № 02.a03.21.0005, 27.08.2013).

References

[1] Volkov N V , Kalin B A, Valikov R A, Yashin A S and Yalcev V N 2012 *Russian Physics Journal* **55** (12-3) 186
[2] Volkov N V and Kalin B A 1991 *Dig. Material science questions of nuclear technology* (M.: Energoatomizdat) p 64
[3] Volkov N V 2011 *Surface. X-ray, Synchrotron and Neutron researches* **3** 40
[4] Kalin B A and Volkov N V 2007 *Dig. scientific papers "Scientific session of the MEPhI 2007"* **9** 44

IOP Conf. Series: Materials Science and Engineering **130** (2016) 012006 doi:10.1088/1757-899X/130/1/012006

Influence of thin oxide layers on tribological properties of E110 alloy tubular specimens under dry friction conditions

B A Kalin, N V Volkov, R A Valikov, A S Yashin and T V Yakutkina

National Research Nuclear University MEPhI (Moscow Engineering Physics Institute) Kashirskoe highway 31, Moscow, 115409, Russia

E-mail: NVVolkov@mail.ru

Abstract. Experiments to simulate wear process of fuel cladding in case its contact with spacing grid. System «sphere-plane» selected as friction pair. Tubular parts of fuel claddings from E110 alloy diameter of 9.15 mm and length of 50 mm (wall thickness 1 mm) were used for investigations. Some claddings were subjected to ion cleaning and polishing under the influence of Ar^+ ion beam with average energy of 3 keV. Samples were oxidized in steam-water conditions (*T*=300° C, *p*=17 MPa, time up to 100 h) to create thin oxide layers with a thickness of 1 mkm on the tubes surface. It is found that wear of the metallic samples takes place in elastically plastic deformation conditions at initial stage (2-5 min). Presenceof thin oxide layer (of thickness up to 200 nm) on the samples surface contributes to reduce wear due to the uniform redistribution its fragments on the friction track, and wear also samples takes place in elastically plastic deformation conditions. Presence of oxide layer with thickness of 700 nm on the samples surface increases wear in conditions of abrasion friction.

1. Introduction

Progress of nuclear energy and nuclear technology has put forward a completely new tasks in the area of tribology. For example, one of the most pressing tasks is to increase the wear resistance (endurance) and service life of tribo-contact pairs, such as fuel cladding - spacing grid.

2. Materials and tools

Tubular parts of fuel claddings from E110 alloy diameter of 9.15 mm and length of 50 mm (wall thickness 0.65 mm) were used for investigations.

Some claddings were subjected to ion cleaning and polishingunder the influence of Ar^+ ion beam with average energy of 3 keV, on the installation ILUR-03 [1] designed for ion beam processing the surface of long tubes with a diameter up to 20 mm (Figure 1). Samples of the tubes in the initial state and after ion treatment were oxidized in a steam environment (T = 300 °C, P = 17 MPa, a time of 10, 20 and 100 hours) for creating on the surface of the tubes of thin oxide layers. Before the experiments, in order to control the process of deterioration, we measured the surface roughness R_a on profilograph TR-200 and the surface microhardness of the PMT-3 (136° angle of a diamond pyramid, the load 90 and 120 g).

Analysis of the conditions of deterioration shells in the case of its contact with the spacer grid shows [2,3] that the closest approximation to real conditions corresponds to deterioration of a pair "sphere-plane" (Figure 2). The motion of a spherical indenter is advantageously carried out around the circumference of the shell, i.e., while rotating the tube around its axis. The indenter motion along tube geneatrix increases influence of local anisotropy of properties.

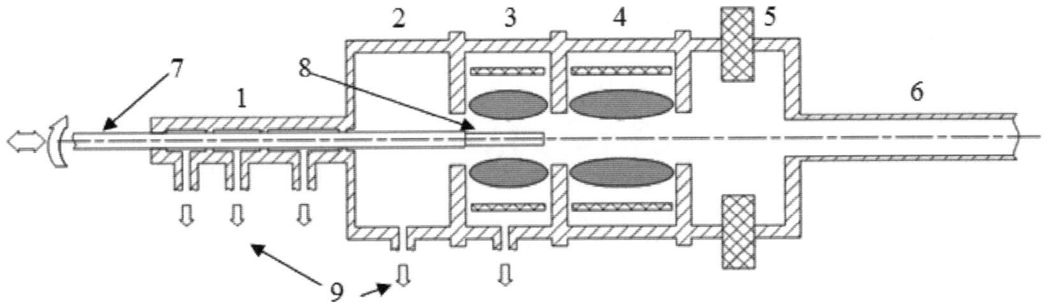

Figure 1. The scheme of the installation ILUR-03: 1 – gateway, 2 – transition chamber, 3 – ion cleaning chamber, 4 – ion polishing chamber, 5 – magnetron chamber, 6 – chamber to exit processing items, 7 – manipulator with a sample holder, 8 – processing item, 9 – vacuum pump connection.

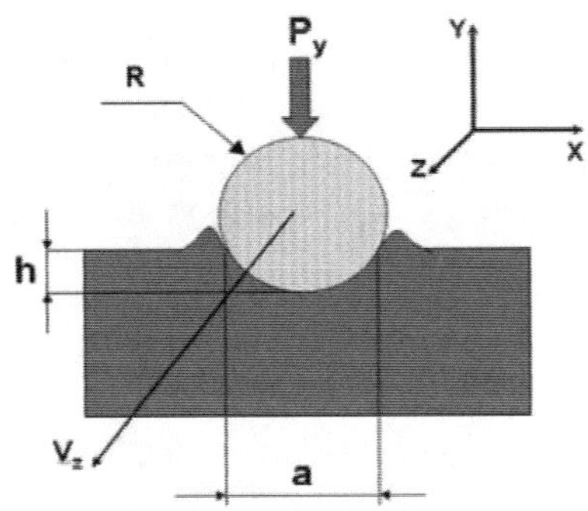

Figure 2. The scheme of deterioration of the friction pair "sphere-plane": 1 - cylindrical sample, 2 - ball indenter of radius R = 0,5 mm, a – width of the friction track, h - depth of the friction track, P_y - load, V_z - speed of movement.

Tribological experiments were conducted on the installation Friction-T (diagram is shown in Figure 3). As the friction pair selected "sphere-cylinder" diameter of the spherical indenter 1 mm, the type of friction - dry, load on the indenter P_y - constant. Selection of load caused by need to ensure the conditions close to the real, i.e., on the indenter at a load of 35 g corresponds to a pressure of 170 MPa, 50 g - 200 MPa.

Deterioration is implemented with a movement of the indenter along the surface with the velocity V_z = 1.4 cm/sec, which corresponds to the actual fuel rod conditions, the oscillation frequency of 0.5 Hz. Time Test - 2, 5, 10, 20, 30, 40, 60 minutes. As the main controlled parameter of deterioration was selected width of friction track – a, which was measured using an optical or scanning electron microscopy [4].

The approximation model of the "ideal indentation" of the ball in the cylindrical shell, friction track depth may be calculated according to the formula $h(t) = R- \{R^2- [0.5a\ (t)]^2\}^{0.5}$ where: h - the depth of the friction track, R - the radius of the spherical indenter of 0.5 mm, a - the width of the friction track.

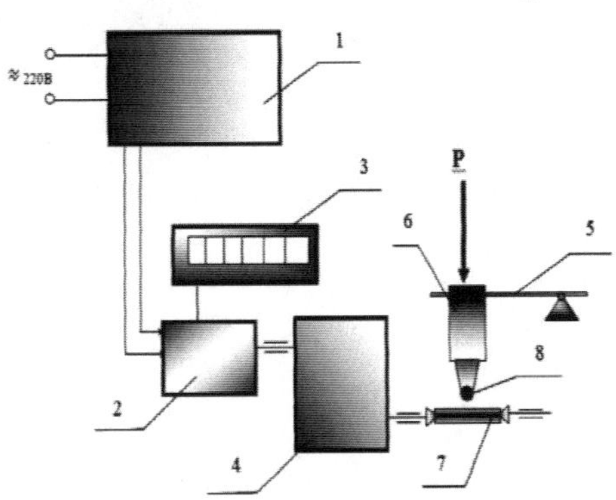

Figure 3. The Installation scheme Friction-T: 1 - power supply, 2-motor, 3 - speed control panel, 4 - precision reducers, 5 - balk of the loading indenter and its movement along the sample, 6 - the holder of a spherical indenter, 7 - cylindrical sample, 8 - ball indenter.

Volumetric wear I_v and linear wear rate I_h friction surface samples was calculated using the measure of h(t), which allows on the basis of tests carried out accelerated depreciation forecast to hold for long term use products [5,6].

3.Results and discussion

The measurements of the surface roughness shells samples indicate that in the normal state (after mechanical polishing), the value of surface roughness value is typically R_a = 0.9-1.5μm after ion treatment is reduced to R_a = 0.5-0.7 μm, and after oxidation in a steam environment is at level: the samples in the normal state - R_a = 1.5-1.0 μm, after ion cleaning - R_a = 0.7-0.9 μm.

Microhardness of samples in the normal state varies in the range $H_\mu = 127 \pm 12$ kg mm^{-2}, and after the ion cleaning there is a slight increase to $H_\mu = 140 \pm 12$ kg mm^{-2}.

In Figure 4 exemplary shows photographs friction track samples in normal condition, after 5 and 30 minutes of testing. From which it is seen that at the initial stage of deterioration (Figure 4a) are formed circular (longitudinal) strain bands (indicated by arrows), are also observed metal particles formed as a result of the effect of setting indenter with subsequent separation from the sample.With the increase in test time (figure 4b) the surfaceof friction track filled metal particles, some of them are covered with an oxide layer and pressed into the surface layer of the sample. At this stage, grinding the surface layer of the material and the stabilization of its restructured.When test time more than 30 minutes there is a further coloring material particles and their subsequent grinding, which contribute to increasing the abrasive component of deterioration, the track profile is formed completely of friction and it is further deepening.

In the case of cylindrical samples at the surface of thin oxide layer process of deterioration is slowing. As seen in Figure 4c width of friction track substantially less (test time of 30 minutes: the layer - 214 μm, in a normal state - 350 μm) compared to a sample placed in the normal state (Figure 4a), annular strips of metal deformation is practically not observed . At higher magnification in a metal layer revealed a large number of oxide particles (size of 50-100 nm), pressed into the surface layer of the sample.

Increasing the test period (Figure 4a) causes a broadening of the friction road which state is characterized by a homogeneous structure consisting of fine particles of the oxide layer (smaller than 50 nm). Areas related to the separation of the material are practically not observed.

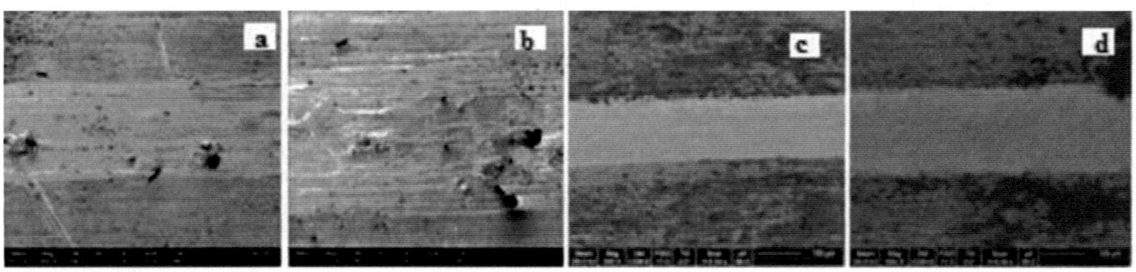

Figure4. Photos of the friction tracks of sample E110 alloy in a normal state and after oxidation in a steam medium: a - the test for 5 minutes; b - the test 30 minutes; c - 5 min test time, the thickness of the oxide layer 200 nm; d - the test 30 minutes, the thickness of the oxide layer 200 nm.

Figure 5 shows experimental data depth change friction tracks of test time h (t) samples in normal condition and the oxide layer (curves 1 and 3), after ion treatment (curve 2), and after ion cleaning with an oxide film (curve 4) that a good approximation to describe the kinetic curves logarithmic following form h (t) = A × Ln (t) + B, where A and B coefficients are selected for each state of the samples.

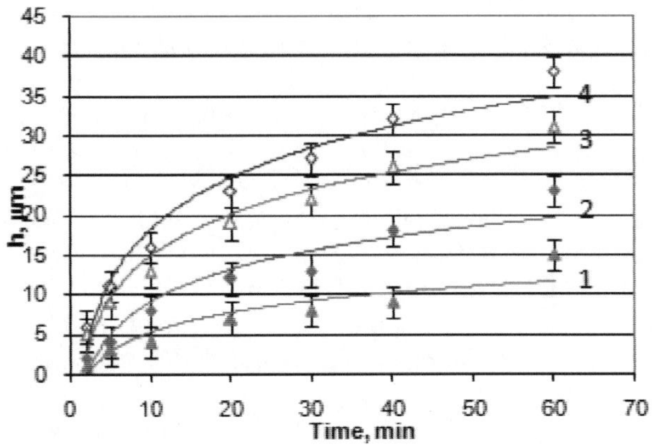

Figure 5. Changing the depth of the friction track of the test time h (t) samples: 1 - sample in normal state; 2 - sample after ion cleaning; 3 - the sample in the normal state with the oxide layer; 4 - sample after ion polishing with the oxide layer.

The presence of the oxide layer on the samples in the normal state and after ion treatment substantially reduces wear. Based on the kinetic dependences of deterioration from test time *h(t)* were calculated loss material volume *V(t)*and the intensity of the linear wear $I_h(t)$.

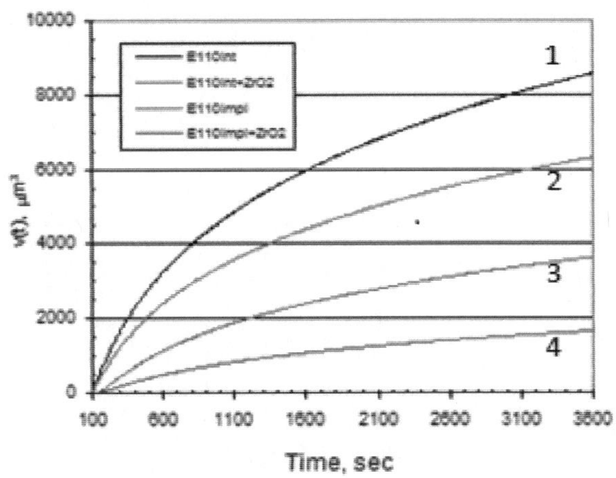

Figure 6. The dependence of volume loss in material from the test time V(t): 1 - the sample in normal state; 2 - sample after ion cleaning; 3 - the sample in the normal state with the oxide layer; 4 - sample after purification by ion oxide layer.

Figure 6 shows the dependence of V(t) the studied samples which show that the minimum loss of volume of the material observed in the samples after ion polishing with oxide layer (curve 4) $V(t) \approx 1500$ $m\mu^3$.

4. Conclusion

Based on the tribological testing of cylindrical samples of zirconium alloys E110 in conditions of dry friction (with a load of 35 g, the rate of movement of the contacting surfaces 2.67 cm/sec) is set as follows:

- at the initial stage of deterioration metal samples (2-5 min) formed with track friction characteristic annular deformation zones, i.e. Wear is under elastic plastic deformation;
- ion polishing reduces abrasive component of wear;
- presence on the surface of samples of thin oxide layer (thickness 200 nm), thereby reducing of deterioration and wear depth approximation depending on the time with a good reliability is described by a logarithmic dependence which is characteristic of friction in terms of elastic-plastic deformation.

Acknowledgements

This work was performed within the framework of the Center of Nuclear Systems and Materials supported by MEPhI Academic Excellence Project (contract № 02.a03.21.0005, 27.08.2013).

IOP Conf. Series: Materials Science and Engineering **130** (2016) 012006 doi:10.1088/1757-899X/130/1/012006

References

[1] Volkov N V, Kalin B A, Valikov R A, Yashin A S and Yalcev V N 2012 *Russian Physics J.* **55** 186

[2] Baranov V M, Gricenko A I, Karasevich A M, Kudryavcev E M, Remizov V V and Sarychev A G 1998 *Acoustic diagnostics and control of fuel and energy complex enterprises* (Moscow: Science) 304

[3] Volkov N V, Ananieva E A, Valikov R A, Kurilkin M O, Oleinikov I V and Phu Vin Tkhein 2010 *Coll. Proc. 9th All-Russia with int. participation scientific and practical conf. "of Rapidly Quenched materials and coatings" BMP* (Moscow: MATI) 226

[4] Kalin B A, Volkov N V and Anan'eva E A 2010 *Journal of Surface Investigation* **4** (3) 538

[5] Khebdy A and Chechevadze A V 1987 *Handbook of Tribology in 3 volumes* (Moscow: Science) **1** 256

[6] Baranov V M, Kudryavcev E M, Sarychev G A and Schavelin V M 1998 *Acoustic emission during friction* (Moscow: Energoatomizdat) 256

Changes of structure and crystallographic texture of cladding tubes from austenitic steel under thermal creep testing

Yu Perlovich, M Isaenkova, V Fesenko, P Dobrokhotov and A Tselishchev

National Research Nuclear University MEPhI (Moscow Engineering Physics Institute) Kashirskoe highway 31,Moscow,115409, Russian Federation

E-mail: yuperl@mail.ru

Abstract. The process of changes in structure and crystallographic texture of cladding tubes from austenitic steel ChS68 under thermal creep testing were studied. Testing of tubes was conducted at the temperature 700 °C in the air by their stretching in axial direction under the stress 160 MPa until rupture. By data of phase and texture analysis a number of processes, accompanying plastic deformation of tubes during thermal creep tests at elevated temperature, were identified. The main texture components of original tube, as well as texture components of different parts of the tested tube are {110}<001> and {112}<11$\bar{1}$>. In the rupture zone the component of the texture of tension with axis <111> along the tube axis becomes stronger. This effect is connected with activation of dislocation slip in the deformed area of tested tube near the new-formed neck. At the same time the character of texture changes in the zone of tube rupture indicates to development of the dynamic recrystallization, conditioned by the total influence of all factors, which control the passage in the tube of thermal creep. In addition, it was revealed the activation of martensitic transformations in the zone of maximal deformation of tube as a result of its creep tests.

1. Introduction

The service period of a fast breeding reactor fuel rods is largely determined by the creep of the cladding material, since its upper part is known to have the highest temperature. Therefore it is necessary to provide the required values of rupture strength and creep rate, which are conditioned by the stability of structure and phase composition of the material during the period of operation. To provide stability of structure it is necessary to conduct systematical study of creep mechanisms of cladding tubes from austenitic steels. Crystallographic texture is sensitive indicator of the processes activated during plastic deformation. Examination of the changes of texture and phase composition along the tested tube reveals features of the processes that are manifested in the high-temperature tests, such as phase transformations, plastic deformation and dynamic recrystallization.

In the current work the regularities of structure and texture changes of cladding tubes from austenitic steel ChS68 as a result of their creep test were studied.

2. Experimental

Cladding tubes 6.9x0.4 mm from ChS68 steel (composition is given in table 1) were examined as follows:

1. Initial state: austenization, 20% cold work (sample 1);
2. Half of a tube after rupture during uniaxial (along the axis of the tube) creep test (sample 2).

Testing proceeded in the air at a tensile stress of 160 MPa, a temperature of 700 °C; time to failure - 6028 hours. According to the measurement of the tube elongation during the test total strain amounts 1.6%. By the measurement of the tested tube diameter (figure 1), minimum residual strain amounts to 0.3%, whereas in the rupture zone strain amounts 2%.

Table 1. Chemical composition of ChS68 steel.

Composition, wt. %													
C	Si	Mn	S	P	Cr	Ni	Mo	Ti	V	B	N	Co	Al
0.05	0.5	1.7	0.006	0.005	16.4	14.8	2.0	0.26	0.1	0.003	0.005	0.01	0.02

Oxide film of the tested tube was removed by electrolytic etching in a mixture of phosphoric acid H_3PO_4 and chromic anhydride CrO_3 at voltage 24V and current value 2.4 A. The thickness of the removed layer is approximately 30-35 µm. X-ray structure analysis and phase analysis were performed for the etched surface of the tested tube. To conduct the study of the crystallographic texture for both initial and tested tube the small rings with height of 3 mm were cut out from the tubes on the electrospark discharge machine. Red lines that points out the cut and red numbers of the corresponding rings used for the texture measurement are placed on figure 1.

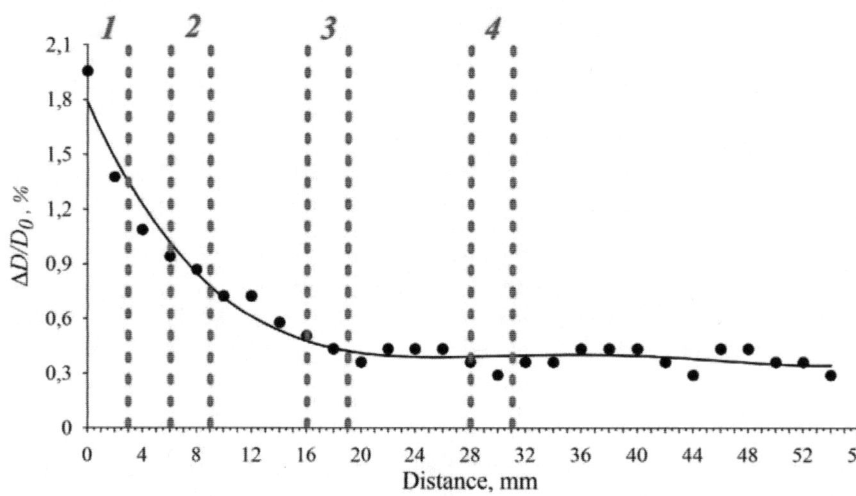

Figure 1. Relative diameter change of cylindrical sample 2 along tested tube. Count starts from the destruction place.

3. Methods

In the current work following methods of structure and texture study were used:

1) X-ray phase analysis method;
2) methods of construction of direct and inverse pole figures.

The investigation of the structure of austenitic tubes was conducted on diffractometer D8 DISCOVER (Bruker, Germany) using filtered radiation CuK_α. For the phase identification Bruker AXS DIFFRAC.EVA v.3.0 software and ICDD PDF-2 database were used.

Quantitative phase analysis of the samples from austenitic steel was not carried out because of the presence of crystallographic texture, which causes significant redistribution of the intensity of diffraction spectrum lines.

The study of crystallographic texture is usually conducted by X-ray measurement of direct pole figures (DPFs) {hkl}, using the tilting of sample [1]. DPF {hkl} is a distribution of a pole density of

IOP Conf. Series: Materials Science and Engineering **130** (2016) 012007 doi:10.1088/1757-899X/130/1/012007

the selected type <hkl> normal in a stereographic projection of the tube sample with its characteristic external axes: a radial *R*, tangential *T* and longitudinal *L*. Pole density is directly proportional to the intensity of the X-ray beam reflected from the respective planes {hkl}. Therefore the registration of the intensity of X-ray beam diffracted from the planes {hkl} for all possible sample orientations relative to the directions of the primary and reflected beams should be conducted for the measurement of DPF. Planes of the selected {hkl} type are reflecting at different orientations of sample. Normal to this planes divides the angle between the primary and reflected beams in two. The amount of the intensity registered is proportional to the volume of grains, taking part in diffracted beam formation.

Texture measurement was conducted on DRON-3 diffractometer, equipped with automatic texture set, with the accumulation of data on the hard drive of a personal computer. Further treatment was carried out with the use of the original software. Incomplete DPF {100}, {110} and {111} were constructed, permitting to determine main texture components of austenitic phase.

Inverse pole figures (IPF) show probable distribution of external axis of interest belonging to the sample relative to the internal crystallographic axis. Plane normal is attributed to the weight proportional to the probability of matching them to the axis of the sample [1].

Direct and inverse pole figures are special cases of orientation distribution function (ODF) [2]. With the use of mathematical instruments needed it is possible to obtain IPFs from ODF. In this paper, the construction of the IPFs carried out by their calculation of ODF, restored on incomplete DPFs using LABOTEX software [3]. This prevented the preparation of texture free reference sample.

4. Results

4.1. Phase analysis results

According to [4] the thickness of the oxide layer on surface of the cladding tubes from austenitic steel, tested under similar conditions does not exceed 7-10 μm. Diffraction spectra were recorded for all fragments of the tube, indicated on figure 1 by red numerals. Recording was performed after removing oxide film from the surface of the tube by etching. Spectra were used to identify phase composition of the samples. Following phases were found:

- γ-phase of the iron, austenite;
- traces of Fe_2O_3 oxide;
- α′-martensite (base centered tetragonal structure) and ε-martensite (HCP structure).

Figure 2 shows that with increasing distance from the rupture area varies the quantitative ratio of the detected phases. The intensity of the peaks, corresponding to austenitic and martensitic phases is proportional to the composition of the respective phase. Figure 3 shows how intensity of (111) peak of austenitic phase increases (the amount of austenite increases), while intensity of (110) peak of martensitic phase decreases (the amount of martensite decreases).

Besides that on spectrum (Figure 2) one can find peaks, corresponding to oxide phase, most intense in locations of the investigated tube, where α′-martensite peaks also have the highest intensity.

Additional etching of 20-25 μm leads to the disappearance of peaks, corresponding to the martensite phase and oxides. Only austenite phase peaks remain in the spectra.

4.2. Crystallographic texture

According to the received data, texture of cold worked and annealed (tested under the temperature of 700 °C) are characterized by a set of approximately the same texture components: {110}<001> and {112}<11$\bar{1}$> (figure 4).

To observe the changes of crystallographic texture of the tested tube it is more convenient to build IPF. In this work IPF were built from DPF {100}, {110}, {111}. Figure 5 shows IPF of *R* and *L* directions of initial tube and of zones 1 and 4 of the tested tube. The appearance of new texture peaks on IPF_R of zone 1 texture near the [[122]], [[113]], [[123]] and [[013]] poles was detected.

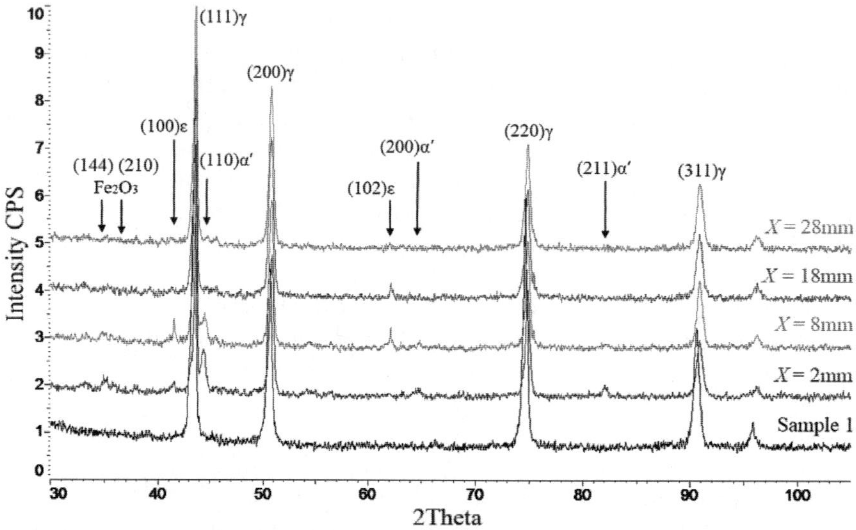

Figure 2. The spectra of different areas of the tested tube, located at various distances X from the rupture place.

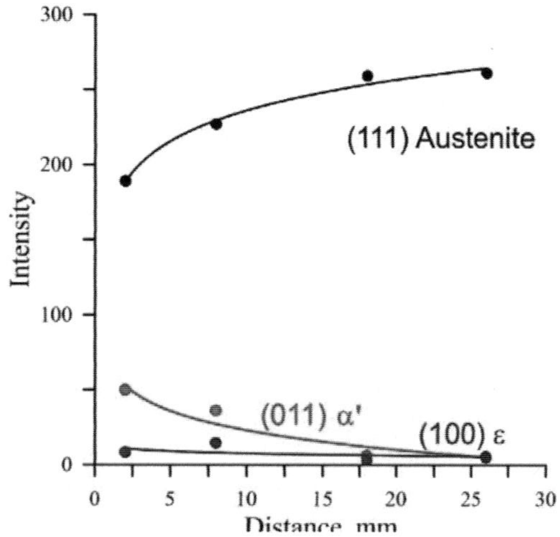

Figure 3. Intensity of the peaks (hkl), corresponding to different phases as a function of the distance from the rupture place, sample 2.

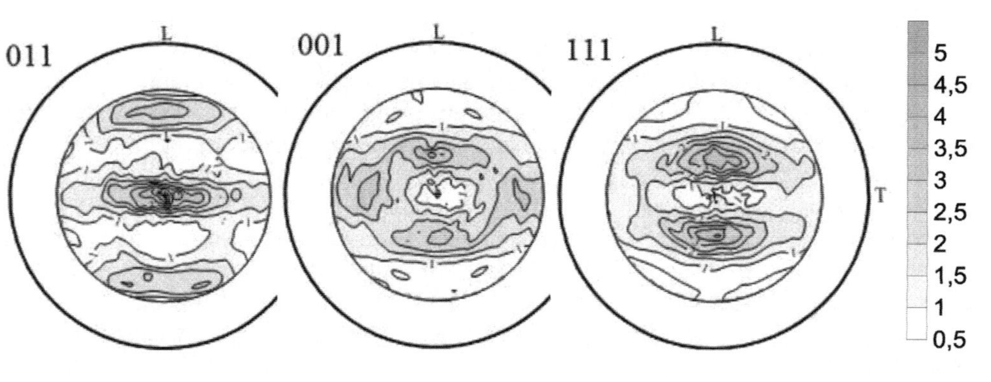

Figure 4. Incomplete direct pole figures {100}, {110}, {111} of initial tube cold worked by rolling to 20% (sample 1).

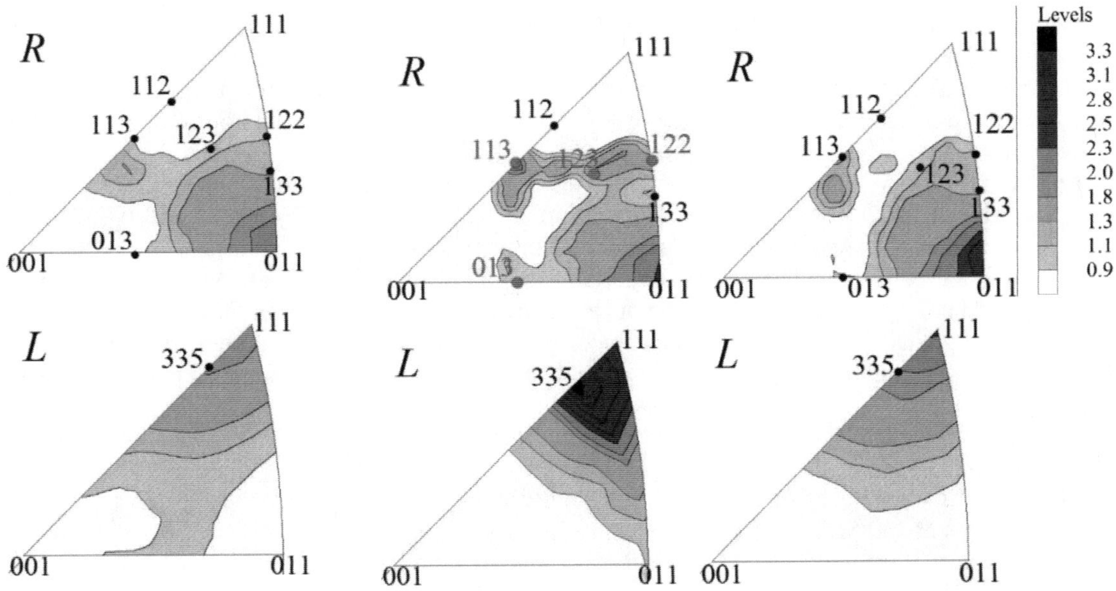

Figure 5. IPF *R* and *L* of sample 1 and of sample 2 (zones 1 and 4)

5. Discussion

5.1. Martensitic formation

There is a clear relationship between the presence of oxide film and martensite peaks on diffraction spectra of sample 2. Peaks of Fe_2O_3 oxide phase on figure 2 corresponds with presence of α'-martensite peaks.

Martensite formation in austenitic steels of different compositions was studied in works [5-7]. During the oxidation process of austenitic steels there is possible the formation of both the oxide film and the propagation of intergranular corrosion, associated with the precipitation of carbides at the grain boundaries. In both cases, the oxidation process is accompanied by a change in the chemical composition of the boundary areas (chrome depletion, oxygen saturation). This leads to the shift of martensite start temperature M_S in a higher values. As a result a martensitic formation in austenitic steels subjected to corrosion is observed.

The works devoted to martensitic transformations are pointing out that the appearance of α'-martensite is possible along with the formation of intermediate ε-martensite by the following scheme: $\gamma \rightarrow \varepsilon \rightarrow \alpha'$ [8-11]. There is a direct correlation between amounts of martensite formed and the level of deformation applied. The formation of ε-martensite is observed only in cases of stress induced martensitic transformations. The decreasing of martensite amount along the tube length starting from the rupture area, as well as the presence of traces of ε-martensite are both suggest the idea of the occurrence of stress induced martensite transformations in the studied steel. Although the absence of additional evidences puts under a question the possibility of stress induced martensitic transformations in the investigated steel.

5.2. The features of texture changes of the tested cladding tube

Initial cladding tube is characterized by FCC rolling texture {110}<001> [12]. Different parts of the tested tube was deformed up to different strain levels. For example, material of zone number 4 (red numeral on figure 1), placed far away from the rupture (28-30 mm) have suffered the lowest diametric deformation, so this zone can be used as test piece, stored at the test temperature for over than 6000 hours.

All zones of the tested tube have the same texture components {110}<001> as for the initial tube. Polygonization and aging of the tube at the test temperature provide intensification of the main rolling texture components.

The difference between the texture of the number 1 zone and the texture of other zones of the tested tube is the presence in the texture of zone 1 of additional peaks {221}, {113}, {123} and {013} (figure 5). These components are specific to the recrystallization textures of FCC materials [12]. Appearance of the mentioned components could testify about activation of the recrystallization process in the rupture area. Thus, deformation of the tube at the temperature of 700 °C promotes the formation and growth of new grains, orientation of which is corresponds to the slopes of texture maximums. This process certainly be accompanied by redistribution of alloying elements in the material of the test tubes.

6. Conclusions

Following results were obtained for the ChS68 steel creep tested at the temperature of 700 °C under the applied stress of 160 MPa:

1. Martensite formation, associated with the change in steel composition as a result of oxidation processes, was detected.
2. The change of martensite amount along the tested tube length indicates a greater thickness of the layer saturated oxygen in the rupture zone.
3. The main texture components of the initial tube and of the different zones of the tested tube are {110}<001>. In the rupture zone the component of the texture of tension with axis <111> along the tube axis becomes stronger. This effect is connected with activation of dislocation slip in the deformed area of tested tube near the new-formed neck.
4. The character of texture changes in the zone of tube rupture indicates to development of the dynamic recrystallization, conditioned by the total influence of all factors, which control the passage in the tube of thermal creep.

Acknowledgements

This work was performed within the framework of the Center of Nuclear Systems and Materials supported by MEPhI Academic Excellence Project (contract № 02.a03.21.0005, 27.08.2013).

References

[1] Taylor A 1961 *X-ray Metallography* (New York: Wiley)
[2] Bunge H J 1982 *Texture Analysis in Materials Science* (London: Butterworth)
[3] Pawlik K 1986 *Phys. stat. sol.* B **134** 477
[4] Simms H G 2011 *Oxidation behaviour of Austenitic stainless steels at high temperature in supercritical plant* M.Res. thesis, University of Birmingham
[5] Butler E P and Burke M G 1986 *Acta Metall.* **34** 557
[6] Fontaine A L, Yena H-W, Trimby P, Moody S, Miller S, Chensee M, Ringera S, Cairney J 2014 *Corrosion Science* **85** 1
[7] Susan D F, Knorovsky G A, Robino C V, Michael J R, Rodriguez M A and Perricone M J 2012 *Science and Technology of Welding and Joining* **17** 321
[8] Lecroisey F and Pineau A 1972 *Metallurgical Transactions* **3** 387
[9] Shirdel M, Mirzadeh H and Parsa M H 2015 *Materials Characterization* **103** 150
[10] de Abreu H F G, da Silva M J G, do Nascimento A M and Freitas F N C 2011 *Materials Science and Technology* **27** 1627-31
[11] Tamura I 1982 *Metal Science* **16** 245
[12] Vishnyakov Ya D, Babreko A A and Vladimirov S A 1979 *Texture Formation Theory in Metals and Alloys* (Moscow: Nauka)

Regularities of crystallographic texture formation in cladding tubes from Zr-based alloys during their production

M Isaenkova, Yu Perlovich and V Fesenko

National Research Nuclear University MEPhI (Moscow Engineering Physics Institute), Kashirskoe highway, 31, Moscow, 115409, Russia

E-mail: MGIsaenkova@mephi.ru

Abstract. This paper summarizes researches of the authors, which are directed on the development of the methodological basis of X-ray studies in the materials science of zirconium and on the systematization of new experimental results obtained using developed methods. The paper describes regularities of the formation of the crystallographic texture and the substructure inhomogeneity of cladding tubes from zirconium alloys at various stages of their manufacturing, i.e. during hot and cold deformation, recrystallization, phase transformations and interactions of the above processes.

1. Introduction

Zirconium stands out among all other metals by extreme importance of the crystallographic texture for the service properties of products from the Zr-based alloys [1-12]. Zirconium-based alloys are used for manufacturing of critical structural elements of nuclear thermal neutron reactors. Their behavior under exploitation is controlled by anisotropy of the radiation growth and creep and by the development of hydride formation, which are predetermined in the significant degree by the features of crystallographic textures formed in products [11-14]. The correlation of corrosion behavior of cladding tubes with their crystallographic texture determined by a misorientation of neighboring grains was also observed in papers [15-17]. Therefore, optimization of manufacturing technologies of products from zirconium alloys requires a knowledge of regularities of the texture formation in them under different types of a plastic deformation and a heat treatment. Only the scientific support and the justification of technological developments at the field of zirconium production can ensure its real progress and effectiveness. In particular, this applies to the systematic study of the texture formation in products at successive stages of their production, and to the analysis of all other crystallographic aspects of the structure formation, one way or another related to the development of the texture.

The development of the deformation texture in a metallic material is consist in following: grains, initially had different orientations, as a result of activation of micromechanisms undergo regular rotation of the crystal lattice and at the orientation space are moving to certain final orientation, stable in relation to the used deformation scheme. At the same time, the substructure forming in a grain depends on its "biography", consisting of initial and final orientations, the trajectory of the grain at the orientation space, strain micromechanisms acting on different parts of its trajectory, the resulting distribution of dislocations and the level of the strain hardening. So, the grains corresponding to different components of the deformation texture obviously differ in their "biographies" and substructure, whereby the metal material with a multi-component texture is, in fact, a composite. The character of the deformed material substructure determines the inhomogeneous development in its recovery processes, recrystallization and phase transformation (PT) during the subsequent heat treatment [10, 18-19].

In this paper, the authors have summarized the results of own researches directed at systematization of the new experimental results, obtained using the developed X-ray methods for investigation of the substructure of textured materials. The processes of structure forming of products are considered in connection with development of their crystallographic texture, what allows to deviate from traditional one-dimensional description of product structure and to characterize it by using of distributions of recorded parameters, giving a much more complete understanding of the studied material.

2. The used methods for studying of the crystallographic texture and substructure of textured materials

Analysis of the crystallographic texture and structure of products from Zr-based alloys was carried by using domestic and foreign diffractometers: DRON-3, DRON-3M, D500/TX (manufactured by Siemens) and D8 DISCOVER with LYNXEYE position-sensitive detector (Bruker's production). Filtered or monochrome radiations of chromium, cobalt and copper were used for measuring texture. The detailed description of the method of crystallographic texture analysis by means of direct pole figures (DPF), features automated recording and calculation of partial and complete DPF, taking into account defocusing of X-ray beam, the estimation of measurement errors and DPF construction are presented in [20]. As a rule, samples by dimensions 3×3 mm^2 or composite samples 17×17 mm2 were made (Figure 1). In the case of measuring the texture of the sample of small size it is assumed that the entire surface is within the primary beam, that allows correcting for defocusing analytically, taking into account only change of the test volume by decreasing the depth of the beam penetration.

For a full description of the orientation of the hexagonal crystallite, it is sufficient to know exactly the orientation of the basal normal [0001] and the angle of rotation of the lattice about it, i.e. to assess the crystallographic texture of a polycrystal the construction of two complete DPF (0001) and $\{10\bar{1}0\}$ or $\{11\bar{2}0\}$ is enough. Taking into account the features of the X-ray spectrum of α-Zr the texture of Zr-based alloy products are characterized by DPF (0001) and $\{11\bar{2}0\}$.

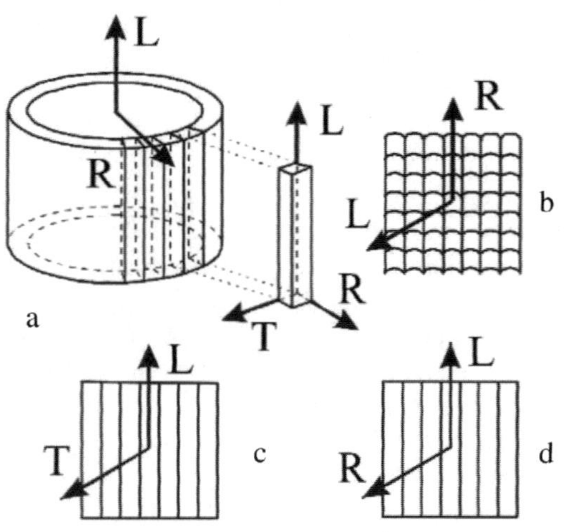

Figure 1. The production of composite samples by size 17×17 mm^2 cut out from tube for X-rays study of its texture: *a* – segment of the tube; *b* – *L*-sample; *c* – *T*-sample; *d* – *R*-sample. The test surfaces of composite sample are perpendicular to the corresponding axes of the tube: radial (*R*), tangential (*T*) and longitudinal (*L*).

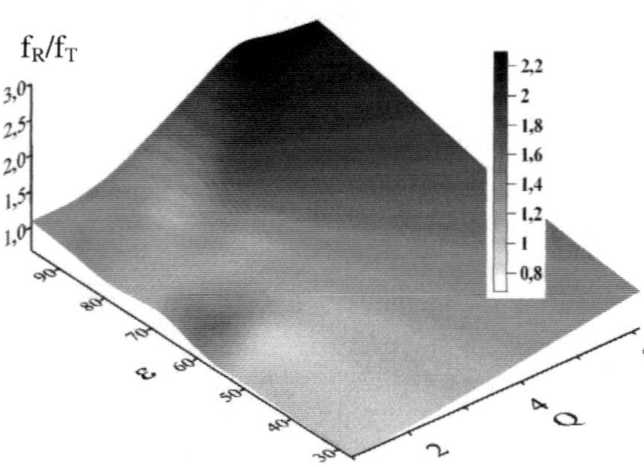

Figure 2. Change of the ratio of integral texture parameters with decreasing of cross-sectional area of tubular billets ε_Σ and varying of Q-factor at their cold rolling

To calculate the integral parameters of texture, also called Kearns parameters or *f*-parameters, it is necessary to build complete DPF (0001). During recording the one section of the sample, only partial DPF can be constructed, due to the geometry of a standard method of texture registration called "on reflection". This disadvantage can be compensated by different methods of constructing complete DPF from partial DPFs, described in detail in [10, 20]. Two main methods are as follows:

1) mutual combination of three partial DPFs obtained for three mutually perpendicular-sections of the product (see Figure 1);
2) extrapolation of partial DPF on uncharted area of stereographic projection.

We used these methods as the most convenient and reliable. They mutually complement each other. The method of mutual combination cannot be used in case of increased layerwise heterogeneity of the investigated products and in the case of impossibility of preparing samples for all three mutually perpendicular cross-sections (see Figure 1). At the correct using of these methods, the CDPFs constructed with their help coincide. In papers [10, 20] the limits of applicability of the different methods of building CDPF (0001) are considered.

For the purpose of study of structure inhomogeneity of the investigated material the generalized direct pole figures (GDPFs), presenting distributions of the diffraction or substructural parameters of subgrains on stereographic projection depending on orientations of the reflecting surfaces, were build [21-25]. The GDPF construction was carried out by registration of X-ray line profile during measuring of texture for each successive sample position by using a scanning motion of a point detector, or by using a position-sensitive detector. Availability of GDPF allows us to build the distributions of volume fractions of α-Zr grains, differing by values of substructural characteristics.

Investigations were carried out at the industrial or model low-alloyed zirconium-based alloys characterized by a small number of additional phases (β- and intermetallic phases). They include alloys of Zr-Nb, Zr-Fe, Zr-Cr, Zr-Nb-Sn-Fe systems, also alloys additionally doped with oxygen. Paper [20] describes how additional phases influences on features of forming rolling texture.

3. The main stages of the formation of the tubes texture

The final texture of tubes is determined by processes of plastic deformation and heat treatment at successive stages of their production, i.e., during hot and cold deformation and intermediate annealings. Processes of the texture formation at different stages of the manufacturing of tubes are considered in [4, 26-28]. The stress state in sheets under rolling are usually described by the tensor, which is characterize by two components - tension in the rolling direction (RD) and contraction in the normal direction (ND). In the case of pressing, forging or rolling of tube billets along with compression in the radial direction (R) of tube and stretching along its axis (L) the component in the tangential or circumferential direction (T) is added, due to the shape of the product. Therefore, for the description of the stress state in the tube during deformation authors of papers [2, 29] introduced Q-factor ($Q = (\Delta t/t)/(\Delta\overline{D}/\overline{D})$, where t – wall thickness, $\overline{D}$ – average diameter of the tube), allowing to estimate the compressive stress along the R and T-directions. The additional component of the stress in the tangential direction of the tube allows controlling the formation process of its crystallographic texture.

The initial stages of the texture formation in sheets, described in detail in [10, 30], are remained in the case of deformation of the tube. Therefore, first of all, we introduce the notation of main texture components observed in the products from zirconium-based alloys:

T0 – basal texture, i.e. when basal axes are parallel to the R(ND)-direction (R – radial direction for tubes, and ND – normal direction for sheets);

T1 – basal axes are deviated from R(ND) to L(RD)-direction on angle ψ=15-20°;

T2 – basal axes are deviated from R(ND) to T(TD)-direction on angle ψ=20-40°;

T3 – basal axes are parallel to the T(TD)-direction;

T4 – basal axes are parallel to the L(RD)-direction.

Texture components T3 and T4 can be detected at the product only in that case, when during technological treatments the PTs $\alpha\leftrightarrow\beta$ occur inside the product material. Texture components T1 and

T0 are observed at the initial stages of plastic deformation and may also appear in the case of high-temperature treatments (thermo-mechanical treatment, involving a PTs). The final texture of tubes is texture T2, the deviation angle of basal axes from radial direction may vary from 20 to 90°, i.e. up to texture T3. Texture component T3 especially prepared at the channel tubes for CANDU-reactor. For this reason, the tubes are manufactured by compression at a temperature within the (α+β)-phase region of the state diagram followed by cold rolling up to 20% of strain in the wall thickness and the long annealing (24 hrs) at temperature 400°C. Such a low degree of cold deformation allows to maintain the texture maximum of DPF (0001) near the T-direction and to eliminate the undesirable maximum in the L-direction.

At the step of hot-pressing or forging the textureless material or material with the texture varying from T2 to T3, as it is observed during manufacturing of CANDU tubes, can be obtained depending on the temperature of its deformation. Common for all textures of the extrusion (the hot deformation) is the location of the basal axes within the zone extending along the T-R-T diameter of the stereographic projection of the sample, which is determined by the symmetry of the used strain schemes. Features of the basal axes distribution within the mentioned above zone depend on the nominal technological parameters and structural characteristics of the material determining the preference activation of slip and twinning or noncrystallographic mechanisms of the grain boundary sliding, and also the dynamic interaction described modes [31].

For the cold rolling of tubes, on the basis of numerous data presented in the form of two-dimensional correlation diagrams [10, 32], we built a three-dimensional graph, which associates the ratio of integral texture parameters f_R/f_T with values of total strain over the cross section of the tubes ε_Σ and Q-factor (Figure 2). According to this diagram, with increasing ε_Σ and Q-factor the ratio f_R/f_T increases. Obtained results help us to systematize manifestations of texture heterogeneity of cladding tubes and to identify the causes of its development.

The initial texture of samples is a major factor by means of which the influence of previous treatment is realized on the texture formation under the α-Zr cold rolling. The complex of original orientation predetermines the fraction of grains, in which twinning is activated during the rolling of sample and those that contribute to the formation of the component T1. At the regions where earlier treatment is associated with fragmentation of grains and distortion of the crystal lattice, the activity of twinning and pyramidal slip is reduced. Difficulty of pyramidal slip weakens the components T1 and reduces the stage of its stability. For intermediate degrees of deformation, this corresponds to some deviation of reorientation trajectories of α-Zr grains in the side of the T-direction.

4. Processes of recrystallization and phase transformations inside deformed tube

In [33] by means of construction of difference diagrams it was able to demonstrate the preferential growth of the grains, basal axes of which are deviate from its stable positions of the rolling texture. It has been found that the α-Zr matrix regions that were deformed with the prevailing participation of twinning are characterized by a reduced tendency to recrystallization. Based on the established laws of textural changes, recrystallization mechanisms operating in the deformed α-Zr matrix have been revealed. The orientation of recrystallized α-Zr grains corresponds to the sides of deformation texture maxima in which the level of the crystal lattice distortion reaches a maximum and centers of recrystallization arising in these regions can, in the process of their growth, be accompanied by a 30° rotation of the matrix about its basic normal. It has been shown that this rotation of the deformed matrix upon recrystallization annealing of samples only occurs in the case of predominant participation of prismatic slipping in the process of deformation, as well as upon fairly slow heating of samples and in the absence of additional texture components in α-Zr, for example, such as T1+T2.

Similar regularities were obtained for tubes (Figures 3-5). Figures 3-4 show the grain size distribution of the dislocation density in the deformed (Figures 3-*a*, 4) and annealed at 580°C (Figure 3-*b*) tubes from zirconium-based alloys. Strain hardening of α-Zr estimated by the physical broadening of X-ray lines varies in a very wide range and is distributed in such manner that as the distance of the grain orientation from the center of the texture maximum is larger so dispersion and/or distortion of the crystal lattice is enhanced (Figure 4-*b*). Therefore, the largest crystallites with the least distorted lattice regardless of its type correspond to the central parts of texture maxima, whereas crystallites corresponding to texture minima are the most dispersed and the most distorted lattice.

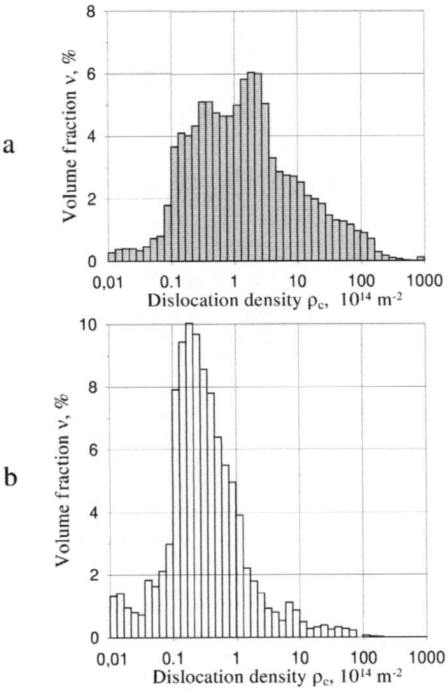

Figure 3. Distributions of volume fractions $\nu(\rho_c)$ for α-Zr grains of Zr-1%Nb tube 9 mm in diameter:
a – deformed state, ε=80%;
b – annealed state, 480°C / 3 h.

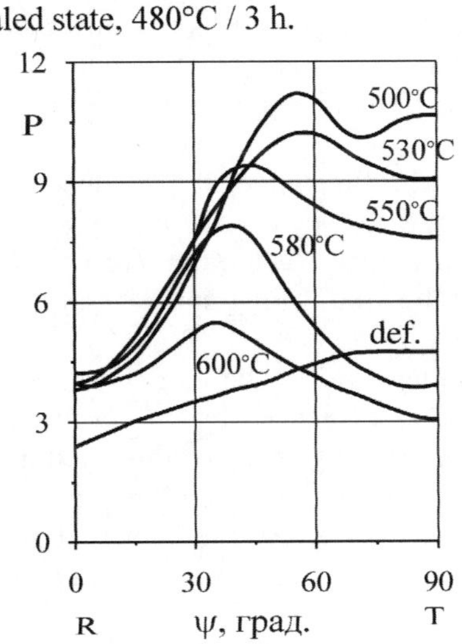

Figure 5. Change of distribution of intensity of X-ray reflectivity (0004) in the R-T-section of tube with increasing of annealing temperature of deformed tube. The annealing temperature is indicated near the corresponding curve.

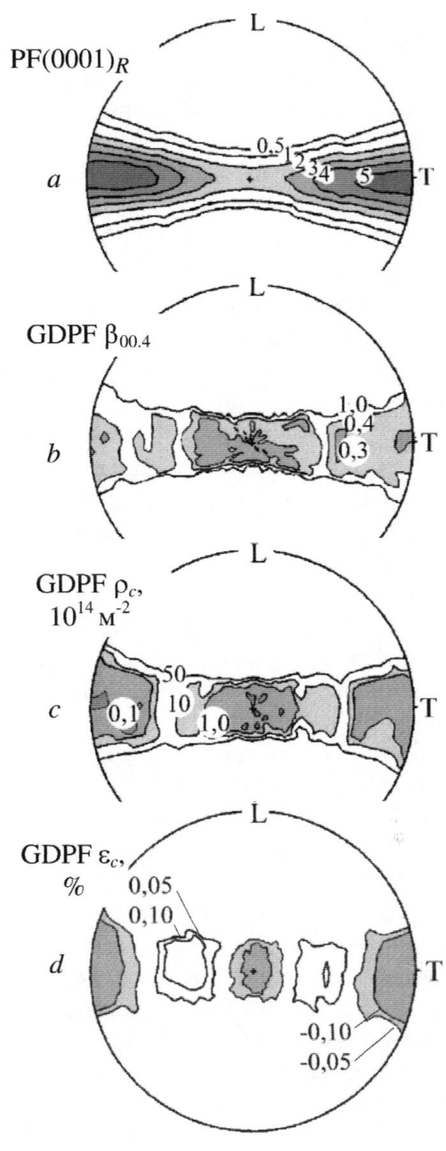

Figure 4. Complete GDPF for α-Zr of rolled Zr-1%Nb tube 20 mm in diameter, ε=65%: *a* - CDPF (0001); *b* - GDPF β_{0004} in deg.; *c* - GDPF ρ_c в 10^{14} m^{-2}; *d* - GDPF ε_c in percent

State of the material is characterized by a wide range of substructure conditions and must be described by the distribution of substructure parameters. Full distributions of *c*- and *a*-dislocation density in the grains of α-Zr plotted for tubes from Zr alloys indicate that the dislocation density depends on the orientations of the grains and varies within a few orders of magnitude, from ~ 10^{12} up to 10^{16} m^{-2} for the rolled tube and up to 10^{15} m^{-2} in the same tube after annealing at 480°C.

Grains with basal axes deviated from the texture maxima are characterized by higher values of the X-ray line broadening $\beta_{(0004)}$ and of the density of *c*-dislocations ρ_c (Figures 4-*b* and *c*). According to a comparison of GDPF ρ_c and GDPF ε_c (Figures 4-*c* and *d*), the elastic strain of the crystal lattice in the tube is distributed in strict accordance with dislocation density: compressive elastic strain is characteristic for regions with the lowest density of dislocations and stretching ones is localized in grains with higher dislocation density.

In [34] it was shown that the reorientation themselves basal axes are observed at the recrystallization of α-Zr in the tubes (annealing temperature 580°C) (Figure 5), resulting in that rolling texture with predominant component T3 becomes texture with predominant components T2 (30-50°). The observed growth of grains, the basal axes of which are deviated from the radial direction at 30-50°, is fully explained by the maximum dislocation density in these grains (Figure 5-*c*).

The most widespread opinion, concerning recrystallization of α-Zr, consists in assertion that the main mode of its lattice reorientation comes to 30°-rotation about basal axes [5, 35]. But, the real situation is more complicated; in particular, under recrystallization of α-Zr its basal axes rotate as well [7]. Recrystallization of α-Zr, connected with the rotation of its crystalline lattice by 30° about basal axes, takes place only in the case, when the prismatic slip participated in previous plastic deformation. The greatest number of recrystallization nuclei forms in α-Zr grains, plastically deformed by operation of different slip systems, as in crystallites, corresponding to stable orientations of the rolling texture. The increase of the heating rate to reach the annealing temperature and the presence of additional texture components, such as T1 + T2 [10, 30], also prevent the rotations of the prismatic axes.

The findings suggest about realization in the material of two fundamentally different mechanisms of recrystallization. Firstly, in the absence of 30 ° rotation of prismatic axes, i.e., at the maintaining of the deformation texture, recrystallization nuclei are formed in the body of the deformed grains with orientation preserved strained matrix. Secondly, the rotation of prismatic axes means nucleation of perfect structure at the junction of neighboring grains, which are then completely absorbed deformed matrix, due to the increased rate of migration of the borders with misorientation angle of 30 ° [35]. Tendency of α-Zr grains to recrystallization, involving a rotation of the crystal lattice around the basal normal depends on the orientation of this normal.

At the final stage of manufacturing of cladding tubes of the finishing size the annealing allows to change the value of the f_R-parameter at 0.05-0.10 by varying the pole density of the basal normal near the R-direction and shifting of texture maxima (Figure 7). Besides, it is possible both increase and decrease f_R-parameter determining by the initial texture of the cold rolling and used at the same time the value of Q-factor [32].

PTs α→β in cold-rolling tubes from zirconium-based alloy also develop non-uniformly, including the mutual absorption of grains with different orientations, as evidenced by the results described in [36]. The grains with orientation deviated from the maximum of initial rolling texture primarily undergo to PT α→β. The essential shift of the texture maxima relatively to their initial position is observed. According to findings, these orientations belong to the areas of the higher strain hardening. A feature of texture changes in the tubes at PFs α→β→α is to strengthen the textural component with the tangential orientation of the basal normal.

At the heating of cold-worked tubes above the temperature of PTs, competition between recrystallization and α→β PT arises in the case of sufficiently high strain hardening. As a result, the heterogeneous structure consisting of grains, which have undergone α→β PT from the deformed and recrystallized states, organizes in materials. The extremely heterogeneous structures could be observed in the weld zone obtained by arc welding of cold rolled sheets [36].

IOP Conf. Series: Materials Science and Engineering **130** (2016) 012008 doi:10.1088/1757-899X/130/1/012008

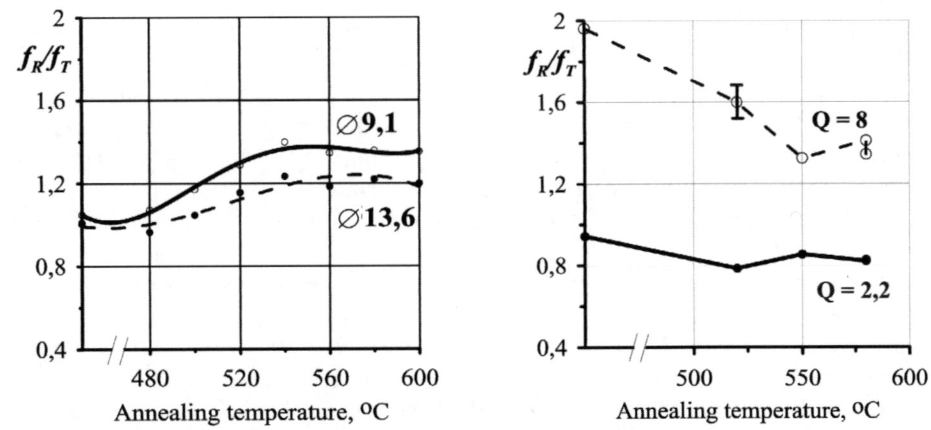

Figure 7. Change of the ratio of integral texture parameters during recrystallization of tubes of different diameters (*a*) and deformed with using different values of Q-factor

5. Conclusions

Creation of a complex of new methods of X-ray studies of reactor zirconium-based alloys, taking into account the existence in them developed crystallographic texture and the associated substructure inhomogeneity, as well as the development of methods of their most complete description of them by distributions of diffraction and substructural parameters allowed to systematize new experimental results:

1) regularities of formation of texture and substructure inhomogeneity in the deformation of zirconium alloys in a wide temperature range, the corresponding β-, (α + β) - and α-region of the phase diagram were identified, and at the same time acting deformation mechanisms were refined;

2) regularities of texture changes during recrystallization and phase transformations in tubes from zirconium alloy with a real polycomponent texture were established;

3) effect of different parameters of technological process on the texture formation was identified.

Acknowledgements

This work was performed within the framework of the Center of Nuclear Systems and Materials supported by MEPhI Academic Excellence Project (contract № 02.a03.21.0005, 27.08.2013).

References

[1] Douglass D L 1971 *The Metallurgy of Zirconium* (Vienna: Int. Atomic Energy Agency) p 360

[2] Zaymovsky A S, Nikulina A V and Reshetnikov N G 1994 *Zirconium alloys in atomic energy* (Moscow: Energoizdat) p 256 [in Russian].

[3] Beskorovainiy N M et al 1995 *Constructional materials for nuclear reactors: textbook for high schools* (Moscow: Energoizdat) **chapter 8** p 491[in Russian].

[4] Tenckhoff E 1988 *Deformation mechanisms, texture and anisotropy in Zirconium and Zircaloy* (ASTM, Special technical publication (STP 966), Philadelphia) p 77

[5] Kocks U F, Tome C N and Wenk H-R 1998 *Texture and Anisotropy. Preferred orientation in polycrystals and their effect on materials properties* (Cambridge University Press) p 676

[6] Woo C H 1985 *J Nucl Mater* **131** 105

[7] Coleman C E et al 2009 Mechanical properties of Zr-2,5%Nb pressure tubes made from electrolytic powder *Zirconium in the Nuclear Industry: 15th Int. Symp.* **ASTM STP 1505** 699

[8] Coleman C E 1982 Effect of texture on hydride reorientation and delayed hydrogen cracking in cold-worked Zr-2.5Nb *Zirconium in the Nuclear Industry: 5th Conference.* **ASTM STP 754** D.G. Franklin (ed) 393
[9] Shamardin V K and Kobylyanskiy G P 1996 *Fizika Metallov i Metallovedenie* **81** (3) 76
[10] Isaenkova M G and Perlovich Yu A 2014 *Regularities of development of crystallographic texture and substructure inhomogeneity in zirconium alloys during deformation and heat treatment* (Moscow: NRNU MEPhI) p 528 [in Russian].
[11] Perlovich Yu A, Isaenkova M G, Konoplenko V P, Novikov V V and Prasolov P F 1982 *Soviet Atomic Energy* **52** (5) 299
[12] Perlovich Yu and Isaenkova M 1994 *Numerical predictions of deformation processes and the behaviour of real materials* (Roskilde, Denmark) p 343
[13] Perlovich Yu A, Isaenkova M G, Kim Yu S and Kim S S 1999 *Problems of atomic science and technology. Physics of irradiation damages and radiation materials science* **77** (2) 58
[14] Perlovich Yu, Isaenkova M and Fesenko V 2008 *Application of Texture Analysis, Ceramic Transactions* **201** 189
[15] Perlovich, Y A, Isaenkova M G, Medvedev P N, Fesenko V A and Thu S S 2015 *Inorganic Materials: Applied Research* **6** (3) 259
[16] Lichter B D, Flanagan W F and Lee D N 2002 *Materials Science Forum* **408-412** 991
[17] Kim H J, Kim T H and Jeong Y H 2002 *J Nucl Mater* **306** 44
[18] Perlovich Yu A, Isaenkova M G and Fesenko V A 2004 *Izvestiya RAN. Seriya physicheskaya* **68** (10) 1462-71 [in Russian]
[19] Perlovich Yu, Bunge H-J and Isaenkova M 2000 *Zeitschrifft fur Metallkunde, Materials Research and Advanced Techniques* **91** (2) 149
[20] Perlovich Yu A, Isaenkova M G and Fesenko V A 2013 *Industrial laboratory* **79** (7) part 1 25
[21] Bunge H-J, Perlovich Yu, Isaenkova M and Fesenko V 1996 *Textures of Materials, Proc. of the 11th Int. Conf. on Textures of Materials* 1455–60
[22] Perlovich Y, Isaenkova M and Bunge H-J 2001 *Materials Science Forum* **378–381** 180
[23] Perlovich Y and Isaenkova M 2002 *Metallurgical & Materials Transactions* **33A** (3) 867
[24] Perlovich Yu, Isaenkova M, Fesenko V and Bunge H-J 2005 *Archives of metallurgy and materials* **50** (2) 303
[25] Isaenkova M, Perlovich Yu and Fesenko V 2007 *Z. Kristallographie* **26** 327
[26] Holt R A and Aldridge S A 1985 *J Nucl Mater* **135** 246
[27] Moulin L, Reschke S and Tenckhoff E 1984 *Zirconium in Nuclear Industry: 6th International Symposium* **ASTM STP 824** 225
[28] Fleck R G, Price E G and Cheadle B A 1984 *Zirconium in nuclear industry: 6th International Symposium* **ASTM STP 824** 88
[29] Konishi T, Matsuda K and Teranishi H 1972 *Can Met Quart* **11** (1) 165
[30] Isaenkova M G, Perlovich Yu A, Thu S S, Krymskaya O A and Fesenko V A 2014 *Tsvetnye Metally* **12** 73
[31] Perlovich Yu A et al 2006 *Physics of Metals and Metallography* **102** (6) 637
[32] Isaenkova M G, Perlovich Yu A, Fesenko V A, Solovev V N and Sergacheva M I 2014 *Tsvetnye Metally* **12** 62
[33] Isaenkova M G, Perlovich Yu A, Fesenko V A, Krymskaya O A, Krapivka N A and Tkhu S S 2014 *The Physics of Metals and Metallography* 115 (8) 756
[34] Perlovich Yu A, Isaenkova M G, Shmeleva T K, Nikulina A V and Zav'yalov A R 1989 *Soviet Atomic Energy* **67** (5) 811
[35] *Recrystallization of metallic materials* 1978 (Moscow: Metallurgiya) Hessner F. (ed.) p 352 [translated into Russian].
[36] Perlovich Yu and Isaenkova M 1997 *Textures and Microstructures* **30** 55

Effect of layerwise structural inhomogeneity on stress-corrosion cracking of steel tubes

Yu A Perlovich[1], O A Krymskaya[1], M G Isaenkova[1], N S Morozov[1], V A Fesenko[1], I V Ryakhovskikh[2] and T S Esiev[2]

[1]National Research Nuclear University MEPhI (Moscow Engineering Physics Institute), Kashirskoe Highway 31, 115409 Moscow, Russia
[2]LLC «Gazprom VNIIGAZ», Moscow region, Razvilka village, Russia

E-mail: YuPerl@mail.ru, OAKrymskaya@mephi.ru, MGIsaenkova@mephi.ru

Abstract. Based on X-ray texture and structure analysis data of the material of main gas pipelines it was shown that the layerwise inhomogeneity of tubes is formed during their manufacturing. The degree of this inhomogeneity affects on the tendency of tubes to stress-corrosion cracking under exploitation. Samples of tubes were cut out from gas pipelines located under various operating conditions. Herewith the study was conducted both for sections with detected stress-corrosion defects and without them. Distributions along tube wall thickness for lattice parameters and half-width of X-ray lines were constructed. Crystallographic texture analysis of external and internal tube layers was also carried out. Obtained data testifies about considerable layerwise inhomogeneity of all samples. Despite the different nature of the texture inhomogeneity of gas pipeline tubes, the more inhomogeneous distribution of texture or structure features causes the increasing of resistance to stress-corrosion. The observed effect can be explained by saturation with interstitial impurities of the surface layer of the hot-rolled sheet and obtained therefrom tube. This results in rising of lattice parameters in the external layer of tube as compared to those in underlying metal. Thus, internal layers have a compressive effect on external layers in the rolling plane that prevents cracks opening at the tube surface. Moreover, the high mutual misorientation of grains within external and internal layers of tube results in the necessity to change the moving crack plane, so that the crack growth can be inhibited when reaching the layer with a modified texture.

1. Introduction

Welded steel pipelines, used for the oil and gas transportation under high pressure, have been known to fail by stress corrosion cracking (SCC), which can be intergranular, transgranular, or a combination of both, depending on the environmental and operational conditions. Recently the problem of SCC has become especially urgent in Russia and other countries, which have extended system of high-pressure main gas pipelines (MGP). The initiation and development of SCC cracks is known to depend on combined effect of three factors: the susceptibility of the material to the SCC, the presence of local tensile stresses exceeding the thresholds and the action of corrosive environment [1-2].

Among features of the material affecting on its susceptibility to the SCC, the structural and textural characteristics should definitely be considered, as well as their inhomogeneity across the tube wall thickness. It is known [3-6] that hot rolling of steel sheets, used for the manufacturing of MGP tubes of large diameter, leads to a formation of layered textured heterogeneity. Thus, parallel to the rolling plane in the external and internal layers of tubes different crystallographic planes are arranged, as well as crystallographic directions, aligned along the axial and tangential directions, vary across the tube wall thickness. Consequently, during the tube operation its corrosion damage in the different layers is accompanied by the crack formation of different types and orientations. Depending on the accepted at that or another plant manufacturing technology the rolling sheets their layerwise textural and structural heterogeneities differ.

The aim of this work was to identify the texture and structure characteristics of tubes material of main gas pipelines, as well as their layerwise inhomogeneity, which affecting their susceptibility to SCC.

2. Experimental data

Different MGP, located in various environmental conditions and consisted of tubes, having different diameters and made from different steel types, were investigated within the frame of this work (Table 1). Different sections were cut out from each MGP: with fixed by eddy current testing SCC defects and without them. Then from each MGP section the samples of two types were prepared:

1) samples, having the studied surface perpendicular to the tube axis, which is the rolling direction of the initial sheet (RD) (labelled ⊥RD), were prepared for registration of X-ray lines;

2) samples, having the studied surface parallel to the rolling plane of the original sheet (denoted ||RD), were prepared for texture pole figures (PF) registration of internal (about half of wall thickness) and external (surface) layers of the tube.

Table 1. Main characteristics of studied MGP.

MGP	Characteristics of pipeline steel / operating period	Diameter × wall thickness of pipeline (mm)	Average operating pressure (MPa)	Depth SCC cracks (detected by eddy current test) (mm)
1	steel K52, thermostrengthened / about 30 years	1220×15.2	5.4	1.0-1.2 not detected
2	steel X70, controlled rolling / about 25 years	1420×1.7	7.4	1.7-2.5 not detected

Assessment of the structural state of samples was conducted by the half-width ($B_{1/2}$) of ferrite's (α-phase) X-ray lines and its lattice parameter, which is the main phase of the tube material. The distributions of these values were constructed across the tube wall thickness. For texture analysis of internal and surface tubes layers by three incomplete PF: {110}, {100} and {112}, – were registered by means of standard procedure [7]. Then the complete PFs were constructed through orientation distribution function (ODF), which in its turn was restored using LABOTEX software [8]. PF is a pole density distribution (P_{hkl}) of normals of the selected type <hkl> in the stereographic projection, with P_{hkl} proportional to the volume fraction of a particular grain orientation.

Figure 1 represents PFs of studied samples, which analysis testifies about the significant texture inhomogeneity of the tubes material from thermostrengthened steel (MGP1). On the tubes external surface the texture is characterized with typical for hot-rolling of bcc metals components {110}<100> and {112}<111>, while in the inner layers of tube the components {100}<110> and {111}<110> are predominant. I.e. the majority of grains in the external layers oriented so, that their crystallographic planes {110} and {112} || to the rolling plane (RP) of the sheet while the directions <100> and <111> ||RD. In the inner layers the crystallographic planes {100} || RP, and the directions <110> ||RD. Therefore, to characterize the texture inhomogeneity along the sheet thickness in this case it is advisable to use the ratios of the integrated intensities $I_{(110)}/I_{(200)}$ and $I_{(110)}/I_{(222)}$ of X-ray lines registered for surface ⊥RD. The distribution of this parameters across the tubes wall thickness is shown in figure 2. The values are normalized to the corresponding ratios for the texture free material. It is evident that in the sample with SCC defects (figure 2-a, black lines) the character of layerwise texture inhomogeneity along the tube wall expressed less sharply than in the sample without cracks. The difference of the texture components ratio in the internal and external layers is higher for the sample without cracks (figure 2-a, red lines). This indicates the higher texture inhomogeneity for this case, which seems to increase the tube resistance to SSC, because during the cracks growth when reaching the layer with a modified texture, their opening is inhibited or stops due to the high mutual misorientation of grains of different layers and the necessity of changing the plane of moving cracks, what requires additional tensile stresses.

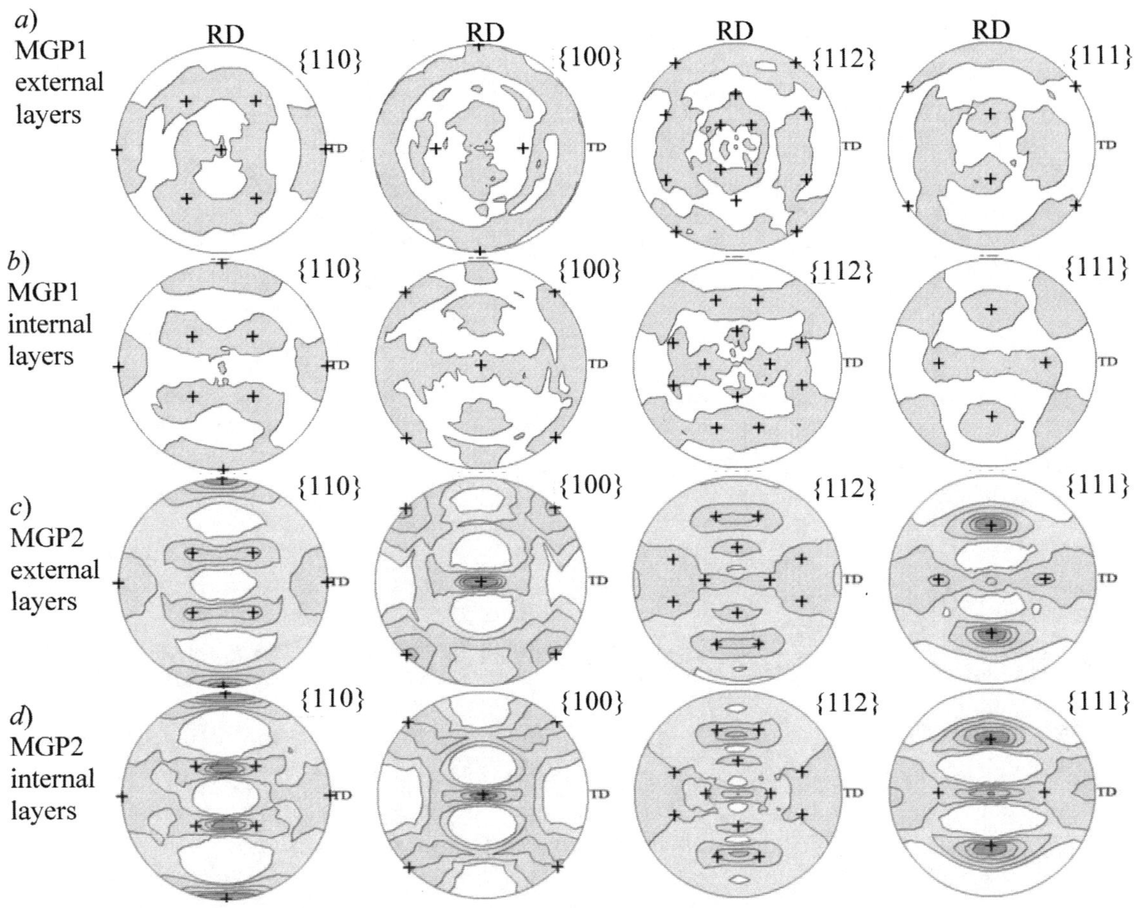

Figure 1. Typical PFs {110}, {100}, {112} and {111} for external and internal layers of thermostrengthened tube (*a*, *b*) and the tube produced using controlled rolling (*c*, *d*).

The texture of tubes made from sheets obtained by controlled rolling (MGP2) is much sharper, its character has a fundamentally different form: the main components in external and internal layers are {100-111}<110> and additional {111}<112>. In this case, the ratios of the X-ray lines intensities $I_{(110)}/I_{(200)}$ and $I_{(110)}/I_{(112)}$ can be used as the characteristics of the texture inhomogeneity along the tubes wall (figure 2-b). This is due to the view of PF {100}, where the number of normals <100>||RD varies depending on the texture features. As one can see the more non-uniform distributions of the texture components ratios are typical for the sample without SCC cracks detected during the defectoscopy. This confirms the suggestion that higher degree of the texture inhomogeneity increases the tube resistance to SSC.

The distributions of the lattice parameters and the half-width of X-ray lines (figure 3) for investigated samples indicate that the content of point defects and lattice distortions on the surface of the tubes is higher than in the underlying layers. This is apparently is a consequence of the saturation

IOP Conf. Series: Materials Science and Engineering **130** (2016) 012009 doi:10.1088/1757-899X/130/1/012009

of the surface layers of the sheet with interstitial impurities, what might occur both in the process of hot rolling on the air, and during operating when there is a contact with the soil electrolyte.

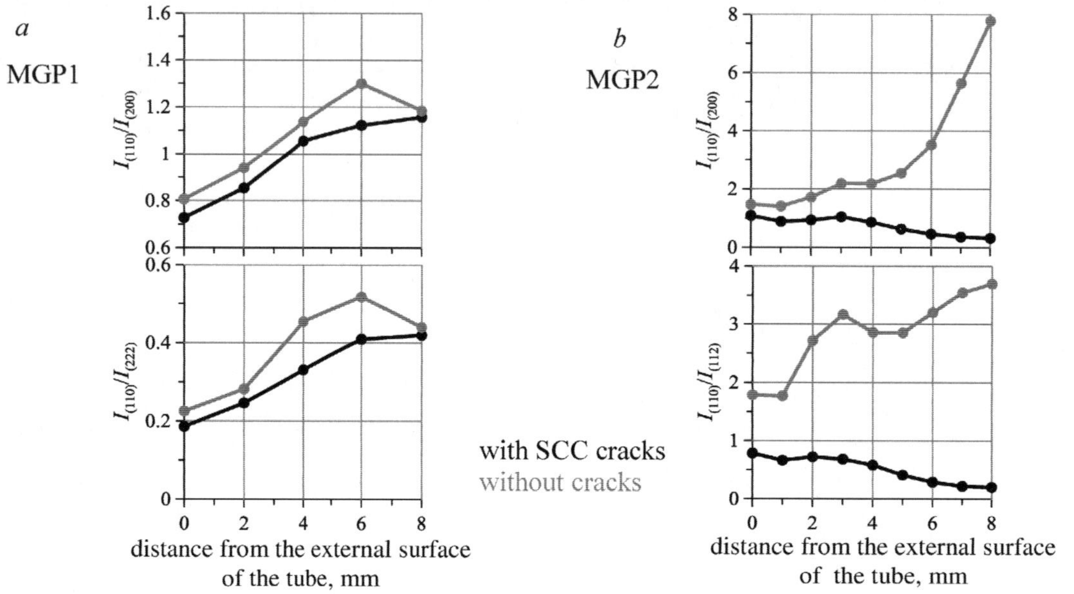

Figure 2. Texture inhomogeneity across the tubes wall thickness of MGP1 (*a*) and MGP2 (*b*).

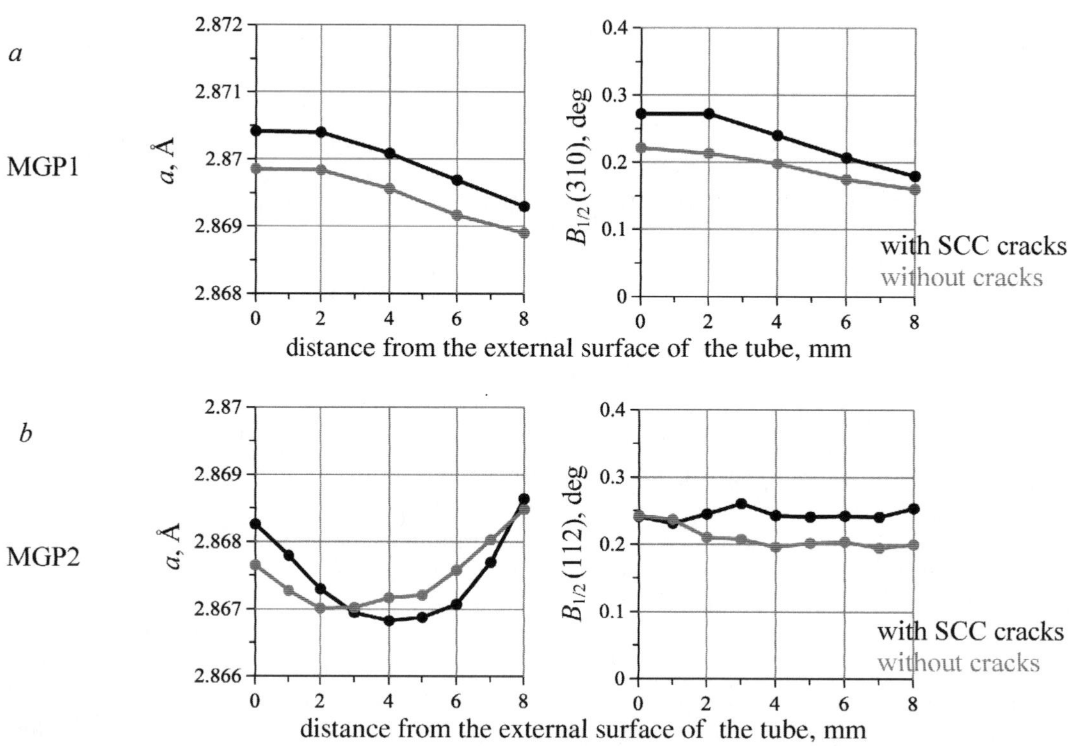

Figure 3. Distributions of the lattice parameter (*a*, Å) and the half-width of X-ray lines $B_{1/2}$ (112) across the tubes wall thickness of MGP1 (*a*) and MGP2 (*b*)controlled rolling (*b*).

3. Discussion

On the basis of obtained data we can see that depending on the accepted production technology of the sheets rolling at certain plant the layerwise textural and structural inhomogeneity differs, because its character is determined by temperature conditions, the percent of reduction, the strain rate and other parameters of thermo-mechanical treatment of the material [3-5]. Moreover, even within the one plant the rolled sheets may have different inhomogeneity due to temperature variation within the length of the sheet.

The higher non-uniform distribution of textural or structural features may increase the resistance of the tube to SCC. This effect can be explained by saturation of the surface layer of the hot-rolled sheet and obtained therefrom tube with interstitial impurities, whereby its lattice parameter will be higher than that of the underlying layers [3, 4]. This leads to the fact that the inner layers have a compressive effect on the surface layer in the rolling plane, preventing the crack development. In addition, the crack opening can be inhibited when achieves the layer with a modified texture due to the high mutual misorientation of grains of different layers and the necessity of changing the moving crack plane, what requires an increasing of applied stresses. Moreover the initiation of axial cracks on the surface of the MGP is promoted by the presence of high-angle boundaries, which sorbent impurity atoms, thus stress concentration at the crack tip increases. Therefore, the presence of sharp multicomponent texture in the surface increases susceptibility of the tube to SCC. Thereby, it is necessary to create a thin surface layer having single texture component and the texture of the inner layers should significantly differ from the texture of the outer layers. The dependency of the structural and textural characteristics upon the parameters of the specific rolling technology, revealed by investigation of the tubes, could allow elaborating recommendations on optimization of rolling regimes preventing the tubes SCC.

4. Conclusion

The presented X-ray study of the tubes samples from different sections of main gas pipelines shows their significant structural and textural inhomogeneity across the wall thickness. Despite the different nature of textural inhomogeneity of the investigated tubes it was shown that the more non-uniform distribution of the textural characteristics leads to an increasing of their resistance to SCC. Internal and external layers of tubes should be sharply different in texture, because the cracks propagation slows down or stops when reach the layer with a modified texture due to high mutual misorientation of grains and the necessity of changing the plane of moving cracks.

Acknowledgements

This work was performed within the framework of the Center of Nuclear Systems and Materials supported by MEPhI Academic Excellence Project (contract № 02.a03.21.0005, 27.08.2013).

References

[1] Arabei A B et al 2009 *Steel in Translation* **39** (9) 725
[2] Zheng W, Elboujdaini M, Revie R W 2011 Stress corrosion cracking in pipelines *Stress corrosion cracking: Theory and practice* ed V S Raja, Tetsuo Shoji (Cambridge: Woodhead Publishing Limited) chapter 19 749
[3] Perlovich Yu, Isaenkova M 2010 *Int J Mater Form* **3** (1) 1143–46.
[4] Perlovich Yu 1994 *Proc. 15th Riso International Symposium on Materials Science*. Ed S I Andersen et al (Riso National Laboratory, Roskilde, Denmark) 445
[5] Jonas J J 2009 *Microstructure and Texture in Steels* ed A Haldar, S Suwas, D Bhattacharjee (London:Springer) chapter 1 3
[6] Kocks U F, Tome C N, Wenk H R 1998 *Texture and anisotropy* (Cambridge UniversityPress) 675 p
[7] Perlovich Y, Isaenkova M, Fesenko V 2013 *Zavodsckaya laboratoriya* **79** № 7 (1) 25 (in Russian)
[8] LaboTex v. 3.0 by LaboSoft (Krakow, Poland) on http://www.labosoft.com.pl

Comparison of microstructure changes and irradiation induced growth of E635 alloy under neutron irradiation influence

V Filippov, D Lauer, Y Lauer and V Petrov

National Research Nuclear University MEPhI (Moscow Engineering Physics Institute), 115409, Kashirskoe shosse 31, Moscow, Russia

E-mail: vipetrov@mephi.ru

Abstract. Model samples of zirconium E635 alloy after rolling, annealing, quenching were subjected to reactor irradiation in BOR 60 up to a fluence ~ 10^{26} m^{-2}. The Mössbauer studies data on the state of iron in the original samples previously subjected to rolling, annealing and quenching, and on the irradiated alloy samples are given. It is revealed that in the initial samples iron is in different intermetallic compounds, and also partly in the solid solution. During irradiation the iron atoms move from one type of intermetallic compounds to another one and in to the solid solution. Data on the relative content of iron in different phases and compounds are compared with the results of irradiation-induced growth strain measurements. The correlation between irradiation-induced grows and iron atoms redistribution in alloys is found.

1. Introduction

Radiation growth of zirconium alloys is one of the important phenomena in the operation of nuclear reactors. To establish the mechanisms of irradiation-induces growth the complex investigations of many alloy parameters must be carry out as before irradiation and also after irradiation. The phase composition changing must be studded too. The necessary information about the origin and evolution of secondary phases and redistribution of elements in the alloy matrix during thermomechanical treatment, as well as information about the phase transformations under the influence of reactor irradiation can be obtained by Mössbauer spectroscopy. This is possible because iron and tin contained in alloy compositions have Mössbauer isotopes. This method allows study the phase composition of alloys with low content of alloying elements. Mössbauer spectroscopy data on iron redistribution in alloys and data on radiation growth are analyzed to find correlation between them.

2. Materials and methods

The alloy of type E635 (Zr-1.2% Sn-0.34% Fe-1.0% Nb-0.03 ÷ 0.05% O) is chosen for investigations. Model alloy samples were prepared with introduced Mössbauer isotopes ^{57}Fe and ^{119}Sn. The alloy composition and method of preparation are chosen close to the composition and method of preparing of industrial alloys. As the heat treatments the quenching, recrystallization and annealing of rolled samples and quenching are used. Before being placed in the fast neutron reactor BOR-60, all the

IOP Conf. Series: Materials Science and Engineering **130** (2016) 012010 doi:10.1088/1757-899X/130/1/012010

samples were autoclaved to create a protective oxide film to prevent the penetration of hydrogen in a sample from lithium in the reactor tests, since such procedure is carried out with real samples.

To investigate changes in the microstructure during irradiation, samples were taken after the first stage of irradiation (up to fluence of $(2 \div 3)10^{25}$ m^{-2}) and after the great irradiation (up to fluence of $11 \cdot 10^{25}$ m^{-2}), which corresponds to a dose of approximately 20 displacements per atom (dpa). Mössbauer studies of quenched, rolled and annealed samples were performed before irradiation and after each of the two stages of irradiation. The technique of sample preparation, modes of radiation tests and Mössbauer studies described in [1, 2].

The data about the phase composition changes of the samples during irradiation obtained by transmission electron microscopy (TEM) and irradiation-induced growth were compared in previous paper [2]. Mössbauer spectroscopy allows obtain more accurate data on the quantitative change in the phase composition of the samples [1]. That's why it is necessary to consider new data. In this paper quantitative data on the phase composition of the irradiated samples obtained by Mössbauer spectroscopy are used for comparison.

3. Analysis of results

The detected by Mössbauer spectroscopy iron containing phases and compoundsand the distribution of iron among these phases and compounds in the alloy E635 [1], as well as relevant data about the irradiation-induced growth (IIG) [2, 3] are summarized in table.

Table. The relative concentrations of the iron-containing phases and compounds and the irradiation induced growth in the samples of the alloy E635.

Treatment	Phase, compound	Before irradiation	1- step of irradiation	IIG (%)	2 - step of irradiation	IIG (%)
Annealing	S.s. Fe in α-Zr	–	32 ± 2	0.06	29 ± 2	0.11
	Zr_3Fe	1.3 ± 1.0	35 ± 3		37 ± 2	
	$Zr(Nb,Fe)_2$	75.7 ± 2,0	16 ± 2		17 ± 2	
	$(Zr,Nb)_2Fe$	23 ± 2	17 ± 4		17 ± 2	
Rolling	S.s. Fe in α-Zr	–	34 ± 2	0.04	30 ± 2	0.18
	Zr_3Fe	1.9 ± 1.5	11 ± 2		38 ± 3	
	$Zr(Nb,Fe)_2$	81 ± 2	33 ± 3		16 ± 2	
	$(Zr,Nb)_2Fe$	17.1 ± 3	22 ± 2		16 ± 3	
Quenching	S.s. Fe in α-Zr	13 ± 2	30 ± 2	0	34 ± 5	-0.01
	Zr_3Fe	21 ± 2	38 ± 4		15 ± 4	
	Zr_2Fe	30 ± 2	–		–	
	$Zr(Nb,Fe)_2$	8 ± 2	16 ± 2		30 ± 2	
	$(Zr,Nb)_2Fe$	28 ± 2	16 ± 2		21 ± 5	

In most cases the presence of three or more microphases were revealed in the alloy with a sufficiently low error. The relative concentrations of individual phases varied from 1.3 to 81% of the total iron content in the alloy. These concentrations were changed under irradiation and the changes in the 1-st and 2-nd stages of irradiation were revealed in different ways. Solid solution was represented by some feature. It was practically absent in annealed and cold worked (rolled) samples, but it was manifested in noticeable quantity after irradiation. Another feature is the disappearance of one of the phases in the quenched sample after irradiation.

Irradiation-induced growth strain values for all the samples did not exceed 0.18% for all types of thermomechanical treatments. Additionally, with increasing rolling deformation the irradiation-induced growth increases. In quenched samples the irradiation-induced growth was virtually nonexistent; by the way, the texture was not observed.

The change of the relative concentration of phases, in which iron is in a different quantitative ratio with other chemical elements, indicates the redistribution of alloying elements between the second phase particles and the solid solution in the alloy under irradiation. Apparently, there is relationship between the evolution of microstructure and irradiation-induced growth. It is believed that the irradiation-induced growth of zirconium is due to two factors: the alignment of oriented loops and climb of dislocations stimulated by irradiation [4]. The direction of flow of point defects during irradiation may occur due to the formation of a large number of opposite point defects and care for their dislocation and other drains. These streams can be weakened by increasing mutual recombination of point defects by radiation.

We can assume that the role in creating or preventing directed flows of point defects plays the process of atoms redistribution. The received data are considered from this point of view.

At the initial stage of irradiation there is a significant redistribution of the iron atoms: intermetallic compounds $Zr(Nb,Fe)_2$ are destroyed in rolled and annealed samples and in quenched samples and iron goes from compounds $(Zr,Nb)_2Fe$ and Zr_2Fe into solid solution and into compound Zr_3Fe, and in the quenched sample into $Zr(Nb,Fe)_2$. In the annealed sample there are more phase content changes compared to rolled samples. An interesting fact is that: after the first radiation exposure step the irradiation-induced growth of annealed samples higher than that of the rolled.

After the second stage of irradiation the phase composition of annealed sample varies slightly, but in the rolled sample the concentration of compound Zr_3Fe increases due to decay of compounds $Zr(Nb,Fe)_2$ and $(Zr,Nb)_2Fe$. We would note that the phase compositions of the annealed and rolled samples after the second step of irradiation are the same. In quenched samples there are the processes of iron atoms redistribution which are in the reverse to rolled sample.

Thus, in rolled samples the redistribution of atoms can facilitates the formation of the directed flows of point defects, whereas in the annealed sample at the second irradiation stage such processes are practically absent. Best irradiative properties of the annealed samples in comparing with the rolled those during the second stage of irradiation can be explained by absence of the atoms redistribution processes. Slightly worse results of irradiation-induced growth strain of annealed samples in the first stage of exposure can be explained by more intensive process of atoms redistribution.

As to quenched alloys, it is necessary to take into account their essential difference from rolled and annealed alloy in a phase composition. This difference is particularly evident in the presence of a large number of non-equivalent state of the iron atoms. Metastability of these states leads to a large number of competing processes of redistribution of atoms under irradiation, which may contribute to the recombination of point defects and prevent the formation of directed streams of point defects and thus slow down the irradiation-induced growth.

4. Conclusion

1. Quantitative Mössbauer data on the iron phases and compounds redistribution in rolled, annealed and quenched samples of E635 alloy before and after two stages of radiation exposure compared with the irradiation-induced growth strain value.
2. The assumptions about the possible correlation of atom redistribution and irradiation-induced growth are made: processes of atoms redistribution in rolled and annealed samples contribute to the formation of directed streams of point defects. So annealed samples have the worst irradiation-induced growth strain value at the first stage of irradiation due to more intensive process of redistribution of atoms, while in the second stage of irradiation, they showed better radiation characteristics and virtually no redistribution of atoms.
3. The presence a large number of metastable non-equivalent states of the iron atoms in quenched samples result in a large number of competing processes of atom redistributions under irradiation, which may lead to recombination of point defects and prevent the formation of directed streams of point defects.

Acknowledgements
This work was performed within the framework of the Center of Nuclear Systems and Materials supported by MEPhI Academic Excellence Project (contract № 02.a03.21.0005, 27.08.2013).

References

[1] Filippov V P, Lauer D E, Lauer Yu A, Petrov V I, Kargin N I and Permyakov Yu V 2013 Investigations of redistribution of iron in a zirconium alloy due to neutron irradiation *Hyperfine Interact* **219** 129-33

[2] Filippov V P, Lauer D E, Lauer J A, Petrov V I, Novoselov A E and Kobylyansky G P 2010 Structural factors irradiation growth alloys Zry-2, NSF, E635, identified by Mossbauer spectroscopy and transmission electron microscopy *Physics and Chemistry of Materials Processing* **2** 18-25

[3] Shishov V N, Peregud M M, Nikulina A V, Shebaldov P V, Tselischev A V, Novoselov A E, Kobilyansky G P, Ostrovsky Z E and Shamardin V K 2002 Influence of Zirconium Alloy Chemical Composition on Microstructure Formation and Irradiation Induced Growth *Zirconium in the Nuclear Industry: Thirteenth International Symposium. ASTM STP 1423* ed Gerry D. Moan ad Peter Ruling (ASTM International, West Conshohocken, PA) pp 758-79

[4] Brik V V and Nekludov I M 2001 The pattern of dislocation structure evolution in self-organizing materials *VANT Series: Physics of radiation damages and radiation materials science (80)* **4** pp 9-13

IOP Conf. Series: Materials Science and Engineering **130** (2016) 012011 doi:10.1088/1757-899X/130/1/012011

Increasing corrosion resistance of carbon steels by surface laser cladding

V I Polsky, V L Yakushin, P S Dzhumaev, V N Petrovsky and D V Safonov

National Research Nuclear University MEPhI (Moscow Engineering Physics Institute) Kashirskoe highway 31, Moscow, 115409, Russia

E-mail: valera_bob@mail.ru, safonovdenis1993@mail.ru

Abstract. This paper presents results of investigation of the microstructure, elemental composition and corrosion resistance of the samples of low-alloy steel widely used in the engineering, after the application of laser cladding. The level of corrosion damage and the corrosion mechanism of cladded steel samples were established. The corrosion rate and installed discharge observed at the total destruction of cladding were obtained. The regularities of structure formation in the application of different powder compositions were obtained. The optimal powder composition that prevents corrosion of samples of low-carbon low-alloy steel was established.

1. Introduction

A large number of machinery parts go out of order a result of abrasion, erosion, corrosion, and so on. Modern technology offers different methods to increase lifetime of details. One such method is laser cladding (LC) [1]. Laser cladding is a perspective and actively developing method of restoring worn parts and create a new application by high corrosion-resistant low-alloy powders on low carbon steel parts. There are manufactured and reducing surfacing [2]. Surfacing is performed by applying a molten metal on the surface of the product has been heated to the melting point or to a temperature reliable wetting liquid weld metal [3]. Cladding layer is integrally formed with the base metal. At the same time the chemical composition of the cladding layer may differ significantly from the composition of the base material. The thickness of the cladding metal (alloy), formed by one or more layers may be different - up to 0.5 mm or more [4].

Besides the above mentioned main goals of LC is improving the corrosion resistance of various parts and assemblies of mechanisms originally made from the cheapest low-carbon low-alloy steels [5]. In this regard, the details of the creation of such steels with LC increase the performance, which in turn is a perspective objective of modern engineering production [6, 7]. The objective of this work is to study the laws of corrosion destruction of samples with different types of LC based on iron and cobalt and choice the optimal composition of the powder and technological conditions applying LC.

2. Materials and Experimental Methods

Experiments on the application of various laser claddings were carried out at two different laser systems: Huffman HC-205 manufactured by HUFFMAN (USA) and "Scanner" (co-production of US-Russia). Huffman HC-205 is a new generation of precision surfacing equipment to the ytterbium fiber laser emission power from 600 to 1000 Watts. Feature Position "Scanner" is the presence on the scanning surface to be treated with the laser head emission capacity from 800 to 1300W. In both installations were used powder dispenser THERMACH Power Feeder AT-1200, which provides precise metering and continuous supply of the powder onto the sample surface at different angles in an argon atmosphere.

Laser cladding was applied to flat substrates made of low-alloy mild steel type St10. Cladding deposited in the form of pure powder brands PR-10R6M5 and 2537 and mixtures. Compositions of powders and their relative proportions in the mixture are shown in tables 1 and 2, respectively.

Table 1. Chemical composition of powders for claddings.

Powder mark	Chemical composition (wt. %)									
	Fe	C	Cr	W	Mo	Si	Mn	Ni	S	Co
PR 10R6M5	80	1	4	6.5	5.2	0.5	0.55	0.4	0.03	-
2537	3	0.9-1.2	27.5-29.5	4-4.8	–	0.9-1.3	–	3	-	57.2-60.7

Table 2. Quantitative ratio of applied powders.

Sample No	Mark of the deposited powder	
	PR-10R6M5 (wt. %)	2537 (wt. %)
0	0	0
1	100	0
2	75	25
3	50	50
4	25	75
5	0	100

Corrosion tests on the intergranular corrosion were carried out of modernized method of AMU (GOST 6032-2003).

In accordance with GOST 6032-2003 test method of AMU conducted by exposing samples of corrosion-resistant steels for 8 hours in boiling aqueous solution containing 50 g / dm^3 $CuSO_4$ and 250 cm^3 / dm^3 H_2SO_4.

Taking into account that the samples of carbon and low-alloy steels for the tests on the intergranular corrosion subjected to considerable degradation, in comparative corrosion study was selected a "soft" mode test: samples were kept for 8 hours in a boiling aqueous solution containing 120 g / dm^3 $CuSO_4$ and 120 cm^3 / dm^3 H_2SO_4.

The calculation of the corrosion rate V_k was performed by their weight loss after exposure to boiling aqueous solution. Samples were weighed with precision ± 0.01 mg of electronic analytical balance GR-202 (A & D, Japan).

Research of topography and surface microstructure, determination of the nature of corrosion interaction and elemental composition of the samples was carried out by scanning microscopy in analytical instrument EVO 50 XVP produced by Carl Zeiss (Germany). The distribution of elements along and across the LC investigated in the wave spectrometer INCA Wave-500, manufactured by Oxford Instruments (UK) with a resolution of 5 eV and a sensitivity of 0.01% (wt.).

Microhardness of samples was measured by the Vickers method on the unit HVS-1000 (China) with automatic loading of the indenter in the largest load for p = 1 N. Time under load was 20 s. The relative error in determining the value of the microhardness of not more than ± 7%.

3. Results and Discussion

It was investigated four types of samples in the work: without cladding (sample No.0) and surfacing of various compositions (samples No.2–No.5). Sample without surfacing is a standard for comparing the corrosion resistance of a pure sample and surfacing samples (samples No.2–No.5).

A study of the microstructure and elemental composition of the samples after corrosion tests by measuring the rate of general corrosion and corrosion buildups, defined the nature of corrosion, depending on the composition of the deposited powder. To select the optimal composition which has the best corrosion resistance, were using the following parameters: the lowest rate of general corrosion and the lowest rate of corrosion cladding, the best coupling of a covering with the substrate (no discontinuities at the substrate-surfacing), minimal heat-affected zone. Analysis of the data (Table 3) showed that the best corrosion resistance has a sample with cladding No.5 (as surfacing material used pure powder brand 2537): the rate of general corrosion of 37.81 g / ($m^2 \cdot$ h), the rate of corrosion surfacing 0.021 g / ($m^2 \cdot$h). This sample has the lowest recorded speed of corrosion, a satisfactory grip surfacing to the surface (no pores, cracks and delamination) and virtually no heat-affected zone (HAZ). Should be noted that all tested samples has satisfactory adhesion to the substrate, and the HAZ is virtually absent, but of the corrosion rate is somewhat higher than the sample №5. The highest rate of general corrosion and corrosion cladding installed in the sample No.2 (as surfacing material was a mixture of powders brand PR-10R6M5 (75 wt.%) and 2537 (25 wt.%)) – 41.97 g / ($m^2 \cdot$h) and 0.42 g / ($m^2 \cdot$h), respectively. The corrosion rate of the reference sample without deposition (sample No.1) is somewhat lower – 41.45 g / ($m^2 \cdot$h).

Table 3. Experimental certain corrosion rate of various samples with LC

Sample No.	The corrosion rate, g/($m^2 \cdot$h)	Corrosion rate of cladding, g/($m^2 \cdot$h)
0	41.45	–
1	39.84	0.25
2	41.97	0.42
3	40.55	0.31
4	39.11	0.17
5	37.81	0.021

It is also established that based on the claddings of powder brand PR 10R6M5 the mechanism of intergranular corrosion is typical, which increases with the content of the powder brand in 2537 is replaced by the mechanism of surface pitting. For the sample without cladding (sample No.0) characteristic is the surface (uniform) corrosion.

Figure 1 shows images of the microstructure of samples No.0 (a), No.1 (b) and No.5 (c) to demonstrate the different types of corrosion damage. In areas like cladding and fine-meshed structure of the substrate is observed. It was found that samples No.1–No.4 character of intergranular corrosion damage, while the sample No.5 – superficial ulcerous. For samples with cladding based on the powder brand PR-10R6M5 boundary cells are observed accumulations of corrosion products and cracked. For samples with a weld overlay based on the brand 2537 (starting from 50 wt.%) no segregations and cracks. Figure 2 is an end to the grinding areas of layered element analysis claddings No.1 (a) and No.5 (b).

As shown by the analysis of the composition of elements in corroded cladding different for samples No.1 and No.5. For sample No.1 is characterized by a large number of isolation oxide mainly iron and chromium - elements that are present in the composition of the cladding.

For sample No.5 no selections were found. It is also noted that when the deposited powder mixture the border of cladding-substrate is wavy, while the deposited pure powder of the same brand – line (figure 2).

IOP Conf. Series: Materials Science and Engineering **130** (2016) 012011 doi:10.1088/1757-899X/130/1/012011

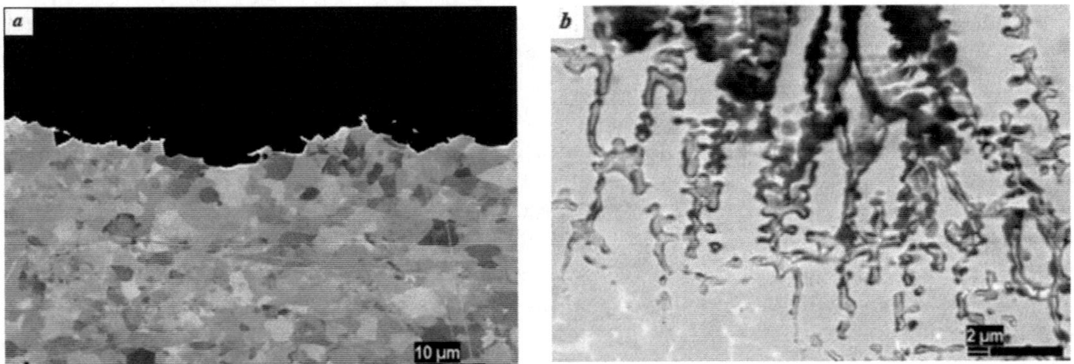

Figure 1. Microstructure corroded substrates LC: a – a sample without LC; b – sample LC based on the powder brand PR-10R6M5 (100%); c – sample LC based on the powder brand 2537 (100%)

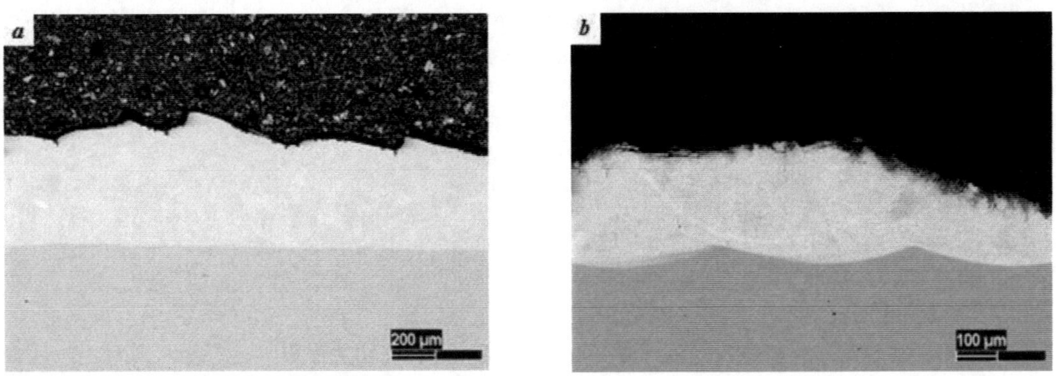

Figure 2. Melting area cladding with the substrate: a – LC substrate based on pure powder of the same brand; b – LC substrate based on a mixture of powders of different grades

Based on researches, the optimal composition of the powder is set for maximum corrosion resistance of samples in a medium boiling aqueous solution containing 120 g / dm^3 $CuSO_4$ and 120 cm^3 / dm^3 H_2SO_4. Revealed that with decreasing amounts of iron and cobalt content increases in the cladding, the corrosion rate decreases. The presence of small amounts of nickel able to slow down the corrosion process, even with high content of iron. The optimum is cladding based on the pure powder brand of 2537 based on the cobalt, can significantly reduce the speed of the corrosion process.

4. Conclusions

It is shown that all the LC have a close-meshed microstructure and rather uniform distribution of the alloying elements on cladding volume.

It is established that at all claddings coupling with a substrate the satisfactory: cracks, peeling and the pores on the boundaries of the substrate-surfacing were not found. All surfacing HAZ is virtually nonexistent.

It is revealed that when fusing one type of powder border of substrate-cladding direct, and when fusing mix of powders (irrespective of a percentage ratio) border of substrate-cladding becomes wavy.

It is revealed that at cladding based on PR-10R6M5 brand powder the mechanism of intergranular corrosion is predominant, and at cladding based on brand 2537 powder is dominated by the mechanism of surface pitting.

The optimum composition of the cladding (pure powder brand of 2537), has the lowest rate of total corrosion and corrosion cladding constituting 37.81 g / ($m^2 \cdot h$) and 0.021 g / ($m^2 \cdot h$), respectively.

Thus, as result of the conducted researches, was established the optimal composition by anti-corrosion properties of LC powders, allowing to increase service life and decrease corrosion of the wares based on low-carbon low-alloy steels by more than 3 orders of magnitude.

Acknowledgements
This work was performed within the framework of the Center of Nuclear Systems and Materials supported by MEPhI Academic Excellence Project (contract № 02.a03.21.0005, 27.08.2013).

References

[1] Chistyakov G V 2010 *New Indust.Techn.* **3** 22
[2] Petrovsky V N, Shtumm V G, Dzhumaev P S and Polsky V I 2012 *Nuc. Phys. and Engin.* **4** 1
[3] Tuominen J, Nokki J and Pajukoski H 2012 *Proc. of 31 st. ICALEO* 192
[4] Shapeev V P, Isaev V I and Chepepanov A N 2011 *Physical mesomechanics* **14** 107
[5] Dzhumaev P S, Petrovsky V N, Polsky V I and Yakushin V L 2015 *Non-ferrous Metals* **1** 50
[6] Turichin G A, Zemlyakov E V, Pozdeeva E Yu and Tuominen J *2012 J.Metal Science and Heat Treatment* **54** (3–4) 139
[7] Popova E V, Tarasova T V, Smurnov I Yu and Sarbaev B S 2012 *Science and education* **3** 23

IOP Conf. Series: Materials Science and Engineering **130** (2016) 012012 doi:10.1088/1757-899X/130/1/012012

Effect of impurities and grain boundaries on the kinetic characteristics of the radiation damage of iron and austenitic steels

D Demidov[1], **E Smirnov**[1] **and A Tsepelev**[1,2]

[1]National Research Nuclear University MEPhI (Moscow Engineering Physics Institute), 31 Kashirskoe highway, Moscow, 115409, Russia
[2]Institute of Metallurgy and Materials Science of the Russian Academy of Sciences (IMET RAS), 49 Leninsky pr., Moscow, 119991, Russia

E-mail: dmitrydem11@gmail.com

Abstract. Austenitic stainless steel AISI 304, 316 and similar in composition, are used to create many elements of the reactor core, such as fuel cladding of fast-neutron reactor. It is known that during the operation, they became subject to such type of radiation damage, as the vacancy swelling and radiation creep. In this paper, was analyzed the effect of alloying elements, impurities and their complexes with radiation defects (RD) on the characteristics of RD and radiation-enhanced diffusion (RED) Parameters of vacancy voids nucleation and growth processes were also studied on the example of Cr18Ni10T steel. Evaluation of the temperature dependence of steady-state concentration of RD for materials with different binding energy complexes "defect - impurity", the effective values of mutual recombination rate RD for materials complexes, influence the formation of impurity complexes and volume density of grain boundaries on the rate of growth of vacancy voids and radiation creep was conducted. The possibility of varying the characteristics of the complex defect - impurity and grain boundary size for the suppression of radiation creep and vacancy swelling was studied

1. Introduction

There exist a lot of ways of decreasing in the vacancy swelling and radiation creep of austenitic steels, but physically all of them are reduced to a decrease in the alloy concentration of point radiation defects (RD).

The first method consists in steel alloying of small amounts of alloying elements (AE) such as Ti, Zr, Si, Mo, Nb, whose atoms form the impurity complexes with vacancies, which reduces the concentration of free vacancies, and increases the radiation-enhanced diffusion (RED), promoting rapid recombination of point defects[1]. The second method is based on increasing the volume density of grain boundaries in the system (decreasing grain size) that give rise to a large amount of the unsaturating sinks for vacancies and of interstitial atoms [2].

2. The calculation part

To assess the role of impurities in reducing the concentration in the RD, it is necessary to use the model of radiation - enhanced diffusion in systems with impurities and the system of kinetic equations annealing RD in steels alloy with impurities [3, 4]:

$$\frac{\partial C_v}{\partial t} = K_0 - K_{iv}C_vC_i - K_{3v}C_vC_I + K_{4v}C_{vI} - K_{5v}C_vC_{iI}$$
$$\frac{\partial C_i}{\partial t} = K_0 - K_{iv}C_iC_v - K_{3i}C_iC_I + K_{4i}C_{iI} - K_{5i}C_vC_{vI}$$
$$\frac{\partial C_I}{\partial t} = K_{4i}C_{iI} + K_{4v}C_{vI} - K_{3i}C_iC_I - K_{3v}C_vC_I + K_{5i}C_vC_{vI} + K_{5v}C_vC_{iI} \quad (1)$$
$$\frac{\partial C_{vI}}{\partial t} = K_{3v}C_vC_I - K_{4v}C_{vI} - K_{5i}C_vC_{vI}$$
$$\frac{\partial C_{iI}}{\partial t} = K_{3i}C_iC_I - K_{4i}C_{iI} - K_{5v}C_vC_{iI}$$

Here K_0 is the irradiation dose per second, K_{iv} is the recombination constant of interstitial atoms and vacancies, C_i is the concentration of interstitial atoms, C_v is the concentration of vacancies, C_I - concentration of impurity atoms, C_{iI}, C_{vI} - the concentration of complexes "interstitial atom - impurity atom" and complex "vacancy - impurity atom, respectively. K_3 and K_4 are the rate constants of formation and decay of appropriate complexes, characterizing the recombination defect in the complex ($i + vI \rightarrow I, v + iI \rightarrow I$). Description of radiation defects concentration depend on various temperatures, and the sink density values, generation rate of defects at different stages of annealing are given in [4, 5]. Annealing of radiation defects at high temperatures and substantial density of sinks is the main phenomenon, realized at the analysis of the RED in metals.

In accordance with [3], process of defects annealing at a time $t = \tau_1 = \dfrac{1}{\sqrt{K_0 K_{iv}^{eff}}}$ completes when defects are accumulated in the process of mutual recombination, where

$$K_{iv}^{eff} = W[D_i + zC_I(D_v exp\frac{E_{iI}^b}{kT} + D_i exp\frac{E_{vI}^b}{kT})] \approx WD_i(1 + D_i exp\frac{E_{vI}^b}{kT}) . \quad (2)$$

The next moment C_v increases and C_i decreases in proportion to the $\sqrt{t}$ and from the time point

$t = \tau_3 = \dfrac{1}{K_{vs}C_s}$ there is an annealing steady state for which the solutions of the equations (5) with low concentration of sinks are [3 - 5]:

$$C_v = \sqrt{K_0K_{is}/K_{iv}^{eff}K_{vs}} = \sqrt{K_0D_i/K_{iv}^{eff}D_v}$$
$$C_i = \sqrt{K_0K_{vs}/K_{iv}^{eff}K_{is}} = \sqrt{K_0D_v/K_{iv}^{eff}D_i} . \quad (3)$$

Solutions based on substantial concentration sinks are given later:

$$C_v = \frac{\rho D_i}{2D_{iv}}\left(\sqrt{\frac{4K_0K_{iv}}{\rho^2 D_i D_v}+1}-1\right)$$
$$C_i = \frac{\rho D_v}{2D_{iv}}\left(\sqrt{\frac{4K_0K_{iv}}{\rho^2 D_i D_v}+1}-1\right) \quad (4)$$

We consider grain boundaries as RD sinks and volume density (grain boundary size) will be varied. The volume density of grain boundaries meant by the ratio of the volume occupied by the grain boundaries to the volume of grains themselves in a dedicated unit volume.

To estimate the effect of alloying elements on the concentration of RD, let us consider their energy characteristics of the interaction with the RD. To do this for a large number of alloying elements and impurities found in austenitic steels, let us analyze the impact of the binding energy in "AE atom - vacancy" on the effective recombination coefficient RD, a decrease which should lead to a significant reduction in free-migrating vacancies.

In addition to increasing K_{iv}^{eff}, the role of the introduction of alloying elements is to bind a large number of free-migrating vacancies in the complexes which gives to changing their diffusion properties, prevents their accumulation in the pores and reduces the intensity of the climb of edge dislocations, responsible for the acceleration of the radiation creep

Table 1. Values of the volume density of grain boundaries and grain size [7]

Volume density of grain boundaries (κ)	Grain size, d (mm)	In figures 3, 5, 6
10^{-1}	$7{,}50\cdot10^{-1}$	1
10^{-2}	$7{,}50\cdot10^{-2}$	2
10^{-3}	$7{,}50\cdot10^{-3}$	3
10^{-4}	$7{,}50\cdot10^{-4}$	4
10^{-5}	$7{,}50\cdot10^{-5}$	5
10^{-6}	$7{,}50\cdot10^{-6}$	6

Later, we addusing vacancies concentration dependences on the binding energy in the complex "vacancy - AE atom", temperature, irradiation dose, AE concentration. From figures 1 – 4 it can be seen that the concentration of RD is significantly reduced with the increase in the binding energy of the complex "vacancy - atom AE" and with increasing volume density of grain boundaries.

One can also note the constant exponential growth of RD recombination with an increase E_{vI}. Graphs of the complexes RD concentration, as well K_{iv}^{eff} dependences on the AE concentration are given at work [1].

Figures 5 and 6 shows that the increase in the volume density of grain boundaries significantly reduces the concentration of free-migratory vacancies in the system and, as a consequence, reduced the rate of growth of vacancy voids and radiation-accelerated creep. The constructed model confirms proven the linear dependence of these effects on the concentration of the RD system.

Also, a significant role to reduce these rates of increase in the binding energy is played in "AE atom - vacancy", as represented in figures 7 and 8. It can be seen that the increase of the binding energy not only reduces the rate of growth of pores and creep but displaces their highest values at higher temperatures.

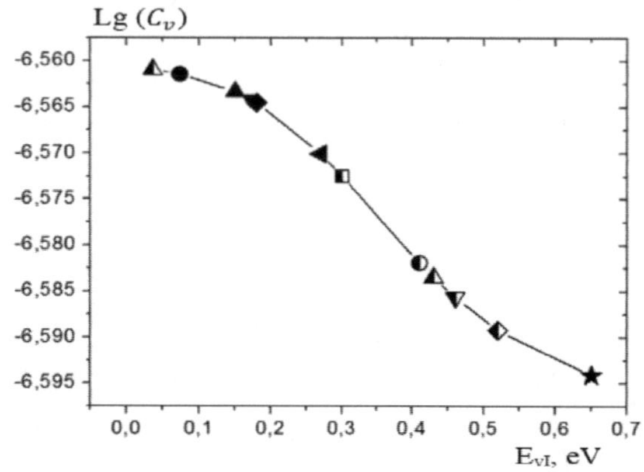

Figure 1. The vacancies concentration dependence in the binding energy of impurity complexes in austenitic steels. Temperature - 973 *K*. The rate of generation of RD: $K_0 = 10^{-6}$ dpa/c, κ = 10^{-3}. AE concentration - 0.4 at. %.

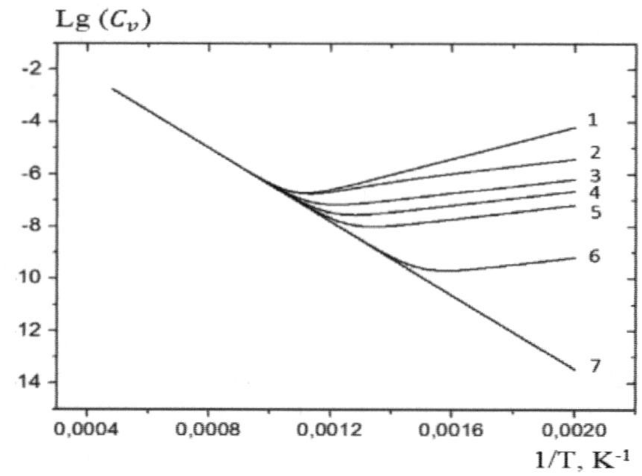

Figure 2. Temperature dependence of vacancies concentration. : AE – Ti, κ = 10^{-3}, 1 – without alloying; $K_0 = 10^{-6}$; 2 – $C_I = 0.01$ at. %, $K_0 = 10^{-6}$; 3 – $C_I = 0.3$ at. % $K_0 = 10^{-5}$; 4 – $C_I = 0.3$ at. %, $K_0 = 10^{-6}$; 5 – $C_I = 0.3$ at. % , $K_0 = 10^{-7}$; 6 – $C_I = 0.5$ at. % , $K_0 = 10^{-6}$; 7– concentration of equilibrium vacancies.

The results obtained in this work, are in good agreement with the results of [8] and also take into account not reviewed in [6] the effect of temperature variation of the shear modulus, which is important to accurately determine the binding energy of the complex "vacancy - atom AE".

Consider the effect of the grain boundaries volume density on the vacancies concentration and complex "vacancy - atom AE" and the impact of the binding energy upon the recombination constant RD - K_{iv}^{eff}.

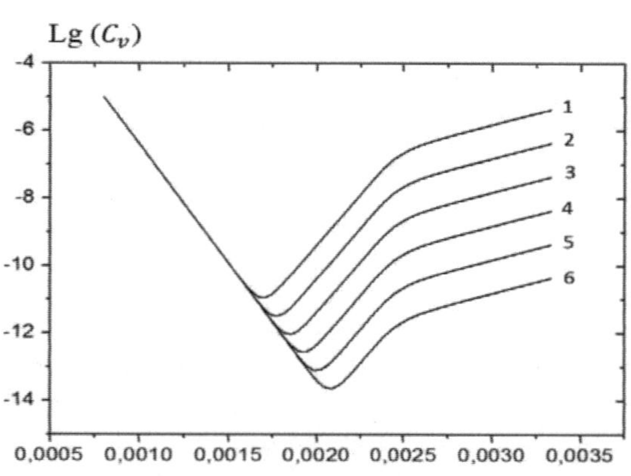

Figure 3. The vacancies concentration dependence on the temperature in austenitic steels for various bulk density of grain boundaries (table 2). AE - Ti, 0.4 at. % , $K_0 = 10^{-6}$ dpa /s.

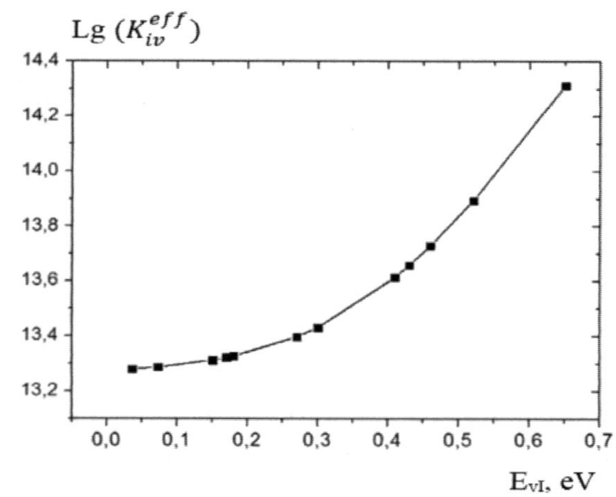

Figure 4. The constant recombination dependence of radiation defects on the binding energy in austenitic steels. $T = 973$, $K_0 = 10^{-6}$ dpa/s, κ = 10^{-3}.

We will consider the effect of the binding energy in the complex "AE atom - vacancy" and the grain boundaries volume density at the rate of vacancy pores growth and radiation creep speed. Let us estimate the values based on expressions (5) and (6), respectively [8]:

$$\frac{dr}{dt} = \frac{\Omega}{r_v}\left[Z_v D_v C_v - Z_i D_i C_i\right] \tag{5}$$

$$\dot{\varepsilon} = \frac{2\Omega L}{9}\left[Z_i D_i C_i - Z_v D_v C_v\right], \tag{6}$$

where Z_i and Z_v are preference factors, Ω is the atomic volume, L is the bulk density of dislocations, r_v is the initial radius of the growing pores. The values of L, Z_i, Z_v, r_v, are taken from work [8]:

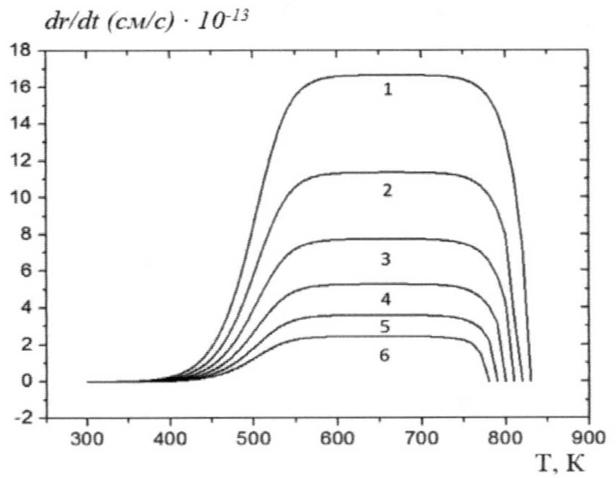

Figure 5. Curves are given different grain boundaries bulk density (table 1). AE - 0.4 at. % Ti. $K_0 = 10^{-6}$ dpa /s. Dependence of vacancy pore growth rate in the austenitic steel pores on temperature.

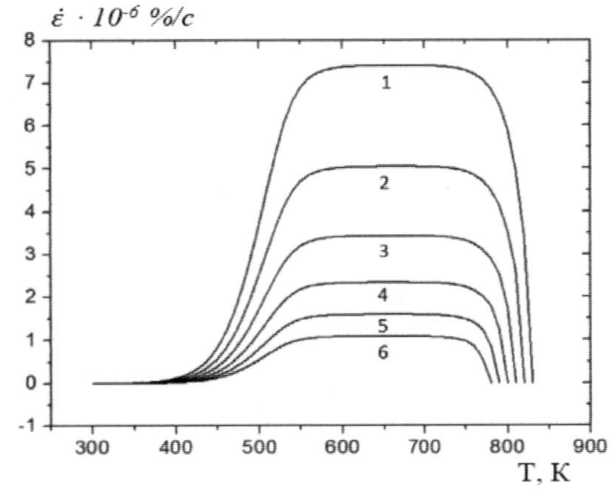

Figure 6. Curves are given for different grain boundaries bulk density (table 1). AE - 0.4 at. % Ti. $K_0 = 10^{-6}$ dpa /s. The velocity of the austenitic steel radiation creep on temperature.

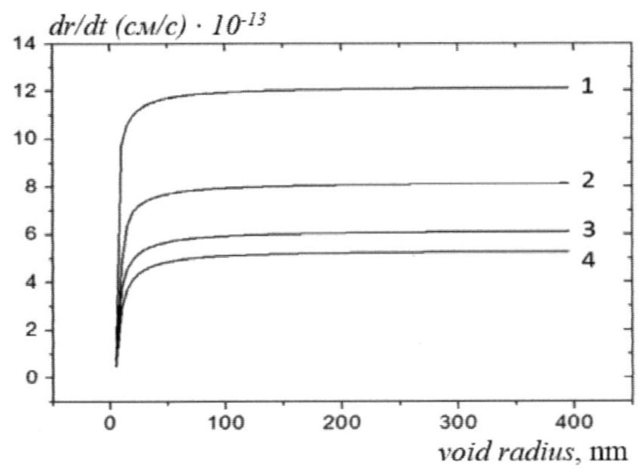

Figure 7. Dependence of vacancy pore growth rate in the austenitic steel pores on pore radius. T = 800 K. The concentrations of the alloying elements are equal to 0.4%. $K_0 = 10^{-6}$ dpa / s. $\kappa = 10^{-3}$. 1 – alloying of P; 2 – alloying of S; 3 – alloying of Nb; 4 - alloying of Ti.

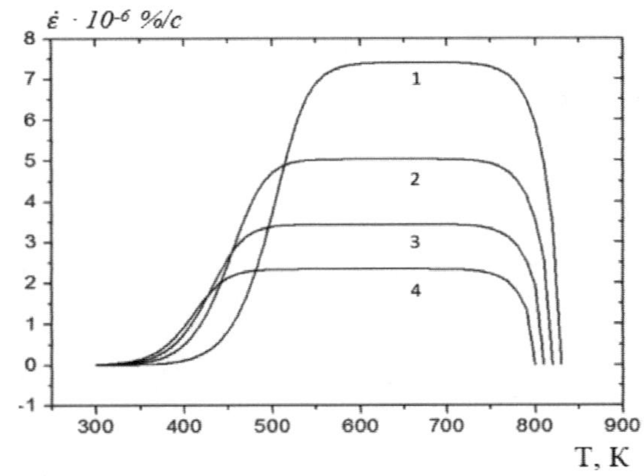

Figure 8. The velocity of the austenitic steel radiation creep on temperature. The concentrations of the alloying elements are equal to 0.4%. $K_0 = 10^{-6}$ dpa/s. $\kappa = 10^{-3}$. 1 – alloying of P; 2 – alloying of S; 3 – alloying of Nb; 4 – alloying of Ti.

3. Conclusion

The paper analyzed the possibility of suppressing the vacancy swelling and radiation-accelerated creep of austenitic steels through the introduction of alloying elements and varying the volume density of grain boundaries. It also analyzes the possibilities of varying the parameters of the alloying elements to reduce the concentration RD in the system.

It has been shown that there is an exponential decrease in concentration RD [1], by increasing the binding energy of the impurity complex and by increasing the concentration of the alloying element in the system. The influence of the radiation defects concentration and share grain boundaries on the vacancy rate of swelling and radiation creep is analyzed.

The dependences of the radiation creep rate and vacancy swelling rate for the various parameters of the alloying element and temperature have been given. The obtained dependences are in good agreement with the results of work [8]. It is shown that there is a significant decrease in these speeds by increasing the binding energy of the impurity in the complex and in the decrease in the average volume of grain.

Acknowledgements
This work was performed within the framework of the Center of Nuclear Systems and Materials supported by MEPhI Academic Excellence Project (contract № 02.a03.21.0005, 27.08.2013).

References

[1] Demidov D N and Smirnov E A 2015 15 *Int. Conf. on Rad. Phys. of Met. and All.: Int. ural seminar (*Kyshtym*)* **11** 15
[2] Zelenskiy V F, Neklyudov I M and Chernyaeva T P 1988 *Radiation defects and swelling of metals* (Kiev: Naukova dumka) p 293
[3] Akhiezer I A and Davydov L N 1985 (Kiev: Naukova dumka) p 225
[4] Smirnov E A 2013 *Metal Physics and Advanced Technologies* **35** (8), 1001–12
[5] Novoselov A N , Smirnov E A and Shmakov A 2013 *Metal Physics and Advance Technologies* 35 (7) 1005-12
[6] Faulkner R G and Song S–H 1996 *Mater. Sci. Technol.* **12** 904
[7] Mehrer H 2007 *Diffusion in Solids* 645
[8] Mansur L K and Yoo M H 1978 *J. Nucl. Mater.* **74** (2) 228

Abnormalities of physics and mechanical properties, behavior of helium and hydrogen in the V-Ti alloys (Overview)

M S Staltsov, I I Chernov, B A Kalin, O N Korchagin and V M Anan'in

National Research Nuclear University MEPhI (Moscow Engineering Physics Institute), Kashirskoe highway 31, 115409 Moscow, Russia

E-mail: m.staltsov@gmail.com

Abstract. The paper presents the results of studies of physical and mechanical properties, helium and hydrogen behavior in vanadium-titanium alloys depending on titanium content. In particular, the results of helium swelling research, thermal desorption studies of helium and hydrogen behavior, results of internal friction measurements, measuring amount of hydrogen retained introduced by various methods. It was shown that the addition of titanium to vanadium have nonmonotonic influence on the behavior of implanted helium and hydrogen, as well as on the physical and mechanical and radiation properties known in literature. It is expected that such an abnormal influence of titanium on various properties of vanadium-titanium alloys occurs because of the interaction of vanadium and titanium atoms with atoms of interstitial impurities.

1. Introduction

The properties of vanadium alloys are of considerable interest to researchers because vanadium considered a promising material for use as a structural material in fast and fusion reactors. Vanadium and its alloys exceed a number of other candidate materials on complex of some properties such as specific strength, drop of induced activity and etc. Of greatest interest are alloys on the base of system vanadium-titanium. A number of investigators has been observed that alloying by titanium nonmonotonically influences some alloys properties [1–9]. It is shown [10] the data the system of V-Ti alloys is abnormal on change of physical-mechanical, radiation and other properties depending on content of alloying element, especially at low concentration area from a few tenths to a few percents of titanium. It is also noted that the mechanical properties, creep, radiation swelling of vanadium alloys significantly depend on the presence and concentration of interstitial impurities (O, N, C etc.).

2. Experimental results

2.1. Internal friction

Measuring temperature dependences of logarithmic decrement damping of oscillations (abbreviation IF for later use to denote of internal friction) is one of the methods for investigation allowing to explain the processes occurring in the interaction of alloy atoms with interstitial impurities. Vanadium and its alloys with 0.3, 0.5, 0.7, 1, 5 and 10 wt.% Ti have been investigated by IF [11]. Measurements were carried out at temperature range from room temperature up to 1123–1173 K. Table 1 shows the contents of oxygen and nitrogen in the initial samples and after measurement of IF. It is seen that in the alloy V-5 wt.% Ti (as well as in other alloys with titanium) the concentration of oxygen is significantly higher than in the initial vanadium. Oxygen and nitrogen content increases during heating at IF measurements.

Table 1. The contents of oxygen and nitrogen in initial V (1), alloy V-5 wt.% Ti (3) and in V after measurement of IF (2) [11].

№	The sample	O_2 (%)	Standart deviation	N_2 (%)	Standart deviation
1	V	0.081	0.0098	0.009	0.0003
2	V	0.098	0.0088	0.012	0.0024
3	V–5% Ti	0.321	0.012	0.0064	0.0003

Investigations of alloys with 0.3, 0.5, 0.7, 1 and 10 wt.% Ti shown that only single peak is at temperature of 498–500 K in IF spectra of vanadium, and it is the oxygen maximum [11]. In alloys with titanium there appear two additional peaks. The peak intensities of IF increase with increasing of Ti concentration up to 0,5 wt.%, decrease with increasing of Ti contents up to 5 wt.% and increase again with increasing Ti concentrations to 10%. All three peaks disappear substantially at a high content of titanium. A small peak is in the region of the maximum II only, but there is a maximum IV at 748 K. That is, titanium nonmonotonically affects the internal friction of alloys. Microhardness and mechanical properties of alloys change at the same regularity depending on titanium concentration in vanadium (Table 2).

Table 2. Mechanical properties at room temperature of several alloys of vanadium [3].

Alloy	$\sigma_в$, МПа	δ (%)	ψ (%)	HV (Мпа)
V–1% Ti	688	~0	~0	2863
V–2.5% Ti	618	12.5	23.7	1873
V–5% Ti	525	26.5	77.9	1687
V–10% Ti	628	27.4	66.5	1900

A number of assumptions were made in the analysis of experimental results [11]. Oxygen contained in the sample can interact only with the vanadium atoms. In so doing each oxygen atom interacts with its closest atom disposed in the center of the BCC lattice of vanadium. The height of maximum I is proportional to number of interacting pairs of O-V, i.e. to oxygen content.

Titanium captures part of available oxygen in solution at presence Ti in vanadium due to a greater affinity for oxygen than that of vanadium (enthalpies of formation of oxides TiO, Ti_2O_3, VO and V_2O_5 are 524.17, 1516.50, 426.38 and 1254.00 kJ/mole, respectively [12]). Whereby the oxygen peak height decreases. Apparently, part of titanium atoms forms a bond with vanadium atoms also creating a pair of Ti-V and a maximum III in the IF spectrum [11]. The reason for maximum III increase in the alloy with 0.5 wt.% Ti compared to 0.3 wt.% Ti is obvious, since a part of oxygen atoms are already occupied, and increase the titanium content leads to an increase in the number of pairs Ti-V.

However, peaks II and III are reduced with further increasing of titanium content, i.e. number double bond reduced. It can be explained reducing of number of pair interactions with increasing titanium content if it is granted that titanium connects the connection of oxygen and vanadium to itself when titanium content greater than 1%, forming ternary complexes V-O-Ti with higher activation energy than binary complexes. In fact, alloy with 5 wt.% Ti gives a small maximum at 748 K. As this takes place, this maximum is increased approximately two times with doubling of titanium concentration [11]. It seems reasonable to say that this maximum created by complex of three atoms of Ti-O-V located in the nearest distances in the titanium lattice [13]. If this is so, the formation of same complexes of atoms can be viewed in different alloys, such as in our case: the same complex Ti-O-V with approximately the same activation energy, but in vanadium-based alloy.

One of the major reasons for non-monotonic dependencies of V-Ti alloys properties on titanium content can be peculiarity of interaction of V and Ti atoms with interstitial impurities. As discussed above, number of binary complexes O-V, Ti-O and Ti-V increases with increase of Ti content and, apparently, their concentrations reach a maximum in the alloy with 0,5 wt.% Ti. A maximum strength (microhardness) of V-Ti alloys is observed approximately for the same concentration of titanium. Heights of IF peaks caused by binary complexes reduced due to formation of an increasing number of ternary complexes V-O-Ti with further increase of the Ti concentration [11]. Accordingly, the strength properties of alloys reduced also. In the alloy V-5 wt.% Ti, peaks due to double bonds are virtually absent, and almost all the complexes are ternary. There is a minimum of strength properties in this alloy (see Table 2). Based on these results, one can conclude that double complexes in V-Ti alloys are more effective detents for dislocations than triple bonds, causing hardening of alloys and decrease of their ductility. Repeated increase the strength properties of the alloy with 10 wt.% Ti caused by doubling the number of ternary complexes as compared with V-5 wt.% Ti alloy as demonstrated an increase two times the peak height of IF in the first alloy [11].

2.2. Regularities of helium behavior

Regularities of gas swelling as well as capture, retention and release of helium were investigated in vanadium-titanium alloys irradiated by helium ions up to fluence of $5 \cdot 10^{20}$ He^{+}/m^{2} at 273 and 923 K [14–16]. It was shown that a nonmonotonic influence of titanium on the helium porosity parameters and gas desorption peaks temperature as on IF and mechanical properties cited above.

In studying the effect of alloying elements on the behavior of implanted helium in vanadium alloys it was supposed [14, 16] that having a larger size titanium atoms ("oversized" atom) create local compression regions of in vanadium lattice, which are sinks for radiation vacancies during low-temperature irradiation (273 K). It was found several high temperature maxima of gas release during investigations the behavior of implanted helium by means of thermal desorption spectrometry (TDS). Their appearance was explained by that vacancies trapping the helium atoms along with simple complexes of type He_mV_n (He is atom of helium, V is vacancy, m and n are their number in the complex, respectively) in pure vanadium form thermally more stable complicated complexes in V-Ti alloys also. This leads to an increase of helium release temperature on doping titanium to vanadium as compared with pure vanadium. All these complexes have different bonding energy and dissociate upon heating to different temperatures accordingly. As shown in [17], thermal stability of complicated complexes of $He_mMe_kV_n$ type are significantly higher in the substitutional alloys with "oversized" atoms than that of simple complexes of He_mV_n type. The results of [14] shows that the temperatures of the main peaks of TDS nonmonotonically depend on titanium content in V-Ti alloys as in the case of above-mentioned dependencies of mechanical properties and IF on concentration of titanium.

It is known [18] that complicated complexes of He_mXV_n type (X is oxygen or nitrogen) may be formed with interstitial impurities in the vanadium also, but their dissociation temperatures and binding energies significantly lower than that for He_mV_n for small values of n in a complex of He_mXV_n. However, cause of non-monotonic dependence of TDS peaks temperature and accordingly of activation energies of gas release on titanium content can be complex processes of various bonds/complexes formation of V-O(N). V-Ti, Ti-O(N) type, disappearance of the double bonds and the formation of ternary bonds of V-O(N)-Ti type only at high concentrations of titanium in V-Ti alloys [11].

It is unlikely formation of simple helium-vacancy complexes in pure vanadium at high-temperature irradiation (923 K) by He^{+} ions. However, thermally stable complex of $He_mTi_kV_n$ type may be formed in V-Ti alloys, i.e. a significant part of helium and vacancies remains in the form of complicated complexes in vanadium alloyed by oversized titanium. As a result highest helium swelling is observed in pure vanadium and swelling is significantly reduced in V-Ti alloys (Table 3). However, the effect of titanium on bubble parameters is nonmonotonic. The bubble sizes and swelling are reduced with increasing of titanium concentrations up to 0.5 wt.%, they are increased when titanium content is 0.7 wt.% Ti and are decreased again in alloys with 1, 5 and 10 wt.% Ti, during which titanium concentration insignificantly affects on bubble density (see Table 3). Minimum size of bubbles and swelling are observed in the alloy V-10 wt.% Ti with a maximum content of titanium and in ternary alloy V-4 wt.% Ti-4 wt.% Cr. It is possible that factors mentioned above play a role here, namely the formation of various double V-O(N), Ti-O(N) and ternary V-O(N)-Ti bonds purifying alloy matrix from oxygen and nitrogen resulting in redistribution of atoms and defects [11, 14].

Table 3. Parametres of helium bubbles in vanadium and its alloys (d_{max} and d_{av} are maximal and average sizes, respectively, ρ is density, S is gas swelling) [14].

Material	d_{max} (nm)	d_{av} (nm)	ρ, (10^{24}, m^{-3})	S (%)
V	~ 22	16	0.017 ± 0.004	4.5 ± 1.1
V–0.1% Ti	~ 3.5	1.9	0.9 ± 0.2	0.5 ± 0.2
V–0.3% Ti	~ 3	1.7	1.3 ± 0.3	1.1 ± 0.3
V–0.5% Ti	~ 2.5	1.5	1.7 ± 0.4	0.4 ± 0.1
V–0.7% Ti	~ 6.5	3.7	1.8 ± 0.5	0.7 ± 0.2
V–1% Ti	~ 3	2.0	1.5 ± 0.4	0.4 ± 0.1
V–5% Ti	~ 3.5	1.6	1.4 ± 0.3	0.5 ± 0.2
V–10% Ti	~ 2	~ 1	3.3 ± 0.8	0.32 ± 0.08
V–4%Ti–4%Cr	~ 2	~ 1	3.6 ± 0.9	0.21 ± 0.05

2.3. Capture, retention and release of hydrogen

It was found nonmonotonic influence of titanium on behavior of hydrogen in alloys V-Ti with different titanium content obtained in TDS studies [19]. The samples were saturated with hydrogen for 2000 h at a temperature of 620 K in an autoclave. All the spectra exhibit low-temperature peak I in the temperature range of 780–790 K, the height of which depends on the alloy composition in contrast to its position on the temperature scale, and high-temperature peak II at 870–880 K, the height and position of which on the temperature scale depend on the alloy composition. An increase in the titanium content in vanadium results in the change in the intensities of peaks I and II: at 0.5 wt.% Ti, the ratio between the heights of these peaks hardly changes; at 1 wt.% Ti, the height of peak I increases and the intensity of peak II decreases; and vice versa, at 5 and 10 wt % Ti, the height of peak I decreases and the intensity of peak II increases, this effect being more pronounced in the range from 5 to 10 wt.% Ti.

It is seen from the dependence of the amount of hydrogen retained on the titanium content, which was determined by the TDS method [19], that 0.5 wt.% Ti reduces the hydrogen concentration by nearly a factor of 3 as compared with that captured by vanadium; at 1 wt.% Ti, the amount of hydrogen retained returns to the initial level; the maximum hydrogen concentration is observed in the V–5 wt.% Ti alloy and it is ~$5.5\cdot10^{20}$ at. H/g (ratio of the number of hydrogen atoms to the number of atoms of the alloy at. H/at. V is ~4.5%). The amount of hydrogen decreases to ~$4.5\cdot10^{20}$ at. H/g (at. H/at. V ~ 3.8%) as the titanium content increases to 10 wt.%. In such a way, the effect of the Ti concentration in V on the amount of hydrogen retained is as nonmonotonic as its effect on the position of peak II in the TDS spectrum.

It was shown in study of the absolute amount of hydrogen retained in V–Ti alloys determined in the RHEN-602 gas analyzer, that the dependence of the amount of hydrogen retained on the titanium content is similar absolutely to that obtained by the TDS method [19].

The regularities of the capture, retention and release of hydrogen and the effect of alloying elements, in particular, Ti to V, on these processes are largely determined by the state in which hydrogen occurs in metals. It is widely believed that hydrogen in materials can exist in one of the following states: atomic (H^0), ionic (H^+ and H^-), molecular (H_2), in compounds with metals (hydrides of M_mH_n type), and in compounds with impurities (e.g., CH_4, H_2S). In addition, hydrogen can form complexes with point defects, for example, with vacancies H_mV_n (V means a vacancy; m and n are the number of hydrogen atoms and vacancies in the complex, respectively) and with vacancies and impurities $H_mX_nV_k$ (X means an interstitial impurity), and other complexes.

Hydrogen was introduced into the samples in the autoclave without causing radiation damage to their structure. The presence of two gas release peaks in the TDS spectra indicates the existence of two main types of traps for hydrogen in vanadium and its alloys. Since the position of low-temperature peak I on the temperature scale is independent of the chemical composition of the material, we have assumed that traps for hydrogen in the temperature range of 780–790 K are such defects, whose binding energy with hydrogen and hydrogen capacity are not affected by the presence of atoms of the alloying elements in solid solution Since vanadium hydrides are thermally unstable, namely, they dissolve already at temperatures from 270 K, such traps can be thermal vacancies [20], dislocations, grain boundaries, twins, discontinuities, and other hereditary defects, which in combination with hydrogen atoms have a positive binding energy. For example, in α-Fe, the binding energy is 0.45–0.53 eV for H–V (V is vacancy), 0.71–0.90 eV for H–V_m (m > 4), 0.24–0.62 eV for H–dislocation, 0.78 eV for H–void, 0.75–0.70 eV for H–helium bubble, and 0.29 eV for H–TiC [21]. Calculations showed that in vanadium single vacancy can hold twelve hydrogen atoms when they are introduced simultaneously and six hydrogen atoms when they are introduced sequentially [20]. It is shown in [22] that in vanadium alloys the chemical composition of the alloy is responsible for the amount of hydrogen trapped. The fact that the intensity (height) of peak I depends on the Ti content may indicate, for example, that titanium atoms change the concentration of this type of traps and/or the binding energy between them and hydrogen atoms.

Traps of the second type, from which hydrogen release upon heating results in the appearance of peak II, are more effective for trapping the hydrogen. In this case, both the intensity of peak II and its position on the temperature scale are determined by the titanium concentration, whose influence on them and the amount of hydrogen retained are nonmonotonic [19]. It can be assumed that in addition to hydrogen captured by the traps of the first type, it is retained in titanium hydrides, or complex hydrides like $Ti_{1-x}V_xH_y$, or γ-TiH_2 titanium hydride, since the standard enthalpy of formation of titanium hydride is less (–144.4 kJ/mol [23]) than that of vanadium hydrides (–39.9 kJ/mol [24]).

It was shown in Table 1, vanadium contained ~0.08% O_2 and ~0.009% N_2, and the V–5wt.% Ti alloy contained ~0.32% O_2 and ~0.0064% N_2. It is established using IF measuring methods and X-ray phase analysis that titanium at low concentrations participates in the formation of various complexes such as Ti–O, Ti–O–V, $VTiO_x$, and others, and nitrides. This occurs because the enthalpy of their formation is significantly less than that of titanium hydride [19], i.e., titanium in an amount of 0.5 wt.% (0.53 at.%) is not involved in the formation of hydrides, but is consumed for binding oxygen and nitrogen, the total concentrations of which in vanadium amounts to 0.58 at.% [11]. It seems that this should not lead to a decrease in the amount of hydrogen retained in the V–0.5% Ti alloy compared with that in vanadium. This reduction in the amount of hydrogen is possible in the case where oxygen and nitrogen in a vanadium solid solution promote retention of hydrogen; for example, they greatly distort the crystal lattice as interstitial impurities. Such interstitial places play the role of vacancy sinks and, hence, they trap hydrogen atoms. The titanium induced escape of oxygen and nitrogen from the solid solution reduces the number of such traps and leads to a reduction in the hydrogen concentration in the V– 0.5% Ti alloy [19].

The amount of hydrogen retained again increases with increasing Ti content to 1 and 5 wt.%, since only a part of titanium is involved in the formation of oxides and nitrides, while the rest forms hydrides with a dissociation temperature of 870 K and above, which becomes clear from comparing the temperature of peak II and the temperature at which γ titanium hydride decomposes [19].

It is shown in [25] that, at a content of ~10% Ti, vanadium begins to take up part of the electron density from titanium. Therefore, a decrease in the electron density of titanium atoms reduces the number of hydrides formed and, correspondingly, reduces the amount of hydrogen retained in the V–10% Ti alloy as compared with that in the V–5% Ti alloy [19].

4. Conclusion

The paper shows that the effect of titanium content on helium swelling, processes of energy dissipation in the crystal lattice (internal friction), capture, retention and release of introduced helium and hydrogen in vanadium-titanium alloys is nonmonotonic, as well as the effect of titanium on some physical, mechanical and high-temperature properties and radiation resistance of V-Ti alloys are known from literature. Analysis of experimental data obtained by different methods, leads to the conclusion that perhaps such an influence occurs due to the interaction of atoms of vanadium and titanium with interstitial impurities, in particular redistribution of O, C and N atoms linkages between vanadium and titanium atoms.

Acknowledgements
This work was performed within the framework of the Center of Nuclear Systems and Materials supported by MEPhI Academic Excellence Project (contract № 02.a03.21.0005, 27.08.2013).

References

[1] Ivanov A N, Krukova L M and Chernov V M 2008 *Voprosy At. Nauki i Techniki: Materialovedenie i Novie Materialy* **2(71)** 61 (in Russian)
[2] Ermolaev G N 2008 *Voprosy At. Nauki i Techniki: Materialovedenie i Novie Materialy* **2(71)** 53
[3] Efimov Yu V, Baron V V and Stavitskiy E M 1969 *Vanadium and its Alloys* (Moscow: Nauka) 254 pp (in Russian)
[4] Loomis B A, Nowicki L J and Smith D L 1994 *J. Nucl. Mater.* **212–215** 790
[5] Chung H M, Loomis B A and Smith D L 1994 *J. Nucl. Mater.* **212–215** 804
[6] Ivanov L I and Platov Yu M 2002 *Radiation Physics of Metals and its applications* (Moscow: Nauka) 300 pp (in Russian)
[7] Kalin B A, Chernov I I, Yakushin V L et al 1984 *Atomic Energy* **57** 173
[8] Wang Y, Kanedome M, Yasuda T. et al 2004 *J. Nucl. Mater.* **329–333** 477
[9] Chen J, Xu Z and Yang L 2002 *J. Nucl. Mater.* **307–311** 566
[10] Kalin B A, Chernov I I and Staltsov M S 2011 *Nucl. Phys. and Engin.* **2(4)** 320 (in Russian)
[11] Anan'in V M, Kalin B A, Korchagin O N et al 2012 *Inorg. Mater.: App. Res.* **3**(3) 243
[12] Novozhzhenov V A 2001 *Introduction to Organic Chemistry* (Barnaul: Altay University) 650 pp (in Russian)
[13] Gupta D and Veining S 1965 In: *Internal Friction And Defects in Metals* / Ed. by V S Postnikov (Moscow: Metallurgiya) 47 (in Russian)
[14] Staltsov M S, Chernov I I, Kalin B A, Kyi Zin Oo, Polyansky A A, Staltsova O S, Aung Kyaw Zaw, Chernov V M and Potapenko M M 2015 *J. Nucl. Mater.* **461** 56
[15] Chernov I I, Staltsov M S, Kalin B A et al 2010 *Atomic Energy* **109(3)** 176
[16] Chernov I I, Staltsov M S, Kalin B A, Kyi Zin Oo, Aung Kyaw Zaw, Statsenko V I, Korchagin O N and Staltsova O S 2012 *Fizika I Chimiya Obrabotki Materialov* **3** 22 (in Russian)
[17] Chernov I I, Kalin B A, Kalashnikov A N and Ananin V M 2002 *J. Nucl. Mater.* **307–311** 362
[18] Nita N, Miyawaki K and Matsui H 2007 *J. Nucl. Mater.* **367-370** 505
[19] Aung Kyaw Zaw, Chernov I I, Staltsov M S et al 2015 *Inorg. Mater.: App. Res.* **6(2)** 138
[20] Gui Li-Jiang, Liu Yue-Lin, Wangb Wei-Tian et al 2013 *J. Nucl. Mater.* **442** 688
[21] Kalin B A, Kalashnikov A N, Chernov I I and Shmakov A A 2011 In: *Proc. 7th Int. School of Young Scientists and Specialists (IHISM'11) "Interaction of Hydrogen Isotopes with Structural Materials"*, Moscow, 10 (in Russian)
[22] Natesan K, Soppet W K 2000 *J. Nucl. Mater.* **283–287** 1316–21
[23] Barin I. 1993 *Thermochemical Data of Pure Substances* Part I (VCH Verlags Gesellshaft, Weincheim)
[23] Golubkov A N and Yukhimchuk A A 1999 In: *Proc. of NATO Advanced Research Workshop on Hydrogen Materials Science and Chemistry of Materials* 2-8 Sept. (Yalta, Ukraine) 255
[24] Votinov S N, Kolotushkin V P and Parfenov A A *Atomnaya Nauka i Promishlennost* (http://bus.znate.ru/docs/index-27154.html?page=3)

Thermal stability and high-temperature deformation of tungsten nanocomposite

D N Ignatiev[1,2], A A Pavlov[1], E S Solntseva[1], M L Taubin[1,2] and A A Yaskolko[1,2]

[1]Federal State Unitary Enterprise "Science Research Institute "LUCH", Zheleznodorozhnaya st 24, Podolsk, Moscow region, 142100, Russia
[2]National Research Nuclear University MEPhI (Moscow Engineering Physics Institute), Kashirskoe highway 31, Moscow, 115409, Russia

E-mail: taubin@sialuch.ru

Abstract This paper presents the result of research of thermomechanical treatment impact on polycrystalline tungsten structure and properties. Samples were exposed to deformation. Two deformation technics were applied: uniaxial rolling and cross direction rolling. After deformation all samples acquired the form of foil. After deformation the relation between the structure of the foils and the annealing time was investigated.
The investigation of high-temperature deformation was conducted using the samples which underwent preliminary heat treatment, and which didn't. The result of heat treatment and tensile strength investigation of pre-machined polycrystalline tungsten samples at elevated temperature is presented in this study. It is shown that annealing at 2670 K after uniaxial rolling could result in significant increase of high temperature creep resistance of tungsten.
The factors that determine the effect of reducing high-temperature deformation of modified tungsten have been identified. As a result of mechanical treatment with subsequent annealing a unique nanoscale structure is formed. The tungsten nanocomposite has been obtained for the first time. During investigation of tungsten nanocomposite thermal stability its structural stability has been confirmed.

1. Introduction

One of the most important task of material science is invention of structural materials resistant to elevated temperature and mechanical treatment into corrosive media. Structural materials of high-temperature and high-loaded systems must have such properties as high tensile strength, high ductility and low creep at elevated temperatures. Tungsten based materials are used as a high temperature material in many fields of science and technology due to the high rates of long-term strength and creep resistance. However, increasing temperature and operation time lead to deterioration of tungsten properties, so it is needed to enhance creep strength at temperatures above 1770 K.

To improve high-temperature strength of tungsten an idea of creation tungsten nanocomposite is suggested. Grid of nanoscale potassium filled pores which forms in its structure should contribute to a slowdown of recrystallization processes. Insoluble nanoscale potassium pore will prevent the formation and growth of recrystallized nuclei.

2. Experimental

The influence of thermo mechanical treatment on tungsten based materials structure and properties was studied using microfoils 100 μm thickness. The foil-shape samples with two types of structure: equal axis grains (after cross-direction rolling deformation) and oblong grains (after one-direction rolling deformation) are used in this research. Thermal stability investigation was conducted in vacuum chamber under pressure $2.6{\cdot}10^{-4}$ Pa. The annealing was carried out by direct drawing a current trough the samples in three stage: the 1- st - 30 min long up to 1670 K with heating rate 300 deg/min and after up to 2670 K with heating rate 50 deg/min; the 2-nd - 90 min long at temperature 2670 K; the 3-rd - 180 min at the same temperature.

To determinate time dependence of high temperature annealing on material structure the samples were marked by diamond paramecium after the 1-st 30 minutes annealing.

The high temperature deformation was studied under the constant load at temperature 2470 K. The investigation of high-temperature deformation was conducted using the samples which underwent preliminary heat treatment, and which didn't.

The experimental result are supplement with X-ray and micro structural analysis by using the JEOL 6460 microscope.

3. Result and Discussion

Different mechanical treatment and annealing at 2670 K of polycrystalline tungsten samples resulted in formation of various recrystallized structure with same chemical composition. After high temperature annealing all modified tungsten foils have a structure with grid of nano-sized pores (from 50 to 150 nm) on grain boundaries. Pores present in specimens with oblong grains and with equal axis grains alike and almost all pores are located on borders of grains.

Samples from the cross-rolled foil acquire equally grain size structure after high-temperature annealing. But samples with uniaxiality rolling have elongated grains structure, the crystallographic orientation of all the grains changed in the same direction [100] (Figure 1). Dislocation density within their grains body decreased, the X-ray spectrum lines half-width is evidence of this phenomena (Figure 2). Long-term annealing does not lead to significant changes in the material structure: clusters of small pores at big grains boundaries are maintained after that. Observed effect has been stable up to 2670 K.

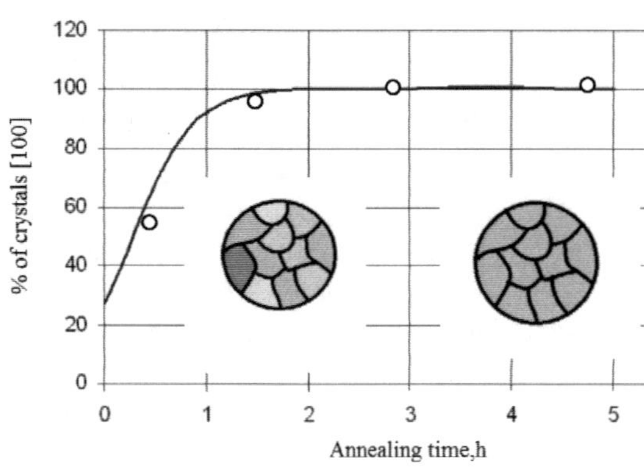

Figure 1. Amount of crystallites with end on direction

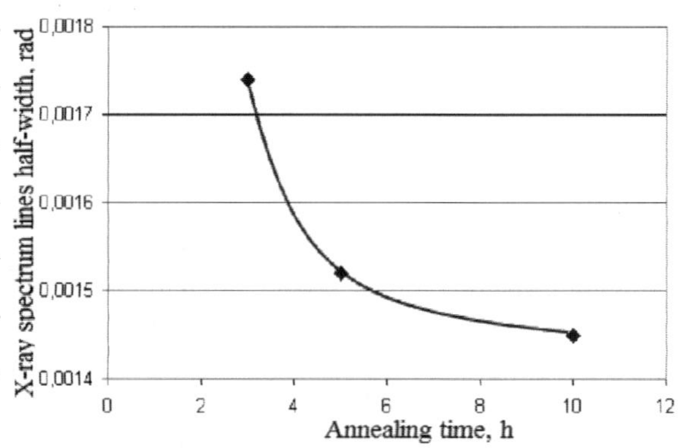

Figure 2. The X-ray spectrum lines half-width

The structure difference is described by various mechanical treatment. In initial material unsolvable potassium is located in stuff pores with effective diameter about 0.1-1.0 microns. One direction rolling forced pores get the elliptic form extended in a direction of rolling. In course of annealing at elevated temperatures this form becomes unstable and breaks up to the pores of the smaller dimension built in rows in a direction of rolling. Pores row hinder grain growth in perpendicular to the rolling direction in process of recrystallization under annealing. It leads to the formation of structures with oblong grains elongated towards the direction of rolling (Figure 1). In such a structure grains slippage considerably decrease.

Specimens with oblong grains structure after heat treatment at 2670 K under constant load high – temperature deformation has shown significant reduction (3 times) of deformation rate, and on the contrary, deformation rate under constant load of samples with equal axis grains structure is accelerated. It is caused by different materials structure which formed after heat and mechanical treatment (Figure 3). The cross-rolled sample from non-heat treated tungsten has recrystallized during creep testing at 2470 K and had a structure with a smaller average grain size than the sample from the pre-annealed material. So its deformation rate was significantly higher than non-heat treated one (Figure 4).

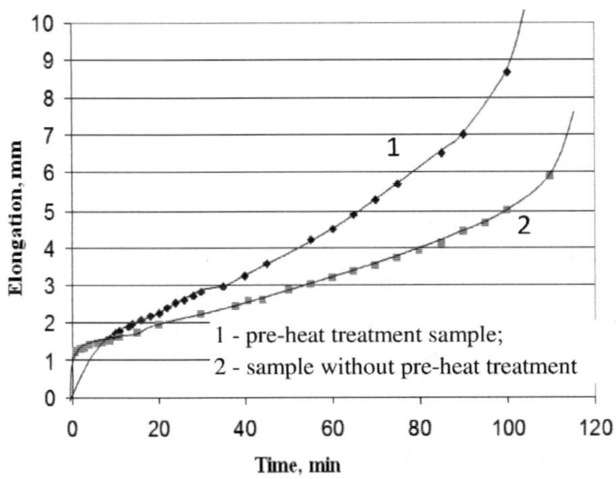

Figure 3. High-temperature deformation of tungsten with equal axed grains

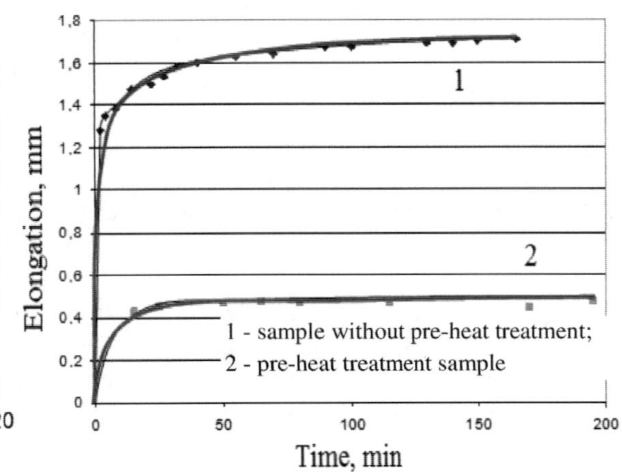

Figure 4. High-temperature deformation of tungsten with oblong grains

Comparing pre-annealing samples with equal axis grains structure and oblong grains structure shows that first one have greater creep rate than materials with oblong grain. The sample with equal axial has considerably large speed especially at the second stage of deformation, as shown in Figure 5. It means that after annealing samples with oblong grains have greater creep resistivity, than without it.

The factor determining a high temperature deformation reduction is a special structure which is formed under uniaxial rolling with subsequent high temperature annealing. This structure is characterized by elongated grains with grid of nano-sized pores at their boundaries which are developed increasing creep resistance [1- 3].

The results of investigations of thermal stability and the factors, which determine the effect of reducing high temperature deformation of modified tungsten, are presented. It is shown that thermomechanical treatment (one direction rolling with subsequent high-temperature annealing) results to formation of tungsten nano-composite unique structure which have an effect of significant reduction of high-temperature deformation [4 - 6]. Such materials as tungsten nanocomposite, which is thermally stable up to 2670 K, obtained for the first time.

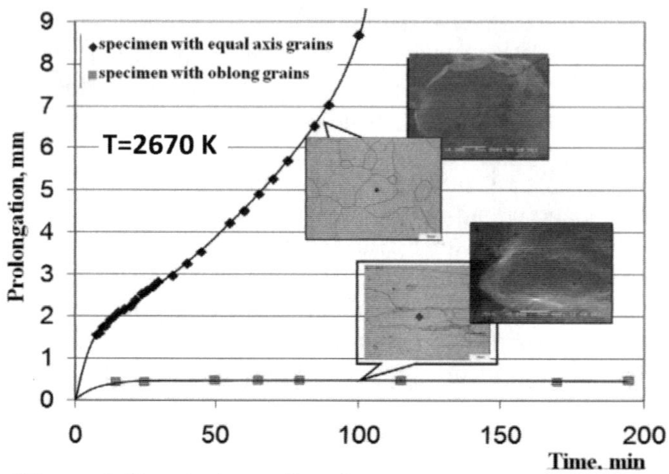

Figure 5. Comparison of pre-heat treatment samples deformation with different structure

4. Conclusion

Research of influence mechanical and heat treatment on structure and properties of polycrystalline tungsten are conducted. It is shown that heat treatment at temperature 2670 K of one chemical compound tungsten after different type of rolling deformation leads to formation of various recrystallized microstructure. One-direction rolling with subsequent annealing initiate textured structure (oblong grains are elongated far the [100] direction), cross-direction rolling with subsequent heat treatment leads to form the structure with huge equal axis grains. However both type of samples have the structure with grid of nanoscale pores (from 50 to 150 nanometers) on grains boundaries.

The high-temperature deformation tests at 2670 K under constant load show significant reduction of oblong grain sample creep (up to 3 times), and in the contrary the raise of deformation speed in equal grain sample after annealing.

The factor defining reduction of high-temperature deformation is the unique structure with oblong grains and grids of potassium nanoscales pores on their boundaries, which form as a result of a one-direction rolling and high-temperature annealing after that.

The tungsten nanocomposite has been obtained for the first time. During investigation of its thermal stability it is shown that tungsten nainocomposite thermally stable up to temperatures 2670 K.

Acknowledgements

This work was performed within the framework of the Center of Nuclear Systems and Materials supported by MEPhI Academic Excellence Project (contract № 02.a03.21.0005, 27.08.2013).

References

[1] Alekseev S, Taubin M and Yaskolko A 2009 *20th Workshop ISTC in Korea* (Seul)
[2] Alekseev S, Taubin M and Yaskolko A 2013 *Proc. of 2nd Biotechnology World Congress* (Dubai)
[3] Alekseev S, Taubin M, Pavlov A and Yaskolko A 2013 *BIT's 4th World Gene Convention* p 301.
[4] Wright P K 1978 *Metallurgical Transactions* **9** 955
[5] Yamazaki S, Ogura S and Fukazawa Y 1978 *High Temperature – High Pressure* **10** 329
[6] Briant C L 1993 *Metallurgical Transactions* **24** 1073

Iron atoms redistribution in oxide films of Zr-Fe, Zr-Fe-Cu alloys during corrosion in autoclave at 350°C

V Filippov and A Bateev

National Research Nuclear University MEPhI (Moscow Engineering Physics Institute), 115409, Kashirskoe shosse, 31, Moscow, Russia

E-mail: vpfilippov@mephi.ru

Abstract. The data on changes of iron atoms state in the oxide films of binary Zr-1.24 mas.%Fe and ternary Zr-1.39 mas.%Fe-0.60 mas.%Cu zirconium alloys are obtained. Alloys are subjected to corrosion tests under autoclave conditions at 350°C temperature in a steam-water environment under pressure p = 16.8 MPa. In initial specimens of the alloys the iron atoms are in the form of intermetallic compounds. In oxide films the decomposition of intermetallic compounds and formation of new compounds occurs with structural phase distortion. In the oxide films metallic the metallic iron particles α–Fe, iron oxide in the form of hematite α–Fe_2O_3, solid solutions of iron ions in ZrO_2 are formed. The phase composition of the oxide films depends on the alloy composition and changes during the growth process of the oxide film.

1. Introduction

Zirconium alloys are used as materials for claddings of nuclear reactors. The extension of operating time of fuel rods up to 6-7 years and increasing the burning of fuel from 35 to 70 MWt days/kg U [1] depend on corrosion resistance of zirconium alloys. To create the new corrosion-resistant alloys it is necessary the understanding of the processes in the oxide films during corrosion of zirconium alloys especially in a steam environment closed to reactor conditions. The same data are needed for understanding the influence of additional doping of different elements. The perspective alloying elements are tin, copper, chromium, iron, tungsten and nickel. Iron has Mössbauer isotope ^{57}Fe and therefore Mössbauer spectroscopy can be used to study the redistribution of iron atoms in oxide films. The aim of this paper is to obtain the data about the redistribution of the iron atom states in oxide films during corrosion in autoclave at the temperature of 350 °C, for binary Zr-Fe and ternary Zr-Fe-Cu. alloys.

2. Experimental

Two zirconium based alloys Zr-1.24 mas.%Fe and Zr-1.39 mas.%Fe-0.60 mas.%Cu were prepared. The iron concentration differs slightly, so any differences in the corrosion behaviour and in the iron atoms redistribution can be caused by the presence of copper in the alloy. The alloy specimens were melted in an electric arc furnace using nonconsumable tungsten electrode in an argon atmosphere, and then subjected to thermo-mechanical treatment with the technology described in [1, 2]. Then the samples were rolled approximately to thickness of 800 microns, and were annealed at 650 °C during 2 hours. Corrosion tests were carried out in stainless steel autoclaves at temperature 350 °C, with 16.8

IOP Conf. Series: Materials Science and Engineering **130** (2016) 012015 doi:10.1088/1757-899X/130/1/012015

MPa pressure. Oxide films were separated from the metal by dissolving the metal substrate in 3% solution of hydrofluoric acid. From the obtained oxide films the absorbers were prepared for Mössbauer studies [2]. ^{57}Co (Cr) was used as a Mössbauer source. The spectra were obtained using the Mössbauer spectrometer operating at constant acceleration mode at room temperature. The spectra were fitted using Mössbauer programs Univem MS. The technique of determining the absolute concentrations of iron atoms described in [4].

3. Results and discussion

3.1. Kinetic of corrosion

The kinetic curves of corrosion tests are shown in Figure 1. During initial stage of corrosion the oxide films on alloys were strongly adhered to the metal and had a black color. For Zr-Fe alloy some whitened grains were detected under microscopic examination on the surface of the oxide film with thickness of 2.8 microns. The number of grains increases with increasing film thickness. The analyze of kinetic curves shows that in initial stage of corrosion the growth of oxide film thicknesses are almost parallel, with approximately equal exponents. The oxide film thicknesses of binary alloy a little more than those of ternary alloy with a copper. Approximately at 6500h the transition to catastrophic corrosion (breakaway) is detected for binary ally. For the ternary alloy, the oxide film thickness increase according to parabolic law until the end of corrosion tests (about 18000 h). This means that the ternary alloy Zr-Fe-Cu has a higher corrosion resistance than binary alloy Zr-Fe. Taking in to account that all the processes for the preparation of samples and corrosion tests were identical and the only differe nce is the presence of copper in the second alloy, one can assumed that there are the effect of copper in this alloy. The effect of copper on the redistribution of iron atoms in oxide films can be explained from the analysis of Mössbauer spectra of oxide films.

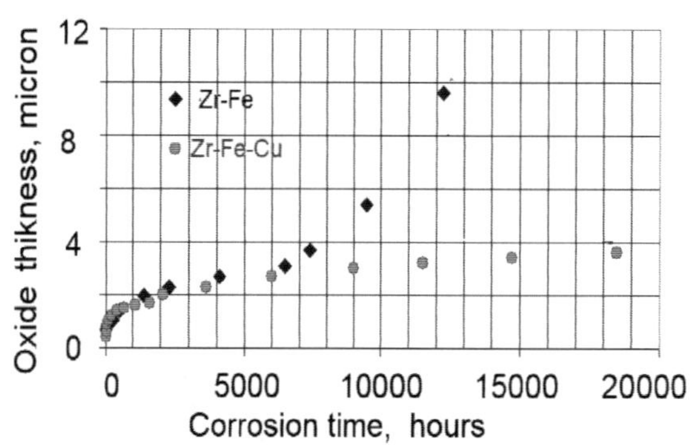

Figure 1. Kinetic corrosion curves of zirconium alloys tested in autoclave at 350 °C temperature, and 16.8 MPa pressure.

3.2 Mossbauer data

Mössbauer spectra of some oxide films are presented in Figure 2. One can see the superposition of the lines of magnetic hyperfine splitting of the magnetic phases and lines of the quadrupole splitting of the paramagnetic phases. Practically all spectral lines are broadened. The broadening of the lines can be caused by the superposition of lines of different phases with similar parameters of the spectra, associated with non-stoichiometry of compounds and defects in the crystal lattice. Analysis of the spectra is made using the previously obtained data [1, 2]. These spectra are fitted with programme MSTools [3] giving the distributions of quadrupole splitting and hyperfine magnetic fields (HMF). The obtained distributions are shown in Figure 3. One can see the differences in spectra because Mössbauer spectroscopy is sensitive to the local and crystallographic environment. Each distinct crystallographic, magnetic and chemical environment for Fe forms different Lorentzian lines in the Mössbauer spectrum. That's why in complex absorbers there are the local distortions and atomic disorder surrounding the Fe atoms, rather than simply reflecting the ideal point symmetries of the relevant sites. The lines of spectra have large width lines because of some distribution of hyperfine magnetic fields and quadrupole splitting. The sextets of the field distribution were constrained to an area 3:2:1 with linewidth of 0.25 mm s^{-1}. The steps of ~0.14 T for the magnetic field distribution were used. The results are shown in Figure 4. The isomer shift was allowed to vary linearly with the magnetic hyperfine field. The magnetic field distribution may reflect a wide distortion in particle size and the presence different form of iron oxides. Intensive pick is in the region of 33.0 T for both spectra (Figure 4) of samples. These picks associate with presence of particles of α–iron in oxide films. The presence of picks with low intensities in the region of ~30 and ~27 T (Figure 4 b) can be explained by influence of Cu atoms in alloy. Such magnitudes of hyperfine magnetic fields can reflect the presence of sold solutions of Cu atoms in the iron particles. Such picks are not observed for sample of binary alloy. For both oxide samples the picks of HMF of ~18.0 T are observed with relative intensities of ~ 0.04.

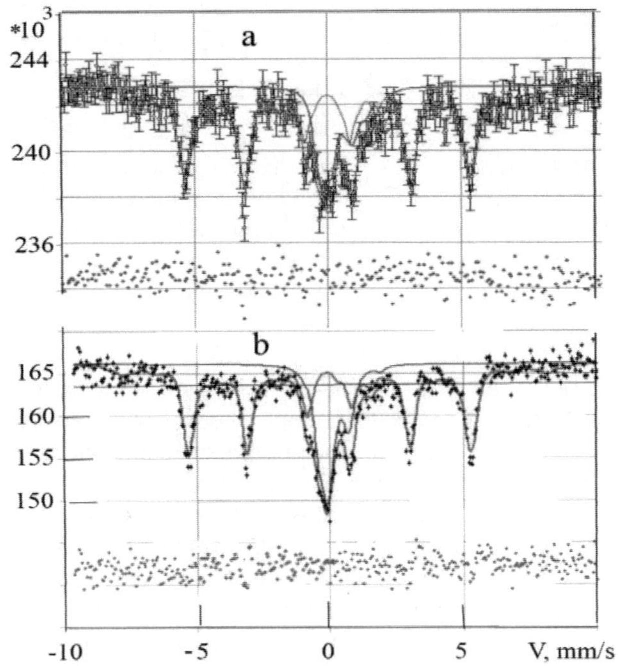

Figure 2. Transmission Mössbauer spectra of ^{57}Fe in: a) oxide film of 2.2 μm on alloy (Zr-1,24 mas.% Fe) after 1136 h of corrosion, b) oxide film of 3.2 μm on alloy (Zr-1,39 mas.% Fe-0.60 mas.% Cu) after 8100 hours of corrosion.

These picks may deal with the presence of intermetallic compound $ZrFe_2$. $ZrFe_2$ can be formed in the process of oxidation of intermetallic compound Zr_3Fe. As Zr is more active in interaction with oxygen the Zr atoms leave intermetallic compound Zr_3Fe and formed crystals of ZrO_2. In the Zr_3Fe particles the deficit of Zr atoms can takes place. This situation promotes to formation of particles $ZrFe_2$. In further process of corrosion these particles are oxidized too. That's why the lines of $ZrFe_2$ particles were not observed in conversion electron Mössbauer spectroscopy (CEMS) spectra. CEMS spectra reflect information from surface layers in which full oxidation is realized. As to pick with magnitude of magnetic field ~ 26.0 T it can be associated with the interfacial region between the nanocrystallites or /and by presence disordered amorphous – like structure of some crystals [5].

The picks with maximums of HMF in the range of 40.0-51.5 T can reflect the presence of iron oxides and hydroxides in oxide films of investigated alloys. The intensities of these picks are small. It means that concentrations of iron in these compounds are small. Correct identification of these compounds at this step of investigations is difficult. More probable the simultaneous presence of iron compounds such as α–Fe_2O_3, Fe_3O_4, β–Fe_2O_3.and in the form of metallic iron and an unknown phase called as Fe^{4+}.

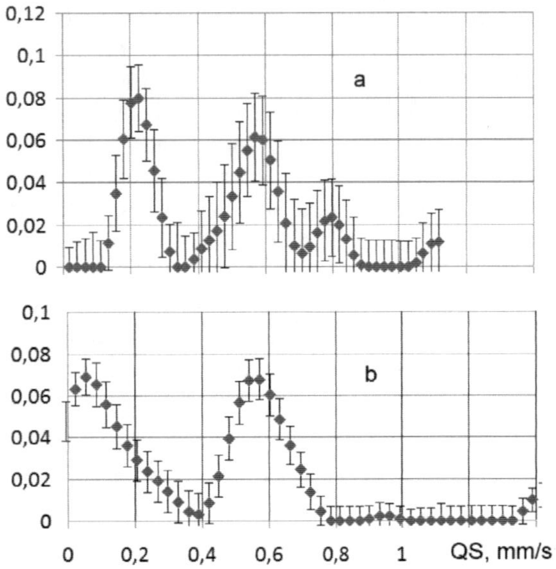

Figure 3. Quadrupole splitting distribution of Mössbauer spectra of ^{57}Fe in: a) oxide film of 2.2 μm on alloy (Zr-1.24 mas.%Fe) after 1136 h of corrosion, b) oxide film of 3.2 μm on alloy (Zr-1.39 mas.%Fe-0,60 mas.%Cu) after 8100 hours of corrosion.

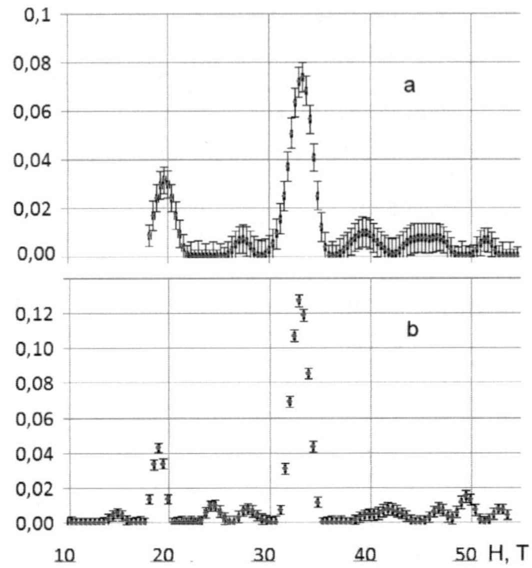

Figure 4. Hyperfine magnetic distribution of Mössbauer spectra of ^{57}Fe in: a) oxide film of 2.2 μm on alloy (Zr-1.24 mas.%Fe) after 1136 h of corrosion, b) oxide film of 3.2 μm on alloy (Zr-1.9 mas.%Fe.0,60 mas.%Cu) after 8100 hours of corrosion.

Quadrupole splitting distributions (QSDs) are a further evolution in modeling Mossbauer spectra of oxides in which there are poorly-resolved quadrupole pairs. The QSDs model the local distortions and atomic disorder surrounding the Fe atoms in investigated samples. The differences in the shape of QSDs can be explained by presence of Cu atoms in Zr-Fe-Cu atoms. For binary alloy (Figure 3 a) three picks are visible with maximums ~ 0.2 mm/s, ~ 0.59 mm/s and ~ 0.8 mm/s. For ternary alloy (Figure 3 b) only two picks are visible with maximums ~ 0.1 mm/s, ~ 0.58 mm/s. For both oxide films the lines with QSDs in range 0.1-0,2 mm/s can be explained by presence of sold solutions of Fe^{3+} ions in ZrO_2. The QSDs with magnitudes ~ 0.58 mm/s and more can reflect the presence of sold solutions of Fe^{2+} ions in ZrO_2. The intensity of third pick in spectra of QSDs of oxide film of binary alloy is in the range of experimental error, but in all fitting it is visible. It is possible that some regions of ZrO_2 have iron ions in sold solution with great gradient of electrical field. This gradient deals with a distortion of structure of ZrO_2. The distortion of structure of ZrO_2 correlates with increasing of corrosion velocity of this alloy. The large width of QSDs for ternary alloy in the range of ~0.1 mm/s can result from probable existence solid solutions of iron ions Fe^{4+} in ZrO_2.

The relative content of iron in the form of so cold Fe^{4+} in the ternary alloy oxide films is greater than for binary alloy. These data allow us to make the assumption that the corrosion resistance is related to the content of iron in the form of metallic iron and probably an unknown phase called as Fe^{4+}. Thus the doping of the zirconium alloy by copper improves the corrosion resistance of the alloy and prevents the formation of large concentrations of compounds α–Fe_2O_3 .

4. Conclusions

1. Iron atoms in oxide film are investigated after corrosion test of binary Zr-Fe and ternary Zr-Fe-Cu alloys in autoclave at 350 °C of temperature and 16.8 MPa of pressure.
2. It is shown that the doping of copper in zirconium alloy with iron increases the alloy corrosion resistance and changes the process of iron atoms redistribution. In oxide films the compounds and

IOP Conf. Series: Materials Science and Engineering **130** (2016) 012015 doi:10.1088/1757-899X/130/1/012015

phase are detected such as 1) α–Fe_2O_3, 2) α–Fe, 2') solid solution Cu in α–Fe, 3) solid solution Fe^{2+} in ZrO_2, 4) solid solution Fe^{3+} in ZrO_2, 5) unknown phase witch can be compound $ZrFe_2$.
3. It is revealed that in oxide films of ternary alloy having a higher corrosion resistance, the relative content of α–Fe is more in comparison with binary alloy. This result in conclusion that time of transition to catastrophic corrosion correlates with the formation of this phase in oxide films .The doping of copper in the alloy prevents the formation of iron oxides and increases the corrosion resistance of the ternary alloy.

Acknowledgements. This study was supported by the RF Ministry of Education and Science, agreement no. 14.575.21.0086, unic project identifier RFMEFI57514X0086.

References

[1] Pecheur D, Filippov V P, Bateev A B, Ivanov Yu Yu 2002 *Zirconium in the Nuclear industry, 13 international symposium ASTM STR 1423*, Cherry D Moan, Eds. ASTM international, PA 135–153
[2] Babikova Yu F, Gruzin P L, Filippov V P and Ivanov A V 1981 *Izvestia VUZov, physic part* **1** 70–74 (in Russian).
[3] Rusakov V S, Kadyrzhanov K K 2005 *Hyperfine Interact.* **164** 87–97
[4] Filippov V P, Petrov V I, Lauer D E, Shikanova Yu A 2006 *Hyperfine Interact.* **168** 965–71
[5] Dino Fiorani 2005 *Surface Effects in magnetic nanoparicles.* Ed (ISM-CNR Area delta Ricerca di Roma Rome, Italy) p 221

IOP Conf. Series: Materials Science and Engineering **130** (2016) 012016 doi:10.1088/1757-899X/130/1/012016

Radiation-induced softening of Fe-Mo alloy under high-temperature electron irradiation

A Tsepelev[1,2]

[1]Institute of Metallurgy and Materials Science of the Russian Academy of Sciences (IMET RAS), 49 Leninsky pr., Moscow, 119991, Russia
[2]National Research Nuclear University MEPhI (Moscow Engineering Physics Institute), 31 Kashirskoe highway, Moscow, 115409, Russia

E-mail: tsep@imet.ac.ru

Abstract. Effect of radiation-induced change of mechanical properties of Fe-5 wt.% Mo alloy irradiated with electrons (2 MeV) at room temperature and 400°C has been investigated. Mechanical properties were estimated by Miniaturized Disk Bend Test technique. Effect of radiation softening of the alloy is ascertained the value of which was increased with temperature rise. With the purpose of separation of thermal and radiation contributions into the effects, the tests were carried out for specimens annealed in the same thermal conditions (temperature and duration of annealing) just as during irradiation. Thermal annealing and electron irradiation at 400°C is found to bring to multidirectional effects of the alloy strengthening and softening respectively. It is concluded that irradiation suppresses the effect of thermal-induced strengthening and stimulates a softening of the alloy due to more significant changes in the structure and phase composition of it.

1. Introduction

The most important characteristics of structural materials used in nuclear power plants is their strength and plastic properties, and stability of these properties under irradiation. There are two main reasons for changes in the mechanical properties of metals and alloys under irradiation — it is the formation of the so-called radiation damages (for example, depleted zones) which are obstacles in the way of gliding dislocations, and changing of structure and phase state of materials caused by radiation enhanced diffusion processes. The first mechanism plays a leading role under cascade-forming irradiation (neutrons, swift heavy ions) and usually leads to radiation strengthening of irradiated materials.

At the same time, radiation-induced structure and phase transformations, caused by an introduction of excess concentration of radiation point defects (vacancies and interstitials) into irradiated materials and controlled by the processes of radiation-enhanced diffusion (RED) and/or radiation-induced segregation (RIS), may stimulate the transition of metastable alloys (which are the most commertial ones) in the more equilibrium state (RED effect), or, contrary, result in to local deviations from the equilibrium state of the system under the given conditions (RIS effect). Such processes can be occurred at sufficiently low temperatures when conventional thermal diffusion is almost suppressed. The upper limit of manifestation of such effects is accepted the temperature about of 0,5 T_{melt} when concentration of radiation point defects becomes comparable with the thermal-equilibrium concentration of vacancies.

It is known that radiation can be used as a tool for directional changes in the properties of materials, however for the practical realization of these opportunities it is required to understand the controlled mechanisms of radiation-stimulated processes in irradiated materials under different conditions of radiation treatment (temperature, flux and fluence of irradiation).

Most of the known data on the effect of irradiation on the mechanical properties of metals and alloys were obtained under neutron or ion irradiation, and they are evidence of radiation strengthening effect occurring due to formation of various radiation damages, resulting in displacement cascades and playing the role of obstacles for moving dislocations. At the same time, the "diffusion" effect of irradiation can lead to both strengthening and softening of irradiated material depending on what kind of structure is formed in the process of accumulation and annealing of radiation point defects. In these mechanisms of structure changing, an interaction of mobile radiation point defects with intrinsical materials "genetic" defects (dislocations, impurity atoms, phase precipitations, boundaries, and so on) plays the main role, and it is possible for any kind of irradiation when the lattice atoms are knocked out from their sites. Note that changes of materials properties due to radiation-induced structure and phase transformations may be masked by other radiation effects, especially at low doses.

In pure form, the "diffusion" effect of irradiation can be observed, for example, under MeV-energy electron irradiation when displacement cascades and defect clusters does not to be created. In this case, irradiation leads to the formation of isolated vacancies and interstitials able to freely migration before annihilation with each other or annealing at sinks (dislocations, interfaces, voids, etc.).

The most convenient way of investigation of diffusion-controlled mechanisms of radiation-induced processes is study of model binary alloys, the number of possible crystal-chemical reactions of radiation point defects with sinks in which is limited.

In this work, the experimental verification of the possibility of changing the mechanical properties of Fe-Mo alloy induced by the enhancement of the diffusion-controlled processes under electron irradiation has been carried out.

2. Experimental technique

Investigations were carried out with the use of Fe-5 wt.% Mo alloy. Accordingly to the phase diagram, the temperature of the structure transition in a two-phase state is below 500°C for this alloy, however on account of a small value of the thermal diffusion coefficient at such temperatures ($\sim 10^{-18}$-10^{-19} cm^2/sec), decomposition of the solid solution does not occurred under normal conditions. The alloy was produced by vacuum casting from 99.9% Fe and 98% Mo. After graduated rolling down to a thickness ~0.2 mm, the resultant foil was annealed in vacuum at 1000°C for 4 hours. The flat specimens of 15×30 mm in size were cut from the rolled foils. Next these specimens were irradiated with 2 MeV electrons (beam current of 200 μA) in air at room temperature and 400°C for 5 hours (fluence of $\sim 10^{18}$ cm^{-2}). The room temperature (20±5°C) was kept by water jet flashing of specimens, elevated temperature (400±10°C) — by compressed air blow-off of it.

Mechanical properties of the alloy were judged by the results of the Miniaturized Disk Bent Tests (MDBT) of disk specimens of 3 mm in diameter which were cut off from irradiated foils [1-3]. Tests were carried out using a special adjusting device with the help of Instron 3382 test machine by means of specimen indentation with a cylindrical punch of 1 mm in diameter with a loading rate of 0.2 mm/min. Registered "load–specimen deflection" curve allows to make a qualitative estimation of the material mechanical properties changes after irradiation because the shape of this curve is very sensitive to structure and phase state of the material [4]. Altogether in five specimens were tested for the alloy in each state (initial, annealed, and irradiated).

3. Results and discussion

Figure 1 shows the loading curves of Fe-5 wt.% Mo alloy disk microspecimens in initial state and after electron irradiation (2 MeV, $\sim 2\cdot 10^{18}$ cm^{-2}) at room temperature and 400°C. For clarity, in one typical curve for each condition is shown on the figure and these curves are displaced relative to each other. As can be seen, irradiation leads to a reduction of a maximum load to specimen's failure.

Comparison of all experimental curves for irradiated specimens (which are not shown in the figure) allows concluding about tendency towards lowering the strength characteristics of the alloy with increasing the temperature of irradiation.

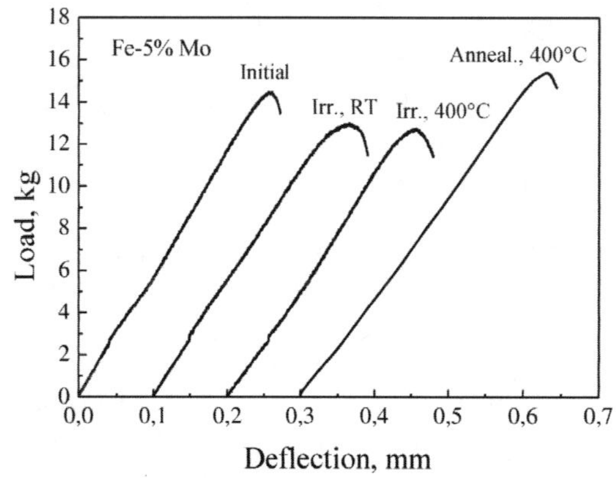

Figure 1. Loading curves of Fe-5% Mo disk microspecimens in initial state (vacuum annealing, 1000°C/4 h), after electron irradiation at room temperature and 400°C (4 h, 10^{18} cm^{-2}), and after additional vacuum annealing at 400°C/5 h.

However, enhancement of the radiation softening effect with increasing of irradiation temperature up to 400°C could be depend not only from irradiation, but also with the structural changes occurring in the alloy at higher temperatures, for example, with the collapse of the solid solution. To test this hypothesis, initial specimens were subjected to additional vacuum annealing at 400°C for 5 hours, which repeats the temperature and time conditions of irradiation. It was found that the additional annealing of the alloy after preliminary high-temperature heat treatment (1000°C/4 h) at a much lower temperature (400°C) not only leads to a softening of the alloy, as in the case of irradiation at this temperature, but conversely, causes its hardening compared with the alloy in initial state. Sensitivity of the alloy to an additional low-temperature annealing indicates clearly that the pre-annealing at 1000°C was insufficient to obtain an equilibrium structure of the alloy. Moreover, it was found multidirectional effect of irradiation and thermal annealing at the same temperature. Apparently, annealing stimulates the structure transformation causing strengthening of Fe-Mo alloy, and irradiation, on the contrary, — it's softening.

More explicitly effect of radiation softening of Fe-5% Mo alloy was manifested during measurements of Vickers microhardness conducted on 402MVD (Woolpert Wilson Instrument) under the load on indenter (diamond pyramid with an apex angle of 136°) of 50 g (Figure 2). The depth of the analyzed surface layer was about 10 microns. Total in 5 measurements on the point was carried out and the average values of microhardness for the alloy in each state were calculated. Accuracy of determining the average value was ±5%. As can be seen, with increase of the temperature of annealing or irradiation, microhardness of the surface layer is decreased, and difference between the average values of microhardness of irradiated and annealed at 400°C specimens greater than the measurement error. At the same time, these data show that annealing at 400°C decreases the microhardness of the alloy, while, according to MDBT measurements, additional annealing at 400°C increases the strength characteristics of the alloy, i.e. strengthening effect takes place.

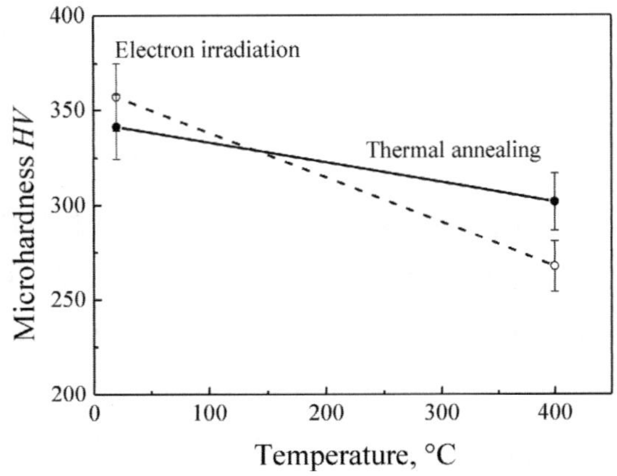

Figure 2. Changing of microhardness of Fe-5% Mo alloy in initial state and after electron irradiation at room temperature and 400°C.

The observed difference in microhardness behavior and changes of the MDBT-curves for irradiated and annealed specimens may be associated with the fact that the microhardness values characterize the strength properties of sufficiently thin surface layer, whereas the MDBT-curves — that of the whole of volume. Obviously, kinetics of radiation point defects annealing near the powerful sink like the surface and in the volume of material will be different and it may have effect on the nature of radiation-stimulated processes occurring near the surface and in the volume.

Despite the low amount of radiation-induced change of mechanical properties of Fe-5 wt.% Mo alloy, the data obtained convincing enough demonstrate the radiation nature of the alloy softening effect, especially at elevated temperature. Undoubtedly, firstly, the reason of change in mechanical properties of Fe-Mo alloy under electron irradiation is radiation-induced structure and/or phase transformations and, secondly, behavior of these processes is different in the volume and near the surface. It may be related to the different kinetics of accumulation and annealing of radiation point defects as well as with different crystal-chemical reactions (e.g., oxidation) on the surface of irradiated specimen as it is cooled by water or air jet.

Acknowledgements
This work was performed within the framework of the Center of Nuclear Systems and Materials supported by MEPhI Academic Excellence Project (contract № 02.a03.21.0005, 27.08.2013).

4. References

[1] *The Use of Small-Scale Specimens for Testing Irradiated Materials* eds Corwin W R and Lucas G E 1986 (Philadelphia, Pa: ASTM STP 888) p 380
[2] Tsepelev A and Poymenov I 1992 *J. Nucl. Mater.* **189** 222
[3] Tsepelev A, Ivanov I and Poimenov I 1995 *Proc. IEA Int. Symp. on Miniaturized Specimens for Testing of irradiated Materials* (Julich, Germany) p 59
[4] Kluech R. and Maziasz P.J. 1990 *Proc. Int. Conf. on Reduced Activation Materials for Fusion Reactors* ASTM STP 1047 (Philadelphia: American Society for Testing and Materials) p7

Multiscale study of the boron and carbon behavior in the fracture zone of the rapidly quenched nickel-based superalloy by the nuclear physics methods of the track and activation autoradiography

A V Shulga

National Research Nuclear University MEPhI (Moscow Engineering Physics Institute), Kashirskoe highway 31, Moscow, 115409, Russia

E-mail: avshulga@mephi.ru

Abstract. The results of the study of the fracture zone of the rapidly quenched nickel-based superalloy after tensile tests are presented. The main attention was attended to the multiscale investigation of boron and carbon distribution in the area of the cracks propagation by the direct nuclear physics methods of the track and activation autoradiography. Tensile tests were performed at the temperatures up to 1150 °C. The nuclear reactor MEPhI IRT-2000 and cyclotron were applied for the autoradiography studies. Significant depleting of boron in the fracture zone which was revealed can be explained due to the intensive moving of the dislocation.

1. Introduction

Nickel based superalloys are the high temperature materials widely applied in critical components of a gas-turbine [1-6]. Microalloying of boron and carbon being used to improve the mechanical properties, in particular, the high temperature strength, ductility and cohesive strength of the grain boundaries [7-11]. Special attention should be given to the role of boron and carbon in the advanced rapidly quenched Ni-based superalloys manufactured by the powder metallurgy via hot isostatic pressing (PM HIP) of the melt-atomized powders. Such features of the rapidly quenched structure as high dispersivity of the microstructure elements, small dendrite arm spacing, super saturation of solid solution and the large interparticle surface in the HIPped compact provide the possibility of the essential peculiarities of boron and carbon behavior in the PM HIP Ni-based superalloys. The features of the crack propagation in Ni-based superalloys, especially in advanced rapidly quenched PM HIP superalloys have the significant meaning for estimation of their mechanical properties.

The important aim of this work is a study of the behavior of boron and carbon in fracture zone of the PM HIP Ni-based superalloy by the high sensitivity direct autoradiography technique. Simultaneously investigation of the specimen structure and the autoradiographs on boron and carbon was performed at a wide range of magnification. The technique of autoradiography allows for a multi-scale analysis of boron and carbon distribution in the various elements of the structure including macro-, metho-, and microstructure. A detailed correlation study of the structure PM HIP Ni-based superalloy compact in the fracture zone and boron, carbon distribution was performed.

2. Experimental details

The composition of the investigated PM HIP Ni-based superalloy EP741NP is listed in Table 1. The specimens with the width and thickness of 3 mm and the working length of 6 mm were cut from the as-HIPped compact of the Ni-based superalloy EP741NP.

Rapidly quenched powders were produced by plasma rotating electrode process (PREP).

Table 1. Chemical composition [in wt.%] of the PM HIP Ni-based superalloy EP741NP

Ni	Al	Ti	Cr	Co	W	Mo	Nb	Hf	C	B
Remaning 56.1	5.1	1.8	9.1	15.7	5.5	3.8	2.6	0.29	0.04	0.012

HIP parameters were the following: 1170 °C, 4 h, 150 MPa. Powder was produced by the plasma rotating electrode process (PREP).The main aim of the tensile tests was to reveal the effect of temperature on the flow behavior and mode of fracture of the PM HIP Ni-based superalloy EP741NP. The tensile tests were carried out at the temperatures 800 – 1150 °C and at 10^{-1}, 10^{-3} s^{-1} strain rates.

The specimens were studied by light metallography, autoradiography on boron and carbon, Scanning Electron Microscopy (SEM), and X-ray Spectral MicroAnalysis (XRSMA). Method of the track autoradiography on boron has been provided the possibility to study macro-, metho-, and micro inhomogeneity of boron and has been used to detect the presence of boron in various elements of the structure such as solid solution in the volume of grain, segregations at the grain boundaries and dispersed boride precipitates. The autoradiography method on boron is based on the use of the nuclear reaction $^{10}B(n,\alpha)^{7}Li$ on thermal neutrons. Fission products, mainly α-particles, were recorded by means of a plastic film detector. This nuclear method has a high sensitivity (the minimum detectable concentration of boron is of about 1 ppm) and allows investigating the multiscale spatial boron distribution with a low (macro) and a high resolution (metho- and microstructure). Dark/bright areas on the negative autoradiograph image correspond to a lower/higher concentration of tracks and therefore to the lower/higher concentration of boron. Separate tracks are detected as small black dots on the positive autoradiographs with various densities depending on boron concentration in solid solution. The coincidence of separate tracks from boron atoms in the boride precipitate results in formation of the spot on the autoradiograph which size depends on the size of boride particle.

An activation autoradiography method was used to obtain the multi-scale distribution of carbon ranging from the macro inhomogeneity in the specimen macro structure to the micro inhomogeneity in the microstructure elements. The autoradiography method is based on the use of the nuclear reaction $^{12}C(d,n)^{13}N$ on carbon atoms with 2 MeV deuterons. Secondary β-particles were detected on the specimen surface by the β-radiometer. This nuclear method has a high sensitivity and includes the operation of measuring the induced activity of the β-radiation due to the nuclear reaction. It must be noted that simultaneously with the more uniform distribution of carbon in the specimens, the induced activity of the β-radiation detected during the autoradiography investigation was increased.

3. Results and discussion

Figures 1 – 3 show macro- and microstructure, activation and track autoradiographs of carbon and boron distribution in the specimen of the rapidly quenched PM HIP Ni-based superalloy EP741NP after the tensile tests. Significant decreasing of boron concentration in the fracture zone of the specimen of the rapidly quenched PM HIP Ni-based superalloy EP741NP compact was observed on the track autoradiograph of boron and was confirmed by the dissolution of boride precipitates detected by scanning electron microscopy and x-ray spectral microanalysis. The length of this boron depletion area is about one millimeter from the fracture surface. Essential boron depletion in the fracture zone under testing at the temperature 1050 °C corresponds to an anomalous high diffusivity of boron (~10^{-9} m^2/s). Interaction of boron atoms with moving dislocations that realize the plastic deformation at the micro crack tips is evidently the main reason of this effect which develops as boron "sweeping out" under tensile test.

Simultaneously in this area has been revealed the nucleation a set of micro cracks evidently at the weak grain boundaries after a small plastic strain.

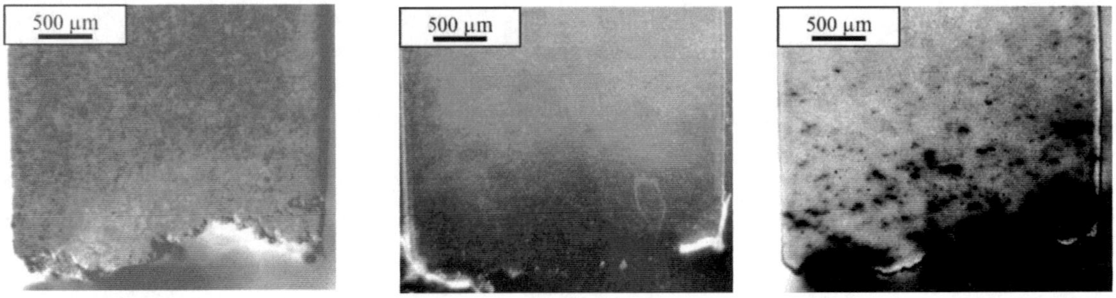

a b c

Figure 1. Macrostructure (a), track and activation autoradiographs of boron (b, negative image) and carbon (c) distribution in the fracture zone of the specimen of the rapidly quenched PM HIP Ni-based superalloy EP741NP after the tensile test at the temperature 1050 °C and strain rate 10^{-3} s^{-1}.

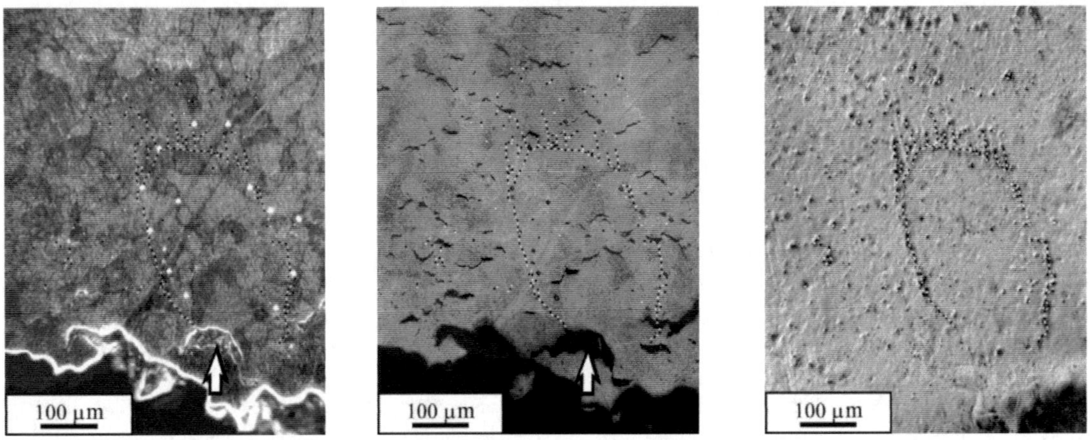

a b c

Figure 2. Microstructure (a, b), track autoradiograph of boron distribution (c, positive image) near the metho crack in the specimen of the PM HIP Ni-based superalloy EP741NP.

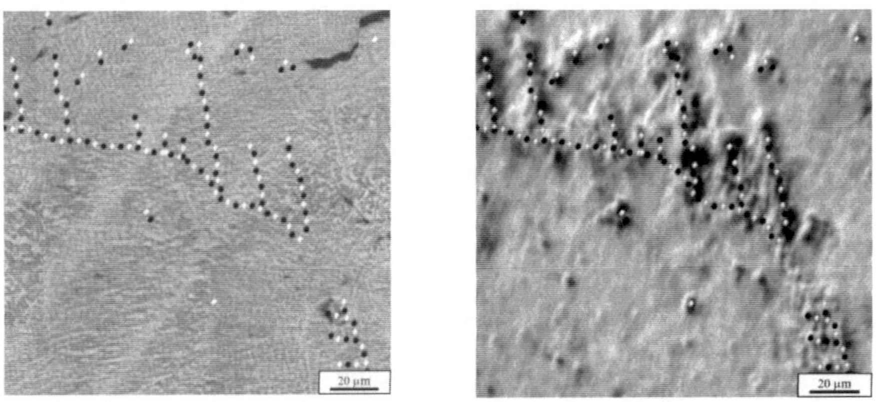

a b

Figure 3. Submicrostructure (a), track autoradiograph of boron distribution (b, positive image) near the metho crack depicted by the black-white dot lines in the specimen of the PM HIP Ni-based superalloy EP741NP at the high magnification.

These micro cracks don't propagate to the fracture of specimen evidently because other grain boundaries are stronger and also owing to plastic deformation at their crack tips. A few larger intergranular (metho) cracks were nucleated during continued plastic deformation at the certain boundaries.

The specimen after the tensile test at the temperature 1050 °C and strain rate 10^{-3} s^{-1} is characterized by the metho crack (indicated by the arrow in Fig. 2) which provides the formation of the oval-shaped trail decorated by boron segregation. An examination of this metho crack, which did not lead to fracture, shows that the crack was initiated in an intergranular fashion. Whereas this crack was not propagated to fracture of the specimen, therefore this confirms that plastic deformation at the crack tip is the main reason of boron segregation. Intensive dislocations glide at the surface of maximum shear stresses near this crack tips and interaction of the moving dislocations with boron atoms evidently results in formation of the oval-shaped trail enriched with boron. Atmospheres of boron atoms on the moving dislocations form well known solute atmospheres produce a drag force to moving dislocations [12, 13].Under considerable deformation, the crack size becomes large enough to cause fracture resulting in the mixed intergranular and ductile tearing fracture mode.

The multiscale study of boron distribution in the fracture zone of the specimen of the PM HIP Ni-based superalloy EP741NP after the tensile test was revealed decreasing the density of borides precipitates of the type (W, Mo)B according to the in the fracture zone simultaneously with depletion of the track density in this area. Decreasing of track density in the fracture zone on the track autoradiograph on boron distribution indicates the decreasing of boron atoms in solid solution. Whereas boride precipitates are characterized by the coincidence of separate tracks from boron atoms on the track autoradiograph with formation of fine spots which correspond to the fine boride particles. Therefore decreasing of this spots density on the track autoradiograph means the decreasing of the boride particles density under tensile test at 1050 °C.

After the tensile tests at the lower temperatures, for example at 800 °C, was detected by the track autoradiography, scanning electron microscopy, and x-ray spectral microanalysis the homogeneous boron distribution without depletion of boron and dissolution of dispersed boride particles of the type (W, Mo)B in the fracture zone of the specimen. Thus, the significantly lower solubility of boron at this temperature evidently is the main reason of this effect.

An activation autoradiography investigation of carbon distribution in the fracture zone of the tested specimens allows detecting the absence of significant change of carbon distribution as it was detected in the case of boron. The main reason of this difference is evidently the essentially higher thermal stability of carbide phases in the nickel-based superalloys when compared with the thermal stability of boride phases. The results of the study of boron and carbon distribution in the fracture zone of the rapidly quenched nickel-based superalloy by the nuclear physics methods of the track and activation autoradiography demonstrates essentially different effect of crack propagation on boron and carbon distribution.

A detailed study of carbide particles distribution by scanning electron microscopy and x-ray spectral microanalysis in addition to the activation autoradiography in the fracture zone was confirmed an essentially higher stability of type (Nb, W)C carbide phase then boride phase (W, Mo, Cr)B type particles distribution under the high temperature tensile tests. Therefore carbon and carbide particles maintain resistance to fracture under the high temperature tensile test in significantly higher degree when compared with boron and boride particles effect.

4. Conclusion

In this work behavior of boron and carbon in the specimens cut from the rapidly quenched PM HIP compact of the Ni-based superalloy EP741NP were investigated. Multiscale study (at macro, metho, micro level) of boron and carbon distribution by autoradiography methods on boron and carbon simultaneously with metallography, Scanning Electron Microscopy (SEM), and X-ray Spectral MicroAnalysis (XRSMA) (nano level) were performed. The significant effect of fracture process on the boron distribution in the fracture zone and the microstructure of the rapidly quenched PM HIP compact of the Ni-based superalloy were determined. It was established the followings:

1. Significant decreasing of boron concentration in the fracture zone of the specimen of the rapidly quenched PM HIP Ni-based superalloy compact after tensile test at 1050 °C and strain rate 10^{-3} s^{-1} was

detected. The length of boron depleting area reaches one millimeter in direction to the fracture surface. Simultaneously in this area has been revealed the nucleation a set of micro cracks evidently at weak grain boundaries after a small plastic strain.

2. Essential boron depletion in the fracture zone under testing at the temperature 1050 °C can be explained as an anomalous high diffusivity of boron. Interaction of boron atoms with the moving dislocations that realize the plastic deformation at the crack tips is evidently the main reason of this effect. Some influence of the dilatation strain in the near tip field occurs evidently on the boron migration.

3. Influence of the test temperature on the features of boron distribution in the fracture zone of the rapidly quenched PM HIP Ni-based superalloy EP741NP was also investigateded. After tensile tests at the lower temperature, for example at 800 °C the homogeneous boron distribution without depletion of boron in failure zone was detected.

4. In the fracture zone was revealed an essentially different behavior of boron and carbon, which are both the microalloying interstitial element. Carbon distribution in fracture zone of tested specimens indicates essentially higher stability when compared with boron distribution.

Acknowledgements

This work was performed within the framework of the Center of Nuclear Systems and Materials supported by MEPhI Academic Excellence Project (contract № 02.a03.21.0005, 27.08.2013).

References

[1] Iljin A A, Stroganov G B, Fatkullin O Kh, Shulga A V and Martinov V N 2008 *Structure and properties of rapidly quenched alloys* (Moscow: publishing house Altex, in Russian) p 588
[2] Shulga AV 2013 *J. Nucl. Mater.* **434** 133
[3] Kitaguchi H S, Li H Y, Evans H E, Ding R G, Jones I P, Baxter G, Bowen P 2013 *Acta Materialia* **61** 1968
[4] Ning Y, Fu M W and Yao W 2012 *Materials Science and Engineering: A* **539** 101
[5] Ning Y, Yao Z, Fu M W and Guo H 2010 *Materials Science and Engineering:* A **527** 6968
[6] Soo S L, Hood R, Aspinwall D K, Voice W E and Sage C 2011 *CIRP Annals, Manufacturing Technology* **60** 89
[7] Zhou P J, Yu J J, Sun XF, Guan H R and Hu Z Q 2008 *Materials Science and Engineering A* **491** 159
[8] Cadel E, Lemarchand D, Chambreland S and Blavette D 2002 *Acta Materialia* **50** 957
[9] Yan B C, Zhang J and Lou L H 2008 *Materials Science and Engineering: A* **474** 39
[10] Song S-H, Zhang Q and Weng L-Q 2008 *Materials Science and Engineering: A* **473** 226
[11] Chen X M, Lin Y C, Wen D X, Zhang J L and He M 2014 *Materials and Design* **57** 568
[12] Hirth J P and Lote J 1992 Theory of dislocations (Second Ed, Florida) p 476
[13] Broek D 1974 *Elementary engineering fracture mechanics* (Leyden) p 368

Requirements to the procedure and stages of innovative fuel development

V Troyanov, L Zabudko, A Grachyov and O Zhdanova

Institution "Innovation and Technology Center by "PRORYV" Project, State Atomic Energy Corporation "Rosatom"

E-mail: tvm@proryv2020.ru

Abstract. According to the accepted current understanding under the nuclear fuel we will consider the assembled active zone unit (Fuel assembly) with its structural elements, fuel rods, pellet column, structural materials of fuel rods and fuel assemblies. The licensing process includes justification of safe application of the proposed modifications, including design-basis and experimental justification of the modified items under normal operating conditions and in violation of normal conditions, including accidents as well. Besides the justification of modified units itself, it is required to show the influence of modifications on the performance and safety of the other Reactor Unit' and Nuclear Plant' elements (e.g. burst can detection system, transportation and processing operations during fuel handling), as well as to justify the new standards of fuel storage etc. Finally, the modified fuel should comply with the applicable regulations, which often becomes a very difficult task, if only because those regulations, such as the NP-082-07, are not covered modification issues. Making amendments into regulations can be considered as the only solution, but the process is complicated and requires deep grounds for amendments. Some aspects of licensing new nuclear fuel are considered the example of mixed nitride uranium –plutonium fuel application for the BREST reactor unit.

1. Introduction

The way from the most attractive idea on modifying nuclear fuel for Nuclear Plant up to its application is very long and stretched for many years. This fact is related to so-called process of licensing the fuel with implemented modifications. The current report will touch the issues of innovative structure of fuel roads and fuel assemblies, fuel licensing requirements and regulations, the status of design-basis justification of BREST-OD-300 fuel rods and assemblies, existing calculation codes and the Comprehensive Program of design-basis and experimental justification of dense fuel as well.

2. Innovative structure of fuel rods and fuel assembly of BREST –OD- 300 Reactor Unit

The task of current report is not the discussion of reasons for developing innovative structure of fuel rods and fuel assembly of the BREST-OD-300 reactor unit. We restrict ourselves to listing the features that distinguish the fuel rods and fuel assemblies from the known and currently used design of fuel assembly of BN-600 reactor and BN-800 as well.

Structural material for fuel rod cladding and end pieces, as well as all fuel assembly elements is a ferritic-martensitic class steel called EP 823. This steel has never been used in fast reactors except for some experiments in the BOR-60 (cladding of several experimental fuel rods with oxide fuel) and BN-600 (fuel assembly hexagonal wrappers in small quantities).

The pellet column is MNUP. Before starting work on "PRORYV" Project it has been tested several MNUP fuel rods in the RR BOR-60 core , but these experimental fuel rods have differences that do not allow considering them fully referent, the test parameters varied from the parameters of operation of the fuel rods in BREST-OD-300. Great experience of uranium nitride fuel in RR BR-10 application is an important fact confirming, in principle, operation of the fuel rods, but it cannot be considered as a referent in relation to the fuel rods with MNUP.

The diameter of the fuel rods of two types (9,7mm and 10.5mm), as well fuel rod pitch in a grid (near 12.7mm), are new and based on requirements to the reactor neutronic characteristics. These characteristics are important for the thermal hydraulics, and require the development of new designs not only the fuel rods but fuel assemblies as well. The initial idea to use ribbed fuel rod cladding with self-spacing function inside hexagonal fuel assembly can was almost unrealizable because that has not been found possible industrial production of fuel rod cladding with high edge which provides a space for fuel rods distancing. It should be noted that from the licensing of ribbed fuel rod cladding point of view, developers would be faced with a set of additional problems in terms of design-basis, and, especially, in the experimental justification of fuel rods because it would be difficult to carry out the reactor testing of fuel rods of such a geometry.

The design of fuel assembly is unwrapped which is similar to fuel assembly TVS-2M type for VVER-1000. But this analogy is very conditional, because the spacer grids made from steel EP 823 only visually are similar to the grids from zirconium alloy E110. Suffice to say that the strength properties of the steel EP 823, in particular the elasticity modulus, which is three times more and the yield strength, which is also 2-3 times higher, make additional difficulties during forming cells and welding of spacer grids and requirements annealed weld points within a short time after welding make difficulties in the manufacturing racks. Thermo-mechanical behavior of unwrapped fuel assemblies during operation being a part of the BREST' core have significant features, and loading factor of fuel assemblies while operating is not completely the same as in VVER-1000.

Regulation requirements [1] do not settle the required scope of design-basis and experimental justification, but specify that justification should be sufficient and included into design and engineering documentation of the developer. ("While developing the technical specification on reactor unit and (or) modifying the reactor core with the help of new structures of fuel assemblies, new fuel composition and updating the reactor control and safety system or other systems important for safety, the necessary reactor and development testing have to be performed. For proving the fulfillment of safety criteria the sufficiency of performed testing has to be shown in the reactor unit specification (cl. 2.1.4)"). In other words, in reactor plant technical specification or any reference materials, the results of design-basis justification as well as experimental ones in "sufficient" scope should be listed. According to [1], for design-basis justification only certified methods and codes can be applied, but the application area of these certified codes should spread to the conditions, structures and materials appropriate to the licensed product.

Experimental justification must be performed on the reference structure when loading parameters (power, mechanical, thermal, irradiation, corrosive loading) are equal to design ones. This requirement is easier to perform, if the changes in existing units are licensed, especially if there are appropriate experimental facilities and prototype or research facilities as well. As an example we may consider the changes in structural material in the fuel rods of operating reactors: there is a technical possibility to carry out experimental work with new fuel rods in identical conditions. If the experimental facilities are absent, as in the case of BREST type reactors (there is no lead-coolant reactor in operation as the basis), implementing the requirements of experimental justification and reference in sufficient scope becomes unfulfilled. Not only in normal operating conditions the justification shall be implemented, but in all accidents, foreseen by the project. The developer is obliged in its design documentation to justify and to set the maximum design limits for fuel rods, safe operation and operational limits. Such limits are set in [1] for RBMK and VVER fuel, as well as for BN reactor oxide fuel.

Limits have to be set and, in addition, put in [1] for MNUP. Otherwise there is a conflict with the regulations, because it is impossible to meet regulations which do not specify the limits.

IOP Conf. Series: Materials Science and Engineering **130** (2016) 012018 doi:10.1088/1757-899X/130/1/012018

3. Codes for fuel rods while operating in normal conditions and their breakdown

For a long time in various industry organizations different calculation codes for calculating the working capacity of container type fuel rods of fast reactor were created, applied and forgotten. Currently, the only code which is practically used for fast reactor fuel rods is "KORAT", developed and operated by JSC "VNIINM", the chief designer and technologist of fuel rod. Code "KORAT" has been certified by Scientific and Engineering Centre for Nuclear and Radiation Safety for calculating fast reactors fuel rods with uranium oxide and MOX fuel in normal operating conditions (stationary, transient and maneuvering modes of reactor operation).

The formulation and the calculation algorithm are using the approaches specified in [2], of which the main ones are the calculation of the plane circular cross-section of a fuel rod under conditions of plane strain, central symmetry and annular layers with different properties, that allow simulating the fuel pellet, the gap between fuel and rod cladding and fuel rod cladding. The loading is monotonous.

Independent calculating the number of plane cross-sections by the height of fuel rod provides the ability to sum the amount of axial deformation and to determine the elongation of the fuel rod. However, this calculation can be very approximate, since the model does not take into account the capacity of axial ratchet in the interaction of the fuel column and the rod cladding, just as is done in models [3].

The "KORAT" code is adapted, as far as it was possible, to calculate the fuel rods with nitride fuel. For this purpose physical and mechanical characteristics of nitride pellet and the cladding made of steel EP 823 were entered. However, in the "KORAT" code the model of the gas swelling of nitride pellet is missing. The model takes into account the temperature, density scales and structure parameters of the fuel pellet, including porosity. The model of gas emission from the fuel pellet is also absent. This is the reason why both gas emission and swelling are set by empirical dependencies in the "KORAT" code.

The "KORAT" code has no certification on calculating fuel rods with nitride fuel.

The "PINCODE" code is designed and certified in JSC "Institute of Physics and Power Engineering IPPE". The code is used for the calculation of thermo-mechanical behaviour of UOX fuel rods in normal operating conditions. The code was developed within the frame of the SVBR Project development, but for calculating fuel rods with nitride fuel the code was not used.

The code "DRAGON" is developed in IPPE. The code is focused on thermo-mechanical behavior of fuel rods with nitride fuel. For this purpose the model for calculating the swelling of the fuel core under the model of spherical gas pores is entered.

This fact distinguishes the "DRAGON code from the others, because potentially in the "DRAGON" code the swelling of the pellet column depends on the following parameters: irradiation, gas composition and gas pressure under the fuel rod cladding and fuel temperature, and the most important parameter is the stress condition in elementary volumes of the fuel pellet. This condition depends on temperature gradients and isostatic pressure caused by the interaction of the pellet column with the fuel rod cladding.

Axial ratchet is missing in the model and the algorithm of fuel rods calculation by the "DRAGON" code, but the task of introducing this model is settled. The algorithm is going to be implemented during the years 2016-2017. Furthermore, the nearest works to improve calculations of fuel rods by the "DRAGON" code cover the algorithms and techniques of statistical calculations which are taking into account the statistical uncertainty of input data and operational parameters, as well as the individual characteristics of each fuel rods in the core in the current time for determining the probability characteristics of fuel rods failures.

The "DRAGON" code is verified at problems with an analytical solution as well as on all available experimental data for testing fuel rods with nitride uranium fuel and MNUP. This code is not certified in national authorities, because for its full certification is not yet sufficient experimental data obtained in studies of nitride fuel rod prototype.

The "DRAGON" and "KORAT" codes are used while developing explanatory and calculation notes as part of the design and engineering documentation on all experimental and prototype fuel rods and fuel assemblies which are being tested in BOR-60 and BN-600 reactors within the framework of the fuel licensing program (see below). Moreover, as neither "KORAT" nor "DRAKON" is certified for MNUP justification, the application of two independent codes and commands is used as a cross-verification of calculations for increasing calculations reliability.

The "BERKUT" code developed by the Nuclear Safety Institute (IBRAE) was initially established as a mechanistic (best estimation) code for calculating nitride fuel rods with the application of precision physical models. However, the lack of fundamental data on characteristics and phenomena in nitride fuel has forced developers to change the concept. The engineering version of the "BERKUT" code has been created, but there "initial statements" are absent. At the same time, in engineering version of "BERKUT" a models of pellet column swelling and fission gas producing emission are missing. The "BERKUT" code hasn't been applied in practice yet, because it does not represent new ideas, methodologies and algorithms. Certification of the "BERKUT" code was not performed for the reasons mentioned above.

All listed calculation codes are intended to calculate quasi-stationary loading and are used to calculate the fuel rods in normal operating conditions and deviation of normal operating conditions. Criteria of fuel rods integrity do not differ from the criteria accepted for normal operating conditions. While developing the technical specification the list of deviation during normal operating conditions is considered, but usually one self-run is made as the most hard deviation.

The calculations of accident modes, including dynamic ones, are a task for the coming period. Also it is needed to develop a methodology of MNUP justification in the modes of power cycling and, in addition, seismic loading.

An independent task is the calculation of the thermo-mechanical behavior of the fuel assembly. The fuel assembly of beam-rod model is developed in IPPE. The model rather borrows approaches and some results of the "TEREMOK" code development [4]. Currently, only engineering calculations and verification according to factory testing of fuel assembly mockups are performed.

There is no model of fuel cladding' corrosive effect and spacer grids in the lead coolant environment, although such attempts are undertaken. Currently corrosive effect is accounted as corrections for the corrosion via the reduction of structural element thickness.

Correction for corrosion is introduced by the results of corrosion experimental studies in lead stands without reactor irradiation. The experimental possibility of determining corrosion in reactor conditions is absent.

4. The fuel rods calculations in accidental situation

Investigations on the accidental situations analysis in BREST OD-300, made by the Chief designer of reactor unit, has shown that the design accidents can lead to increasing of fuel cladding up to 1100°C . In general the "DRAGON" code could be used for calculating the stress-strain state of fuel rods in these conditions. However, these calculations are faced with the lack of information on physical and mechanical properties of fuel cladding and pellet column in same temperature level and total absence of experimental data to verify these calculations.

According to [1] it is required to introduce the so-called maximum design limits (MDL) for fuel rods which are needed for justification accidental situations on reactor units. In case of increasing the limits fuel rods will be failured. Such MDL for BREST fuel rods are developed on the basis of data on long-term strength of fuel cladding at high temperatures.

The loading conditions of the fuel rod are accepted as typical, based on calculations of gas release from the fuel. Calculations are made according to the worst combination of parameters. The following data is characterized MDL: temperature 800 °C – t_1, 900 °C – t_2 , 1000 °C – t_3. In addition, MDL on MNUP maximum temperature of and the cladding deformation is introduced as well.

5. The program of experimental justification of fuel rods

The structure of works on MNUP justification is specified in Figure 1.

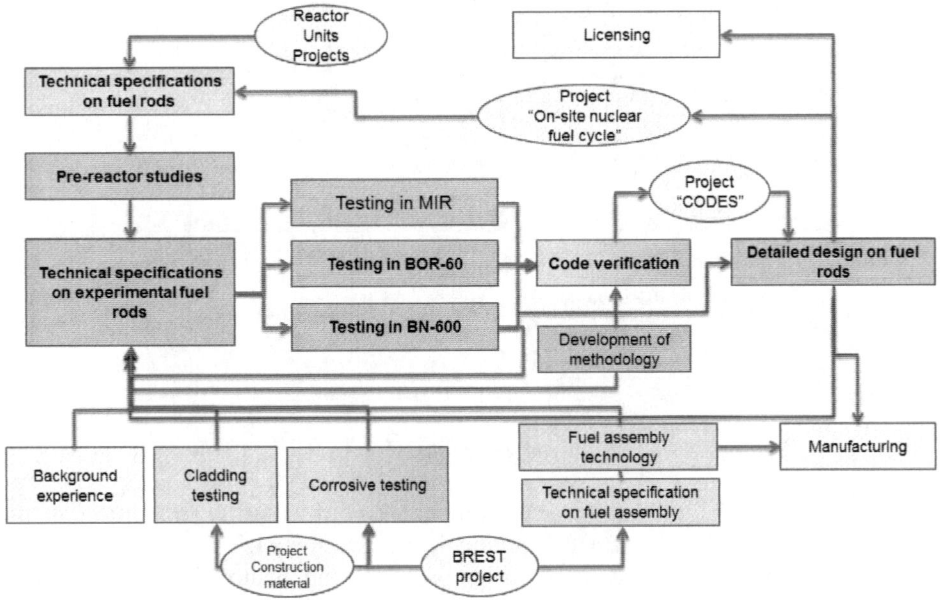

Figure 1. The structure of works on MNUP justification.

The Comprehensive Program of design-basis and experimental justification of dense fuel [5] was developed and approved on December 31st, 2013 by Dr. Pershukov, the Deputy Director General of State Corporation "Rosatom". As a result of executing the mentioned Program the working capacity of fuel rods with MNUP under main operational parameters of BREST-OD-300 and BN-1200 reactor's cores should be explained.

Part of the Comprehensive Program of design-basis and experimental justification are the following:

- pre-reactor examination of MNUP composition;
- methods and equipment for pre-reactor studies;
- development of MNUP carbothermal synthesis and basic technologies for pellet manufacturing;
- the manufacture of experimental fuel rods and fuel assemblies;
- reactor testing in BOR, MIR and BN-600 reactors;
- post-reactor examination;
- modeling and codes (corrosion, fuel rod, thermo-mechanical characteristics of active zone);
- development of structure and basic technologies of BREST fuel assembly;
- alternative structures and technologies;
- technical specifications of fuel rods (standard and experimental).

Schedule for testing MNUP in RR BOR-60 for the period 2012-2016 is given in Figure 2.

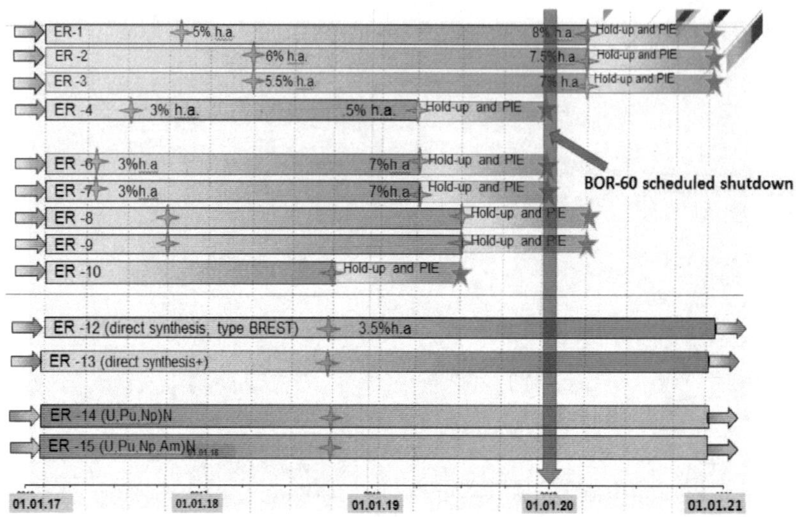

Figure 2. Schedule for testing MNUP in RR BOR-60.

In BN-600 the more meaningful testing conditions in terms of length of irradiation at steady-state operating parameters with minimum influence of transient mode and variations in power are provided.

To ensure the testing presentability in terms of burning and neutron dose and fission density experimental fuel assemblies are placed on the core periphery of BN-600 reactor. This fact leads to significant peaking factor in experimental fuel assembly (EFA). The irradiation parameters of standard fuel rods have a significant range depending on the location of the fuel assemblies. Thus, the results of experimental fuel assemblies' irradiation in conditions of significant heat gradient along the radius will provide data for reasoning the working capacity of fuel rods with various levels of linear heat generation rate, but with required fission rate.

Schedule for testing MNUP in BN-600 is given in Figure 3.

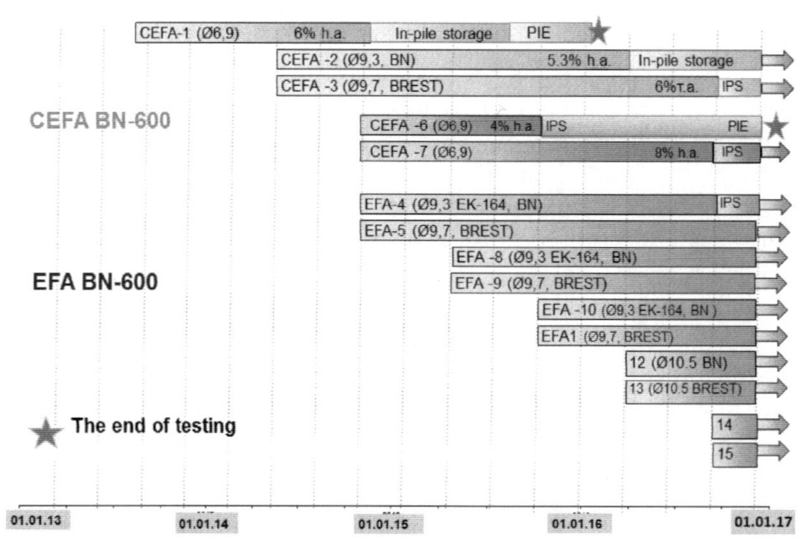

Figure 3. Schedule for testing MNUP in BN-600.

Post-irradiation studies of EAF fuel rods of BN-600 will provide the following additional data on the properties of the mixed nitride:

• by the MNUP swelling rate depending on temperature and burning;

• by the fission gas release from MNUP depending on temperature and burning;

• by the chemical interaction between MNUP with claddings EK 164 hd and EP 823 depending on burning and temperature.

The experimental data will be obtained as a result of works. The data is needed to verify the calculation of temperature and stress-strain state of a fuel rod with MNUP for smooth cladding.

Obtained results will give an opportunity to prepare design-basis justification for making decision related to maximum valid capacity of fuel rods of BN-1200 type reactor and BREST with MNUP as well. The main problem of arranging the experimental fuel irradiation in reactor BN-600 is the need to obtain a permit for testing in Russian Federal Service for Ecological, Technical and Atomic Supervision. But the permit application means an availability of sufficient justification of EFA working capacity.

6. General approach to licensing fuel rods of Reactor Unit BREST-OD-300

The above shows that an innovative type of nuclear fuel cannot be fully experimentally proved, because it is impossible to obtain the full reference during testing the fuel rods. At the same time the question of "sufficiency" of testing in the context of the amount and testing statistical assurance appears.

Moreover, the study should be performed for all operational parameters of reactor unit based on the factors of the statistical uncertainty of operating parameters (so-called factors of overheating). This means that experimental verification should also consider the factors of overheating, but to fulfill it requirements in the experiments methodologically impossible.

The fact that such statistical deviations in testing is impossible to measure and set, thereby to confirm that they were realized in testing, because to definite the rate of these overheating factors often neither set nor measure is not possible. That is, it is impossible to experimentally confirm that these overheating factors were implemented in testing and the experimental result is corresponding to factors.

The only way of new fuel licensing is basic-design justification with the help of certified calculation codes, which verification and certification are extended to the whole range of design and operational parameters of the justified items. For this purpose, the experimental results should cover the range of required parameters; extrapolation of results must be justified and carried out with the help of proven models. In this case, it is possible separately to investigate the properties and behaviour of structural materials and pellet column, to prevent individual differences in the design of experimental items from the licensed ones and to have variations in operational parameters, so as the effect of these variations can be mathematically modeled and described by the physical models.

For example, the length of the pilot fuel rods is not a factor of primary importance, if the behavior model of plane sections of a fuel rod is verified, and the neutron flux density in experimental fuel rods may differ from the neutron flux density in a licensed item, if the effect of neutron flux on the structural material has been studied in other experiments which are equal to by this parameter to the licensed items. Thus, the physical models are formed the basis of such hypothesis. In physical models the influence of the neutron flux density is important for describing the behavior exactly of structural material, and the behavior of the pellet column is not due to the neutron flux density, but the fission density in fuel material, its temperature and structural parameters.

While verifying the behavior models of the pellet column firstly it is required to ensure the adequacy of the linear heat rate and identity patterns of the fuel composition, wherein to receive the rates of energy producing (or burn-up) which correspond to real rate.

This ideology forms the basis of the Comprehensive Program of design-basis and experimental justification, which were discussed in section 5. The influence of each factor on fuel rod performance is determined in a pre-planned testing either in the research reactor BOR-60 or in the commercial reactor BN-600.

The combination of these studies allows verifying and certifying the calculation methodology and calculation code, such as "DRAGON". The limiting parameter which will decrease the area of application (certification) of the “DRAGON” code is achieved heat generation rate (rate of burn-up). The reason is the absence of established model, which prove the influence of this exact factor on the performance of fuel rods with MNUP due to lack of data about processes associated with the evolution of the structure and accumulation of fission products, and the insufficient certainty of the extrapolation of pellet column properties.

The models associated with cracking of the pellet column and fuel rod operating in modes of power cycling will remain not fully verified. The requirement for power cycling when operating in a wide range is specified in the technical specification on BREST reactor, but targeted experimental studies in this direction have not conducted yet. This is due to the fact that neither in RR BOR-60 nor in BN-600 where MNUP is testing, such power changes are not implemented while operating. Technical capability to hold such a reactor testing is not available, and the possibility of justification of such regimes will appear only after the construction and launch of BREST-OD-300 reactor.

Thus, the lack of research reactor unit with lead coolant eliminates the possibility to verify calculation models and programs on some important operation parameters, such as corrosion of structural materials in lead under radiation or fuel rods and assemblies in modes of cycling or power transients. But the experimental demonstration reactor unit with BREST type reactor is constructing to prove the feasibility of idea of a lead reactor with nitride fuel, isn’t it?

References

[1] 2007 *The rules of nuclear safety for reactor units NP-082-07* **4**.
[2] Likhachyov Yu and Pupko V 1975 *The strength of fuel assemblies of reactor units* (Moscow)
[3] Popov V, Khmelevsky M 1980 *Evaluation of axial elongation of fuel rod of RBMK reactor under cyclic change of power* (Obninsk).
[4] Troyanov V, Likhachyov Yu and Folomeev V 2002 *News of Higher Educational Establishments Nuclear Industry* **3** 19
[5] Troyanov V 2014 *J. Rosenergoatom* **11** 28

IOP Conf. Series: Materials Science and Engineering **130** (2016) 012019 doi:10.1088/1757-899X/130/1/012019

Development of lithium and tungsten limiters for test on T-10 tokamak at high heat load condition

I E Lyublinski[1,2], A V Vertkov[1], M Yu Zharkov[1], V A Vershkov[3], S V Mirnov[2]

[1]Joint Stock Company "RED STAR", Elektrolitny proezd, 1a, Moscow, 115230, Russia
[2]National Research Nuclear University MEPhI (Moscow Engineering Physics Institute), Kashirskoe highway, 31, Moscow, 115409, Russia
[3]National Research Center "Kurchatov Institute", Kurchatov sqv., 1, Moscow, 123182, Russia

E-mail: lyublinski@yandex.ru

Abstract. Application of a complex of powerful (up to 3 MW) ECR plasma heating in T-10 tokamak is pulled down with a problem of the strong plasma pollution at power input more than 2 MW. For the solution of these problems the new W and Li limiters is developed and prepared to implementation. As it is supposed, application of W as a plasma facing material will allow excluding carbon influx into vacuum chamber. An additional Li limiter arranged in a shadow of W one will be used as a Li source for plasma periphery cooling due to a reradiation on Li that will lead to decrease in power deposition on W limiters. Parameters and design of limiters are presented. Plasma facing surface of a limiter is made of capillary-porous system (CPS) with Li. Porous matrix of CPS (W felt) provides stability of liquid Li surface under MHD force effect and an opportunity of its constant renewal due to capillary forces. The necessary Li flux from a Li limiter surface is estimated for maintenance of normal operation mode of W limiters at ECRH power of 3 MW during 400 ms. It is shown, that upgrade of limiters in tokamak T-10 will allow providing of ECR plasma heating with power up to 3 MW at reasonable Li flux.

1. Introduction

The main undermined problem in the tokamak reactor is the problem of plasma interaction with the reactor wall. Graphite diaphragms and divertor plates was used in the tokamak reactors for many days long. However high physical and chemical erosion and tritium accumulation in the graphite powder are made to refuse of using this material. Tungsten was offered as an alternative of graphite. Tungsten doesn't interact with hydrogen by chemical way and it has low sputtering ratio. Jet experiments had shown the falling of the hydrogen sorption by an order of magnitude. However tungsten has the big nuclear charge, what leads to significant radiation loss, even its small concentrations. Main radiation loss proceed from hot zone because of neoclassical tungsten ions accumulation in the centre. That's why it's desirable to develop the method of reduction for the tungsten entrance into the plasma. It is the target of the T-10 experiment. Tungsten ring diaphragm (with ring's radius 30 cm) and the mushroom rail limiter (with limiter radius 30 cm) will be the main diaphragms at the experiment. Lithium mobile diagram which is made at the capillary structure base is introduced of tungsten reduction into the T-10. It's assumed to choose the optimal position of lithium diaphragm for tungsten prevention into the discharge. It can happen as a result of two processes: the closing of tungsten surface by the lithium and the cooling of cord's periphery cause of its intake. As expected, tungsten application as a material turned to plasma will allow excluding the graphite intake and the cooling of plasma periphery [1-4] will lead to energy flux reduction at the tungsten diaphragms. Nowadays it's difficult to assess the efficiency of these movements. It's the aim of the future T-10 experiments.

2. Lithium limiter

Lithium limiter – is the source is located in the top vertical port of tokamak's chamber T-10. It can move relatively last closed magnetic surface (LCMS). Limiter's moving in the SOL plasma region allows to regulate energy flux into the plasma. Surface limiter that faced to plasma is manufactured from capillary pore system (CPS) with lithium. CPS material (tungsten felt with radius of pores of 30 microns) provides lithium limiter surface stability under MHD forces influence and surface opportunity on account of capillary forces. The main limiter technical characteristics list is at the table 1.

Table 1. Characteristics of lithium limiter

Parameter	Value
Max heat load, MW/m^2	5
Diaphragm overall dimension H×L×W, mm	95×450×48
The length of working section, mm	323
The diameter of working section, mm	34
The area of opened lithium surface, cm^2	324
The movement range of the lower diaphragm (r1-r2), mm	150
Working temperature, °C	200-550
The heater power, W	1000
The lithium mass, g	50

The lithium limiter construction (figure 1) includes the system of mount/movement and receiving element. The receiving element represents the tubular construction with lithium supply tank. The receiving element forms by supporting tube 1, which manufactured from molybdenum and plays the heat accumulation role.

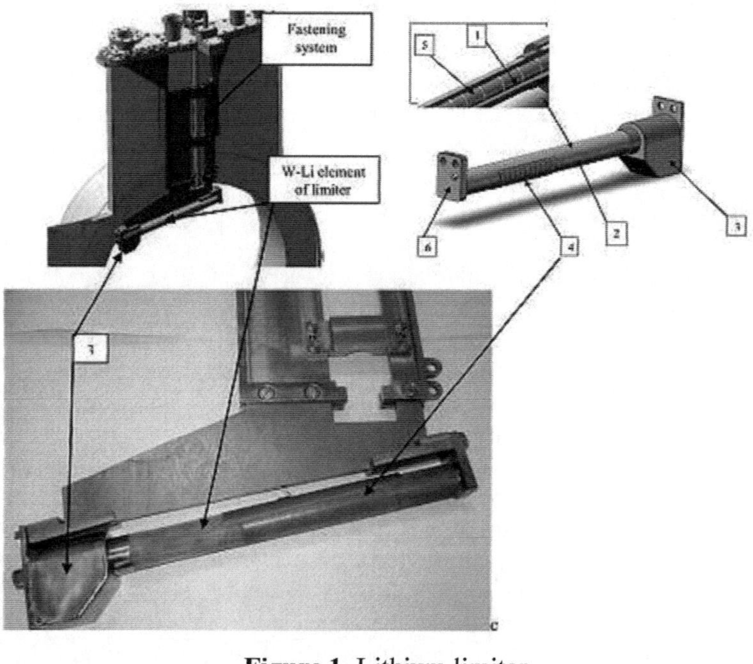

Figure 1. Lithium limiter

The CPS 2, which has hydraulic connection with liquid lithium at the tank 3, installs at the receiving element outer surface. The molybdenum mesh with the pore's radius of 75 microns – is the foundation of the CPS limiter. The elements of pored tungsten additionally installed in direct contact places of receiving surface with the plasma. The limiter's heating to the lithium melting temperature is carried out by electric heater 5, which located inside the tube. The receiving element electrically isolated from the tokamak structure, what allows to change its electric potential relatively plasma.

The receiving element's construction lithium mobile limiter operability estimated magnitude ratings of the lithium surface temperature at adjusted temperature coming heat flow no more than 5 MW/m^2. If the element is located at the LCMS position at the ECRH heat power 3 MW, that max flow is ~ $9MW/m^2$, and the heat load distribution at the surface takes the form that represents at the figure 2.

For the provision of the heat load magnitude 5 MW/m^2 top mobile limiter must be at the shadow of low mobile tungsten limiter. The temperature stabilization at the receiving element surface at the 500-550 °C level reaches on account of the heat removal from the surface to the internal construction elements at these conductions. Temperature evolution of lithium limiter surface at plasma discharge with heat flux 5 MW/m^2 presented on figure 3. The effect of shielding layer lithium steam formation and reradiation in it the part of coming energy didn't take in account in the calculation.

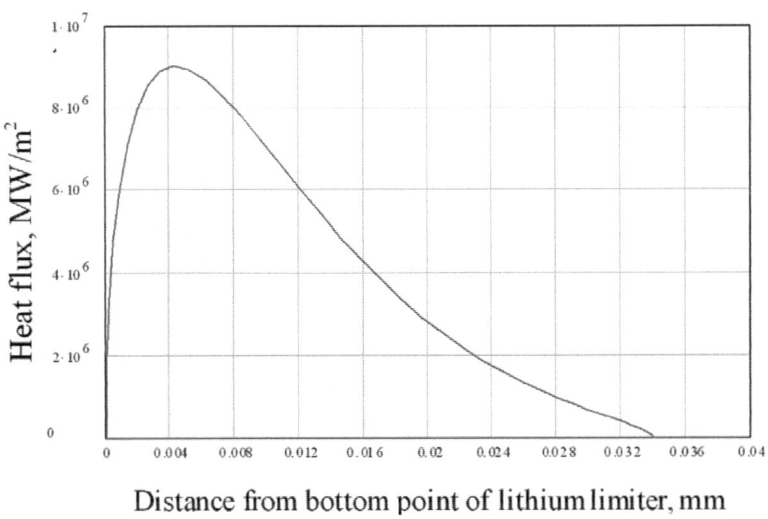

Figure 2. Heat load distribution on lithium limiter surface at power input 3 MW

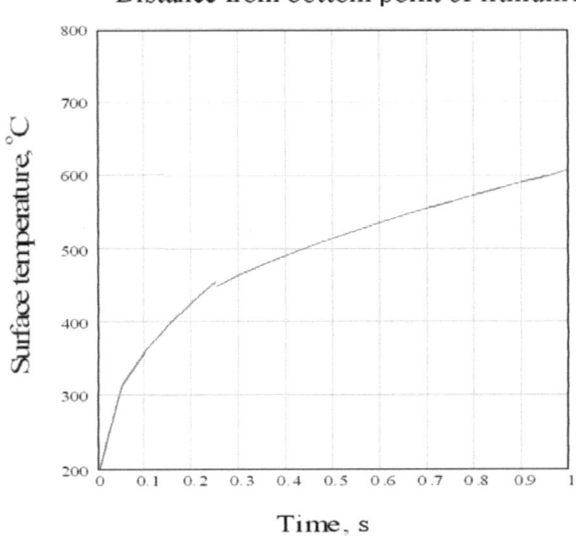

Figure 3. Temperature evolution of lithium limiter surface at plasma discharge with heat flux 5 MW/m^2

3. Complex of tungsten limiters

The tungsten limiter's complex (figure 4) includes fixed circular and mobile "mushroom" (rail) limiter. The circular limiter is the ring with external diameter of 760 mm and internal diameter of 660 mm. The ring plate is perpendicular to the direction of the longitudinal tokamak T-10 magnetic field. The ring strengthens at the wall of vacuum chamber T-10 with the keeping electrical contact. 50 removable tungsten elements with the sizes 50×70×20 mm, that have profiled surface with the optimum tilt angle to magnetic field, forms the receiving limiter surface.

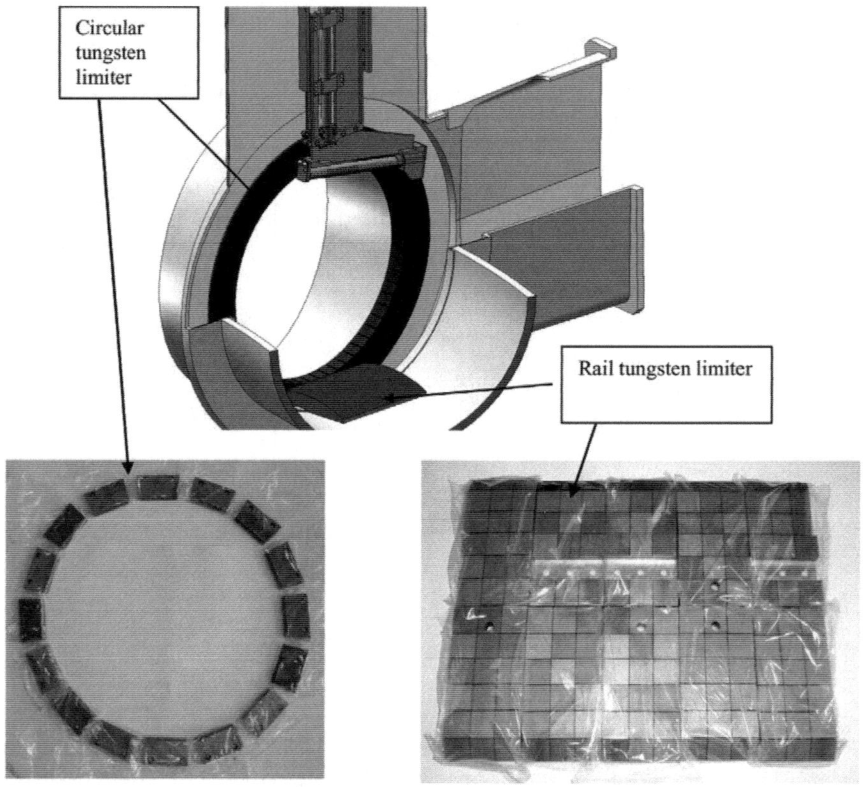

Figure 4. Complex of lithium and tungsten limiters of T-10 tokamak

The lower mobile limiter is the "mushroom" construction that sets the radius of the last closed magnetic surface (the boarder of plasma filament) with an opportunity of changing this radius in the range r = 200-300 mm. The construction consists of fixed foundation which fixed at the flange of low vertical nipple and the mobile cap. The cap is lined by tungsten elements and electrically isolated from the foundation and the section T-10. The cap's top operating surface is made gable with the incline's line of separation which locates perpendicular to longitudinal magnetic field. The incline slope angle to longitudinal magnetic field is 10-15°. The cap's overall dimensions are 300×300 mm at plate, parallel to magnetic field and 50 mm at radial dimension. The overall dimension of tungsten elements are 150×50×20 mm.

The diagnostic system of limiters includes: quick-detachable diagnostic tungsten elements for fast analysis of changing the surface states of tungsten limiters; Langmuir probes are established in the holes at the surface of tungsten elements and represented the information about contact plasma parameters; thermocouples, which measure heat state the limiter complex.

4. Conclusion

The executed values required lithium flow from the lithium limiter surface into the plasma for providing conditions for normal operation tungsten limiters at ECRH power heating 3 MW for 400 ms are shown that required flow reaches at the temperature of 500-550 °C. In this case Z_{eff} of plasma won't exceed the magnitude 2. Moreover, the lithium flow estimated at the base of researching tokamak T-11M results at the tokamak wall doesn't exceed ~ 10 g at 1000 plasma shots, what comparable with lithium influx at using lithium evaporator for the wall T-10 conditioning at the previous experiments. In this way are shown that the limiters modernization at the tokamak T-10 will allow to provide ECRH heat modes with power to 3 MW at moderate lithium fluxes.

Acknowledgements

This work was performed within the framework of the Center of Nuclear Systems and Materials supported by MEPhI Academic Excellence Project (contract № 02.a03.21.0005, 27.08.2013).

References

[1] Evtikhin V A, Lyublinski I E, Vertkov A V, Mirnov S V, et al. 2002 *Plasma Phys. and Contr. Fusion* **44** 955-966
[2] Vertkov A V, Lyublinski I E, Mirnov S V, et al. 2015 *J. Nucl. Mater.* **463** 1156-1159
[3] Mirnov S V, Lyublinski I E, Vertkov A V, et al. 2013 *J. Nucl. Mater.* **438** S224–S228
[4] Mirnov S V, Belov A M, Lyublinski I E, et al. 2015 *Nucl. Fusion* **55** 123015 (11 pp)

IOP Conf. Series: Materials Science and Engineering **130** (2016) 012020 doi:10.1088/1757-899X/130/1/012020

Gas emission from the UO_2 samples, containing fission products and burnable absorber

V P Kopytin, V G Baranov, M A Burlakova, A V Tenishev, R S Kuzmin, S A Pokrovskiy and V V Mikhalchik

National Research Nuclear University MEPhI (Moscow Engineering Physics Institute), Kashirskoe highway 31, Moscow, 115409, Russia

E-mail: Onil709@mail.ru

Abstract. The process gas released from the fuel pellets of uranium fuel during fuel burn-up reduces the thermal conductivity of the rod-shell gap, enhances hydrogen embrittlement of the cladding material, causes it's carbonization, as well as transport processes in the fuel. In this study a technique of investigating the thermal desorption of gases from the UO_2 fuel material were perfected in the temperature range 300-2000 K for uniform sample heating rate of 15 K/min in vacuum. The characteristic kinetic dependences are acquired for the gas emission from UO_2 samples, containing simulators of fission products (SFP) and the burnable neutron absorber (BNA). Depending on the amount of SFP and BNA contained in the sample thermal desorption gas spectra (TDGS) vary. The composition of emitted gas varies, as well as the number of peaks in the TDGS and the peaks shift to higher temperatures. This indicates that introduction of SFPs and BNA alters the sample material structure and cause the creation of so-called traps which have different bonding energies to the gases. The traps can be a grid of dislocations, voids, and contained in the UO_2 matrix SFP and BNA. Similar processes will occur in the fuel pellets in the real conditions of the Nuclear Power Plant as well.

1. Introduction

In connection with the development of fuels for safe and profitable ensuring operationing of the nuclear energy sector and the creation of advanced nuclear power plants, we have created a modified oxide nuclear fuel, which will allow:
- to compensate reactivity by placing Gd in the surface layer;
- to increase the thermal conductivity at the pellet peripheral region by using an $AlGdO_3$ stable phase;
- to slow down the rim-zone formation by using depleted uranium in the surface layer of the pellet.

Using the combined (two-band) pellets based on modified oxide nuclear fuel in WWER allows to solve the problem aiming to improve the fuel efficiency at high burnup [1,2]. The developing of the experimental method for studying uranium dioxide samples thermal desorption and UO_2 samples gas release investigating, containing different amount of simulators of fission products (SFP) and the burnable neutron absorbers (BNA) at a constant heating rate in a vacuum was the task of this researching [3,4]. The problem is actual because the process gas evolving from uranium fuel pellets during burn up reduces the thermal conductivity of the core-shell gap, enhances hydrogen embrittlement of the cladding material, induces its carbonization, and transport processes in the fuel. The composition of the process gases may comprise: CO_2, CO, N_2, H_2, hydrocarbons, water vapor, and others. Using the method of thermal desorption is caused by its large abilities which allow to receive in a wide range of temperatures the information on number, composition and mobility of gases, in dependence to structural parameters of the material.

2. Experiment

The investigation of the gas release from the UO_2 samples, containing SFP and BNA kinetics was performed with a high-vacuum mass spectrometer instalation. Washed in ethanol and dried UO_2 samples were placed in a loading system having a form of a disc with vertical holes and rotating around its axis. The heating was provided by a programmable MINITE-RM400 (made in Russia). The electrical resistance heater is made of a molybdenum tube and allows to heat the sample at a constant rate from room temperature to 2000 K. In this study heating is carried at 15 K / min in the temperature interval 300 - 2000 K. The evacuation of the vacuum system is carried out by the pump station Hi Cube 80 and allows to obtain a vacuum of about 10^{-5} Pa.

Gases released from the test sample as a result of its thermal heating mass spectrum is obtained by means of a mass spectrometer MX-7304, with its sensor connected to the analysis chamber through the valve DU-25 (made in Russia). The sensitivity to hydrogen at a linear heating is about $3{\cdot}10^{-14}$ molecules or concentration of $6{\cdot}10^{-5}$ at. % (0.01 ppm). The thermal desorption mass spectrum is recorded on a PC using the software package Labview 2012.

3. Results

Figure 1a shows typical kinetic dependences of the UO_2 gas samples emission, non- SFP and BNA. Herein a single peak M2 (hydrogen) at 600 K and a peak at 1100 K is presented.

Figure 1b shows typical kinetic dependences of the UO_2 gas samples, containing 8% SFP emission. In this figure, there is one M2 peak (hydrogen) at 670 K and two M28 peaks - at 880 K and 1320 K, respectively.

Figure1c shows the experimental values for the UO_2 sample, containing 11% SFP. There are three M2 peaks (at 670 K, 1020 K and 1320 K). For this sample the M28 peak is observed with wide plateau in the temperature range 950 - 1200 K. There are two M44 peaks - at temperatures of 1050 K and 1300 K. Also there are two M12 peaks at 1050 K and 1300 K and one M16 peak at 1300 K.

The dependencies for UO_2 sample with 6,9% Gd_2O_3 are shown in Figure 1d, which contains five M2 peaks (580 K, 680 K, 820 K, 970 K and 1130 K), three M28 peaks (830 K, 1130 K, 1430 K) and single M44 peak at 1130 K.

Figure 1e shows the data for UO_2 with 8,9% $AlGdO_3$ (6% Gd). In this figure, there is one M2 peak (at 660 K), three M28 peaks (850 K, 1150 K, 1450 K) and single M44 peak at 850 K.

Figure 1f shows the typical kinetic dependences of the gas emission from a UO_2 sample with 8,9% $AlGdO_3$ (6% Gd) and 11% SFP. Figure 1f shows four M2 peaks (560 K, 670 K, 800 K, 1300 K), one M28 peak at 1450 K, five M44 peaks (670 K, 770 K, 910 K, 1050 K and 1370 K), one M12 peak at 1370 K and single M16 peak at 1370 K.

4. Discussion

Based on the analysis of the UO_2 samples gas evolving kinetic dependences it can be concluded that M2 and M28 gassing from the UO_2 samples occurs from the fuel pellet volume. The typical shape of the kinetic dependences indicates that H_2 and CO gassing occurs by diffusion, and the binding energy of the M2 and M28 is different in the sample. The UO_2 samples with 8% of SFP (Figure 1b) also exhibit diffusion gassing of M2 and M28.

However, it is seen that the M2 peak shifts to a temperature of 670 K that indicates the increase in the M2 binding energy within the sample volume and on the kinetic dependence of M 28 gassing a peak shifts to a temperature of 1320 K and formation of an additional peak is observed at 880 K. This indirectly indicates that at the introduction of the UO_2 sample additional 8% SFP defects are formed, which segregate and gases have larger binding energy with them.

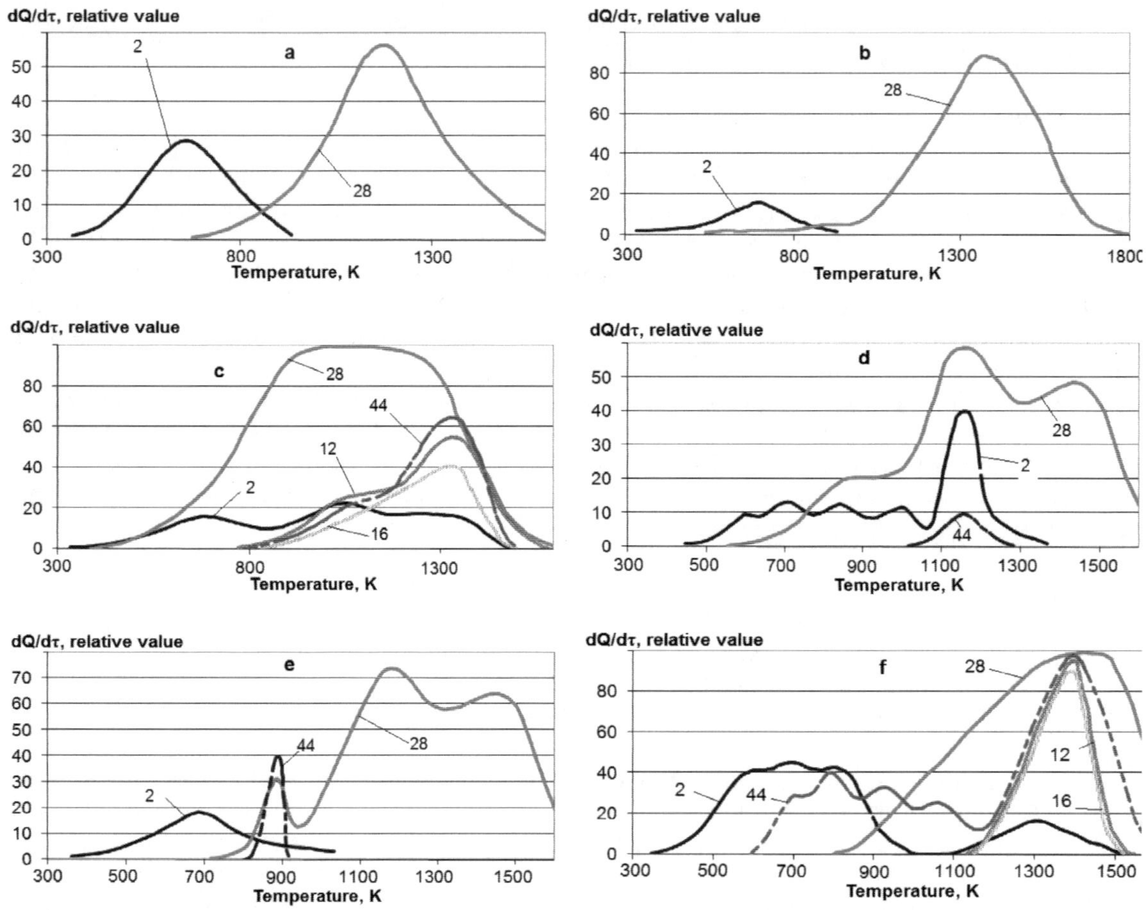

Figure 1. Thermodesorption spectra of evolved gases for the samples:
a – UO_2; b – UO_2+8% SFP; c – UO_2+11% SFP; d – UO_2+6,9%Gd_2O_3; e – UO_2+8,9%$AlGdO_3$;
f – UO_2+8,9% $AlGdO_3$+11% SFP

The kinetic release of gases dependences of UO_2 samples containing 11% SFP (Figure 1c) have a saturated range of gas emission peak emission both in the number of peaks and the amount of generated mass emissions (M2, M28, M44, M12 and M16). It can be suggested that the more complex system defects (dislocations, dislocation networks, dislocation loops, gas bubbles) which can trap gases is formed into the UO_2 fuel pellet with 11% of SFP. However, by to the characteristic emission of gas kinetic dependences type can be said that the gassing takes place by diffusion. The kinetic gas emission dependences at UO_2 samples with content 6,9% Gd_2O_3 also are having several peaks of the sample gas outlet: at the gas M2 there are five peaks, at the M28 - three peaks, and M44 - one peak. In this case M2, M28 and M44 peak is observed at one temperature - 1130 K. The presence of multiple peaks gassing also implies the presence of several places trapping gases in the UO_2 sample, and the existence of peaks gassing with varying mass (M2, M28, M44) at the same temperature 1130 K suggests that the yield of gases takes place in one way, possibly at gas bubbles chain at the grain boundaries. For samples with UO_2 $AlGdO_3$ 8,9% (6% Gd) also exhibits several peak gassing M28 are observed at the kinetic gas emission dependences what suggests the presence of several types defects by trapping gases. And the coincidence of peaks gassing M28 and M44 at a temperature of 850 K, suggests gas exit of the same mechanism, namely, at chain of gas bubbles. In this case the peak of gas output from sample has a steep look. For UO_2 with UO_2 $AlGdO_3$ with 8,9% (6% Gd) and 11% of SFP samples with kinetic dependence gassing (Figure 1f) are even richer spectrum peaks gassing as in the number of peaks, and as the amount of generated gas mass (M2, M28, M44, M12 and M16). This implies that the sample has formed a complex system defects that can trap gases. The presence of the kinetic dependences of the gas emission with masses of the M12, M16 in Figure 1c and Figure 1f allows to suggest that precipitated mass M28 corresponds to CO, and the mass of M44 corresponds to CO_2. The kinetic dependence of the release of gas from the M2 corresponds to H_2.

The presence of sharp peaks of the emission of gas M28 and M44 (Figure 1e) in a narrow temperature range about 850 K, and M2, M28 and M44 (Figure 1d) in the temperature range of about 1130K said suggests that the yield of gases occurs through the chain of gas bubbles at the grain boundaries. In another case, flatter curves indicate that the yield of gas occurs through the diffusion of atoms of the gas molecules. These atoms move to the surface of the material by grain boundary diffusion and then combined into the molecular compounds.

5. Conclusion

In this study the technique was perfected for carrying out experiments on the vacuum thermal desorption of gases from UO_2 fuel pellets, containing SFP and BNA. It was found that the thermal desorption spectrum of gases released depends on the concentration of SFP and BNA in UO_2 fuel pellet. This is due to the fact that the defects are created in the UO_2 samples which are effective gas traps to a sufficiently high temperature (1500 K). The gas composition released from the fuel pellet is the following: H_2, CO, CO_2.

Acknowledgements

This work was performed within the framework of the Center of Nuclear Systems and Materials supported by MEPhI Academic Excellence Project (contract № 02.a03.21.0005, 27.08.2013).

References

[1] Zinkle S J, Terrani K A, Gehin J C, Ott L J and Snead L L 2014 *J. Nucl. Mater.* **448** (1) 374
[2] Durazzo M, Oliveira F B V, Urano de Carvalho E F, Riella H G 2010 *J. Nucl. Mater.* **400** 189
[3] Westphal B R, Bateman K J, Lind R P, Howden K L, Del Cul G D 2005 *Fission product removal from spent oxide fuel by head-end processing* (Tsukuba: Proc. Global 2005)
[4] Kang K H, Song K C, Park H S, Moon J S, Yang M S 2000 *Metals and Materials* **6** (6) 583

Rate control sintering of the uranium dioxide

R S Kuzmin, V G Baranov, M A Burlakova, V V Mikhalchik and A V Tenishev

National Research Nuclear University MEPhI (Moscow Engineering Physics Institute), Kashirskoe highway 31, Moscow, 115409, Russia

E-mail: Onil709@mail.ru, VGBaranov@mephi.ru

Abstract. The application of rate controlled sintering (RCS) technique for fabricating oxide fuel pellets in industrial conditions is investigated. Green pellets of UO_2 (with no additives) which underwent industrial-type compacting were sintered in reducing Ar – 8% H_2 medium using dynamic and isothermal RCS modes with exposure at 1600 °C during 8 hours. Decreasing shrinkage rate resulted in growth of sintered density for the dynamic RCS mode. Opposite to the results of earlier works [1,2], decreasing heating rate did not result in reducing sintering rate. Isothermal mode allowed sintering rate to be maintained between 0.1– 0.15 %/min. Temperature–time modes obtained using RCS are close to the operation mode of through-type furnace used for mass production.

1. Intoduction

Currently oxide nuclear fuel based on UO_2 is essential for atomic energy. To maintain product quality, security requirements and to improve economic indicators a continuous improvement of the technological aspects of manufacturing powders, molding powders and fuel compacts based on uranium dioxide is a relevant issue. The sintering process "raw" compacts receives considerable attention due to the strong economic performance of this process step, with a single cycle can be more than 40 hours. At this step the basic parameters of the sintered fuel pellets are formed: geometry, density, average grain size, porosity, pore distribution, etc. To obtain a ceramic fuel based on stoichiometric uranium dioxide a reducing atmosphere of hydrogen or mixtures of hydrogen and nitrogen is used as medium. Another important characteristic of the sintering medium is humidity, it's control is urgent to obtain stable characteristics of the fuel pellets.

In papers [1-3] the authors presented the results of the analysis of the sintering process UO_2 pellets and (U, Gd)O_2 with different composition when changing the temperature mode and sintering medium. In the papers a preset constant heating rate technique is used. To get more information on the features of the fuel pellets sintering and possible effects on sintered structure the rate control sintering (RCS) method is considered useful.

The RCS method due to changes in the rate of heating is known for a long time. With it, in particular, a very fine grain structure is received, while maintaining a high density. To do this it is necessary to divide the two competing processes - pore coalescence and growth, and grain growth, which is achieved by providing a special densification rate profile [4] (figure 1).

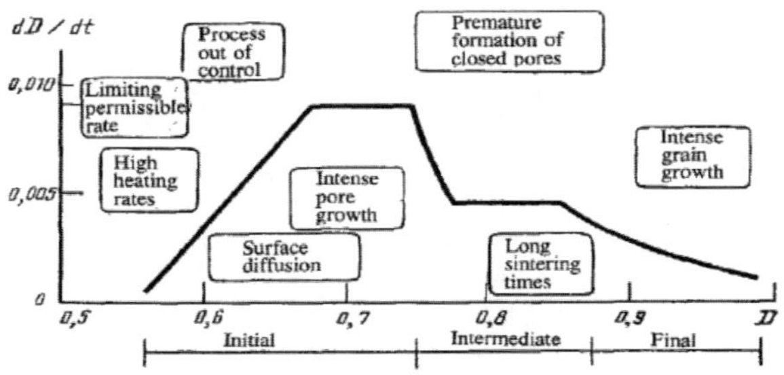

Figure 1. Typical dependence of the sintering rate to shrinkage using RCS with denoted processes, limiting the densification rate [4].

The data on a use of the RCS method to manage the final microstructure of uranium dioxide is absent in literature. In addition, for an oxide nuclear fuel an urgent problem is to increase the grain size using a limited amount of additives that activate grain growth or its complete rejection. Thus, application of the method for uranium dioxide based pellets sintering enables not only a more effective control of structural parameters, but improve the nuclear purity of the fuel.

2. Materials and equipment

In the paper oxide nuclear fuel compacts produced in industrial conditions are investigated. UO_2 powder was obtained by ammonium diuranate (ADU) process. The compacts were produced by the method based on preliminary briquetting with a liquid plasticizing component and then grinding to press powder. The density of the obtained compacts was about 50% of theoretical density (TD).

To investigate the size change kinetics of compacts during of its sintering a horizontal dilatometer Netzsch DIL 402 C (Germany) is used, using its software application a constant heating rate sintering and RCS is possible. Sintering is performed in a dynamic medium of Ar-8% H_2 using a gas flow rate of 50 ml/min.

The microstructure of the sintered UO_2 pellets is investigated by the scanning electron microscope JEOL JSM-6610LV (Japan).

3. Dilatometic measurements

As part of the works on sintering of oxide fuel pellets in modes that simulate industrial production (under laboratory conditions the maximum sintering temperature is 1600 °C), a series out of experiments varying the temperature of isothermal exposures was carried out. Figure 2 shows the dilatometric curves for UO2 pellets, sintered with regular program (hereinafter referred to as the "basic" program – curve 1) and the program with a higher temperature of 4 and 5 isothermal exposures ("optimized" program – curve 2). With 4 and 5 isothermal exposures at 100 °C higher the sintering rate (i.e. the time derivative of length change) was nearly constant.

For the "basic" program sintering takes place in four steps:

1) initial - at temperatures of 800-1000 °C;
2) slow - at temperatures of 1000-1100 °C;
3) accelerated - from 1100 ° C and up to a maximum temperature of 1600 °C;
4) fading - throughout the isothermal holding.

For the "optimized" program, the sintering process rather quickly reaches a constant rate (average 0.07%/min) until the isothermal exposure. Thus, instead of slow 2nd and 3rd stages fast one is observed at an average speed.

From Figure 3 it is evident that during sintering with the "basic" program at the 2nd step the sintering rate is less than 0.02%/min, and at the 3rd - changes along a curve with a maximum of 0.27%/min; in the case of the "optimized" program sintering rate is generally in the range of 0.05-0.10%/min and is less than 0.14%/min.

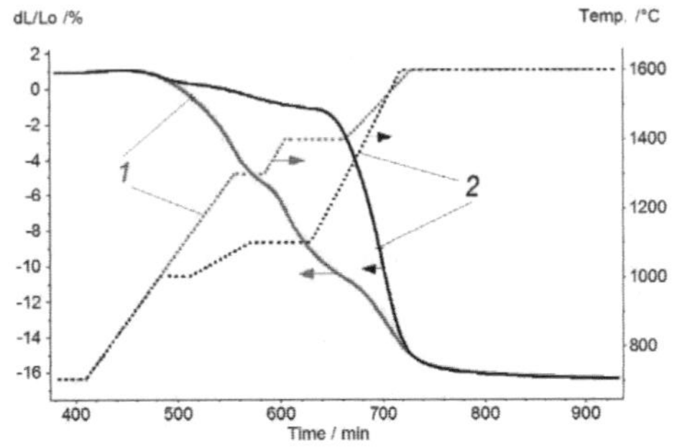

Figure 2. Dilatometric sintering curves (—) with the temperature program (····), simulating a shortened production program with a 12 min beam cycle: 1 - basic, 2 - optimized.

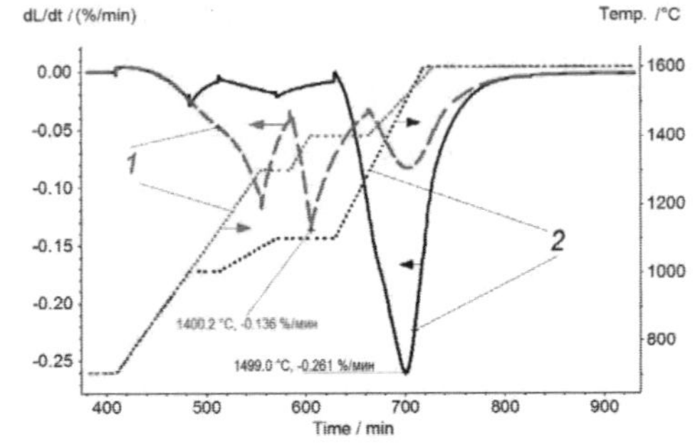

Figure 3. The time dependence of the sintering rate (— and ---) for temperature program (····), simulating a shortened production program with a 12 min beam cycle: 1 - basic, 2 - optimized.

The sintered microstructures analysis, that is shown in figure 4, shows that the shrinkage rate close to constant gives a more homogeneous structure and a smaller pore size.

From these results, it is seen that the fuel pellets produced with "optimized" program have pores not exceeding 2 μm in diameter and a grain size of 5-12 μm.

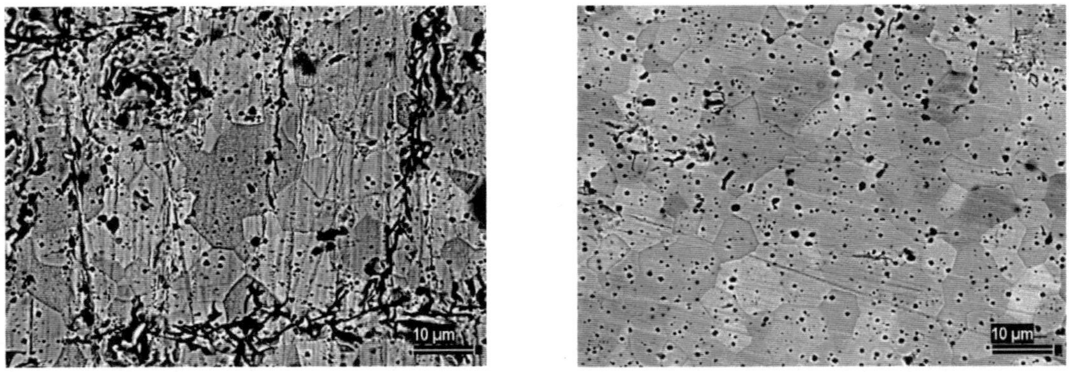

a) б)

Figure 4. The microstructure of UO_2 pellets, sintered with program simulating a shortened production program with a 12 min beam cycle: a) basic, b) optimized.

As mentioned above, the use of RCS can significantly affect on the microstructure parameters. As for ceramic nuclear fuel grain size is important from the standpoint of performance, one of the aims is to eliminate the second sintering stage (figure 1), which suppresses the grain growth. For this purpose, a series of experiments were carried out with the preset shrinkage rates. From figure 5 it is evident, that it is difficult to receive the stable value of the shrinkage rate in the long time interval in the heating process, however, it becomes possible to local time intervals (20-40 min) with reducing the shrinkage rate to 0.05%/min.

The reason it was not possible to maintain a desired sintering rate can be seen on the curve 3, wherein the two peaks are distinctly different. With this a second one, that is more acute, most likely corresponds to the enabling of a new, more rapid sintering mechanism. Such a mechanism may be a plastic material flow. At figure 6 showing the shrinkage and shrinkage rate to temperature it can be seen how quickly the process proceeds almost instantaneously accelerating the sintering in the range of 1325-1350 °C.

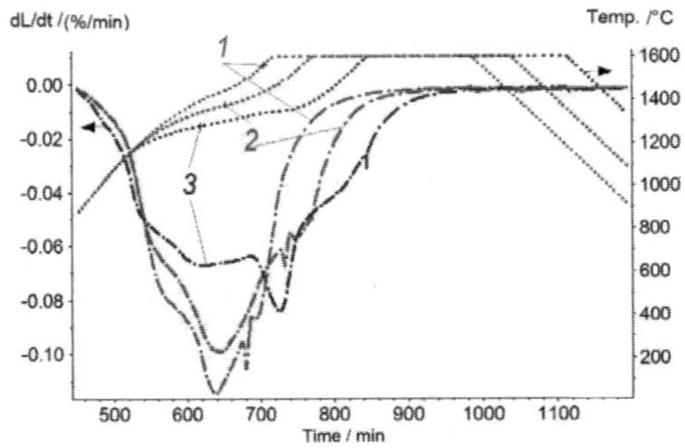

Figure 5. The time dependence of the sintering rate (- · -) with the temperature program (····) using RCS in dynamic mode with preset rate of 1 – 0.1; 2 – 0.07; 3 – 0.05 % / min

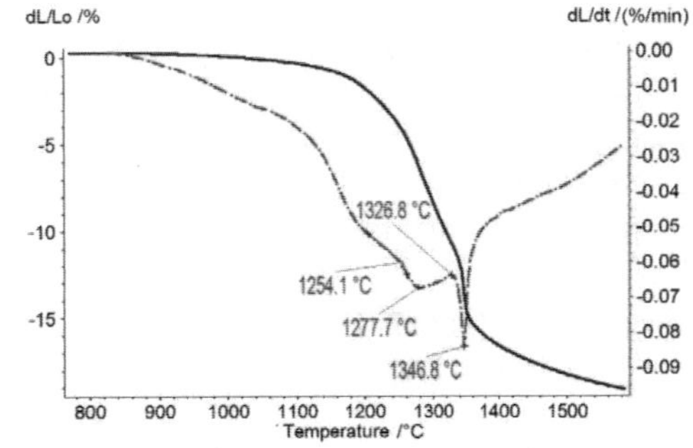

Figure 6. Temperature dependence of shrinkage (—) and sintering rate (- · -) using RCS in the dynamic mode at a preset rate of 0.05% / min

When using a low shrinkage rate of 0.004% / min (figure 7) receiving a constant shrinkage rate also failed. The reason is the limitation of the method and software part in the definition of the value of the derivative at the current time, by which the temperature program is adjusted. However, at the shrinkage rate curve the two peaks corresponding, as supposed, to the two different predominant sintering mechanism, which are activated at these temperatures, are also clearly distinguished. The second peak, showing sintering acceleration finishes at 1327 °C, which is close to the previous case (figure 6).

Thus, obtaining a constant sintering rate of 0.05 or 0.07% / min by feedback method is very difficult due to the rapid sintering acceleration by enabling of new mechanisms when increasing temperature.

However, providing of dynamic mode sintering is only possible with laboratory equipment. Therefore, from the standpoint of practical implementation more preferable is use sintering programs, which are based on the formation of uniform heating segments and some isothermal exposures. Thus, the use of the RCS technique to simulate industrial production conditions is possible only in the isothermal mode.

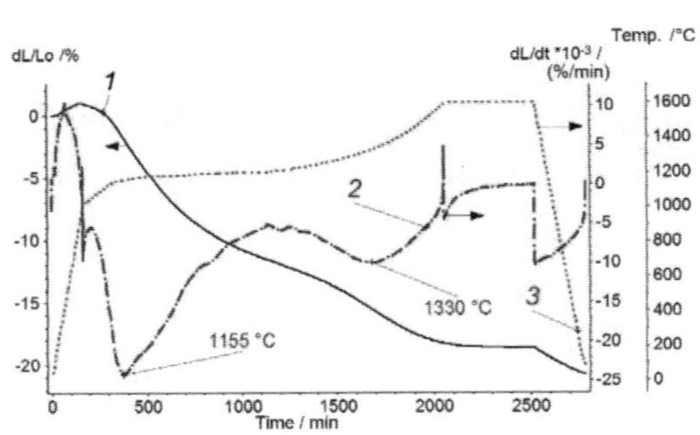

Figure 7. Time dependence of shrinkage (1) and sintering rate (2) with a temperature program (3) using RCS in the "start / stop" mode at a preset rate of 0.004 %/min

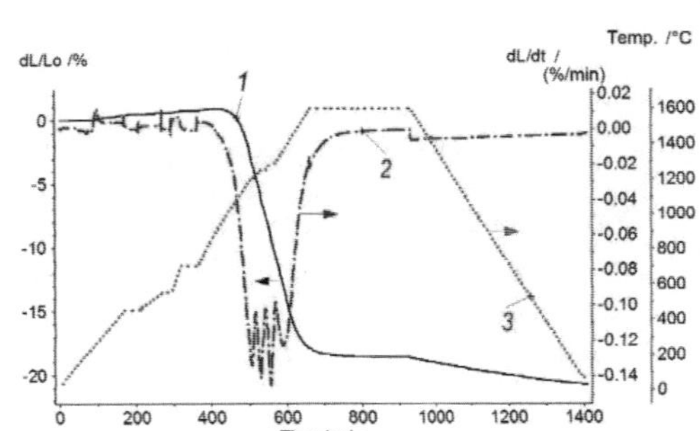

Figure 8. Time dependency of shrinkage (1) and sintering rate (2) when sintering with a temperature program (3) using RCS in isothermal mode at a preset rate interval 0.1-0.15 %/min

Based on the data obtained from previous experiments, limitations to the shrinkage rate were selected of 0.1-0.15% / min. To a temperature of 900 °C, the heating was conducted in the mode simulating the manufacturing sintering (see figure 2). On Figure 8 it is seen that shrinkage is linear during intensive sintering, and the shrinkage rate lies within preset limits.

The resulting temperature program comprises three isothermal exposures at high temperatures and is similar to a program simulating industrial production with higher temperature of 4 and 5th isothermal exposures (figure 9). Thus, the task of improving the profile of temperature for getting the structure with more uniform grain size distribution and fewer microcracks can be solved by maintaining shrinkage rate at a preset level.

The sintered density of the resulting samples was measured by hydrostatic weighing. Accuracy of the method ± 0,03 g/cm^3. The results are shown in Table 1. It is seen that the density values were similar within the measurement accuracy for all temperature programs.

4. Sample structure

Sintered samples microstructure were analyzed by scanning electron microscope.

General view of the microstructure of sintered samples were compared on ×400 magnification. The sample sintered with the "optimized" program has less regions of increased porosity than pellets obtained with RCS and smaller in size, except the pellet sintered under isothermal conditions (figure 10). For this one only small regions with high porosity can be seen. The presence of such regions in all the pellets is due to the lower temperature of the final exposure, compared with the industrial production sintering conditions.

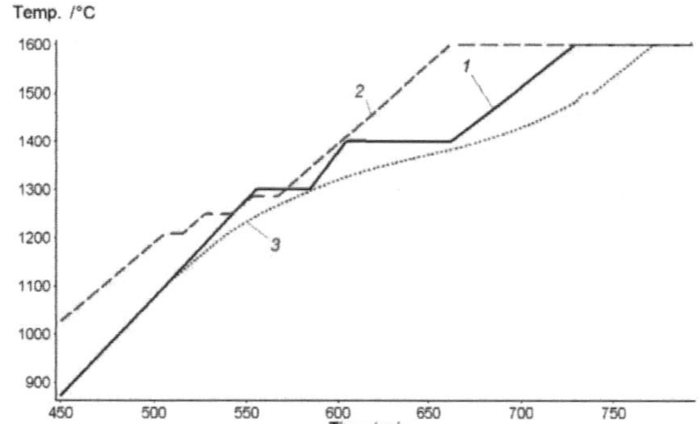

Figure 9. Time-temperature sintering programs for UO_2 pellets:
1 – "optimized",
2 – isothermal mode RCS 0.1-0.15% / min,
3 – dynamic mode RCS 0.07% / min

Table 1 Sintered samples density

№	RCS sintering mode	Shrinkage rate threshold value (%/min)	Density (g/cm^3)
1	No ("basic" program, see figure 2)	No	10.42
2	No ("optimized" program, see figure 2)	No	10.40
3	Start/stop	0.004	10.43
4		0.1 (control factor 25)	10.20
5	Dynamic	0.07 (control factor 25)	10.42
6		0.05 (control factor 40)	10.49
7	Isothermal	0.1 – 0.15	10.39

As a result of the hydrostatic weighing, the sample sintered in dynamic mode at a rate of 0.05%/min has the highest density. But on its microstructure porous portions is seen that are larger and more prominent (figure 11). That fact is due to the large grain size for this sample and the presence of larger pores (15 μm) in the sample that sintered in the isothermal mode.

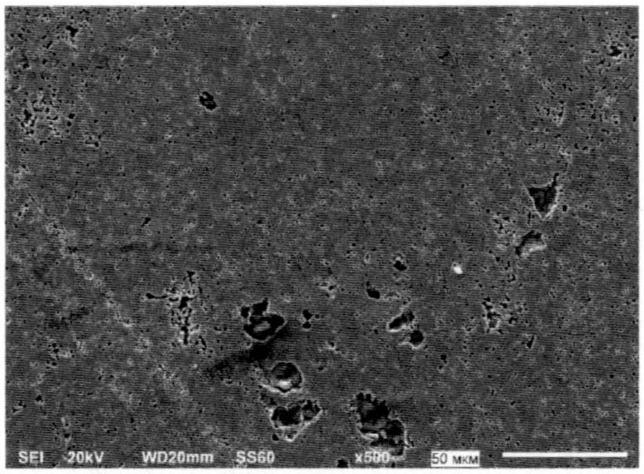

Figure 10. Microstructure of sintered pellets using RCS in isothermal mode.

Figure 11. The microstructure of sintered pellets using RCS in dynamic mode 0.05% / min.

When comparing the microstructures of samples sintered with RCS in dynamic mode at different preset sintering rates a greater number of porous regions and larger pores were found for the shrinkage rate of 0.07 %/min, then that for 0.05 %/min. Microstructure of the sample sintered at 0.1 %/min, does not satisfy the preset requirements. In the central portion of the pellet connected porosity larger than 100 μm is observed. The reason may be a temperature program with fast reaching of a large shrinkage rate, as well as the pressing defect. These dilatometric data for dynamic conditions are consistent with the density measurements: the decrease in the sintering rate increases the density of the samples.

On high magnification images, in which the grain structure is visible, the following features are identified: a sample sintered in isothermal RCS mode has smallest average grain size of less than 3 μm, for other RCS samples that value is about 5 μm. This is due to a high shrinkage rate in the isothermal mode. The most uniform grain size is obtained for the sample with "optimized" program and RCS in isothermal mode. In the dynamic mode, a decrease in the shrinkage rate caused a slight increase in grain size.

Thus, maintaining shrinkage rate at a constant level more uniform grain size can be obtained. If the temperature profile of the shrinkage rate is a curve with maximum larger grain size is available, but the grain size distribution becomes less uniform.

5. Conclusion

Dilatometric study of the sintering process in three programs was carried out: simulation of the temperature profile of the industrial band oven; optimized temperature profile, providing nearly constant shrinkage rate; heating programs, providing a controlled rate of shrinkage. Different modes of RCS were used: start / stop, dynamic and isothermal. With the isothermal mode it was possible to get a close to constant sintering rate (0.1-0.15% / min) and most uniform grain size distribution. Sintering with a dynamic mode at different preset shrinkage rates shows that the total shrinkage increases with decreasing the sintering rate. In the dynamic mode, a constant shrinkage rate was not available. The probable cause is a sharp acceleration of sintering in the temperature range of 1320-1350 °C. Temperature profiles obtained with RCS in the dynamic mode (at the rate of 0.05% / min) and isothermal mode are close to the profile of the "optimized" industrial production program.

The microstructure of sintered samples was analyzed. In general, the samples have similar microstructures and have similar average grain size. For samples sintered in a dynamic mode a decrease in the shrinkage rate caused a slight increase in the grain size. In the isothermal mode a more uniform microstructure, but with a smaller grains is observed. Sintering temperature profile for optimized program provides reduction in the number of large pores in the sintered pellet.

The use of RCS can be used to control the final structure of oxide nuclear fuel pellets and the creation of optimal temperature conditions for band ovens used in industry.

Acknowledgements
This work was performed within the framework of the Center of Nuclear Systems and Materials supported by MEPhI Academic Excellence Project (contract № 02.a03.21.0005, 27.08.2013).

References

[1] Baranov V G, Ivanov A V, Kuzmin R S, Petrov I V, Tenishev A V, Timoshin I S and Khlunov A V 2011 *Atomic Energy* **110** (3) 172
[2] Baranov V G, Kuzmin R S, Tenishev A V, Tarasov B A, Timoshin I S, 2011 *Nuclear Physics and Engeneering* **2**(4) 309 (in Russian)
[3] Baranov V G, Kuzmin R S, Tenishev A V, Khlunov A V, Ivanov A V, Petrov I V and Timoshin I S 2012 *Atomic Energy* **112** (4) 303
[4] Ragulya A V, Skorokhod V V. 1994 *Powder Metallurgy and Metal Ceramics* **32** 3 (in Russian).

High temperature behavior of simulated mixed nitrides

V G Baranov, A V Lunev, V V Mikhalchik, A V Tenishev and D P Shornikov

National Research Nuclear University MEPhI (Moscow Engineering Physics Institute), Kashirskoe highway 31, Moscow, 115409, Russia

E-mail: Onil709@mail.ru

Abstract. Specimen of uranium-based mixed nitrides were synthesized by high-temperature nitriding of metal powder. To investigate thermal stability, samples were annealed at high temperature in a helium atmosphere. During these experiments, the effect of increasing the exposure temperature is studied. Raising the exposure temperature results in a multifold increase of mass loss. A comparison with data on pure uranium nitride shows that increasing the complexity of the nitride systems also results in higher mass loss. Later microscopic investigation of test samples revealed that metal precipitates may be found only on the surface of test samples. Electron probe micro-analysis indicates these precipitates to be uranium metal. Nevertheless, compared to pure uranium nitride, uranium-based mixed nitrides exhibit active evaporation at lower temperatures

1. Introduction

Nuclear fuel based on mixed uranium-plutonium nitride has been considered as perspective for fast nuclear reactors with recycling fuel cycle [1]. Nevertheless, there is a serious drawback in processing and usage of uranium-based nitrides at high temperatures – namely, thermal decomposition [2, 3]. This process hinders the sintering of pellets to a high density, also leading to dissociation and evaporation of uranium and plutonium at 1700–1900 °C. Review of previous works [4] has shown that, despite the large number of theoretical and experimental data on the subject, the dissociation still remains understudied. A detailed study of thermal decomposition in uranium-based mixed nitrides is required.

Conducting experiments with fuel containing plutonium and minor actinides requires special working conditions, i.e. gloveboxes with inert atmosphere. This circumstance greatly increases the complexity of research [5]. At a certain point, one may consider using simulated fuel samples instead where active elements are replaced by chemically identical simulators.

This paper aims to provide further information for the kinetics of thermal decomposition in simulated mixed nitrides.

2. Experimental

2.1 Materials and specimens

Powder of uranium mononitride (UN) was produced by nitridation of hydrogenated uranium metal. Cold pressing was performed then to obtain green pellets. Due to the reactive nature of the metal and nitride powders, all materials are handled and stored in argon-filled gloveboxes, which provided an inert atmosphere. Samples had minimal amount of secondary oxide chemical phases. Microstructure of the uranium nitride green pellets shows open porosity channels with no cracks.

For tests at 2100 °C with mixed nitrides, plutonium mononitride (PuN) is replaced by yttrium and neodymium mononitrides (YN and NdN) which provide similar nitrogen vapor pressure. Thus, YN, NdN and Ce are selected as simulators for PuN.

Fission products are simulated by Nd, Ce, Mo, Zr, Ba, La, Sr, Y, Pr, Ru. To simplify the alloying procedure, Nd, Ce, La, Pr, and Y are substituted with an equivalent 10 and 20 wt. % amount of mischmetal (Mm). MmN was produced by hydrogenation of mischmetal prior to nitridation. The product was then mechanically admixed to uranium mononitride powder.

2.2. Thermal stability tests

Tests using Netzsch STA 449 F1 apparatus with QMS 403 C quadruple mass-spectrometer are performed to determine the threshold temperature of thermal decomposition. Measurements are conducted in high purity helium (6.0 grade) with flow rate of 100 ml/min. Additional helium refinement to < 1 ppb is performed using heated catalytic MonoTorr_filter on the basis of metallic getter. The following parameters are monitored during the thermal gravimetric analysis: sample mass change; the rate of sample mass change or the derivative of mass change with respect to time, and the time dependence of sample's temperature. Thermal stability mode was enabled which means that the exposure occurs as long as the sample does not lose a predetermined mass.

Microstructural investigation of uranium mononitride samples is performed with scanning electron microscope (SEM) JEOL–6610LV equipped with EDS (energy-dispersive X-ray spectroscopy) for X–ray microanalysis.

3. Results

3.1. Uranium nitride

Mass loss of a uranium nitride sample heated up to 2100°C is performed (Figure 1). The sample exposed for 35 min lost a total 2 % of its mass, starting losing mass already at the stage of heating (1600–1700 °C). After reaching the isotherm, mass loss happened in two stages: the average rate of mass loss increases up to ≈0.09 mg/min, and later during exposure mass loss accelerates up to ≈0.012 μg/min. Prior to acceleration, the mass of the samples decreases by ≈2.5 mg, and the total mass loss of sample is about 3.5 mg (1.9 % of the total mass of the sample). At 2100 °C the average rate of mass loss increases after 10 min exposure. Mass loss data performed over other UN samples at this exposure temperature show similar kinetics. If thermal exposure is reduced to 2000°C, the acceleration of mass loss begins after 50 min of exposure (Figure 2). At isothermal exposure the average rate of mass loss increases up to ≈0.03 mg/min, then mass loss accelerates up to ≈0.05 μg/min. Decreasing of exposure temperature results in slowdown of losing mass. The average rate of mass loss becomes smaller.

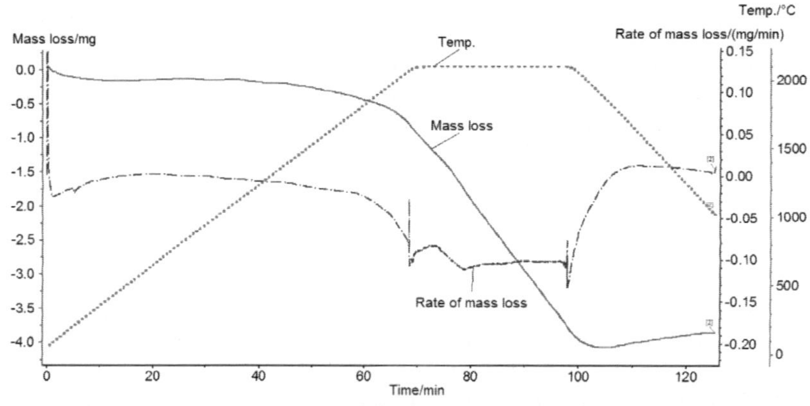

Figure 1. Dependence of mass loss and rate of mass loss on temperature and isothermal exposure duration (2100 °C) for uranium nitride sample.

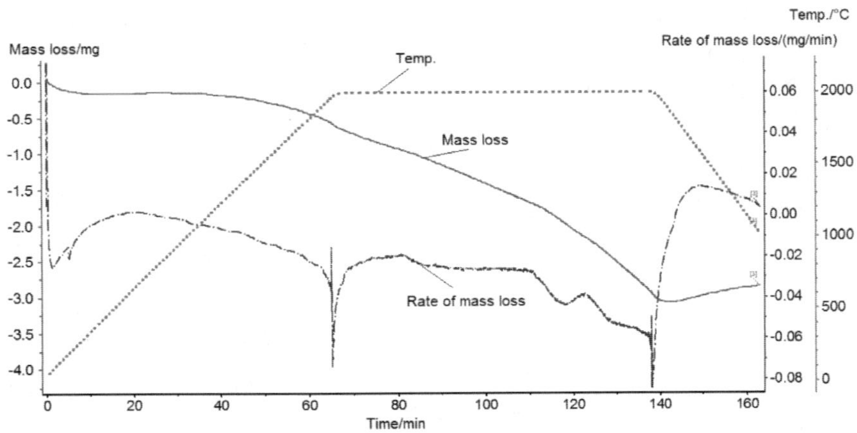

Figure 2. Dependence of mass loss and rate of mass loss on temperature and isothermal exposure duration (2000 °C) for uranium nitride sample.

After thermal stability tests, UN samples were additionally studied using scanning electron microscopy (Figure 3). Electron probe micro-analysis showed that uranium-rich precipitates were formed. The size of precipitates varied from 0.1 to 6.0 μm. Decreasing exposure temperature from 2100 °C to 2000 °C resulted in lower precipitate sizes.

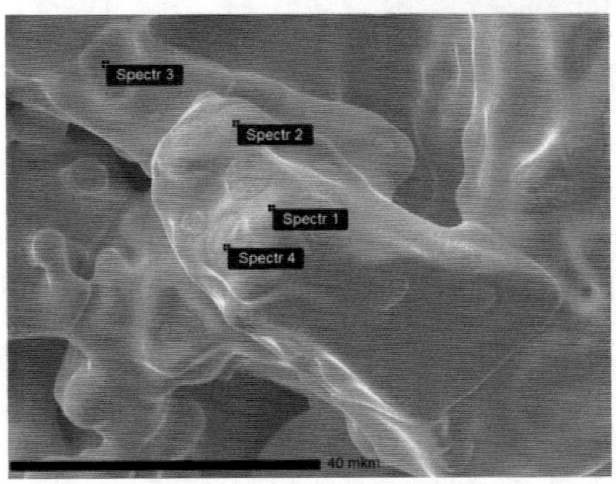

Figure 3. Microstructure and electron probe micro- analysis of uranium mononitride sample exposed at 2100 °C:
Spectr 1 (N – 48.9 at.%, U – 51.1 at.%);
Spectr 2 (U – 100.0 at.%);
Spectr 3 (N – 46.6 at.%, U – 53.4 at.%);
Spectr 4 (U – 100,0 at.%).

Summing up facts mentioned above, the second stage of mass loss may be explained (i.e. severe increase in rate of mass loss during exposure at high temperature – see Figure 1,2). Evaporation of uranium mononitride leads to nitrogen depletion: the N/U ratio of uranium mononitride gradually decreases, and dissociation begins when composition reaches equilibrium at given temperature and partial pressure of nitrogen in the gas atmosphere which then leads to precipitation of uranium metal. The precipitated phase remains in liquid state, since test temperature exceeds the melting temperature of uranium. Moreover, vapor pressure of uranium is large enough to initiate its evaporation and yet increases the rate of weight loss.

The precipitated phase was distributed evenly over the non-restrained surface of the sample. However, no precipitates were found on sample's surface which was in contact with the crucible during tests. Evidently, the crucible restrained access of helium to the surface of the sample in these areas. This resulted in a higher partial pressure of nitrogen over this surface and stabilized UN.

3.2. Uranium-based mixed nitride

Mass loss of a UN+20 wt.%MmN sample heated up to 2100°C is performed (Figure 4). In (U,Mm)N samples mass loss due to evaporation begin at temperatures above 1480°C. Comparing to uranium mononitride, the rate of mass loss is significantly increased in mixed nitrides: 0.12 mg/min for UN,

0.75 mg/min for UN+10 wt.% MmN, and 0.91 mg/min for UN+20 wt.%MmN. During isothermal exposure, the maximum rate of mass loss is about 20% lower. Acceleration of evaporation, nevertheless, does not lead to dissociation.

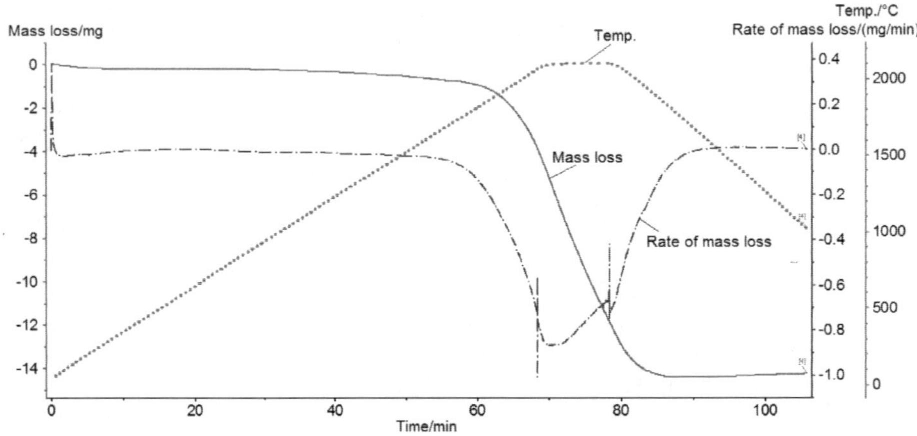

Figure 4. Dependence of mass loss, rate of weight loss for UN+20% MmN samples on temperature and duration of isothermal exposure.

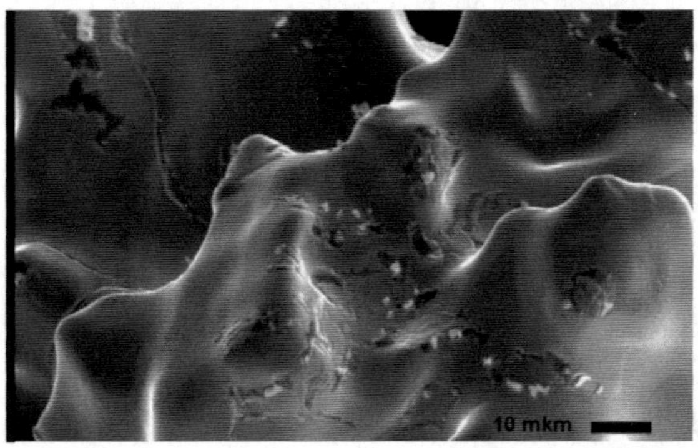

Figure 5. Microstructure of UN+20% MmN sample exposed at 2100 °C.

The resulting microstructure of (U,Mm)N after tests is shown on Figure 5. The distribution of elements in both samples is similar. Ce, La and Nd were evenly distributed along the examined surface.

X-Ray analysis showed that lines are split in two: UN and (U,Me)N meaning that MmN dissolved in UN (Figure 6.). Precipitates of MmN were not observed.

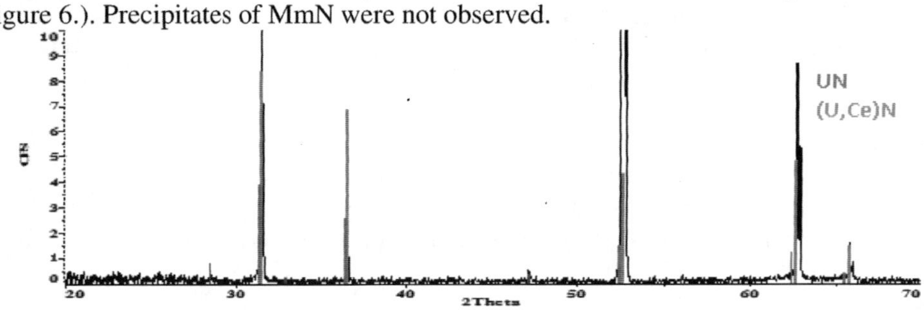

Figure 6. X-ray pattern of UN+20 wt.% Mm sample.

4. Discussion

Evaporation of uranium mononitride leads to nitrogen depletion: the N/U ratio of uranium mononitride gradually decreases, and dissociation begins when composition reaches equilibrium at given temperature and partial pressure of nitrogen in the gas atmosphere which then leads to precipitation of uranium metal. The precipitated phase remains in liquid state, since test temperature exceeds the melting temperature of uranium dioxide. Moreover, vapor pressure of uranium is large enough to initiate its evaporation and yet increases the rate of mass loss.

Mass loss due to evaporation occurs at a much higher rate in mixed uranium nitride samples with 10–20 wt.% of MmN. However, no dissociation happens since precipitates of uranium metal or mischmetal nitrides are not found on the surface of samples.

Both evaporation and dissociation of uranium nitride is strongly dependent on the partial pressure of nitrogen. Low partial pressure of nitrogen in helium leads to dissociation at test temperatures.

Summary

This study addressed one of the key problems with using mixed uranium-based nitrides in closed nuclear fuel cycle: namely the influence of minor actinides and fission products on high-temperature sintering of recycled green pellets. Thermal stability tests conducted both with pure uranium nitride (UN) samples and simfuel samples containing 20 wt.% mischmetal (Mm) show that samples with complex chemical composition are sufficiently less stable than pure UN. For samples with 20 wt. % Mm dissociation temperature drops by 200-300°C. Therefore, it is extremely important to carefully choose the high-temperature processing mode for this fuel so dissociation can be prevented. Both fission products and minor actinides will hinder the performance of mixed nitride fuel. Further work is needed, possibly aimed at selecting proper additives which could increase the dissociation temperature of mixed uranium-based nitrides.

Acknowledgements

This work was performed within the framework of the Center of Nuclear Systems and Materials supported by MEPhI Academic Excellence Project (contract № 02.a03.21.0005, 27.08.2013).

References

[1] Arai Y 2012 *Comprehensive Nuclear Materials* **3.02** 41
[2] Butt D P and Jaques B 2009 *Synthesis and Optimization of the Sintering Kinetics of Actinide* (Boise: Boise State University) p. 191
[3] Uno M, Nishi T and Takano M 2012 *Comprehensive Nuclear Materials* **2.03** 61
[4] M. Pukari 2013 *Experimental and theoretical studies of nitride fuels* (Stockholm: Royal Institute of Technology) Doctoral Thesis (2013) p. 74
[5] Streit M and Ingold F 2005 *J. of the European Ceram. Soc.* **25** 2687–92

IOP Conf. Series: Materials Science and Engineering **130** (2016) 012023 doi:10.1088/1757-899X/130/1/012023

Interaction UMo fuel with Fe and FeCr

S N Nikitin, D P Shornikov, B A Tarasov and V G Baranov

National Research Nuclear University MEPhI (Moscow Engineering Physics Institute) Kashirskoye highway 31, Moscow, 115409, Russia

E-mail: mephi200809@yandex.ru

Abstract. Uranium-molybdenum alloys are perspective nuclear fuel for fast reactors. In this work, a study was conducted of the interaction of uranium-molybdenum alloy with iron and chromium steel at an elevated temperature of 750 °C for 5 hours. It was found that the constant rate of the interaction layer growth for diffusion couple UMo/FeCr is about $5.4 \cdot 10^{-12}$ m^2/s at 750 °C. The phase composition of the interaction layers for the both diffusion couples was determined. The interaction comes along the grain boundaries, there are not interacts UMo alloy grain in the structure of the diffusion zone.

1. Introduction

During the operation in a nuclear reactor metallic fuel comes into close contact with the cladding of stainless steel, because of swelling which contributes to diffusive reacted fuel cladding at a high temperature to form a liquid phase and results in the destruction of the fuel element [1-4]. One of the possible ways to improve compatibility and reduce the swelling of metallic fuel is uranium molybdenum alloying [5-7]. In the case of uranium-zirconium alloy there is a redistribution of the alloy components, migration of the uranium, zirconium, reducing the local temperature of the melting point [3]. Based on the enthalpy of formation of binary mixtures of Mo and Zr with Fe, which is an approximate measure of the reactivity between the two elements shows that Mo has a lower enthalpy (in absolute value) than Zr. That said that Mo is less active in the interaction with a cladding compared to Zr [4]. However, at present there is no clear understanding of the processes occurring in the interaction of uranium-molybdenum fuel with material cladding of fast reactor made of stainless steel. Therefore, the main objective of the work is the experimental determination of the effect of alloying molybdenum nuclear fuel on the nature and kinetics of its interaction with iron and chromium steel.

2. Experiment

Alloys U10 wt.% Mo, Fe, Fe13 wt.% Cr hereinafter U10Mo, Fe13Cr respectively prepared by melting a batch of metal chips in an arc furnace with non-consumable tungsten electrode in an atmosphere of purified argon. To prepare the alloys used pure uranium (99.98), distilled molybdenum, iodine chromium, carbonyl iron and high purity Al (99.999). The batch melted several times to improve the homogeneity of the ingot. All alloys were annealed in vacuum at 1000 °C for one day for homogenizing. The obtained alloy was cut into tablets of 3 mm thickness at electric discharge machine, then subjected to grinding followed by polishing with diamond paste to 1 μm. Follows from the obtained "tablets" samples were taken for analysis of homogeneity and to determine the exact composition of the obtained alloy. To carry out the diffusion annealing, uranium alloy tablets and stainless steel were clamped together between the molybdenum discs, followed by annealing in a quartz ampoule. Before annealing to prevent possible oxidation of the getter vial laid chip zirconium iodide. Then the vial drops diffusion assembly and vial is sealed, evacuated. Diffusion annealing is conducted in a tube furnace at temperature 750 °C, for 5 hours. After completing annealing time the quartz ampoule is placed in a bath of running cold water. Then diffusion couples were cut perpendicular to the direction of diffusion in the electric spark machine, the use of which will allow to cut the diffusion couples without destroying the fragile interaction layer (IL). The next step is to pour a few of diffusion couples into the resin. The resulting sample was further passed metallographic preparation. The microstructure of the surface and the elemental composition of the samples was investigated by scanning electron microscope JEOL 6610LV using energy-dispersive spectrometer Oxford Instruments INCA x-ACT. Phase composition layer interaction in diffusion pairs was determined by quantitative relations (U+Mo)/(Fe+Cr) in phase.

3. Results

3.1. The interaction between U10Mo and Fe

Analysis of the structural-phase state of the interaction layer in diffusion couple U10Mo/Fe after annealing at 750 °C for 5 hours showed that the reaction layer may be divided into the following areas (Figure 1 a-d):

1. $(U,Mo)Fe_2$;
2. $(U,Mo)Fe_2 + (U,Mo)_6Fe$;
3. binary eutectic $(U,Mo)_6Fe + (U,Mo)Fe_2$;
4. U_6Fe + 90wt.% Mo + binary eutectic $(U,Mo)_6Fe + (U,Mo)Fe_2$;
5. binary eutectic $\{(U,Mo)_6Fe + (U,Mo) Fe_2\} + U_6Fe$;
6. binary eutectic $\{(U,Mo)Fe_2 + (U,Mo)_6Fe\}$ + grain U10Mo + U_6Fe.

At the initial stage of interaction comes at the expense iron diffusion along grain boundaries U10Mo, which is confirmed by unreacted grains of uranium-molybdenum alloy (Figure 1 d). However, due to the fact that the temperature of isothermal annealing at 750 ° C which is 25 °C above the temperature of formation of eutectic $U_6Fe + UFe_2$ for binary system uranium-iron character of interaction is different from below the eutectic to form a eutectic phase of rapid interaction begins. Maximum width of the interaction layer region from 200 to 300 µm is U_6Fe + 90 wt.% Mo + binary eutectic $(U,Mo)_6Fe + (U,Mo)Fe_2$. A special feature of this area are the particles enriched more 90wt.% Mo, the amount of which increases gradually with the advance deep into the alloy fuel. One of the main characteristics of the interaction of the shell material with fuel is a penetration depth of the fuel cell in cladding which in of diffusion couple U10Mo/Fe is about 200 µm at 750 °C for 5 hours.

3.2. The interaction between U10Mo and Fe13Cr

As can be seen from the microstructure of the interaction between the layer of chromium steel and uranium-molybdenum fuel (figure 2 a-d), occurs due to diffusion of elements of the shell U10Mo grain boundaries to form ternary eutectic $(U,Mo)Fe_2 + \mu + UFe_6$, where µ 80% U-6% Mo-11% Fe-3% Cr. However, from the fuel alloy there is a change in the mechanism of grain boundary volume, as a result we observe the so-called "feathers" (figure 2 d), while in the border zone is observed only UFe_6. All interaction layer can be divided into five areas:

1. $(U,Mo)(Fe,Cr)_2$;
2. $(U,Mo)(Fe,Cr)_2 + \mu$;
3. ternary eutectic $UFe_6 + (U,Mo)Fe_2 + \mu$;
4. grain U10Mo + ternary eutectic $(U,Mo)Fe_2 + \mu + UFe_6$;
5. UFe_6 + U10Mo.

Chromium rich layer is not formed due to the high temperature annealing. The thickness of the reaction layer in a diffusion couple is about 700 microns at a temperature of 750 °C for 5 hours of isothermal annealing.

IOP Conf. Series: Materials Science and Engineering **130** (2016) 012023 doi:10.1088/1757-899X/130/1/012023

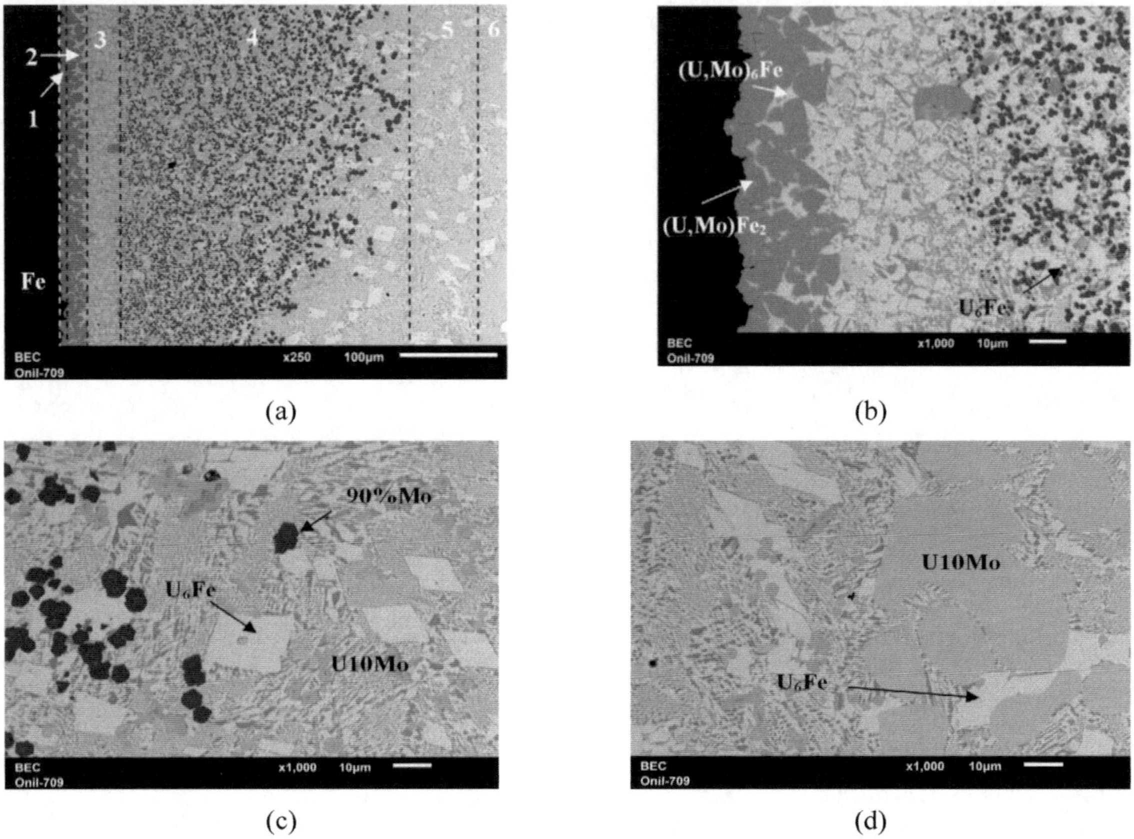

Figure 1. (a) The microstructure IL for diffusion couple U10Mo / Fe after annealing at 750 °C for 5 hours; (b) limit of Fe/IL; (c) the boundary between the areas 4 and 5; (d) between the boundary areas 5 and 6.

The depth of penetration of the fuel components in the shell of diffusion couple U10Mo/Fe13Cr is about 60 microns. Based on the concepts of a parabolic law of growth of the interaction layer ($X^2=Kt$, where X - thickness interaction, t-annealing time) for the first time defined the growth rate constant (K) for diffusion couple U10Mo/Fe13Cr, which is about $5.4{\cdot}10^{-12}$ m^2/s at 750 °C. Comparison of the results is impossible due to lack of data on high-temperature interaction U and UZr with FeCr.

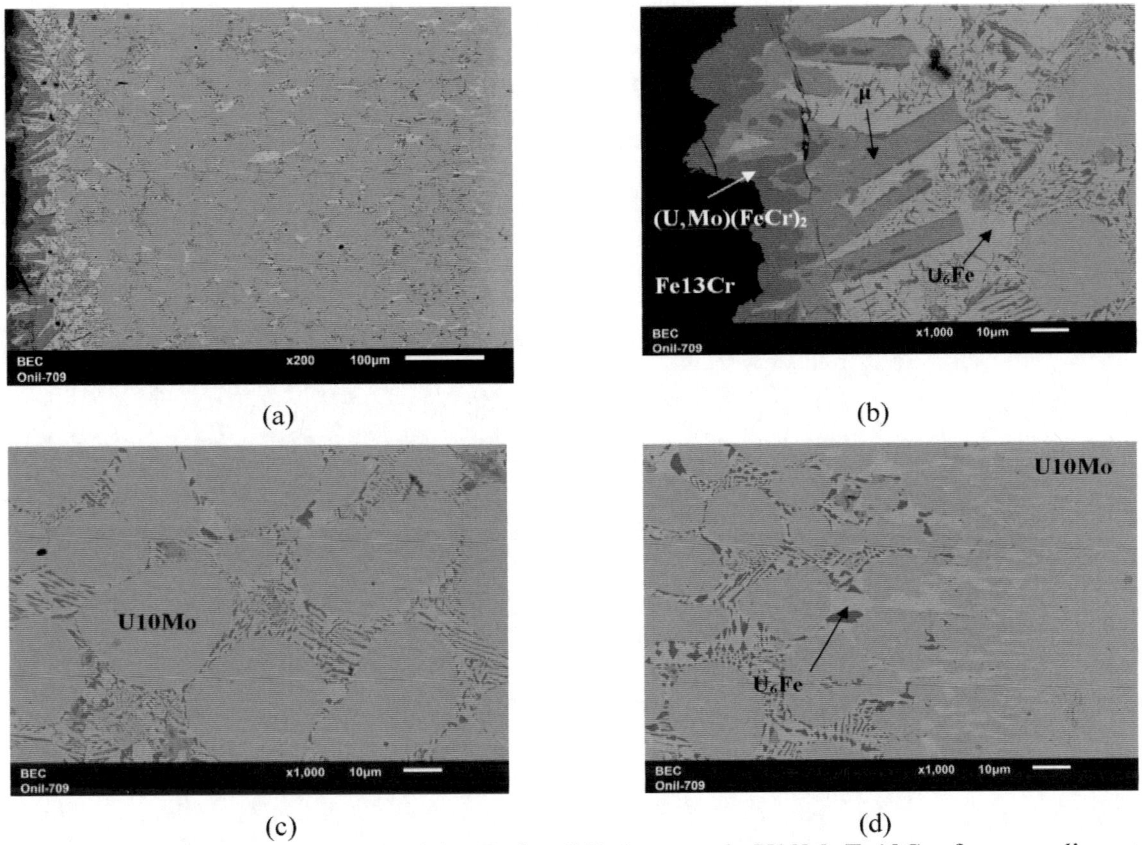

Figure 2. (a) The microstructure of the IL for diffusion couple U10Mo/Fe13Cr after annealing at 750 °C for 5 hours; (b) the border Fe13Cr/interaction layer; (c) the line between 3 and 4 areas; (d) the border area between 4 and alloy U10Mo.

4. Conclusions

In this work, research was conducted of the interaction of uranium-molybdenum alloy with iron and chromium steel at an elevated temperature of 750 °C for 5 hours. It is found that the rate constant of the interaction layer growth for diffusion couple UMo/FeCr is about $5.4 \cdot 10^{-12}$ m^2/s at 750 °C. The phase composition of the interaction layers for the both diffusion couples was determined. The interaction goes along the grain boundaries, as evidenced not interacts UMo alloy grain in the structure of the diffusion zone.

Acknowledgements

This work was performed within the framework of the Center of Nuclear Systems and Materials supported by MEPhI Academic Excellence Project (contract № 02.a03.21.0005, 27.08.2013).

References

[1] Otsuka S, Kaito T, Ukai S and Inoue M 2011 *J. Nucl. Mater.* **441** 286
[2] Nikitin S N, Shornikov D P, Tarasov B A and Baranov V G 2015 *Tsvetnye Metally* **3** 36 (in Russian)
[3] Hofman G L, Walters L C and Bauer T H 1997 *Progress in Nuclear Energy* **31** (1/2) 83
[4] Huang K, Park Y, Zhou L, Coffey K R, Sohn Y H, Sencer B H and Kennedy J R 2014 *J. Nucl. Mater.* **451** 372
[5] Cohen A B, Tsai H and Neimark L A 1993 *J. Nucl. Mater.* **204** 244
[6] Aitkaliyeva A, Madden J W, Miller B D, Papesch C A and Cole J I 2015 *J. Nucl. Mater.* **464** 28

[7] Jia Jian-ping, Wang Zhi-gang, Chen Miao, Wang Xi-sheng, Zhang Peng-cheng and WU Sheng 2013 *Atomic Energy Science and Technology* **47** (2) 295
[8] Kim Y S, Hofman G L and Yacout A M 2013 *J. Nucl. Mater.* **441** 520

Influence of a doping by Al stainless steel on kinetics and character of interaction with the metallic nuclear fuel

S N Nikitin, D P Shornikov, B A Tarasov and V G Baranov

National Research Nuclear University MEPhI (Moscow Engineering Physics Institute, Kashirskoye Highway 31, Moscow, 115409, Russia

E-mail: mephi200809@yandex.ru

Abstract. Metallic nuclear fuel is a perspective kind of fuel for fast reactors. In this paper we conducted a study of the interaction between uranium-molybdenum alloy and ferritic-martensitic steels with additions of aluminum at a temperature of 700 ° C for 25 hours. The rate constants of the interaction layer growth at 700 °C is about $2.8 \cdot 10^{-14}$ m^2/s. It is established that doping Al stainless steel leads to decrease in interaction with uranium-molybdenum alloys. The phase composition of the interaction layer is determined.

1. Introduction

Metallic nuclear fuel is a promising fuel for fast reactors because of its high thermal conductivity, and the ability to achieve high burnup. However, the use of this type of fuel is limited by its strong swelling and physico-chemical interaction fuel with the stainless steel , whereby the fuel element is out of order [1-4]. Currently, the main type of metal fuel is uranium-zirconium alloys. However, in the case of uranium-zirconium alloy there is a redistribution components of the alloy, migration of the uranium, zirconium, reducing the local temperature of melting [1]. A more reasonable way to improve interoperability and reduce the swelling of metallic fuel is uranium molybdenum alloying [5-7]. Molybdenum has a lower enthalpy of formation of binary mixtures with Fe (in absolute value) than Zr. This indicates that Mo is less active in the interaction with a cladding compared to Zr [6].

Also great interest are the high –chromium steel with the addition of aluminum, the main purpose of which is the confrontation corrosion in lead coolant [8-10], through the formation of interlayers of aluminum oxide. Corrosion tests of this kind of steel in lead showed good resistance. The mechanical properties of such alloys can improve by dispersion strengthening. Also noted that such steels inherent low swelling under irradiation damaging dose of 200 to dpa. In addition, tests of chromium steels alloyed with aluminum U-Zr fuels demonstrated the formation of a liquid phase at a higher temperature compared with undoped chromium steel [8]. Increase compatibility is due to the stabilization of α-Zr layer at the interface reaction layer, as well as a partial substitution of aluminum by iron lattice sites, which has a low diffusion mobility of the alloy in the fuel, thereby reducing interdiffusion. Also, introduction the aluminum contributes to the suppression the influence of fission products on the interaction [8] .Also alloying aluminum allows to obtain protective coatings by nitriding the inner surface of the cladding of fast reactor [3]. In this regard, interest testing aluminum-containing chromium steel with U-Mo fuel.

2. Experiment

Alloys U-10 wt.% Mo, Fe-13 wt.% Cr-4 wt.% Al (further designated U10Mo and Fe13Cr4Al) respectively prepared by melting of pure metal chips in arc furnace with non-consumable tungsten electrode in atmosphere of purified argon. To prepare the alloys used pure uranium (99.98), distilled molybdenum, iodine chromium, carbonyl iron and high purity Al (99.999). The batch melted several times to improve the homogeneity of the ingot. All alloys were annealed in vacuum at 1000 °C for one day for homogenizing. The obtained alloy was cut into tablets of 3 mm thickness at electric discharge machine, then subjected to grinding followed by polishing with diamond paste to 1 μm. Follows from the obtained "tablets" samples were taken for analysis of homogeneity and to determine the exact composition of the obtained alloy. To carry out the diffusion annealing, uranium alloy tablets and stainless steel were clamped together between the molybdenum discs, followed by annealing in a quartz ampoule. Before annealing to avoid possible oxidation laid getter of the chip zirconium in the ampoule. Then into the ampoule is lowered diffusion assembly and sealed ampoule, evacuated. Diffusion annealing was performed in a tube furnace, at the temperature 700 °C, the duration of excerpts of 25 hours. Upon completion of annealing time quartz ampoule was placed in a bath of cold running water. Further diffusion couples were cut perpendicular to the direction of diffusion in the electric spark machine, the use of which will allow to cut the diffusion couples without destroying the brittle interaction layer (IL). The next stage is to pour a few pairs of diffusion into the resin. The sample was subsequently passed metallographic preparation. The microstructure of the surface and the elemental composition of the samples was investigated by scanning electron microscope JEOL 6610LV using energy-dispersive spectrometer Oxford Instruments INCA x-ACT. Phase composition of interaction layer in the diffusion couple was determined by quantitative relations (U+Mo)/(Fe+Cr) in the phase.

3. Results

Figure 1 (a-f) shows the microstructure of the interaction layer in the diffusion couple U10Mo/Fe13Cr4Al after annealing at 700 °C for 25 hours. From the analysis of the microstructure can be seen that the eutectic UFe_2 + U_6Fe offline, and the thickness of the interaction region (X) is 50 microns. Proceeding from concepts of a parabolic law of growth of the interaction layer ($X^2=K{\cdot}t$, where t-time annealing) first determined the growth rate constant of the interaction layer $K=2.8{\cdot}10^{-14}$ m^2/s. Based on the distribution of elements over the width of the interaction layer in the diffusion couple U10Mo/Fe13Cr4Al (figures 1b-f, and 2) it can be concluded that the whole interaction layer represents the layered structure comprising four areas:

1. Enriched chromium and aluminum layer Cr_2Al;
2. $(U,Mo)(Fe,Cr,Al)_2$;
3. Enriched molybdenum layer;
4. α-U+ U10Mo.

Areas enriched chromium and molybdenum with aluminum, is only about 10 microns and are the diffusion barriers, preventing further interaction the fuel with cladding material. Rich molybdenum layer is composed of one-third of uranium. As can be seen from the microstructure of the diffusion zone (figure 1 a), at the boundary layer interaction/U10Mo located two phase region. However, EDX analysis of this area did not reveal it in the presence of components of the material cladding. From maps of the distribution of molybdenum to the width of the interaction layer (figure 1 e) shows that in this area are observed phase with a low content of molybdenum in some places.

Molybdenum-rich layer is formed due to lower solubility of molybdenum in the γ-uranium with increasing chromium content. Also due to the low mobility of the chromium alloy is formed in the fuel rich chromium and aluminum layer, which is a Cr_2Al with some solubility of iron and uranium.

Comparison of the kinetics and the nature of the interaction FeCrAl with uranium-zirconium alloy shows the advantage of molybdenum-uranium alloys. However, co-doping with aluminum and chromium does not solve the problem of the interaction of the fuel with the cladding at a high

IOP Conf. Series: Materials Science and Engineering **130** (2016) 012024 doi:10.1088/1757-899X/130/1/012024

temperature at which a liquid phase and diffusion processes take place in a few hours. Therefore, by increasing the appropriate compatibility at high temperatures is a protective coating on the inner surface of the cladding.

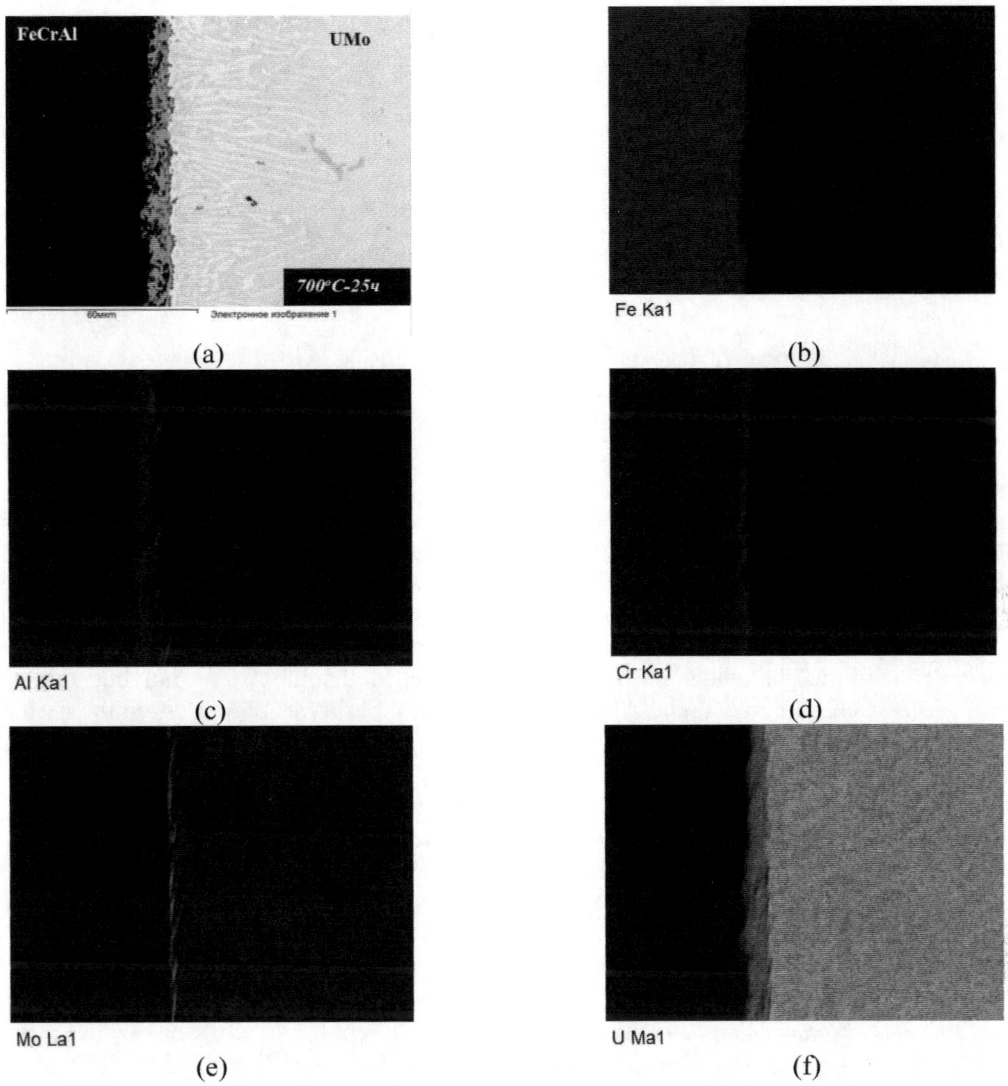

Figure 1. The microstructure of the interaction layer in the diffusion couple U10Mo/Fe13Cr4Al after annealing at 700 °C for 25 hours (a); map the distribution of elements across the width of the interaction layer for iron (b), aluminum (c), chromium (d), molybdenum (e), uranium (f).

One of the promising methods of coating is a chemical-thermal treatment of the inner surface of the cladding with the saturation of the gas phase with oxygen and nitrogen. [3].

As a result, the nitriding process on the steel surface is formed layer of small-disperse intermetallic compounds, whose thickness is about 25 microns. Annealing at 850 °C for 5 hours showed absence of interaction between the metal fuel and the cladding fast reactor.

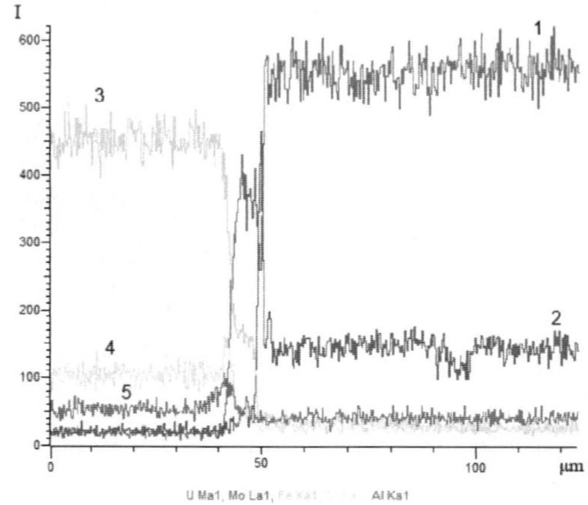

Figure 2. The distribution of elements across the width of the interaction layer: 1-U Ma1, 2-Mo La1, 3-Fe Ka1, 4-Cr Ka1, 5-Al Ka1.

4. Conclusions

In this paper, conducted a study peculiarities of diffusion interaction between uranium-molybdenum alloy and ferritic-martensitic steels with additions of aluminum at a temperature of 700 °C for 25 hours. Determined the rate constants of the interaction layer growth at 700 °C, constitutes $2.8 \cdot 10^{-14}$ m^2/s. It is established that the stainless steel with aluminum doping leads to lower interaction with uranium-molybdenum alloys. It is shown that the interaction UMo/FeCrAl at 700 °C for 25 hours in the diffusion zone formed molybdenum rich layer, due to reduced solubility of molybdenum in the γ-uranium at increase chromium content. Also, due to the low mobility of chromium in the fuel alloy, a thin intermetallic layer Cr_2Al formed.

Acknowledgements

This work was performed within the framework of the Center of Nuclear Systems and Materials supported by MEPhI Academic Excellence Project (contract № 02.a03.21.0005, 27.08.2013).

References

[1] Hofman G L, Walters L C and Bauer T H 1997 *Progress in Nuclear Energy* **31** (1/2) 83
[2] Huang K, Park Y, Zhou L, Coffey K R, Sohn Y H, Sencer B H and Kennedy J R 2014 *Journal of Nuclear Materials* **451** 372
[3] Nikitin S N, Shornikov D P, Tarasov B A and Baranov V G 2015 *Tsvetnye Metally* **3** 36 (in Russian)
[4] Cohen A B, Tsai H and Neimark L A 1993 *Journal of Nuclear Materials* **204** 244
[5] Aitkaliyeva A, Madden J W, Miller B D and Papesch C A, Cole J I 2015 *Journal of Nuclear Materials* **464** 28
[6] Kim Y S, Hofman G L and Yacout A M 2013 *Journal of Nuclear Materials* **441** 520
[7] Jia Jian-ping, Wang Zhi-gang, Chen Miao, Wang Xi-sheng, Zhang Peng-cheng, WU Sheng 2013 *Atomic Energy Science and Technology* **47** (2) 295
[8] Otsuka S, Kaito T, Ukai S and Inoue M 2011 *Journal of Nuclear Materials* **441** 286
[9] Kimura A, Kasada R, Iwata N, Kishimoto H, Zhang C H, Isselin J, Dou P, Lee J H, Muthukumar N, Okuda T, Inoue M, Ukai S, Ohnuki S, Fujisawa T and Abe T F 2011 *Journal of Nuclear Materials* **417** 176
[10] Moriyasu Tokiwai, Masaaki Horie and Kenji Kako, Masayuki Fujiwara 1993 *Journal of Nuclear Materials* **204** 56

Thermal properties of nonstoichiometry uranium dioxide

R Kavazauri, S A Pokrovskiy, V G Baranov and A V Tenishev

National Research Nuclear University MEPhI (Moscow Engineering Physics Institute)
31 Kashirskoe highway, Moscow, 115409, Russia

E-mail: Onil709@inbox.ru

Abstract. In this paper, was developed a method of oxidation pure uranium dioxide to a predetermined deviation from the stoichiometry. Oxidation was carried out using the thermogravimetric method on NETZSCH STA 409 CD with a solid electrolyte galvanic cell for controlling the oxygen potential of the environment. 4 samples uranium oxide were obtained with a different ratio of oxygen-to-metal: O / U = 2.002, O / U = 2.005, O / U = 2.015, O / U = 2.033. For the obtained samples were determined basic thermal characteristics of the heat capacity, thermal diffusivity, thermal conductivity. The error of heat capacity determination is equal to 5%. Thermal diffusivity and thermal conductivity of the samples decreased with increasing deviation from stoichiometry. For the sample with O / M = 2.033, difference of both values with those of stoichiometric uranium dioxide is close to 50%.

1. Introduction

The ratio of oxygen-to-metal (O / U) is the most important parameter of the oxide fuel, which determines the behavior of the fuel in the reactor core. Deviation from the stoichiometric composition (x = O / U - 2) influences on the basic properties of the fuel such as thermal diffusivity, thermal conductivity, ductility, melting point, diffusion coefficient, chemical interaction with the cladding and, as a consequence, the final fuel burning. Thus, the existence of preparation procedure of oxide fuel samples with different deviation from stoichiometry is important for the development and improvement of modern fuel nuclear power reactors.

The standard tablet without central hole was sintered in Ar-atmosphere with 8 % H_2 at 1600 °C for 8 hours. After sintering, the tablet was cut into several samples with the form of discs. Characteristics of the resulting samples are shown in table 1. The density of the samples was measured by hydrostatic method.

Table 1. Density of the samples.

№ sample	Density of sample (g/cm^3)	Mass (g)	Thickness (mm)	Diameter (mm)
1	10.346	0.97854	1.7332	8.411
2	10.339	0.97991	1.7372	8.418
3	10.340	0.98064	1.7378	8.412

IOP Conf. Series: Materials Science and Engineering **130** (2016) 012025 doi:10.1088/1757-899X/130/1/012025

2. Preparation of samples with given deviation from stoichiometry

Changing the oxygen-to-metal ratio of uranium dioxide samples was carried out on the NETZSCH STA 409 CD by thermogravimetry method. To control and change the oxygen potential of the gas flowing through the plant was used system, consisting of the oxygen sensor and the oxygen pump. The sensor measures the oxygen potential of the gas blown through the sample, and the pump is able to pump in or pump out oxygen from purging gas.

After sample preparation, was determined the initial deviation from the stoichiometry by solid electrolyte galvanic cell method (O / M = 2.002).

Using expressions (1) and (2) was calculated final weight of the sample that needs to be obtained by the sample oxidation to obtain the desired O / U [1].

$$\frac{O}{U}=\frac{(100-U_S-Z-m)A}{15.999\cdot U_S}, \quad (1)$$

where U - the proportion of uranium in the sample, at. %; O - the proportion of oxygen in the sample, at. %; U_S - the proportion of uranium in the sample, wt. %; Z - total numbers of non-volatile impurities, wt%; m - moisture content and volatile impurities, wt. %; A - average atomic weight of uranium isotope mixture; 15,999 - the atomic weight of oxygen.

The atomic weight of isotope mixture calculated as:

$$A=\sum F_i A_i, \quad (2)$$

where F_i - mass fraction of the i-th isotope of uranium, A_i - the atomic weight of the i-th isotope of uranium.

The sample was placed on the Al_2O_3 sample holder and heated to 1300 ° C in a stream of helium. Then oxygen pump begins to add oxygen into helium to increase his oxygen potential until the sample mass comes up to the previously calculated. After cooling, the weight of the obtained sample were refined on the scales with an accuracy ± 4 mg and using the expressions (1) and (2) sample deviation from stoichiometry was received. A typical thermogravimetric curve of sample oxidation is shown in figure 1.

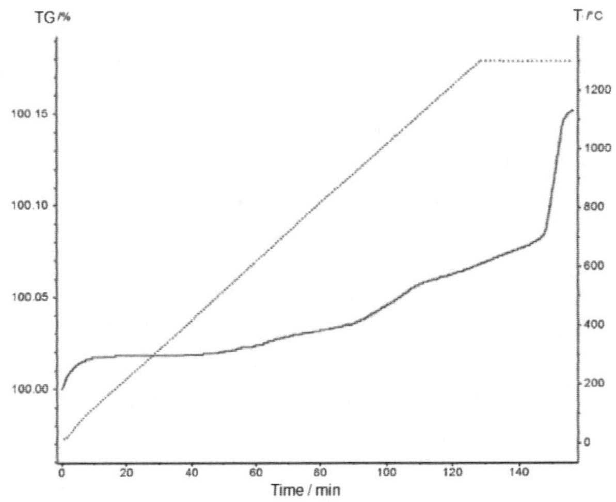

Figure 1. A typical thermogravimetric curve of the uranium dioxide sample oxidation.

Thus, following samples were obtained: O / U = 2.005, O / U = 2.015, O / U = 2.033.

3. Determination of the thermal diffusivity and thermal conductivity of the obtained samples

Measurements of the specific heat and thermal diffusivity were held at Netzsch LFA 427 by laser flash method in a vacuum.

The temperature dependence of heat capacity is shown in figure 2.

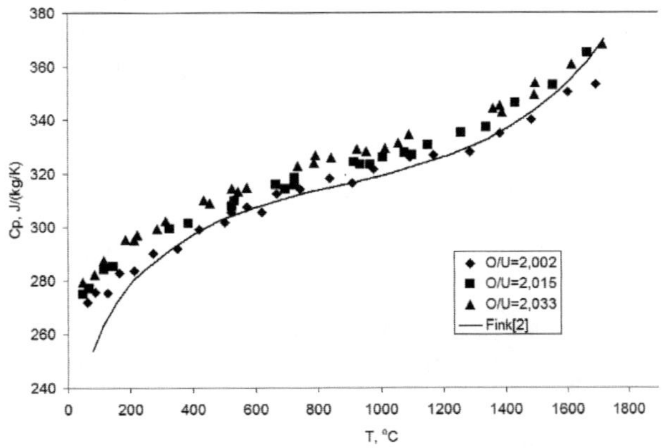

Figure 2. The heat capacity of UO_2 samples.

The heat capacity of the samples increases with temperature from 270 up to 370 J/(g·K) and does not differ more than 5% from each other. The heat capacity of the sample with O / M = 2.002 in good agreement with the results presented in the paper [2].

From obtained data of thermal diffusivity, specific heat and density using the expression (1) were calculated temperature dependence of the thermal conductivity.

$$\lambda(t) = a(t) \cdot C_p(t) \cdot \rho(t) \cdot \tag{3}$$

The obtained data are recalculated to 95% TD by the expression

$$\rho_p = \rho_{100}(1-p)^{2,5} \text{ [3].} \tag{4}$$

Thus calculated temperature dependences of the uranium dioxide thermal conductivity are shown in figure 3.

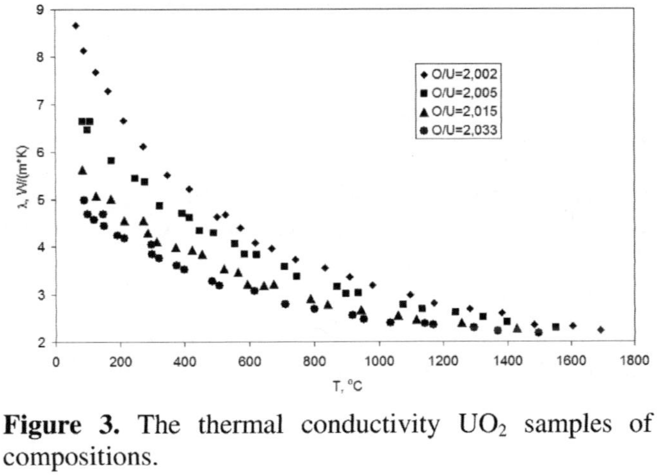

Figure 3. The thermal conductivity UO_2 samples of all compositions.

As is well known the thermal conductivity of ceramic materials above room temperature is determined by the phonon component. In Clemens model [4] it is assumed that the thermal conductivity of the material depends of phonons by phonons and by lattice defects scattering. In hyperstoicheometric uranium dioxide oxygen ions are located in the interstices of the anion sublattice and stand as phonon scattering centers. In addition to each embedded oxygen atom two uranium ions should change their valence from U +4 to U +5, the atomic radius of the ion U +5 (r = 88.0 pm) smaller than U +4 radius (r = 101.1 pm), thus pentavalent uranium ions are also centers of phonon scattering.

Calculated from the obtained thermal conductivity values, thermal resistance values for all samples are shown in figure 4. It can be seen that the curves are linear, but with increasing of stoichiometry the temperature for which linearity is preserved decreases, such as for O / U = 2.033 linearity is preserved only to 1200 °C.

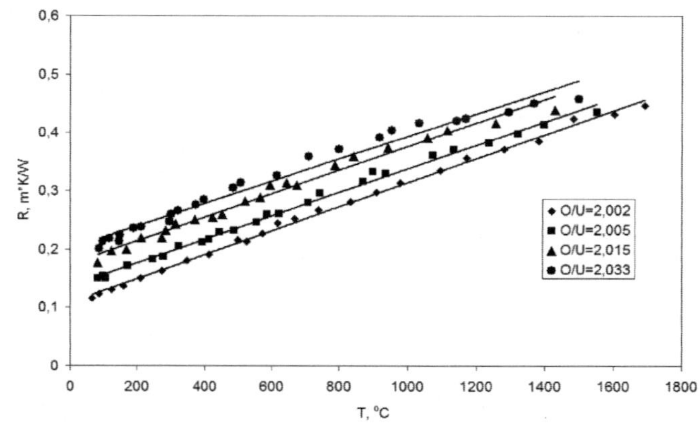

Figure 4. Thermal resistance UO_2 samples of all compositions.

The linearity of the thermal resistance means that the main component of the thermal conductivity is a phonon, and it can be described by the expression (5).

$$R = \frac{1}{\lambda} = A + BT \tag{5}$$

The coefficients A and B, determined by least squares method are reported in table 2.

Table 2. Coefficients in thermal resistance equation.

	$UO_{2.002}$	$UO_{2.005}$	$UO_{2.0015}$	$UO_{2.0033}$
A, 10^{-2}, m·K/W	4.626±0.014	7.311±0.022	10.697±0.032	13.048±0.039
B, 10^{-4}, m/W	2.121±0.006	2.134±0.006	2.067±0.006	1.998±0.006

The experimental data was processing to obtain the following expression for the thermal conductivity of uranium dioxide, depending of the stoichiometry and temperature (6).

$$\lambda_{95}(T,x) = \frac{1}{A(x)+B(x)T} + \lambda_{electron} =$$
$$= \frac{1}{(4,627+0,55\cdot x^{1/2})\cdot 10^{-2} + (2,120-3,3\cdot x)\cdot 10^{-4}\cdot T} + \frac{6400\cdot 1000^{5/2}}{T^{5/2}}\exp(-\frac{16350}{T}) \tag{6}$$

4. Conclusions

In this paper, was developed a method of oxidation pure uranium dioxide to a predetermined value O / U by thermogravimetric method on NETZSCH STA 409 with a solid galvanic cell for controlling the oxygen potential of purging gases. For samples obtained after oxidation were determined temperature dependence of the specific heat and thermal diffusivity by laser flash method. Then thermal conductivities were counted for each deviation from stoichiometry.

Thermal conductivity is strongly dependent on the deviation from the stoichiometry because of the formation of interstitial oxygen ions and pentavalent uranium ions that are centers of phonon scattering. From obtained dependences of thermal conductivity were calculated the thermal resistance of uranium dioxide samples of all compositions. Processing of these curves yielded the expression (6) to calculate the thermal conductivity in dependence of temperature and deviations from stoichiometry.

Acknowledgements
This work was performed within the framework of the Center of Nuclear Systems and Materials supported by MEPhI Academic Excellence Project (contract № 02.a03.21.0005, 27.08.2013).

References

[1] 2011 *ASTM Designation* C1453-00 ASTM International
[2] Fink J K 2000 *J. Nucl. Mater* **279** 1
[3] Lassmann K 2001 *ITU Activity Report* EUR 20252 16
[4] Klemens P G 1960 *Phys. Rev* **119** 507

IOP Conf. Series: Materials Science and Engineering **130** (2016) 012026 doi:10.1088/1757-899X/130/1/012026

Thermal properties of $(Al,Gd)O_3$ doped uranium dioxide

S A Pokrovskiy, V G Baranov and A V Tenishev

National Research Nuclear University MEPhI (Moscow Engineering Physics Institute) 31 Kashirskoe highway, Moscow, 115409, Russia

E-mail: Onil709@inbox.ru

Abstract. In this paper were defined thermal characteristics (thermal diffusivity up to 2000 °C, heat capacity up to 1250 ° C and thermal conductivity of up to 1250 °C) of oxide nuclear fuel with additives of burnable neutron absorber in the form of $(Al,Gd)O_3$. Also for comparison samples UO_2 with gadolinium Gd_2O_3 and pure uranium dioxide were studied. It is shown that the thermal diffusivity and thermal conductivity of UO_2 with dopant of gadolinium in the form of (Al, Gd) O_3 is considerably higher than those of Gd_2O_3 –doped uranium dioxide, due to the fact that (Al, Gd)O_3 does not form a solid solution with UO_2.

1. Introduction

In order to achieve high burnup of WWER fuel requires using burnable neutron absorbers (Gd for WWER fuel), to reduce the reactivity in the initial stages of TVEL action. But the injection to the fuel gadolinium in the form of Gd_2O_3 leads to a significant reduction of uranium dioxide thermal conductivity [1,2,3], and as a consequence an increase of the temperature gradient between fuel and cladding. To solve this problem were offer to inject in uranium dioxide gadolinium as a perovskite phase $AlGdO_3$, which improves the thermal conductivity of the fuel without sacrificing Gd amount in it [4].

To carry out the work following uranium dioxide samples were made: R - pure uranium dioxide, G - uranium dioxide with 6.9% Gd_2O_3 and AG - uranium dioxide with 8.9% $AlGdO_3$. The density of the samples obtained are given in table 1.

Table 1. Density of the sample investigated batches

Batch	Theoretical density (TD) (g/cm^3)	Density ρ (g/cm^3)	% of TD
R	10.96	10.56	96
G	10.80	9.78	91
AG	10.41	9.62	92

2. Determination of heat capacity

The heat capacity in the temperature range 35-1215 °C determined by differential scanning calorimetry method in accordance with standards DIN 51007, ASTM E 1269 in an atmosphere of high purity argon. Specimens have a diameter of 6 mm and a thickness of 1.5-1.6 mm. The heat capacity measurements were performed on the synchronous thermal analyzer Netzsch STA 409 CD. As a reference were used disk of sapphire 6 mm in diameter and 1 mm thick. Relative error of the method is less than 5%. The results are shown in figure 1.

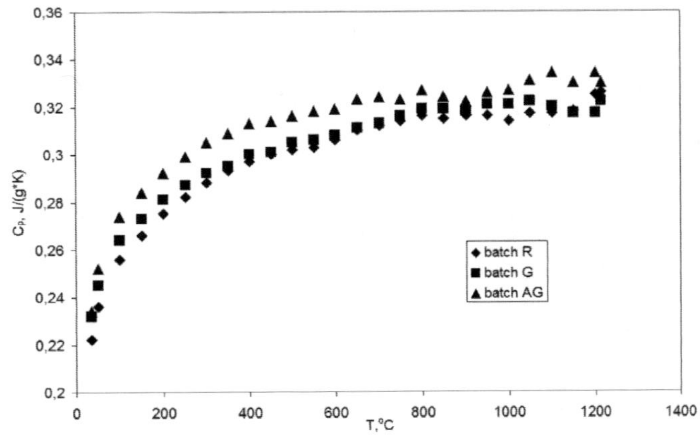

Figure 1. The heat capacity of UO_2 samples of all batches.

It should be noted that because of the small amounts of additives difference of the specific heat values of samples from different batches is small and does not exceed the measurement error.

3. Determination of thermal diffusivity

Measurements were carried out on Netzsch LFA 427 by laser flash method in a high purity argon atmosphere. The samples had form of discs 10 mm in diameter, 2 mm thick.

Uranium dioxide thermal diffusivity values of R, G and AG batches shown in figure 2.

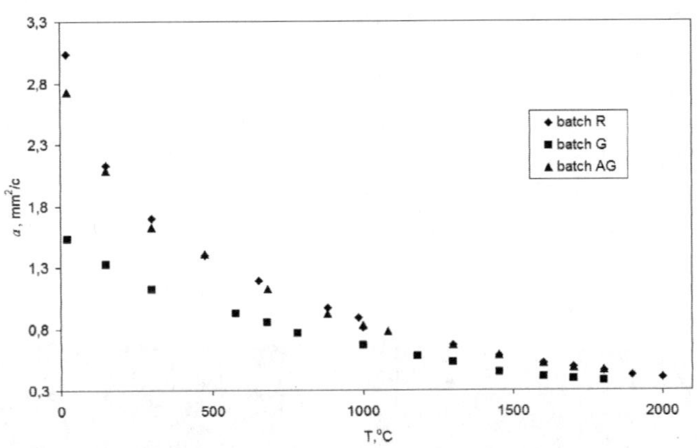

Figure 2. Thermal diffusivity UO_2 samples of all batches.

From these results, it's clear that thermal diffusivity of the samples decreases from 3 to 0.3 mm^2/s while temperature increasing. The thermal diffusivity of the samples with gadolinium additives is lower than that of pure uranium dioxide, but it is clear that a large influence exert form of burnable poison (Gd) is injected into the fuel. In samples of batch G gadolinium injected as Gd_2O_3, and during sintering it dissolves in the matrix and forms a solid solution with UO_2, which as known negatively affect on the thermal conductivity. In the samples of batch AG gadolinium is in the form of a perovskite phase $AlGdO_3$, and does not introduce distortion into UO_2 lattice, resulting in only a slight reduction in thermal conductivity compared to pure uranium dioxide in figure 2.

4. The calculation of the thermal conductivity

From the obtained data of thermal diffusivity, specific heat and density using the expression (1) were calculated temperature dependence of the thermal conductivity.

$$\lambda(t) = a(t) \cdot C_p(t) \cdot \rho(t). \quad (1)$$

Obtained data are recalculated to 95% TD by the expression

$$\rho_p = \rho_{100}(1-p)^{2,5}. \quad (2)$$

The results are shown in figure 3.

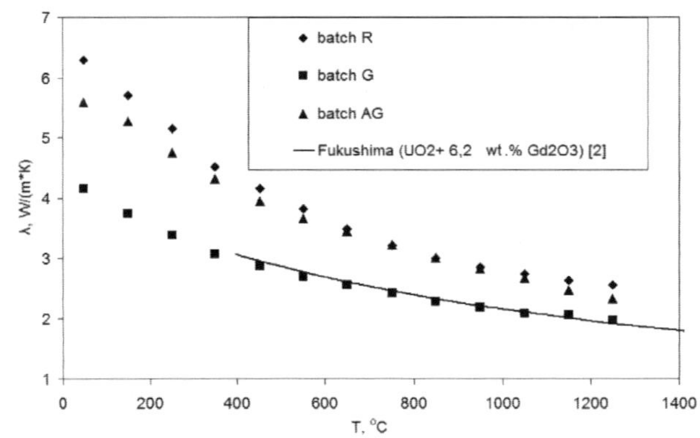

Figure 3. The temperature dependence of the thermal conductivity of samples UO_2.

The thermal conductivity of the samples is in the range 2 - 6.5 W/(m·K). Also we can see that dopants of $AlGdO_3$ much weaker affect on the thermal conductivity UO_2 than Gd_2O_3 one.

5. SEM-study UO_2 samples

The state of gadolinium in the samples G and AG confirmed by SEM (figures 4, 5 and tables 2, 3).

In Figure 4 - the microstructure of the sample AG in secondary electrons, UO_2 looks light gray, $AlGdO_3$ - dark gray. From the distribution map Gd, and gained spectra shows that all gadolinium is in $AlGdO_3$ form. From Figure 5, which shows the microstructure in the secondary electron and gadolinium distribution for sample batch G seen that gadolinium is evenly distributed throughout the matrix UO_2.

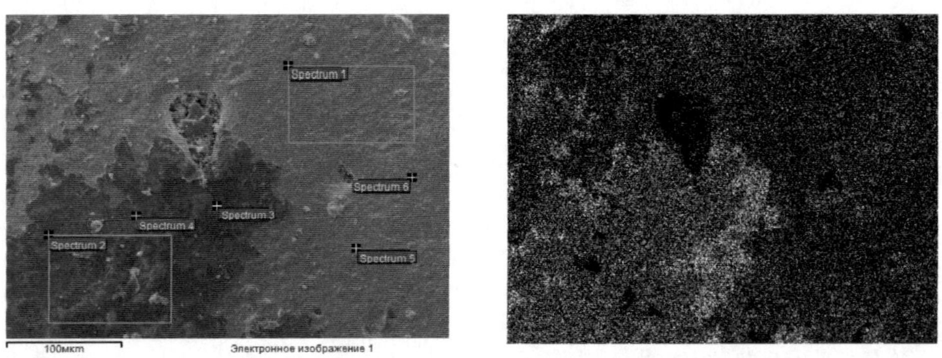

Figure 4. SEM image and distribution map of gadolinium in the sample batch AG.

Table 2. Composition data of the sample batch AG (wt.%)

	O	Al	Si	Gd	U
Spectrum 1	12.42		0.47		87.10
Spectrum 2	32.29	31.82	2.28	12.31	21.30
Spectrum 3	31.27	35.16		12.92	20.65
Spectrum 4	33.86	27.83		8.58	29.72
Spectrum 5	12.73				87.27
Spectrum 6	12.53				87.47

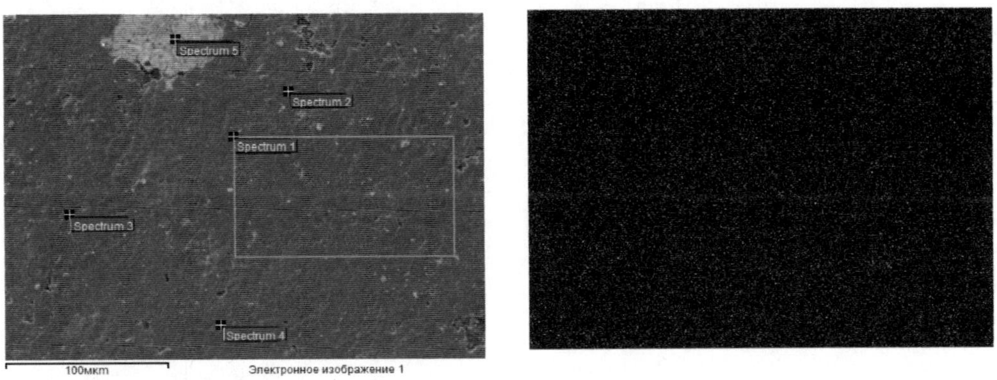

Figure 5. SEM image and distribution map of gadolinium n the sample batch G.

Table 3. Composition data of the sample batch G (wt.%)

	O	Al	Si	Gd	U
Spectrum 1	11.76			4.77	83.48
Spectrum 2	16.94	4.71		6.86	71.49
Spectrum 3	6.49	1.35	0.51	4.90	86.75
Spectrum 4	9.16			5.16	85.69
Spectrum 5	27.30			4.39	68.30

6.Conclusions

In the present work measured basic thermophysical characteristics oxide fuel with additives burnable neutron absorber in the form of a perovskite phase (Al, Gd) O_3 and in the form of a solid solution UO_2 - Gd_2O_3. State of gadolinium in the samples is confirmed by SEM-studies (figures 5, 6).

It is shown that the thermal conductivity of the samples decreases with increasing temperature and is in the range of 2-6.5 W / (m·K), moreover serious influence exert gadolinium state in UO_2: formation of a solid solution UO_2 - Gd_2O_3 causes distortion of UO_2 crystal lattice and as a result, a strong decrease (35%) of thermal conductivity. Gadolinium it the form of perovskite phase does not make distortion into the lattice and thermal conductivity decreased less than 5% in the entire investigated temperature range.

Acknowledgements

This work was performed within the framework of the Center of Nuclear Systems and Materials supported by MEPhI Academic Excellence Project (contract № 02.a03.21.0005, 27.08.2013).

References

[1] Baranov V G, Tenishev A V, Mikheev E N and Pokrovskiy S A 2011 *Atomic energy* **110** 36

[2] Fukushima S, Ohmichi T, Maeda A and Watanabe H *J. Nucl. Mater* **105** 201
[3] Masaki A, Mutsumi H, Toshio K and Yoshiaki K 1996 *J. Nucl. Mater* **231** 29
[4] Kim H S, Joung C Y, Lee B H, Kim S H and Sohn D S 2008 *J. Nucl. Mater* **372** 340

Manufacturing and STA-investigation of witness-samples for the temperature monitoring of structural steels under irradiation

O N Sevryukov, V T Fedotov, A A Polyansky, S A Pokrovski and R S Kuzmin

National Research Nuclear University MEPhI (Moscow Engineering Physics Institute)
31 Kashirskoe highway, 115409 Moscow, Russia

E-mail: sevr54@mail.ru

Abstract. The object of investigations was alloys based on lead and cadmium used as fuse monitors to control the maximum irradiation temperature (fuse temperature monitors, FTM) of samples from structural steels under irradiation in a research reactor IR-8. The result of the work was selected and tested initial materials for production of alloys. A technological scheme of the production of alloys for FTM has been developed and experimental studies of the properties of these alloys have been carried out.

1. Introduction

To develop the technologies of increasing the service life of heat resistant steels, operated for a long time under extreme conditions as a part of critical constructions on the basis of an improved complex technique of certification of the service properties conditioned by the formation and transformation of the nanostructure during irradiation, it was necessary to develop compositions and the manufacturing technology of alloys for FTM in the form of rods with a diameter of about 3 mm and a melting temperature changing in the 298–314 °C range[1,2]. A technological scheme providing the production of rods with a specified chemical composition was proposed and the technique for measuring melting temperatures, which allows certifying FTM-alloys with a required accuracy, was improved.

2. Development of a method to produce alloys with a specified composition

The technological scheme of producing alloys is shown in figure 1. A vertical furnace with resistive heating was used as the basic unit of the equipment. The protection of the melt from oxidation is provided by an argon atmosphere and the melting temperature is controlled by a thermocouple that is connected with a power regulator (not shown in the scheme).

The filling rate of the quartz feeder is regulated by a rotameter, which allows choosing the optimum speed of solidification of the melt. The operability of the technological scheme at the stage of development was confirmed by obtaining samples of rods from a lead-indium alloy with a specified composition Pb–(8–10 wt. %)In

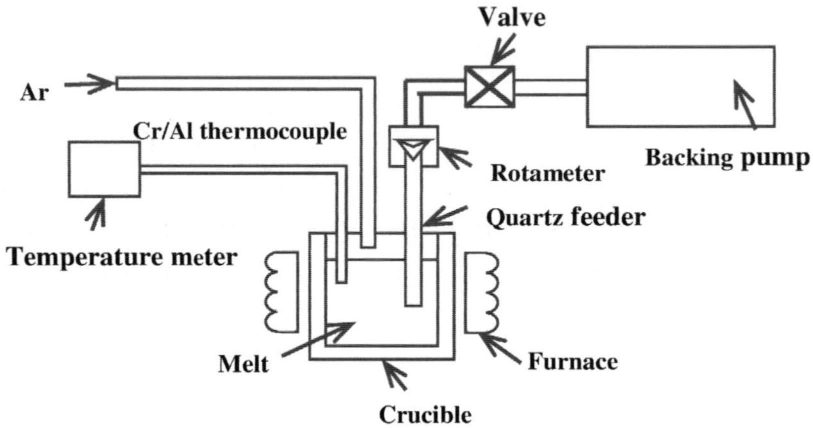

Figure 1. The technological scheme for production of alloys with a melting temperature of 250–350 °C.

The geometric and thermo physical characteristics of the alloy obtained correspond to specified requirements. Lead of analytical reagent grade (ARG), cadmium (ARG), indium of the In0 grade, and tin of very high purity (VHP) were used as initial materials to produce a pilot batch of FTM-alloys [3]. To avoid contamination of the alloys by lead oxides, the melt of lead was preliminarily purified in an argon atmosphere. Alloys of six compositions: No.1 - Cd-(5–10)In; No.2 - Pb-(5–8.5)In; No.3 - Pb-(8–10)In; No.4 - Pb-(2–5)In; No.5 - Pb-(0.5–1.5)Sn and No.6 - Pb-(1–2)In, were made.

3. Measuring the melting temperatures and determining the chemical composition of experimental alloys.

To determine the melting temperature, an installation of the synchronous thermal analysis STA 409 CD of the Firm "Netzsch" (Germany) was used. To measure melting temperatures, a furnace with a Pt-Rh heater and a sample holder TG / DSK with a thermocouple of the type S (Pt–Pt-10%Rh) were used. The heating rate was 10 K/min. Tests were conducted in a dynamic helium atmosphere at a flow rate of 70 ml/min. Samples were placed in an Al_2O_3 crucible with "pierced" lids, and an empty crucible with a lid was used as the "standard".

To determine the influence of various factors (the contact with the crucible and the mass of samples) on the results of the experiment, preliminary measurements on samples of the alloy Pb-5%In were carried out. Three discs of various thicknesses were cut from the initial cylindrical rod of the alloy. Masses of the samples were 21.78, 52.18 and 101.50 mg. The melting of each sample was performed three times.

The results obtained show that the melting temperature shifts towards the area of lower values with increasing the sample mass. Additionally, for samples with large masses after the first melting, there is apparently a change of the contact with the crucible and the temperature of the extrapolated start for repeated meltings shifts towards lower values. Thus, to increase the measurement accuracy, it is necessary to preliminarily prepare (polish) the sample surface to ensure a good contact with the bottom of the crucible. It is also necessary to use samples with similar masses and perform a temperature calibration of the installation using standards with similar geometry and mass.

Determination of the melting temperatures was performed using alloys of six compositions. To preliminarily estimate the stability of the compositions obtained, several rods of each alloy were randomly selected for each composition. Further, samples for tests were produced from an arbitrary place of the rod.

For example, the alloy No. 6 shows a good reproducibility of the melting temperatures of samples of the same composition. Thermograms of the tests of five samples are presented in figure 2.

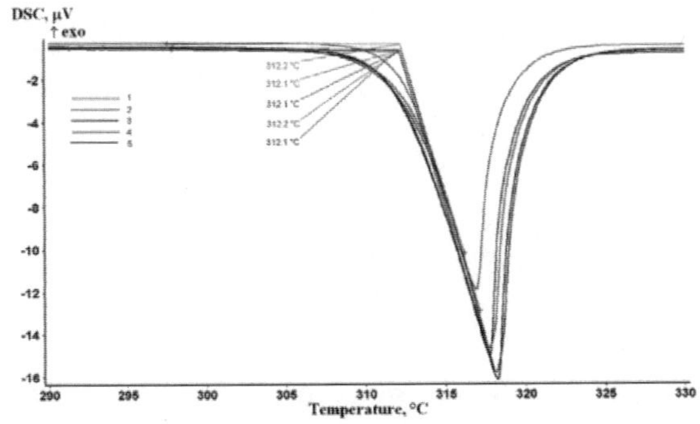

Figure 2. Thermograms of samples of the alloy No. 6 with masses of: 1, 44.71 mg; 2, 63.50 mg; 3, 63.79 mg; 4, 76.60 mg; 5, 78.22 mg.

Based on these results, the stability of the rest of alloys was evaluated by three samples. The test results of alloys No. 2, 3, 4 and 5 are shown in figures 3-6, respectively.

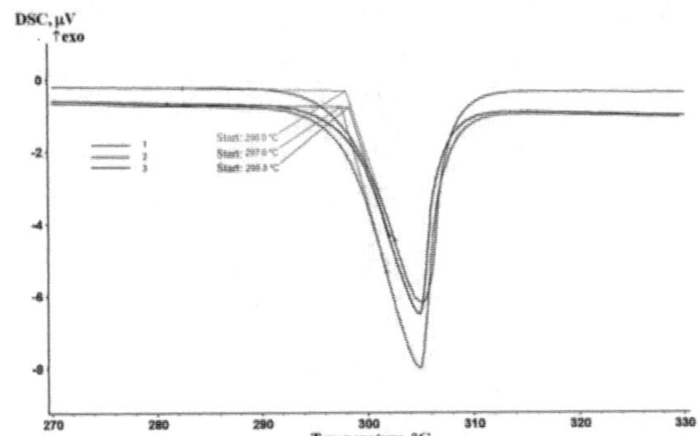

Figure 3. Thermograms of samples of the alloy No. 2 with masses of: 1, 36.78 mg; 2, 40.51 mg; 3, 30.80 mg.

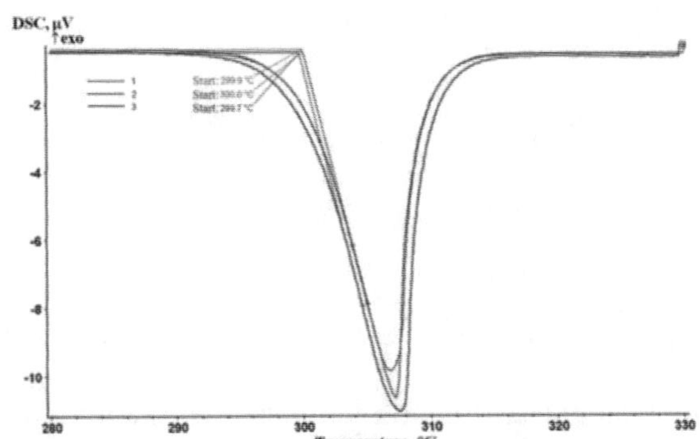

Figure 4. Thermograms of samples of the alloy No. 3 with masses of: 1, 53.85 mg; 2, 58.11 mg; 3, 65.58 mg.

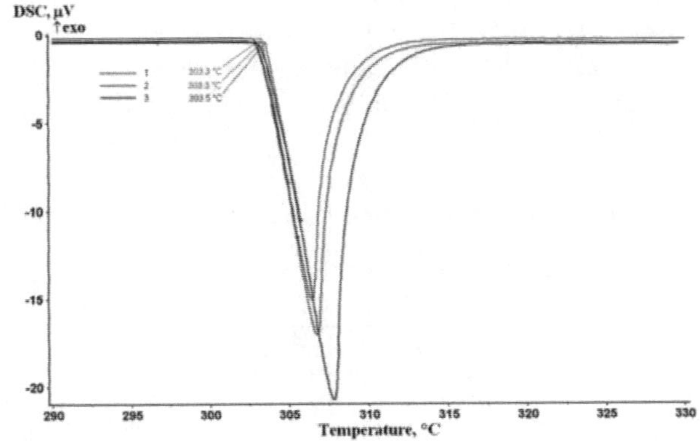

Figure 5. Thermograms of samples of the alloy No. 4 with masses of: 1, 34.32 mg; 2, 41.75 mg; 3, 58.28 mg.

IOP Conf. Series: Materials Science and Engineering **130** (2016) 012027 doi:10.1088/1757-899X/130/1/012027

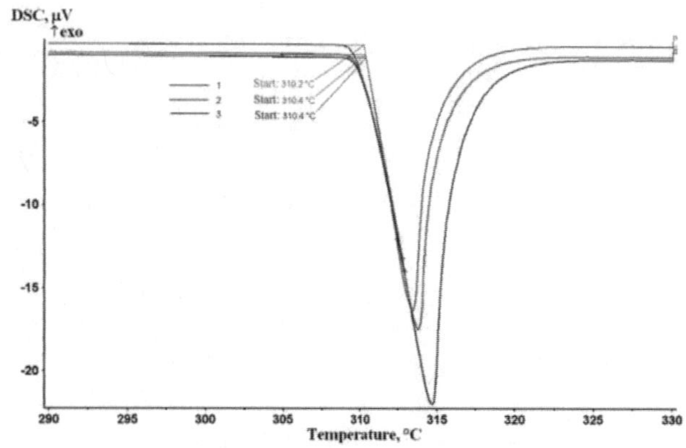

Figure 6. Thermograms of samples of the alloy No. 5 with masses of: 1, 42.33 mg; 2, 45.79 mg; 3, 66.39 mg.

These results testify the stability of the composition and the manufacturing technology of alloys. The average difference of the melting temperatures for various samples of alloys of the same composition does not exceed 0.3 K, except for the alloy No. 2, where it reaches 0.7 K.

Special attention is given to the alloy No. 1. The results of its differential thermal analysis are presented in figure 7.

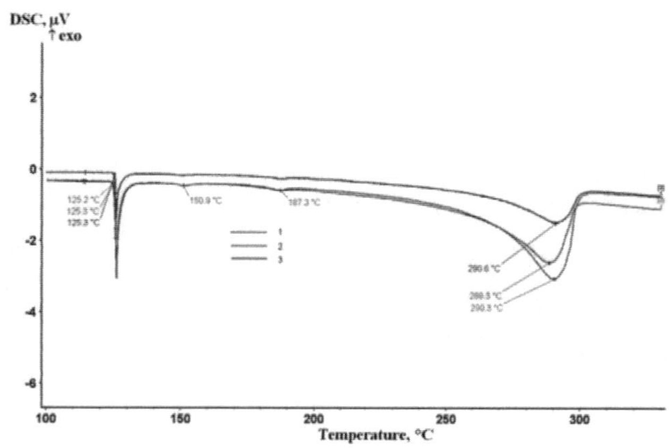

Figure 7. Thermograms of samples of the alloy No. 1 with masses of: 1, 37.81 mg; 2, 28.97 mg; 3, 13.84 mg.

As shown in the thermograms (figure 7), the alloy No. 1 undergoes several transformations at temperatures of 125, 160 and 187 °C that terminate at the temperature of 290 °C at which, presumably, the whole sample transfers into the liquid phase.

In this case it is difficult to accurately determine the complete transformation temperature because it will be significantly influenced by the heating rate and the sample mass. Considering the above said and the fact that the liquid phase appears in the alloy already at the temperatures of ~ 130 °C, the choice of the alloy of the given composition as a temperature indicator is questionable.

To ensure the best measurement accuracy, the thermocouple of the sample holder was calibrated under specified conditions using the melting temperatures of standard substances such as In, Bi and Zn. The calibration results are presented in Table 1.

Table 1. The results of determining the melting temperatures of standard materials.

Material	Mass (mg)	Measured value of the melting temperature (°C)	Standard value of the melting temperature (°C)
In	23.44	157.1	156.6
Bi	54.64	272.0	271.4
Zn	15.26	420.8	419.5

It follows from Table 1 that the melting temperatures measured for standard substances are slightly overrated and, as follows from figure 9, the dependence between the measured and standard values is linear.

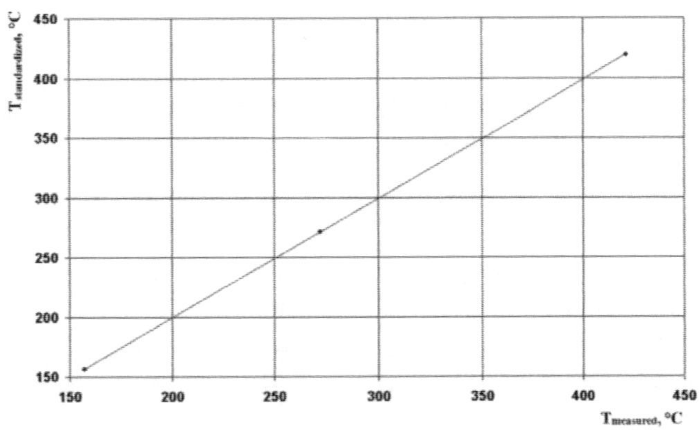

Figure 9. The dependence between the measured and standardized melting temperatures of standard substances.

Using the calibration obtained, repeated measurements of the melting temperature of one sample for each alloy were made. The resulting melting temperatures of the alloys are shown in table 2.

Table 2. The melting temperatures of alloys.

No.	$T_{melting}$, °C
Alloy No. 1	289 ± 2
Alloy No. 2	297 ± 1
Alloy No. 3	299 ± 1
Alloy No. 4	302 ± 1
Alloy No. 5	309 ± 1
Alloy No. 6	311 ± 1

4. Conclusion

Compositions have been selected in the systems of lead-indium, cadmium-indium and lead-tin and a technological scheme for the production of rods with a diameter of 3 ± 0.1 mm has been proposed. The melting points of alloys have been measured by a synchronous thermal analysis. Based on a complex of the investigations performed, the compositions of alloys have been clarified and alloys in the form of rods, providing the temperature control in the temperature range of 289–311 °C with a precision of ± 1–2 °C, have been made.

Acknowledgements

This work was performed within the framework of the Center of Nuclear Systems and Materials supported by MEPhI Academic Excellence Project (contract № 02.a03.21.0005, 27.08.2013).

References

[1] Rabu B, Pagano C, Boucenna A, Addes et al 1999 *Journal of Aerosol Science* **30** (1) 109
[2] Xiao H, Long C, Chen L and Liang B 2013 *Proc. 21st International Conference on Nuclear Engineering, ICONE 2013 (Chengdu, China)* **1** article number V001T02A027
[3] Dubourg R, Austregesilo H, Bals C et al 2010 *Progress in Nuclear Energy* **52** (1) 108

IOP Conf. Series: Materials Science and Engineering **130** (2016) 012028 doi:10.1088/1757-899X/130/1/012028

Neutron-absorbing amorphous alloys for cladding coatings

O N Sevryukov, V T Fedotov and A A Polyansky

National Research Nuclear University MEPhI (Moscow Engineering Physics Institute), Kashirskoye highway 31 , Moscow, 115409, Russia

E-mail: sevr54@mail.ru

Abstract. This paper shows developed compositions of neutron-absorbing cladding alloys based on nickel and containing such elements as B, Gd, Hf, and Mn. The techniques for application of coatings from these alloys on the surface of structural steels have been improved. It has been shown that the amorphous neutron-absorbing coating is more uniform than the crystalline one. The experimental data on the adhesion of cladding coatings with a steel substrate and their neutron-absorbing capacity have been obtained.

1. Introduction

One of the instruments to ensure nuclear safety in the transportation of spent nuclear fuel (SNF) is to create a high-strength transport container that will effectively absorb the radioactive radiations of fission products. In this regard, the problem arises of creating of corresponding structural materials.

It is known that the corrosion-resistant boron steels with a boron concentration of 1 to 2.4 wt. % are sufficiently technological and have satisfactory physicomechanical properties. The steels alloyed with gadolinium are also prospective absorbing structural materials. However, both the first and the second materials have a number of disadvantages. Firstly, it is a small concentration of the absorbing element and the related need to manufacture containers with thick walls. Secondly, the reduced ductility and impact toughness of these materials due to alloying significantly reduce the manufacturability of the product and create an increased risk of breaking the container during its transportation.

In this connection, it is very promising to use an absorbing metal coating on the surface of a high-strength corrosion-resistant steel that is well developed in the industry. However, ensuring the effectiveness of such a coating is possible only at a high concentration of neutron-absorbing elements in the cladding layer. From this point of view, it seems appropriate to use amorphous metal foils obtained by rapid solidification. This method allows to: a) significantly (up to 2 and more times) increase the concentration of sparingly soluble elements in the metal matrix; b) obtain a thin (10-100 μm) plastic foil from practically non-deformable alloys, which provides a high manufacturability of applying a cladding material on the surface of products of complex configuration. In the light of the above-said, the work was performed on the choice of compositions for cladding, improvement of the technology of obtaining amorphous alloys by rapid solidification, the cladding of corrosion-resistance steel samples, metallographic analysis of the coatings obtained and measurement of the absorbing characteristics of the clad samples.

2. The choice of the compositions of amorphizing cladding alloys

In accordance with the known data and the previous studies [1,2], a boron containing nickel alloy (in wt.%, Ni (basis)-3.0Fe-7Cr-3B-4Si) was taken as a base material for investigations. This alloy easily

IOP Conf. Series: Materials Science and Engineering **130** (2016) 012028 doi:10.1088/1757-899X/130/1/012028

becomes amorphous, it is free-running and it well wets the chrome-nickel steels. Its melting point (T_m) is 975 °C. In accordance with the recommendations, it was proposed to add gadolinium, hafnium and manganese as alloying elements and, moreover, to increase the boron concentration up to 5%. To increase the corrosion resistance of the alloy, the chromium concentration should be increased to 14 wt.%.

Thus, further studies were carried out with the following compositions:

- Ni (basis) - 7Cr - 3Fe - 4Si - 3B;
- Ni (basis) - 7Cr - 4Fe - 4Si - 5B;
- Ni (basis) - 14Cr - 4Fe - 4,5Si - 4,5B;
- 45Fe - 40Ni - 7Cr - 4Si - 4B;
- Ni (basis) -3Fe-7Cr-4B-4Si-1Gd-2Hf;
- Ni (basis) -3Fe-7Cr-5B-4Si-5Gd;
- Ni (basis) -3Fe-7Cr-5B-4Si-4Mn.

Besides the above-listed cladding materials, the concentration of boron in the cladding layer was additionally increased by the ligature Ni + 15% B (NiB). To improve the physicochemical properties of the ligature, it was also obtained in the form of an amorphous ribbon by rapid solidification of the melt. For cladding, a standard amorphous alloy Ni + 7% Cr + 4% Fe + 4% B + 4% Si in the form of a ribbon was used. The alloy was layer-by-layer alternated with a ribbon of the NiB ligature and placed on the surface of a plate. The ratio of the both cladding alloys was of such a value that the total boron content in the combined layer after their application on the surface of the steel was not less than 8-9%.

3. The manufacturing technology of ingots

The following materials were used as the initial charge materials for the production of ingots: Ni of the grade H-0 (GOST 849-70), electrolytic Cr, carbonyl Fe, iodide Hf (TsMTU 05-177-69), distilled Gd, semiconductor Si (GOST 19658-74), Mn of the grade Mr00 (GOST 6008-75), amorphous B, and Ni-powder of the grade PNE1 (nickel electrolytic powder, GOST 9722-71).

To ensure a uniform chemical composition of the ingots, metalloids were added in the form of Ni-15%Si and Ni-10%B ligatures. For convenience of the addition of refractory Hf into the alloy, it was added using the Ni-35%Hf ligature obtained in an arc furnace MEPhI-9 with no less than a sixfold remelting. An ingot of the alloy 85%Gd-15%(Ni+Fe) was also melted in an arc furnace.

For the experiments on rapid solidification, ingots with a mass of 250-800 g were casted in an induction furnace using quartz crucibles and graphite molds.

4. The manufacturing technology of rapidly quenched ribbons

Experiments on the improvement of technological regimes of manufacturing amorphous alloys were carried out, using a high-vacuum installation "Crystal-702", in conditions of a controlled gaseous environment, the temperature regime of casting and the solidification rate of the melt [2, 3]. The installation scheme is shown in figure 1.

5. The results of experiments on rapid solidification

The following materials were used for the investigation.

– Ni-7Cr-3Fe-3B-4Si, Ni-7Cr-4Fe-4Si-5B and Ni-14Cr-4Fe-4.5Si-4.5B alloys; a plastic ribbon made from the alloy Ni-45Fe-7Cr-4Si-4B with a thickness of 50 μm and a width of 20 mm; the structure of the alloy is amorphous; the start melting temperature (T_m) is 975, 980 and 990 °C, respectively.

– Brittle fragments of a ribbon made from the alloy Ni-7Cr-3Fe-2Hf-1Gd-4B-4Si with a thickness of 50 μm and a width of 10 mm; the alloy structure is amorphous-crystalline; T_m = 980 °C.

– Brittle fragments of a ribbon made from the alloy Ni-7Cr-3Fe-5Gd-5B-4Si with a thickness of 50 μm and a width of 10 mm; the alloy structure is amorphous-crystalline; T_m = 980 °C.

– A brittle ribbon made from the alloy Ni-7Cr-3Fe-4Si-5B-4Mn with a thickness of 50 μm and a width of 20 mm; the alloy structure is amorphous-crystalline; T_m = 975 °C.

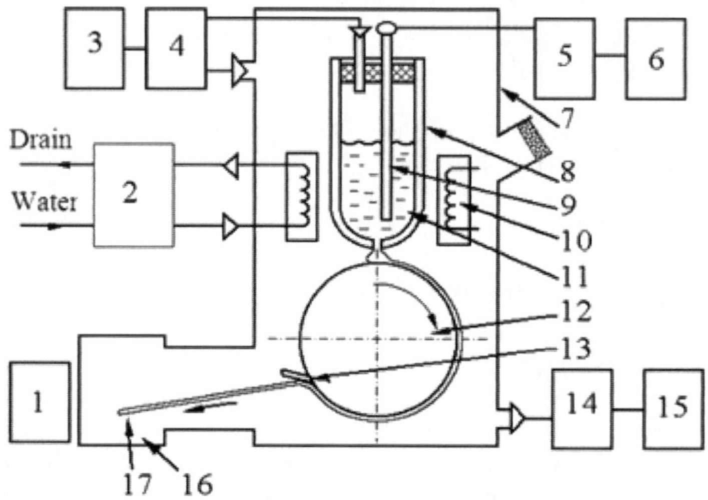

Figure 1. A scheme of the installation "Crystal-702":
1- remote control; 2 - high-frequency generator; 3- cylinder with an inert gas; 4 - gas inlet system; 5, control device such as R111; 6 - potentiometer KSP-4; 7 - camera body; 8 - quartz or ceramic feeder; 9 - thermocouple; 10 - high-frequency inductor; 11- melt; 12 - copper disc for quenching; 13 - ribbon remover; 14 - vacuum system with a thermocouple vacuum gauge VT-2A; 15 - vacuum unit AVZ-20D; 16 - ribbon receiver; 17 - rapidly quenched ribbon.

6. The manufacture of powder alloys

To manufacture amorphous powders containing neutron-absorbing elements, a technological chain was developed. It includes the following steps: embrittlement of a rapidly quenched ribbon without loss of amorphism at the temperature of no more than 500 °C; preliminary grinding of the ribbon into fragments no larger than 5×10 mm^2; grinding in a ball mill in an inert gaseous medium (argon); sifting using a sieve with a mesh size of 100 to 20 μm. To remove grease and dust contaminations containing in the resulting powder, it was washed in acetone with a fivefold change of the solvent and dried on filter paper.

Study of the powder fragments using a microscope showed that the powder can be related to the fragment type and the form of its particles is both isometric and needlelike. The particles can reach 1 mm in length. The grinding technique was the same for all the nickel alloys.

7. The cladding of steel samples by rapidly quenched alloys

Plates with sizes of $40\times40\times5$ mm^3 made from the stainless steel 12X18H10T were used as the samples for cladding. The surface to be cladded was purified from oxides, made even by grinding and then degreased. The cladding alloy was applied in the form of either a powder or equal segments of the ribbon so that the cladding layer thickness was from 150 to 200 μm. The powder was used in cases when the manufacture of geometrically regular fragments of the ribbon was not possible. After laying the cladding alloy on the surface of a steel plate, the plate was placed into the working chamber of a vacuum electric resistance furnace of the SShV-25 type. The furnace was evacuated to a residual pressure of 10^{-3} Pa and heated to the temperature which was 100 °C above the starting melting temperature T_m. The temperature was controlled by chromel-alumel thermocouples that were in contact with the steel plate, as well as visually by the moment of a full spreading over the plate

surface. The heating rate was ~ 20 °C/min, the holding time at the temperature of a full spreading was 5 min, after which the heaters were turned off and the sample was cooled in the furnace.
To estimate the possibility of using the existing technological equipment for applying of coatings, works were carried out on cladding the steel by an amorphous powder at the installation "Kiev-7" under stationary conditions.

Optimal parameters were determined which are necessary to obtain a dense uniform coating with a thickness of ~ 1 mm on a cylindrical sample of the steel 12X18H10T. The spraying distance was 300 mm and the rotation frequency was 50 rev/min. The resulting coating has a good adhesion to the steel [4]. When the sample was treated by a turning lathe, no peeling, flaking or spalling of the sprayed layer was observed. A roughness is marked on the cladding layer surface that is a consequence of the needlelike form of individual particles of the powder that do not have enough time to be melt in the plasma flow. However, the basis part of the coating has a dense microcrystalline structure and a porosity of no more than 10%.

8. The mechanical properties of coatings from rapidly quenched materials

The mechanical tests of ring-shaped samples with applied coatings were performed using the test machine FPZ 10/1. The samples with a width of 10 mm, an inner diameter of 60 mm and an outer diameter of 66 mm (the outer diameter of the coated pipe was 70 mm) were exposed to a compression test.

The distance between the grippers L_{tr} at the time of appearance of a crack in the coating was registered during the loading. The moving rate of the active gripper was 1.0 mm/min. Three samples of each type of the coating were exposed to tests. The results of tests are shown in table 1.

Table 1. Test results on the deformation capacity of neutron-absorbing coatings on the steel 12X18H10T.

Number of a sample	Amorphous coating	Crystalline coating
	L_{tr} (mm)	
1	60.4	68.6
2	63.2	67.3
3	62.1	67.9

As follows from the data presented, the deformation ability of the coating obtained using an amorphous alloy is significantly higher (L_{tr} = 61.9 mm) than that obtained using traditional crystalline powders (L_{tr} = 68.0 mm).

The microhardness of the coatings and the substrate was measured using a micro-hardness testing machine PMT-3. The measurement results are shown in table 2.

Table 2. The microhardness of clad samples of the steel X18H10T.

Alloy composition, wt.%	Microhardness (MPa)
Ni (basis)-3Fe-7Cr-5B-4Si-4Mn	
Amorphous	7500
Crystalline	9600
Ni (basis)-7Cr-4Fe-4Si-5B;	7500
Ni (basis)-14Cr-4Fe-4.5Si-4.5B;	7800
45Fe - 40Ni - 7Cr - 4Si - 4B.	6900
Steel X18H10T	3500

Thus, the results obtained show that the coating plasticity when using amorphous alloys is higher than that in case of crystalline ones [5].

IOP Conf. Series: Materials Science and Engineering **130** (2016) 012028 doi:10.1088/1757-899X/130/1/012028

9. The test of clad samples on the neutron-absorbing capability

To perform comparative studies, four steel samples (No. 1, 2, 3, and 4) with sizes of 30×30 mm were prepared. The sample No. 1 is a 3 mm thick tube fragment of a corrosion-resistant steel 12X18H10T. The sample No. 2 is a tube fragment with a neutron-absorbing coating about 1 mm in thickness. The sample No. 3 is a sheet steel 12X18H10T with a thickness of 3.5 mm and a volume content of boron of 1.5 wt. %. The sample No. 4 is a sheet steel 12X18H10T with a thickness of 5 mm.

Investigation of the neutron-absorbing capability of materials was carried out using a spectrometric installation consisting of a neutron source (Pu-Be, with an activity of 2.04×10^6 n/s), a neutron moderator (140 mm thick organic glass), a cadmium filter with a thickness of 0.5 mm, and a semiconductor thermal neutron detector (DTP 2Kh603B Ts23.394, 022TU).

The neutron-absorbing capability of materials was determined by a coefficient of neutron absorption:

$$k_n = (N_o - N_1)/N_o, \quad (1)$$

where N_o is the number of neutrons recorded without the sample; N_1 is the number of neutrons recorded with the sample.

The results of determining the neutron absorption coefficients are presented in Table 3 which shows that the samples No. 2 (with coating containing neutron-absorbing elements) and No. 3 (containing boron) have the best neutron-absorbing capability.

Table 3. The effectiveness of neutron absorption by steel samples.

Sample number	1 (without a coating)	2 (with a coating)	3 (a borated sheet)	4 (a sheet)
k_n	0.16±0.03	0.46±0.05	0.54±0.05	0.23±0.03

It is obvious that the neutron-absorbing layer decreases the penetration of neutrons by several times and, even in the illustrated variant of the cladding alloy (with a boron concentration in the layer of 3.5%), its effectiveness is highly competitive with that of the volume-alloyed stainless steel.

10. Conclusion

– The compositions and the technology for manufacturing of rapidly quenched alloys based on nickel and containing such neutron-absorbing elements as B, Gd, Hf, and Mn have been developed and amorphous and amorphous-crystalline alloys have been obtained.

– The elements of the cladding technology of steel details using both a rapidly quenched ribbon and powders of amorphous alloys with complex composition have been improved.

– The optimum temperature-time parameters of the application of a coating required to obtain a uniform layer with a thickness of 1.0-2.0 mm have been chosen. Special studies have showed that the coating obtained has a good adhesion to the steel.

– It has been shown that the amorphous neutron-absorbing coating is more uniform in comparison with the crystalline one.

– The mechanical compression tests have shown that the deformation ability of the coating obtained using an amorphous alloy is significantly higher than that obtained using traditional crystalline powders.

– It has been shown that the coating with a thickness of no more than 1 mm from the viewpoint of its neutron-absorbing capability is equivalent to 3.5 mm thick borated steel.

– The experimental data obtained on the adhesion strength of cladding coatings with a steel substrate and their neutron-absorbing capability give grounds to speak about the prospects of using of amorphous materials and the technologies developed for cladding of structural elements of the container as an alternative variant of the biological protection from the penetrating radiation of a spent nuclear fuel.

Acknowledgements
This work was performed within the framework of the Center of Nuclear Systems and Materials supported by MEPhI Academic Excellence Project (contract № 02.a03.21.0005, 27.08.2013).

References

[1] Choi Y, Moon B M and Sohn D S 2013 *Nuclear Engineering and Technology* **45** (5) 689
[2] Kalin B A, Fedotov V T, Sevryukov O N, Grigoryev A E, Skuratov L A, Plyushchev A N et al. 1996 *Svarochnoye proizvodstvo* **1** 15
[3] Peng Y, Zhang C, Zhou H and Liu L 2013 *Surface and Coatings Technology* 218 (1) 17
[4] Xie C X, Yang Y Z and Chen M Z 2013 ***Proc.** 4th International Conference on Manufacturing Science and Engineering, **ICMSE 2013 (Dalian, China)*** 70
[5] Inoue A, Shen B L and Chang C T 2004 *Acta Materialia* **52** (14) 4093–99

The corrosion resistance and neutron-absorbing properties of coatings based on amorphous alloys

O N Sevryukov and A A Polyansky

National Research Nuclear University MEPhI (Moscow Engineering Physics Institute), Kashirskoye highway 31, Moscow, 115409, Russia

E-mail: sevr54@mail.ru

Abstract. The object of the present study was the corrosion-resistant amorphizing alloys with an increased content of boron for cladding the surface of metals, rapidly quenched alloys without boron for protective coatings on a high-boron cladding layer, as well as steel samples with a protective coating with a high content of boron and without boron. The aim of the work is to investigate the corrosion resistance of a coating in water at the temperature of 40 °C in conditions of an open access of oxygen for 1000 h, as well as the features of the microstructure of clad samples before and after the corrosion tests. New data on the corrosion resistance of Cr18Ni10Ti steel samples with a protective layer from a rapidly quenched alloy Ni-19Cr-10Si (in wt.%) on a high-boron coating have been obtained.

1. Introduction

Studies on the development of high-boron neutron-absorbing coatings [1] confirmed the correctness of the chosen direction. However, the leaching of boron [2] in an aqueous medium at elevated temperatures and an open access of oxygen and, respectively, increasing the boron concentration in water are observed. Therefore, this work was carried out to estimate the corrosion resistance of coatings clad by a protective layer before and after the tests in an aqueous medium.

2. Production of samples with a high-boron coating with and without a non-boron protective layer

The methodology of coating the steel was the principle of "a high-boron layer - a layer without boron" in one technological cycle. The "high-boron component - nickel boron alloy Ni-7wt.%Cr-4wt.% Fe-4.5wt.% Si-4.5wt.%B" mixture in the ratio of 1:1 was used as the initial material for cladding. The homogenization of the powder was performed for 30 min in a planetary ball mill with addition of ethanol to prevent its oxidation. The resulting mixture was discharged from the container, slightly dried and applied in the form of a pasty mass to the surface of plates from the steel Cr18Ni10Ti with a diameter of 25 mm and a thickness of 3 mm. The plates had a side board of 1.5 mm in height and 2 mm in width. Two or four layers of a ribbon of the protective microcrystalline alloy Ni-Cr-Si was applied on the surface of a part of the plates that was welded to the plate by a spot electrocontact welding. Then the plates were placed in a vacuum furnace SShV-1.25/25I1-1000 and exposed to cladding in a vacuum of $5x10^{-5}$ mm Hg at the temperature of 1100±25 °C. The temperature was controlled by a chromel-alumel thermocouple and visually by the time of a full spreading of the melt over the surface of a clad sample. The heating rate was about 20 °C/min, then the sample was held at the temperature of full spreading for 15 min. After that, the temperature was lowered to 600-700 °C and the sample was held for 1 h, then the heating was turned off and the sample was cooled with the furnace[3]. A part of the samples with and without a cladding protective layer was exposed to corrosion tests; a part of the samples was prepared for a microstructural and X-ray microspectral analysis.

3. Corrosion tests of samples with a protective cladding layer

To determine the corrosion resistance of the cladding layer, two samples from each part were placed in glass flasks with a capacity of 0.5 l that were filled with distilled water and loaded into a thermostatic drying oven PE-4610. The temperature in the working space of the thermostat was maintained at 40 ± 0.5 °C. The duration of tests was 1450 h. The boron concentration in the solution was measured by a mass-spectrometer. These data are presented in table 1.

Table1. Results of the corrosion tests of clad samples.

Batch number	Characteristics of the clad coating	The weight of samples before the tests (g)	The weight of samples after the tests (g)	The concentration of boron (μg/l)
1	High-boron coating	13.2543	13.2546	33.78
		13.6903	13.6894	
2	High-boron coating + two layers of a ribbon	13.1251	13.1252	28.60
		13.1144	13.1155	
3	High-boron coating + four layers of a ribbon	13.8037	13.8184	22.02
		13.6552	13.6629	

4. Study of the microstructure of samples before and after the corrosion tests

Investigation of the distribution of elements in the "coating-substrate" contact area was carried out using a scanning electron microscope "CamScan - 4DV" equipped with a wave-crystal diffraction spectrometer WDX-3PC. To study the contact zone structure, the reflected *electron imaging mode* was used at an accelerating voltage of 15 kV. Since the electron reflection coefficient increases with the atomic number of the element, a lighter contract of images in reflected electrons corresponds to areas with a larger average atomic number. The results of metallographic studies are presented in figures 1 - 3.

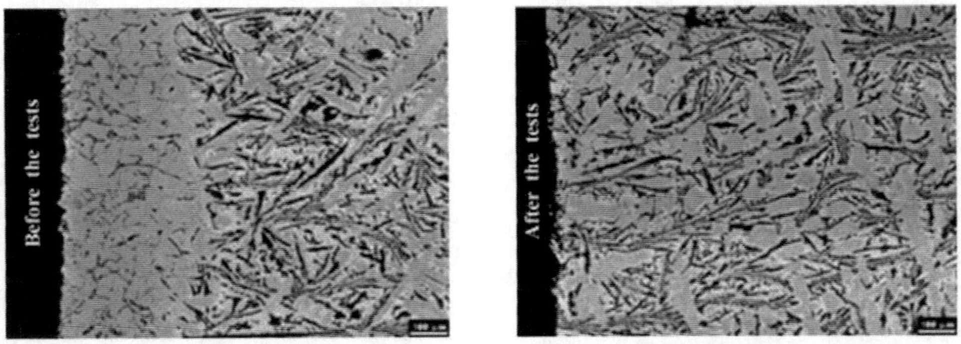

a) b)

Figure1. Microstructures of samples with a two-layer protective layer: a) before the corrosion tests; b) after the corrosion tests (the marker size in the lower right corner of the images is 100 μm).

A protective layer with a thickness of about from 200 to 300 μm is clearly visible at the sample transverse section before the corrosion tests (figures 1a, 2a). This layer is absent at the sample after the tests (figure 1b). Assuming that there was dissolution of the protective layer, the weight loss of the sample with a diameter of 25 mm should be about 0.3-0.5 g, whereas in reality it is 0.0011 g. The transverse section of the sample before the tests was done along its diameter, the section of the sample after the tests was done approximately in a distance of 1/3 of the diameter. Since the high-boron plating powder fills the space inside the cylindrical sample with less than 100% of density and the melt wets well the steel, the formation of a concave meniscus in melting takes place due to capillary forces.

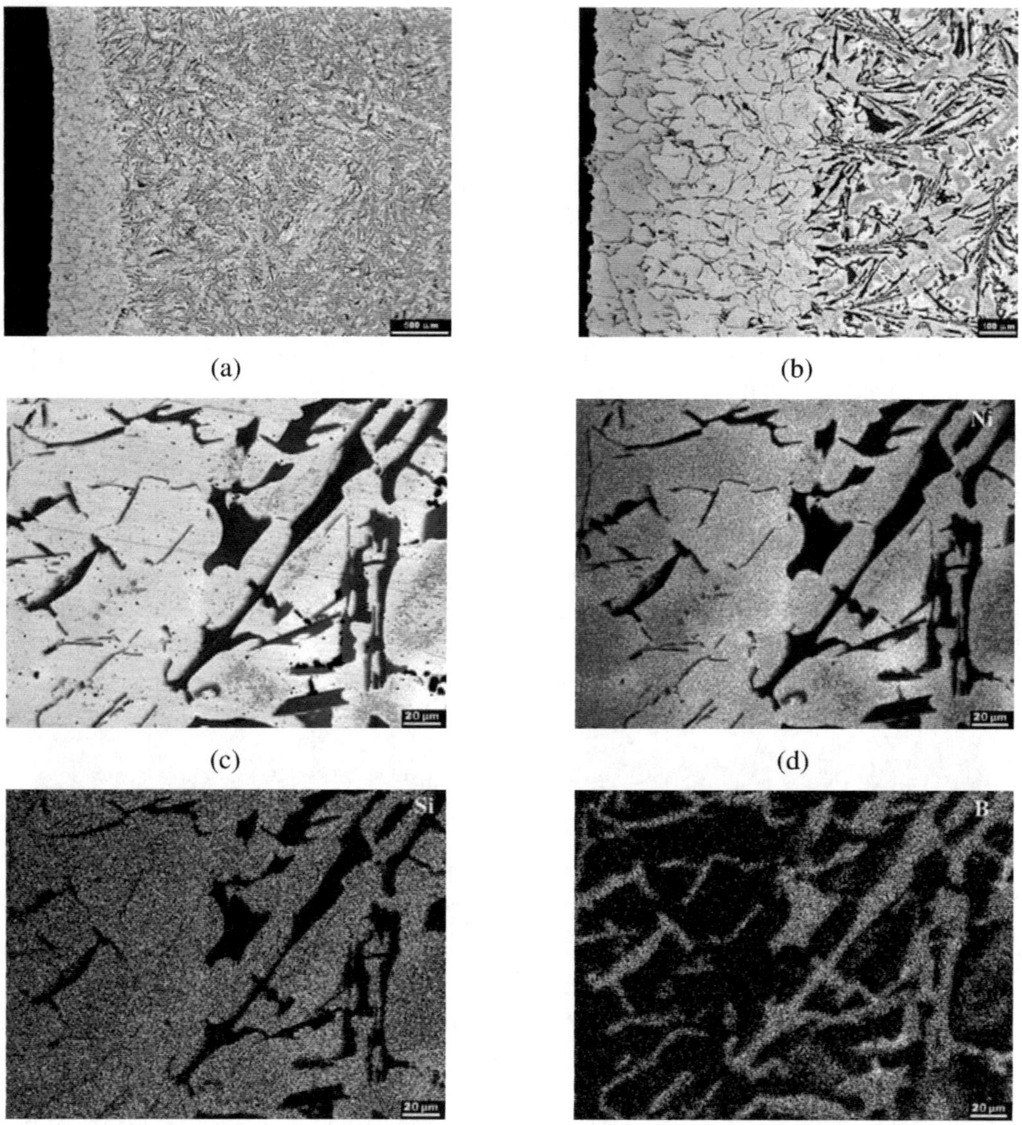

Figure 2. Photomicrographs of the sample cross section with a protective layer before the tests: a, b) in reflected electrons; c) in backward electrons; d, e, f) in a characteristic X-ray radiation of the corresponding element.

As the protective layer melts a little bit earlier than the layer containing boron and the temperature gradient is directed from the edges of the sample to its center, it is quite probable that the central parts of the sample are enriched by a non-boron melt and its content at the peripheral parts of the sample can be little up to its complete absence. This is indirectly confirmed by the fact that the change in the weight of all the samples during the corrosion tests is insignificant and no sharp difference in the boron concentration in water after the tests (see table 1) is observed. It seems appropriate in further works to divide the processes of application of a high-boron coating and a protective layer into two separate stages.

Figures 2 and 3 show data on the distribution of the basic and alloying elements at the "protective layer - cladding layer" border of the sample before the tests and near its surface after the tests

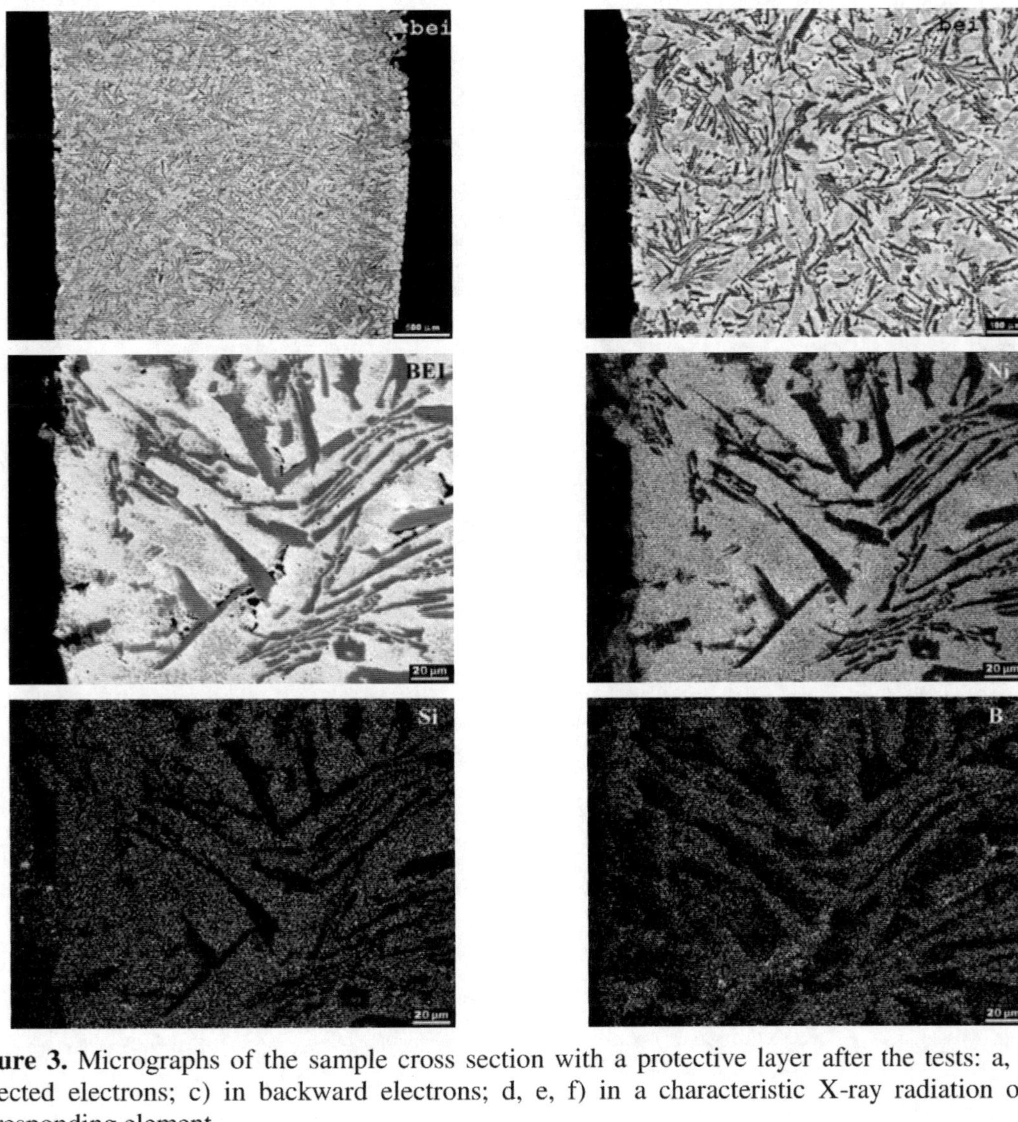

Figure 3. Micrographs of the sample cross section with a protective layer after the tests: a, b) in reflected electrons; c) in backward electrons; d, e, f) in a characteristic X-ray radiation of the corresponding element.

As follows from these data (figure 2a-c), no discontinuities, pores, or foreign inclusions are observed at the "protective layer - cladding layer" border. The distribution of elements (figure 2d-f) shows an active chemical interaction of the melts of the protective and cladding layers without allocation of foreign phases. During the corrosion tests, the microstructure (figure 3a-c) and the distribution of elements (figure 3d-f) did not undergo significant changes, which indicate the stability of the cladding layer even in the absence of a protective coating.

5. Conclusion

- A laboratory technique for the joint application of a protective layer and a high-boron coating onto the surface of corrosion-resistant steel samples has been developed.
- Samples have been made and investigation has been performed of the corrosion resistance of the coating in an aqueous medium at an open access of oxygen and at the temperature of 40 °C for 1450 h. It has been found that the two-layer coating in the form of a 50 μm thick ribbon applied on the surface of a high-boron layer decreases the concentration of boron in water by 1.2 times, and the four-layer coating by 1.5 times.

Acknowledgements
This work was performed within the framework of the Center of Nuclear Systems and Materials supported by MEPhI Academic Excellence Project (contract № 02.a03.21.0005, 27.08.2013).

References

[1] Farmer J C, Haslam J J and Day SD et al 2007 *J. Mat. Res.* **22** (8) 2297–11
[2] Yang Y, Zhang C, Peng Y, Yu Y and Liu L 2012 *Corrosion Science* **59** 10
[3] Liu L, Zhang C 2014 *Thin Solid Films* **561** 70

IOP Conf. Series: Materials Science and Engineering **130** (2016) 012030 doi:10.1088/1757-899X/130/1/012030

Structural-phase state and creep of mixed nitride fuel

I I Konovalov, B A Tarasov and E M Glagovsky

National Research Nuclear University MEPhI (Moscow Engineering Physics Institute) Kashirskoe Highway 31, 115409 Moscow, Russia

E-mail: BATarasov@mephi.ru

Abstract. By the analysis of thermal creep data in conjunction with structural-phase state the most likely mechanisms of UN creep are considered. An equation relating the thermal and radiation creep of nitride fuel with such important parameters as plutonium content, porosity, grain size, the content of impurities of transition metals and oxygen, the carbon content has been suggested. At stationary operating parameters in reactor the creep of nitride fuel with technical purity is defined by the thermal component at mechanism of intergranular slip and by the radiation component, which plays a significant role at temperatures below 1100 °C. Both types of creep in a first approximation have a linear dependence on the stress.

1. Introduction

Mixed uranium-plutonium nitride is a promising fuel for nuclear fast reactors. Thermal creep of nitride defines its technological properties (e.g., sintering, hot pressing of pellets) and radiation creep determines characteristics of mechanical interaction of the fuel with cladding.

The complexity of the problem, in addition to the lack of systemic research on uranium nitride creep, is that the required range of parameters is outside the scope of the parameters that are usually studied experimentally. For example, for designers of fuel elements it's important to know the parameters of fuel creep at strain rates $\dot{\varepsilon}$ from 10^{-7} to 10^{-6} h^{-1} (rate of core deformation caused by swelling) on the basis of several years at temperatures typically below 1400 °C.

Creep deformation is quite a complex process that depends on many factors. To describe such multivariable processes the phenomenological approach was used. To reduce the uncertainty of the parameters in the equation for UN creep, the correlation between self-diffusion and creep has been used.

In this paper, based on general theoretical approaches the equation was suggested relating thermal and radiation creep of nitride fuel with such important parameters as plutonium content, porosity, grain size, the content of transition metal impurities and oxygen, the carbon content.

2. The correlation between thermal creep and self-diffusion

The equation of creep from common positions should contain in addition to traditional factors such as stress, temperature, grain size also other factors associated with structural-phase state of the material.

Data analysis at approximation of the thermal creep rate dependence on the stress σ and the temperature T by the Arrhenius equation $\dot{\varepsilon} \sim \sigma^n \exp(-E/kT)$ indicates that the uranium nitride corresponds to the general laws for the isomorphic compounds: monocarbide uranium [1-3] mononitrides Zr, Ti, and other metals [4].

In order to describe the apparent change of n (from 1 to 6), and E (from 0.4 to $1.2E_{SD}$) two mechanisms of thermal creep were selected for the range of assumed operating parameters of nitride fuel T<1400 °C (0.55Tm) and $\dot{\varepsilon}=(10^{-7}\text{-}10^{-6})h^{-1}$.

First mechanism is the traditional mechanism with power law on stress with $n \sim 4.5$ and creep activation energy E equal to that of the volume self-diffusion E_{vSD}. Second mechanism is the "linear" creep by grain boundary sliding, such as creep by Gifkins [8], with the activation energy equal to that for boundary self-diffusion $E_{bSD} \approx 0.5E_{SD}$.

Based on data analysis for the second mechanism were selected the following values: $n = 1$ and the dependence from characteristic grain size L_G^{-1}. However, this mechanism is also considered for the boundaries between the matrix and the precipitations of second phases.

On this basis, the interrelation between self-diffusion and thermal stationary creep is described by the equations:

$$\dot{\varepsilon}_0 = a_1\sigma^n \exp\left(-\frac{E_{vSD}}{kT}\right) + a_2\sigma\frac{\psi^0}{kT}\exp\left(-\frac{E_{bSD}}{kT}\right) \tag{1}$$

$$D_{SD} = D_{vSD} + \psi^0 D_{bSD} = b_1\exp\left(-\frac{E_{vSD}}{kT}\right) + b_2\psi^0\exp\left(-\frac{E_{bSD}}{kT}\right), \tag{2}$$

where: a_1, a_2, b_1, b_2 – constants, σ – stress, n – exponent (for pure compounds is usually in the range of from 4 to 5), ψ_0 – the total volume fraction of the boundaries, E_{vSD} and E_{bSD} – respectively the activation energies of volume and boundary self-diffusion of less mobile atoms, in our case U and Pu, k – Boltzmann constant, T – temperature in K.

Compilation of experimental data on the temperature dependence of self-diffusion shows that the data can vary by three orders of magnitude for nitride with different purity. Trend is the acceleration of self-diffusion in "dirty" mononitrides while reducing the apparent activation energy E and the exponent n.

In [8] self-diffusion coefficient of U (cm^2/s) in the single crystal of high purity UN is defined by the Arrhenius equation:

$$D_{vSD}^{Th} = 5.4\cdot10^{-3}\exp\left(-\frac{5.0}{kT}\right). \tag{3}$$

For uranium nitride the activation energy of boundary diffusion D_B is unknown, so the data for the isomorphic compound UC were used.

For UC measured ratio of the activation energies of the boundary and volume self-diffusion is 0.53 [12], which is within the interval of typical values for metal compounds $E_{bSD} = (0.5\text{-}0.6)\,E_{vSD}$.

Using the defined ratio for pre-exponential factors for volume and boundary diffusion of UC we obtain the probability boundary diffusion coefficient for UN in cm^2/s:

$$D_{bSD}^{Th} \approx 1.4\cdot10^{-4}\exp\left(-\frac{0.53E_{vSD}}{kT}\right). \tag{4}$$

At temperatures below the temperature of (U,Pu)N dissociation (<1700 °C) Pu redistribution over the cross section of the fuel pellets was not observed. So, it was decided that the self-diffusion of U and Pu in the temperature range of interest differ insignificantly, and to describe the creep one may consider the average value for self-diffusion. To determine the average value depending on plutonium content C_{Pu}=Pu/(U + Pu) the linear correlation E_{vSD} in eV with the melting temperatures of PuN (2760 K) and the UN (3030 K) was used:

$$E_{vSD} = 5.0f_{Pu}, \quad f_{Pu} = 1 - 0.09C_{Pu}. \tag{5}$$

The total volume fraction of boundaries ψ^0 formed by grains and impurity precipitates of second phases and included in (2) can be represented as:

$$\psi^0 = \psi_{gb} + \psi_O + \psi_C + \psi_{tm}, \tag{6}$$

where ψ_{gb} – effective volume fraction of grain boundaries; ψ_O, ψ_C and ψ_{tm} – respectively effective volume fractions of boundaries formed by the precipitates of uranium and plutonium oxides, uranium carbides at (N + C) content above stoichiometry and transition metal compounds (TM) with nitrogen or uranium.

With an average width of the high angle boundaries ~ 0.5 nm, the volume fraction of grain boundaries will be:

$$\psi_{Gb} = \frac{1.5\cdot 10^{-3}}{L_G}, \qquad (7)$$

where L_G – characteristic grain size in μm.

To evaluate other components in ψ^0 it was assumed that the density of spherical precipitates of the second phase is about of 10^{10} cm^{-3} (the observed density of the oxide phase in UN at the oxygen content ~ 1300 ppm [13]) and accepted effective width of the diffusion boundary between the phases with different lattice parameters ~ 1 nm.

If we assume the solubility limit of oxygen in the UN ~ 650 ppm [13] then the volume fraction of the boundaries formed by dioxide precipitates will be:

$$\psi_O \approx 1.1\cdot 10^{-7}(C_O - 650)\text{, at } C_O < 650 \text{ ppm } \psi_O = 0. \qquad (8)$$

Since carbon is soluble in the UN, we must take into account the concentration of carbon above stoichiometry (N + C) / U = 1 which leads to the formation of higher carbide. With the same parameters for UC_2 as for the oxide phase, the volume fraction of the boundaries at the carbon content above the stoichiometry C_C^* in ppm will be:

$$\psi_C \approx 0.8\cdot 10^{-7} C_C^*. \qquad (9)$$

Contamination of transition metals TM mainly occurs by elements belonging to the industrial metal equipment (Fe, Cr, Ni, refractory metals) and ceramic materials (Al, Si, etc.). The volume fraction of diffusion boundaries formed by precipitates of TM (C_{tm} is concentration in ppm) is estimated by relation:

$$\psi_{tm} \approx 0.5\cdot 10^{-7} C_{tm}. \qquad (10)$$

3. The equation for thermal creep of uranium mononitride

In general terms the equation of thermal creep for uranium nitride when correlated with the activation energy of self-diffusion has the power and linear components:

$$\dot{\varepsilon}_0^{Th} = a_1\sigma^{(4.5+\Delta n)}\exp\left(-\frac{(5.0+\Delta E)f_{Pu}}{kT}\right) + a_2\psi^0\frac{\sigma}{kT}\exp\left(-\frac{2.65 f_{Pu}}{kT}\right) \qquad (11)$$

$$\dot{\varepsilon}^{Th} = \dot{\varepsilon}_0^{Th}\exp(0.2P), \qquad (12)$$

where Δn and ΔE – respectively change in the exponent and activation energy with the dispersion hardening by impurity phases.

The porosity factor P, % in (12) is typical for sintered ceramic materials within the porosity of up to 20% [14].

Coefficients in (11) are determined by correlation with refractory ceramic: $a_1/(a_2\psi_0) \sim 10^2$ (for $\psi_0 \sim 10^{-3}$, $a_1/a_2 \sim 0.1$). For $\dot{\varepsilon}^{Th}$ in h^{-1} and σ in MPa the coefficients a_1 and a_2 has the order of 10^6 and 10^5 respectively.

In the dispersion-strengthened carbides and nitrides the activation energy may increase on (1-2) eV, while the exponent n reaches the value of 6-7.

Figure 1 shows a comparison of calculated (P = 10%, L_G = 10 μm) and experimental data. As can be seen from figure 1b dispersed phases formed by impurities may have both reinforcing and softening effect (indicated by arrows in figure 1b).

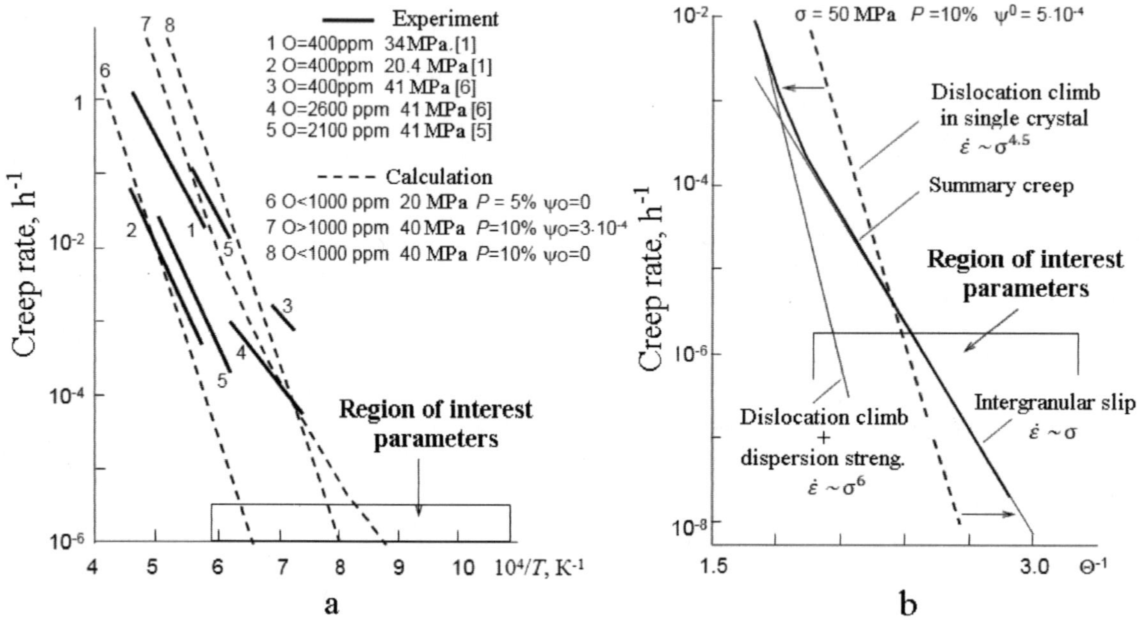

Figure 1. Comparison of the calculated and experimental data on thermal creep of UN (a) and deformation mechanisms (b) depending on the homologous temperature $\Theta = T/T_m$.

4. Comparison of thermal and radiation creep

For determining the contribution of thermal creep under stationary operating conditions of nitride fuel, it's sufficient to compare the linear component of thermal creep $\dot{\varepsilon}^{Th}$, defined by parameter ψ_0, with radiation creep $\dot{\varepsilon}^{R}$, determined by the fission rate G.

According to data the dependence of radiation creep $\dot{\varepsilon}^{R}$ in h^{-1} from fission rate G in cm^{-3}s^{-1} and σ in MPa for «dirty» UN (P = 7%, UO_2 ~ 10 vol.%, L_G = 15 μm, the estimated value $\psi^0 \sim 7\cdot10^{-4}$) taking into account (12) can be written as:

$$\dot{\varepsilon}^{R} = (2.5-3.3)\cdot 10^{-22} G\sigma \exp(0.2P). \qquad (13)$$

Linear thermal creep component, according to (11) and (12) is defined by:

$$\dot{\varepsilon}^{Th} = 10^5 \psi^0 \frac{\sigma}{kT} \exp\left(-\frac{2.65}{kT}\right) \exp(0.2P) . \qquad (14)$$

Figure 2 shows the estimated radiation and thermal creep of UN of nitride samples with mentioned above characteristics (a) and shows the probable temperature dependence for nitride fuel under parameters of operation in reactor (b).

In the range of interests parameters (T < 1400 °C, σ < 50 MPa, $\psi_0 \sim 3\cdot10^{-4}$) creep of nitride fuel will be determined by the thermal and radiation components both having the linear stress dependence.

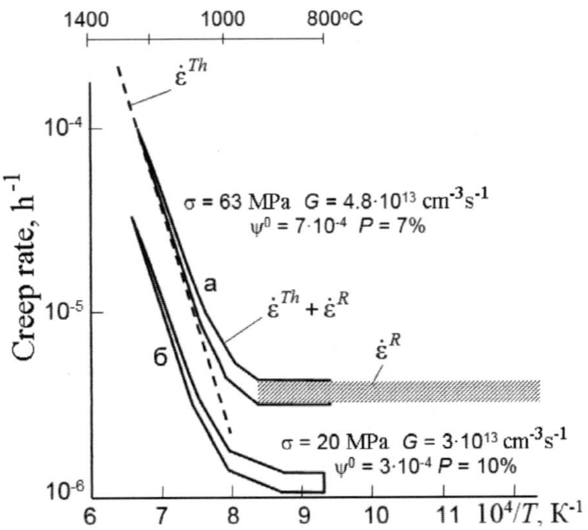

Figure 2. Comparison of the calculated thermal (dotted line) and radiation (shaded area [15]) creep of UN.

5. Conclusions

Data on thermal creep of mononitride fuel can be described by two components: the creep caused by climb of dislocations - the power law creep, and creep due to intergranular slip - linear creep. The temperature of transition from one creep to another depends on the nitride structure and the level of stress (strain rate).

For a correct forecast of thermal creep under the proposed approach the following minimum set of data is required: plutonium content, porosity, grain size, the content of impurities of transition metals and oxygen, the carbon content above stoichiometric ratio (N+C)/(U+Pu) = 1.

At stationary operating parameters of nitride fuel ($T < 1400$ °C) creep is defined by a thermal component with the mechanism of intergranular slip and radiation component, which is beginning to play a significant role at temperatures below 1100 °C. Both types of creep in the first approximation should have a linear dependence on stress (strain rate from 10^{-7} to 10^{-6} h^{-1}).

Acknowledgements

This work was performed within the framework of the Center of Nuclear Systems and Materials supported by MEPhI Academic Excellence Project (contract № 02.a03.21.0005, 27.08.2013).

References

[1] Routbort J L and Singh R N 1975 *Journal of Nuclear Material* **58** 78
[2] Tokar M 1971 *High Temperature Compressive Creep and Hot Hardness of Uranium-Plutonium Carbides* Report LANL LA-4704
[3] 1981Gemlin Handbook of Inorganic Chemistry. *U Uranium: The Behavior of Uranium Fuels in Nuclear Reactors*. Springer Verlag Berlin
[4] Zubarev P V 1985 *Zaroprochnost faz vnedreniya* (*Hot strength of interstitial phases*). Moscow. Metallurgia (in Russian)
[5] Masaaki, Uchida, and Michio 1973 *Journal of Nuclear Materials* **49** 91
[6] Fassler MH et al. *Compressive Creep of UC and UN*. 1965 US Report PWAC-482. Part 1
[7] Kutty T R G, Kaity S and Basak C B, 2012 *Nuclear Engineering and Design* **250** 125.

[8] Gifkins R C 1968 *Journal of the American Ceramic Society* **51** 2

[9] Bradbury M H, Michael H and Matzke Hj 1978 *Journal of Nuclear Materials* **75** 68

[10] Bradbury M H and Matzke Hj 1980 *Journal of Nuclear Materials* **91** 13

[11] Singh R N 1976 *Mechanical-property and diffusion data for advanced carbide and nitride fuels.* Argonne National Lab. USA. Report ANL-AFP-15

[12] Delvoye F 1988 *Etude des nitrudes d'uranium et de plutonium a faibles teneures en oxigene et en carbone*. Rapport CEA (France), CEA-R-5434

[13] Bakynov V S 1994 Refractories **35** V 10 317

[14] Hayes S L et al. 1989 *Development of creep rate porosity correction factor*. Report Texas A&M University ANFL-10-O

IOP Conf. Series: Materials Science and Engineering **130** (2016) 012031 doi:10.1088/1757-899X/130/1/012031

Investigation of nanostructural and monocrystalline materials for units of high intensive X-ray sources

A A Pavlov[1], M L Taubin[1,2], D A Chesnokov[1] and A A Yaskolko[1,2]

[1]Federal State Unitary Enterprise "Science Research Institute "LUCH", Zheleznodorozhnaya st, Podolsk, Moscow region, 142100, Russia
[2]National Research Nuclear University MEPhI (Moscow Engineering Physics Institute), Kashirskoe highway 31, Moscow, 115409, Russia

E-mail: taubin@sialuch.ru

Abstract The concept of increase of characteristics x-ray tubes by use nano-and monocrystal materials is offered in the report.

1. Introduction

Results of own works of the authors and works of the Russian, Japanese, Korean and American experts in refractory metals nanocompozite with increased strength characteristics for cathodes and anodes are analyzed in this report. It is shown, that though monocrystals have shown the better characteristics in comparison with traditional polycrystals both as anodes and as cathodes, more radical way of increase of operational characteristics of the x-ray tubes is application refractory nanostructural materials.

The results of experiments showed that high-temperature sample annealing resulted in a sharp increase (by a factor of more than 3) in material creep resistance at a temperature of 2200°C.

The study of emission characteristics shows that single-crystal tungsten with crystallographic orientation of emitting plane (111) has higher emission characteristics (work function equals 4.43 eV) than polycrystalline tungsten. But best results has nanoscsale W.

In this work results of research of evaporation of tungsten are presented. It is shown that the monocrystal has essential advantages on this parameter. It is important for materials working at temperatures more 2000°C.

An increase in the emission characteristics of a cathode and the strength characteristics of an anode at the same time can be achieved by modifying standard tungsten powders (for cathodes), tungsten and molybdenum powders (for anodes) by nanoscale powders of refractory materials (grain size is 10-100 nm). W, Mo, Nb, Re nanopowders are proposed to be used for producing anodes based on tungsten and molybdenum. It must increase allowable unit loads of an X-ray tube by reducing porosity, increasing strength and plasticity of anode material. Prerequisites for these proposals are some results of work where an increase in density, strength by 10-30%, and plasticity by a factor of 1.5 - 2 was achieved due to modification of steels by ultra-disperse powders of refractory compounds.

The authors carried out some experiments to determine the creep of tungsten alloyed by potassium on samples as foils 100 μm thick, used when manufacturing X-ray tube cathodes.

The results of these experiments showed that high-temperature sample annealing resulted in a sharp increase (by a factor of more than 3) in material creep resistance at a temperature of 2200°C.The authors think that this effect results from forming nanoscale little bubbles filled with potassium at the grain boundaries.

Results of this work can be used when development x-ray tubes with the increased operational characteristics.

2. Emission properties of tungsten

The problem of X-ray tubes at the present stage of developing medical equipment consists in the necessity of producing X-radiation of high intensity when providing small-size focal spots.

It is of particular importance for X-ray tubes of digital scanning X-ray units and mammography. Cathode problems are key problems for providing required roentgen optical characteristics of X-ray tubes.

The main task for having a focal spot of necessary size is electron focusing. Today, the most promising way of producing small-size focal spots is applying the design of cathodes with flat emitting surface.

A flat-shaped emitter is more efficient as compared with a spiral as it has more uniform electric field distribution and, also, large area which is important for producing X-radiation of high intensity [1].

A decrease in work function for increasing emission current is only possible when using new materials.

The results of many-year investigations of materials conformably to their application as emitters of thermionic energy converters [2, 3] showed that tungsten with single-crystal structure could also be promising material for X-ray tube emitters. The results of these investigations showed higher emission characteristics, lower and uniform (as compared with polycrystalline) tungsten evaporation from surface and higher dimensional stability at high operating temperatures due to absence of recrystallization (a considerable disadvantage of emitters made of polycrystalline) up to melting point as well. Proceeding from the above-mentioned, we think that producing a cathode with a flat emitter made of single-crystal tungsten is one of the promising trends in the area under consideration.

The use of single-crystal tungsten as emitter material allows, firstly, reducing the emitter working temperature by 70°C as the work function of the crystallographic face [111] of single-crystal tungsten is (4.4±0.05) eV, being lower by 0.15 eV than the work function of polycrystalline tungsten (4.55±0.05) eV, secondly, making easier electron focusing due to lower emission contrast range of a single crystal.

The results of studying the emission characteristics (Figure 1) show that single-crystal tungsten with crystallographic orientation of emitting plane (111) has higher emission characteristics (work function equals 4.43 eV) than polycrystalline tungsten. However, the best results have nanoscale W (Figure 1).

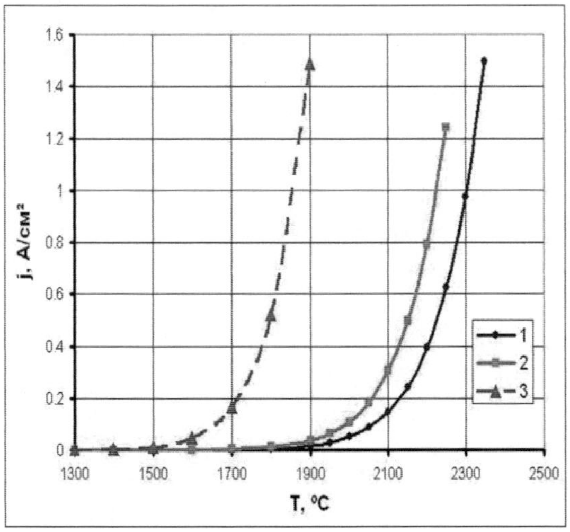

Figure 1. Emission current density of temperature dependence
1 - polycrystalline tungsten;
2 - single-crystal tungsten from plane [111];
3 - nanoscale W.

3. Compare of damageability of poly- and monocrystalline tungsten

The second task for increasing the output characteristics of X-ray tubes is anode problems, i.e. a decrease of its failure.

Doping of tungsten by rhenium is not always optimum, as to mammography it is not acceptable. In this connection, the authors have offered manufactured and tested single-crystal molybdenum (for mammography) and single-crystal molybdenum coated with single-crystal tungsten (for other applications) as anode material.

It is shown that the dependence of reducing in X-ray tube dose rate for a single-crystal anode (Figure 2) differs slightly from the typical dependence for a polycrystalline molybdenum anode. After reducing the dose power by 10% during the first 1500 switches, stabilization takes place, providing large service-life of an X-ray tube.

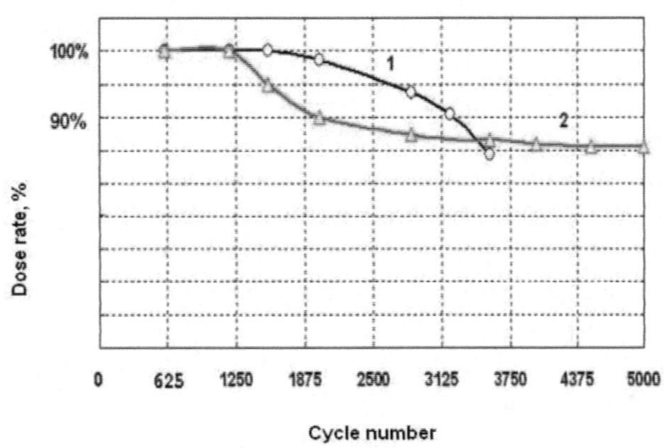

Figure 2. Dose power dependence on the number of switches 1 - polycrystal tungsten doped by rhenium; 2 - single-crystal tungsten.

In work results of research of evaporation of tungsten also are presented. It is shown that the monocrystal has essential advantages on this parameter (Figure 3). It is important for materials working at temperatures more 2000°C

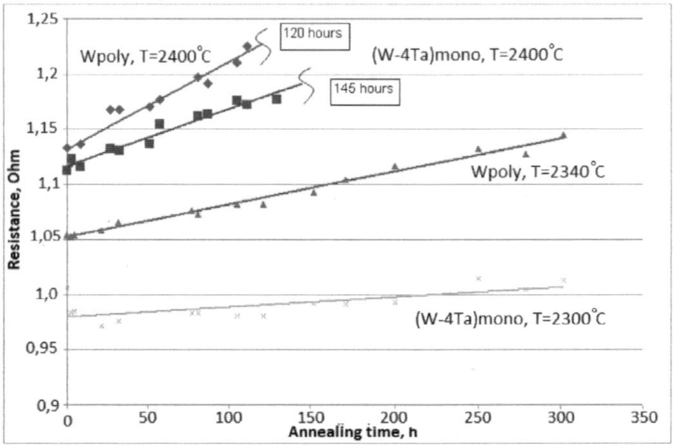

Figure 3. Evaporation of poly- and monocrystalline tungsten.

4. Advantage of nanomaterial for using in X-ray units

Though single crystals showed better characteristics both as anodes and as cathodes, the application of nanomaterials may be a more radical way of increasing the operating characteristics of X-ray tubes.

For instance, the application of nanopowders (Mo, Nb) for producing anodes based on molybdenum or molybdenum alloy with niobium must exclude the influence of the technological disadvantages above-described and increase permissible unit loads to the focal spot due to a decrease in porosity, an increase in strength and plasticity of material.

Some investigation results are known [4] where the work function of nanocrystal tungsten was lower by 0.8 eV than that of coarse tungsten. This phenomenon can be explained by the fact that when forming nanomaterial, current tubes with lower work function appear in the range 10 nm wide near grain boundaries. This effect can considerably increase X-radiation intensity at the same cathode temperatures or decrease the cathode temperature at the same X-radiation intensity by 400°C. Estimations show that at the temperature at which the effect of an increase in the emission characteristics for this material is still available (T ≤1500°C), one may rely on obtaining the density of emission current equal to 10 mA/cm^2.

A number of publications have appeared lately where some results of developing X-ray tubes with the use of carbon nanotubes (CNT) as cathode materials are described.

The development engineers from the University of South Caroline together with the other researchers presented materials on developing a scanning source of X-radiation with a cathode made of CNT. The anode voltage was 40-60 kW, electric field strength - 2.3-3.1 V/cm, emission current density reached 1 A/cm^2.

In parallel with the improvement of the characteristics of cathode made of CNT, the application of nanodimensional dopes may also increase strength properties of tungsten used conventionally.

5. High temperature deformation of tungsten

We obtained the effect of an increase in tungsten strength characteristics at high temperature annealings of samples as 100 μm thick foils of BA type tungsten (tungsten with silicon-aluminum and alkali additives) (Figure 4).

Investigations in the structure of this material samples after annealings (Figure 5). Allowed finding out nanodimensional (≤ 200 nm) inclusions along the grain boundaries of tungsten subjected preliminarily to heat treatment, resulting in a sharp increase (by a factor of more than 3) in material creep resistance at a temperature of 2200°C.

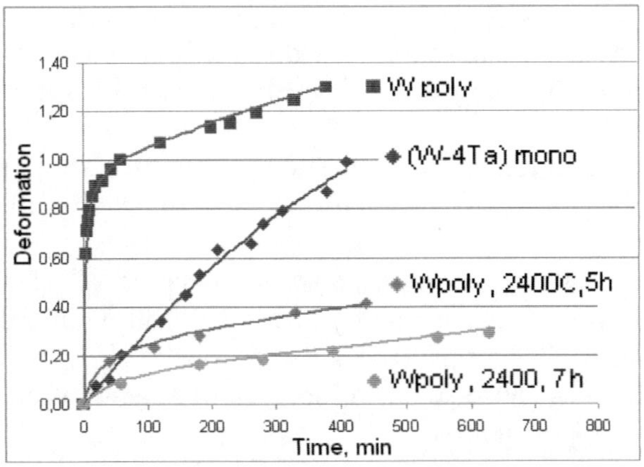

Figure 4. Deformation dependence on time at creep tests. 1 - polycrystalline tungsten without heat treatment; 2 - polycrystalline tungsten preannealed at a temperature of 2500°C for 5 hours. 3 - polycrystalline tungsten preannealed at a temperature of 2500°C for 7 hours. 4 - monocrystalline tungsten.

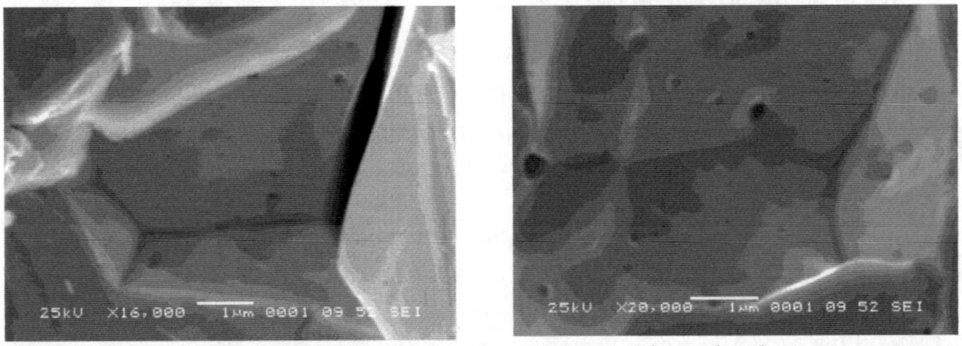

Figure 5. Microstructure of polycrystal tungsten subjected to heat treatment at a temperature of 2500°C for 5 hours.

As a whole, a qualitative increase in the operating characteristics of X-ray tubes with the use of nanomaterials and monocrystal materials can be achieved by a simultaneous increase in the density of electron current (a flat cathode of nanotungsten or tungsten nanocomposite) and by an increase in load-carrying capacity of an X-ray tube (anode made of nanocomposites of molybdenum with niobium or tungsten with rhenium) by providing a great number of interactions between electrons of increased density and a great number of atoms of anode target materials.

For an establishment of influence of duration of annealing on thermal stability of structure of a material on samples, of annealing within 30 minutes, the diamond pyramid puts marks. Then, these samples were annealing at temperature 2400°C within 1.5 hours and 3 hours.

Thus, temperature stability of this effect as shown is at high (up to 2400°C) temperatures. The similar effect of formation nanosizes pores in wires of modified tungsten is described in a number of works [4-6].

6. Conclusions

Comparison of structure and properties of samples from foils with equal and nonequal structure, past preliminary heat treatment and without it is carried out. Thus, for the first time it is received nanocompozite of tungsten in the form of a foil. Structure of this material is stable up to temperatures 2200 °C.

The effect of an increase in tungsten strength characteristics at high temperature annealing of samples as 100 μm thick foils of BA type tungsten (tungsten with silicon-aluminum and alkali additives) obtained.

From experimental data it is visible, that speed of creep of a sample past preliminary thermal processing is lower, than for sample without preliminary thermal annealing. The similar behaviour of a material is connected to distinctions in structure of the tested samples.

The results presented are a cause for developing a new generation of X-radiation sources with increased diagnostic opportunities.

Acknowledgements

This work was performed within the framework of the Center of Nuclear Systems and Materials supported by MEPhI Academic Excellence Project (contract № 02.a03.21.0005, 27.08.2013).

References

[1] Alekseev S, Taubin M and Yaskolko A. 2009 *20th Workshop ISTC in Korea Nanomaterals and nanotechnology* (Seul)
[2] Alekseev S V, Taubin M L and Yaskolko A A 2008 *Proc. 2 All-Russia national congresses on radiology*
[3] Alekseev S V, Taubin M L and Yaskolko A A 2008 *Medical physics* **2**
[4] Wright P K. 1978 *Metallurgical Transactions* A **9A** 955
[5] Yamazaki S, Ogura S, Fukazawa Y and Hatae N. 1978 *High Temperature - High Pressure* **10** 329
[6] Briant C L Potassium 1993 *Metallurgical Transactions A* 24A. 1073
[7] Klinkov A E, Nikolaev Yu V and Vybyvanetz V I 1991 *Proc. The second Inter.Confer. on Nuclear Power Enginering in Space. Physics of Thermionic Energy Converters* (Sukhumi) 37

High temperature dilatometry of simulated oxide nuclear fuel

A V Tenishev, V G Baranov, R S Kuzmin and S A Pokrovskiy

National Research Nuclear University MEPhI (Moscow Engineering Physics Institute)
31 Kashirskoe highway, Moscow, 115409, Russia

E-mail: avtenishev@mephi.ru

Abstract. High temperature dilatometry of model systems based on uranium dioxide with additives of burnable neutron absorbers both as Gd_2O_3 and as $AlGdO_3$, and fission products simulators (FPS) was performed. It shown that in some cases instead of high temperature samples shrinkage there is a sharp transition to the expansion, which is associated with an increase of the samples volume due to the formation of liquid phases. The beginning of a complex composition eutectic melting starts at temperatures from 1950 to 2250 °C in the uranium dioxide samples containing significant amounts of Al, Gd, and FPS. Thus, in the analysis of oxide nuclear fuel behavior at high temperatures should be considered that the formation of liquid phases is possible at a temperature of 1000 °C lower than a melting point of pure stoichiometric uranium dioxide if its initial composition became more complex.

1. Introduction

Currently, the VVER core forms from fuel assemblies containing oxide nuclear fuel based on enriched by U^{235} up to 4 - 5% uranium dioxide with burnable neutron absorber (BNA) (for example Gd_2O_3) that allow compensate the initial reactivity, align the energy release in a core volume and maintain the temperature reactivity coefficient for a given level, thereby, increasing the degree of nuclear fuel burn-up. The nuclear uranium dioxide (UO_2) or UO_2 with gadolinium oxide (Gd_2O_3) fuel pellets also contain small amounts of dopants (D) in the form of Al and Si oxides. Presence of the alumina-silicate phase, gadolinium oxide and fission products accumulated in a pellets during the reactor operation which may form the substitutional solid solutions with UO_2 or distinct phases reduces the thermal conductivity, increases complexity of phase composition, and hence, increases the fuel temperature and makes possible a phase transitions. The presence of phase transitions is one of the limitations of nuclear fuel performance as they leads to changes in volume and properties that can have a significant impact on the reactor fuel assemblies (FA) performance. In this paper analysis of a phase transitions presence in the model nuclear fuel samples was decided to hold with a help of high-temperature dilatometry.

2. Samples

Studies were performed on the model nuclear fuel (MNF) samples produced on the basis of uranium dioxide (table 1). All samples contained D in the following estimated amounts 0.005% Al + 0.0025% Si + 0.006% Cr. As the BNA additive Gd_2O_3 which forms a solid solution with UO_2 or $AlGdO_3$ compound in the form of self-phase was used. To simulate the phase composition of the fuel with burn up additives of fission products simulators (FPS) was used which total amount is equivalent to the 11 at. % burn up. The type and content of the specific FPS are presented in table 2.

Table 1. Features of MNF samples.

№ batch	Composition	Density (g/cm^3)
1	UO_2 + D + 6.9% Gd_2O_3	9.78 ± 0.03
2	UO_2 + D + 8.9% $AlGdO_3$	9.62 ± 0.03
3	UO_2 + D + 11% FPS	9.47 ± 0.03
4	UO_2 + D + 11% FPS + 8.9% $AlGdO_3$	9.45 ± 0.03

Table 2. Composition and content of FPS.

FPS	Y	Zr	Sr	Ba	Ce	Mo	Nd	Ru	La
Content (wt. %)	0.13	1.1	0.24	1.55	2.9	1.36	1.54	1.92	0.73

Samples containing D and BNA were prepared by mixing powders of the components with an intermediate briquetting and rubbing through a sieve. FPS addition was performed on the stage of preparation of uranium dioxide powder by means of joint co-precipitation to ensure uniform elements distribution. Subsequently the powders were pressed with a force of 250 MPa and sintered at 1750 °C in a dry hydrogen atmosphere for 3 hours.

Microstructure and composition studies of the model fuel samples individual phases was performed using a scanning electron microscope JEOL 6610LV, equipped with energy dispersion spectrometer (EDS) element analysis system.

The microstructure of the batch 1 samples represents a uniform grains distribution with an average size of 10 - 12 micrometers. Porosity in turn evenly distributed. Increased porosity observed in local areas with length from 30 to 100 micrometers. Outside of these areas there are individual pores or their absence. Gadolinium is unevenly distributed, but all located in solid solution in the UO_2 matrix.

The microstructure of the batch 2 samples shows unevenly size distribution of the uranium dioxide grain with Al_2O_3 and $AlGdO_3$ inclusions. Part of Gd is dissolved in the UO_2, thus released Al reacts with excess oxygen and form Al_2O_3. Alumina is unevenly distributed on the microsection surface. The matrix grain structure has a bimodal distribution with small (0.2 - 1.5 micrometers) and large (5 - 10 micrometers) grain. Fine grains are mainly formed around the second phase precipitates.

For the batch 3 samples mean grain size was 2 - 3 micrometers. Element analysis indicated that the uranium content in the matrix almost not varies from one grain to another, but the FPS addition distributed less evenly. Ba and Mo which must be allocated in separate phases, sometimes occur in solid solution. Electron microprobe analyzes allow us to conclude that the sample contains from three phases: fluorite phase having the average composition $U_{0,904}La_{0,078}Y_{0,004}Zr_{0,014}O_2$, metallic phase based on Mo (Ru), and perovskite-type phase $BaZrO_3$.

The microstructure of batch 4 samples is generally similar to batch 3, wherein $AlGdO_3$ is also in the form of self-phase precipitates.

Samples phase composition is also confirmed by X-ray data.

3. Equipment and methods

A high-temperature dilatometer DIL 402 E/8 «Pyro» (Netzsch, Germany) was used for the sample length measurements. The initial sample dimensions were measured with an electronic TESA micrometer calibrated with a set of standards.

To protect the graphite elements of dilatometer the test samples of MNF was placed in tungsten containers consisting of a tube with a diameter of 8 mm and two 1.5 mm thick spacers tight-fitting the container form the both sides. Use of the protective container imposes a limitation on the maximum temperature. The dilatometer is not suitable for measurements where graphite sample holder is in contact with the tungsten at temperatures above 2400 °C, so that the maximum temperature of the measurement was 2390 °C.

IOP Conf. Series: Materials Science and Engineering **130** (2016) 012032 doi:10.1088/1757-899X/130/1/012032

To avoid oxidation of samples, measurement of thermal expansion was carried out in an atmosphere of helium class is not worse than 5.0 (99.999%) with an additional heated MonoTorr gas filter.

The following temperature program was used: preheating to 1200 °C at a rate of 50 °C / min and then heating at a heating rate of 20 °C / min to a maximum temperature of 2390 °C, and then cooling without exposure in reverse order. Writing data of a linear displacement sensor while cooling was carried out to 800 °C.

4. Results and discussion

Figure 1 shows the dilatometric data for sample batch 1. It is seen that changes in the size of the sample at the initial stage of heating is due to thermal expansion. When approaching the sintering temperature of samples (1750 °C), the rate of expansion is slowing, due to densification. However, up to the maximum test temperature the expansion prevails over the shrinkage. At temperatures above 1956 °C small deflection starts, and at 2097 °C noise appears on the dilatometric curve which may be due to a liquid phase formation as will be shown below. It should be noted that the effects at these temperatures are repeated at the cooling of sample too, that indicates the constant composition of the phases formed. In the process of cooling the sample has a smooth decrease in its size, while in the high temperature part of the curve of expansion there spikes.

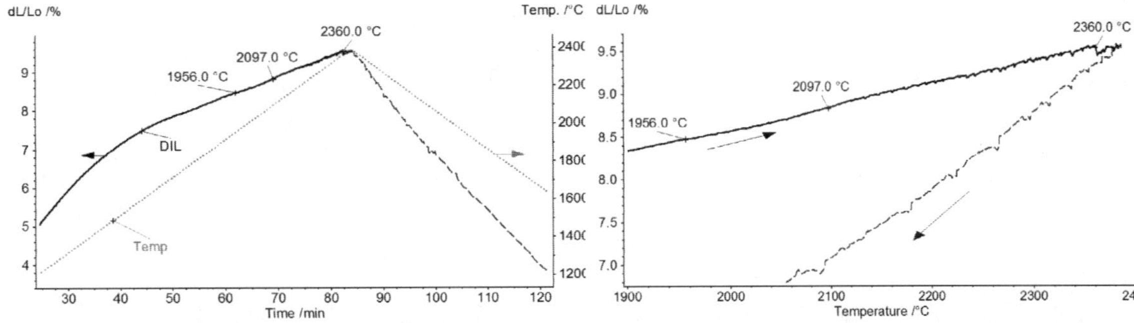

a b

Figure 1. Time (a) and temperature (b) dependencies of the batch 1 sample length (UO_2 – 6.9% Gd_2O_3).

The microstructure of the batch 1 samples after the test remained virtually unchanged. It is still a solid solution with a grain size increased to 20 µm with reduced porosity and uniform distribution of Gd. The presence of other phases has not been observed.

According to the UO_2-Gd_2O_3 phase diagram [1] in equilibrium for a given gadolinium content it must be a solid solution without forming any other phases. A possible reason for the appearance of features on the extension curve at high temperatures can be a phase transitions associated with the formation of eutectic with low melting temperature in the solid-solution and dopants system. For example, according to [2] in the UO_2-Al_2O_3 system at temperatures above 1930 °C two immiscible phases can exist and the eutectic formed at this temperature (figure 2).

The reason that these phases are not detected in the analysis of the microstructure may be a very small amount of dopants used in the manufacture of nuclear oxide fuel.

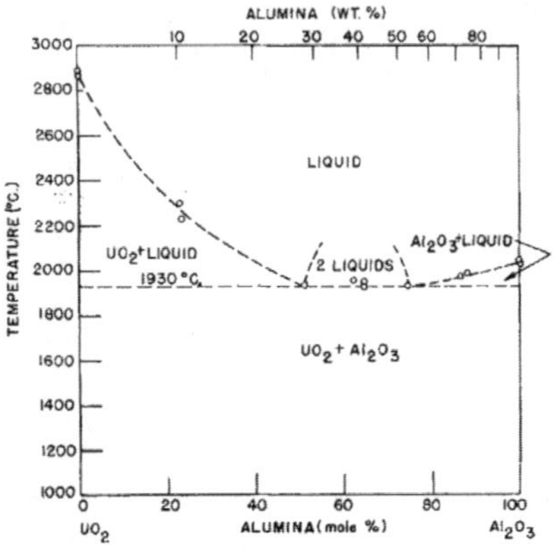

Figure 2. Diagram of phase equilibrium of UO_2-Al_2O_3 [2].

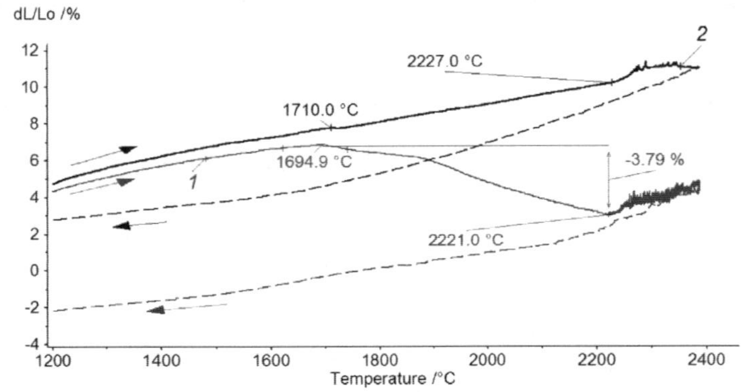

Figure 3. Temperature dependence of the batch 2 sample length: 1 - the first heating, 2 - the second heating, --- – cooling.

Dilatometric curves obtained for the batch 2 samples, are shown in figure 3. On the curve 1, corresponding to the first heating shrinkage about of 4% associated with densification is visible. Unlike of the batch 1 samples in this case shrinkage dominates expansion since the 1695 °C but at 2220 °C dependence drastically reversed and the noise appears on the curve. Curve 2 in figure 3 corresponds to reheat of the same sample after re-polishing of its end surfaces. There is practically no traces of shrinkage, but some features are also appears close to above mentioned temperatures. Apparently a sharp transition from the expansion to shrinkage and noise appearance on the curves at 2220 °C is associated with an increase in the sample volume due to the formation of significant amounts of liquid phase. This confirms by the analysis of the microstructure on the end surfaces of the sample (figure 4), which showed the presence of the phase that was in the liquid state during tests. Element analysis shows that the melted phase is a eutectic type Al_2O_3 - UO_2 with small additions of gadolinium and there is almost no aluminum in the body of the matrix grains.

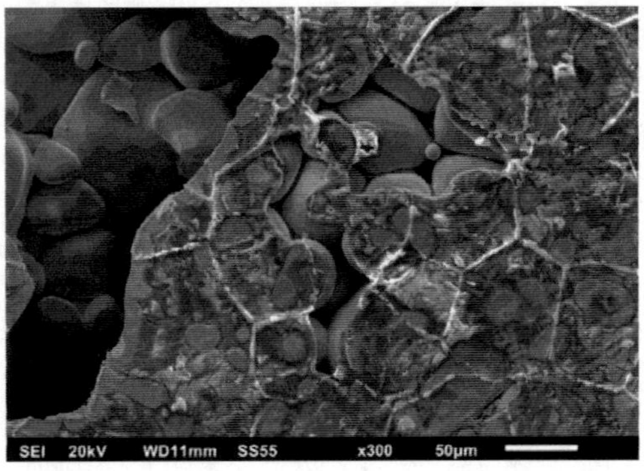

Figure 4. The microstructure of the batch 2 sample surface. The new phase that melts during test.

Thus, in the UO_2 + D + 8.9% $AlGdO_3$ system very small amounts of low temperature melting eutectics is formed at a temperature of about 1700 °C due to the high content of aluminum oxide [3] that results in a significant densification acceleration due to the liquid phase formation on the grains boundary. Further, the composition of this phase is complicating, and at a temperature of about 2220 °C the formation of considerable amounts of liquid phase leads to a noticeable increase in sample volume. The analysis of the cooling and reheating curves should take into account that the composition of the melting phase may be changed in the case of long-term measurements as well as

IOP Conf. Series: Materials Science and Engineering **130** (2016) 012032 doi:10.1088/1757-899X/130/1/012032

the part of the molten phase may evaporate at high temperatures, so the phase transition temperatures may vary.

Results of dilatometry of the batch 3 sample are shown in figure 5. The amount of shrinkage associated with densification of sample was 4.3%. Melting of significant amounts of the substance has not been detected.

The microstructure of batch 3 sample after the test is shown in figure 6. According to the quantitative element analysis the uranium and fission products simulators content is somewhat different in different areas: on the grain boundaries uranium content is slightly smaller, and additives - a few more. Some places with the metallic phase on the basis of molybdenum, ruthenium and tungsten were found on the sample surface that was in contact with tungsten spacer. The presented microstructure shows that, firstly, the process of thermal etching occurred, and part of the material was evaporated, and secondly, there are submelting traces on the surface, but liquid phase separation was not observed as in samples with a high aluminum and gadolinium content.

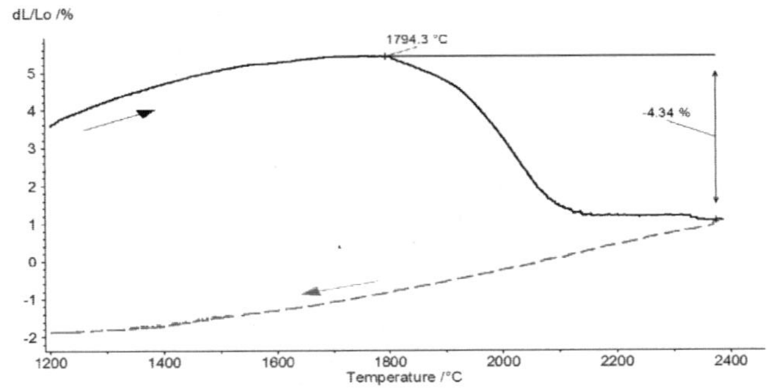

Figure 5. Temperature dependence of the batch 3 sample length (UO_2-11% FPS).

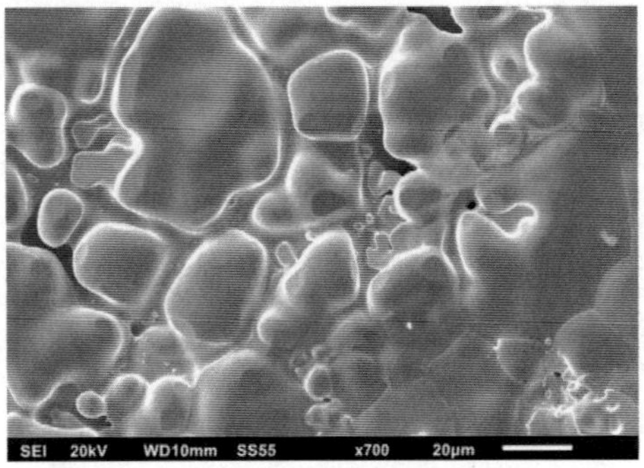

Figure 6. The microstructure of the batch 3 sample end face heated up to 2390 °C in a high temperature dilatometer.

Figure 7 shows the temperature dependence of the length change of the batch 2 and 4 samples, which composition differs only by the FPS presence. Curve for batch 2 is displaced downward for drawing readability, the ordinate axis corresponds to the curve for the batch 4. It can be seen that the introduction of FPS slightly increases the temperature of shrinkage over expansion predominance, but the temperature of a significant quantity of liquid appearance is reduced.

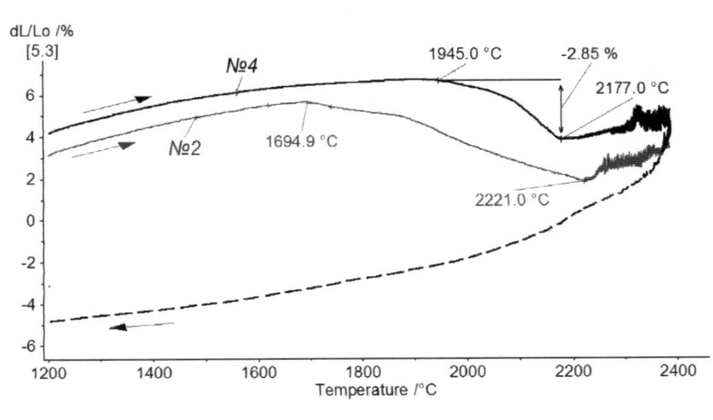

Figure 7. Temperature dependence of the batch 2 (UO_2 + D + 8,9% $AlGdO_3$) and 4 (UO_2 + D + 8,9% $AlGdO_3$ + FPS) samples length change

Microstructure analysis of the batch 4 samples end faces showed the presence of more complex eutectic composition containing Al_2O_3, UO_2, Gd and some FPS similarly to batch 2 samples.

5. Conclusion

1. It is shown that high temperature dilatometry can be used to detect phase transitions and to determine the liquid phase formation temperature in the oxide nuclear fuel samples with FPS and a BNA additions.

2. Shrinkage of samples mainly associated with they sintering occurs in the 1800-2000 °C temperature range.

3. In samples of uranium dioxide containing dopants and BNA in the form of Gd_2O_3 or $AlGdO_3$ at temperatures from 1950 to 2250 °C complex composition eutectics melting is observed both on the sample surface and in the bulk. Most low temperature melting phases composed mainly of Al, Gd and Si oxides and contained a small amount of uranium and FPS. Samples microstructure and composition element analysis of the individual phases after the tests confirmed the melting of some components of the structure.

4. The liquidus line has not been reached for all batches of samples in the presently investigated temperature range (T_{max} = 2400 °C), i.e. there always remains an unmelted frame of uranium dioxide or a solid solution based on it.

5. For the final interpretation of the results, it is beneficial to measure the melting points of the individual components of the structure.

Acknowledgements
This work was performed within the framework of the Center of Nuclear Systems and Materials supported by MEPhI Academic Excellence Project (contract № 02.a03.21.0005, 27.08.2013).

References

[1] Kang K W, Yang J H, Kim J H, Rhee Y W, Kim D J, Kim K S, Song K W 2007 *Thermochimica Acta* **455** 134
[2] Lamberston W A, Muller M H 1953 *J. Amer. Ceram. Soc.* **36** (10) 329
[3] Kolitsch U, Seifert H J, Aldinger F 1997 *Journal of Alloys and Compounds* **257** 104

IOP Conf. Series: Materials Science and Engineering **130** (2016) 012033 doi:10.1088/1757-899X/130/1/012033

Evolution of pore size distribution during sintering of oxide nuclear fuel

V G Baranov,Y N Devyatko, A V Tenishev, V V Mikhalchik and O V Khomyakov

National Research Nuclear University MEPhI (Moscow Engineering Physics Institute) Kashirskoe shosse 31, Moscow, 115409, Russia

E-mail: homaykov_o_v@rambler.ru

Abstract. Uranium dioxide pellets were sintered at various temperature routes and atmospheres with different oxygen content. Statistically calculated pore size distribution of the sintered pellets and distribution function was obtained. It is shown that the average pore size is almost unchanged at intermediate stage of sintering while the total number of pores reduced.

1. Introduction

Nowadays uranium dioxide is the main fuel for commercial reactors. Due to the high melting temperature of uranium dioxide (3120 °C), the most effective way to obtain fuel pellets is high temperature sintering (at temperatures 1700-1750 °C). Industrial manufacturing of uranium dioxide has several features that can be affected on the final properties. At powder stage UO_2 is obtained with hyperstoichiometry composition (O/U> 2). Then powder was compacted by pressing at 200-250 MPa to a green density $\rho_0 = 10.96$ g/cm^3. Sintering is carried out in reducing atmosphere to obtain fuel with stoichiometric composition. At this work sintering of uranium dioxide pellets was conducted in Ar+8%H_2 atmosphere with small additives of oxygen.

The sintering process is the most difficult to describe, and many aspects are very poorly studied. Also this fact is applies to pore formation. Green compact of uranium dioxide has a very complex inner surface. Structure of the green pellets consist of individual powder particles (or microcrystallites with typical size is about 200-300 nm) and large agglomerates of this ones (Figure 1).

Figure 1. Structure of UO_2 green pellet.

During sintering the powder particles are combined and grain boundaries are formed. The inner structure is restructured. Very important that the process of grain growth and dissolution process are occurred simultaneously but these processes unrelated to each other directly.

The main characteristic of sintering process is a sample shrinkage during which occurs dissolution of pores inside the compact. The relative change of the total pores volume is connected with changing of pellet volume. Therefore we need to know law of pores evolution which will allow to describe the sintering process and kinetics of this one.

Most of the papers [1-7] devoted to the theoretical description of sintering process is very often don't take into account the pore size distribution, only average pore size. All authors suppose that characteristic size of all pores are the same. In such approximation the law of change of average size isolated pore is $\Delta\langle R\rangle/\langle R\rangle = g(t, T)$, where $\langle R\rangle$ – the average pore size, t – time, T – temperature, g – function of time and temperature. This type of formula corresponds with formula of linear sample dimension: $\Delta L/L = N\cdot\Delta\langle R\rangle/\langle R\rangle = N\cdot g(t, T)$, where L – linear dimension of the sintered samples, N (= Const) – amount of pores in the pellet. This is equivalent to the choice of the pore distribution in the form of Dirac delta-function $f(R, t) = N\cdot\delta(R-\langle R\rangle)$ centered at a point which is equal to the average pore size. Actually, the pore distribution of pellets is not narrow and during sintering it is occurs both reducing size of pores and their total amount ($N \neq$ Const). Moreover, the average pore size does not coincide with a maximum of size distribution. Hence, the Dirac delta function is a very rough approximation for the pore size distribution function. The kinetics of nuclear fuel sintering and the value of its densification depend on many factors: oxygen potential, the annealing temperature, heating rate [8]. Evolution process of the pore size distribution function may depend on same factors.

The purpose of this work is an experimental investigation evolution of pore size distribution function during the oxide nuclear fuel pellets sintering.

2. Calculation of pore size distribution function

To identify the porosity of uranium dioxide the samples were cut and polished with diamond paste. 2D pore size distribution function was calculated by means of Scanning Electrone Microscope images. Images with different magnifications covered different areas on the section (such as 100, 500 and 1000 μm^2). Image areas were chosen such a manner that pores of a certain size were clearly visible on the image. The total area of the pores then calculated on the images. The obtained data were presented in two columns: the areas of pores and quantity of pores with given areas. This array was based on the calculation of 2D - porosity: S: $P_i = N_i/S$ (N_i – amount of pores, S_i – the total image area, i – group of pores with area S_i.

Usually there are the same dimensions of the pore areas on the different images ($i = k$). To account these "crossing" areas we used the following algorithm. If two images with different areas $S^{(1)}$ and $S^{(2)}$ had groups of pores with equal size, formula of the porosity P_k will:

$$P_k=(N^{(1)}+N^{(2)})/(S^{(1)}+S^{(2)}),$$

where $N^{(1)}$ and $N^{(2)}$ – the amount of pores with the same areas on the images with different magnification.

Thus, we can calculate 2D discrete porosity over the entire range of pore size. The average area of the pores, average equivalent pore diameter and average porosity calculated based on discrete 2D - porosity by formulas:

$$\langle S\rangle = \sum_i S_i P_i \Big/ \sum_i P_i;\ \langle D\rangle = \sqrt{\frac{4}{\pi}} \sum_i \sqrt{S_i} P_i \Big/ \sum_i P_i;\ P = \langle S\rangle \sum_i P_i\,.$$

The most obvious form of pore size distribution is histogram. Where ordinate is the amount of pores in interval of pores diameters [D_k, D_{k+m}] relative to the total amount of pores. Abscissa is the equivalent pore diameter.

$$\Delta N/N = \sum_{i=k}^{k+m} P_i \Big/ \sum_i P_i$$

3. Pore size distribution during sintering with different heating rate

Samples of uranium dioxide were sintered with next temperature route: heating (with different heating rates 4, 6 and 8 K/min), then annealing at 1600 °C during 465 min. Temperature sintering route are shown in Figure 2. There are sintering processes of samples: duration of sintering is decreased with decreasing of heating rate. Table 1 shows the initial and final density of samples sintered with different heating rate. The densities of samples in Table 1 were determined both by means of hydrostatic weighing in water and by means of direct calculation using measured weight and volume of sample. From table 1 it is followed that the final densities of sintered pellets of uranium dioxide are equal within inaccuracy of measurements.

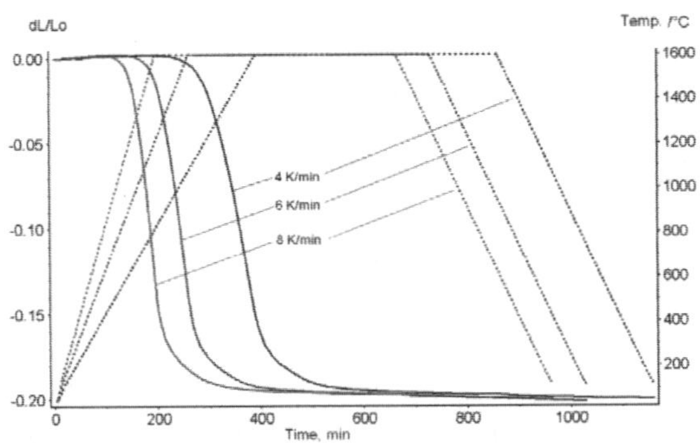

Figure 2. Sintering routes of uranium dioxide samples and their sintering curves.

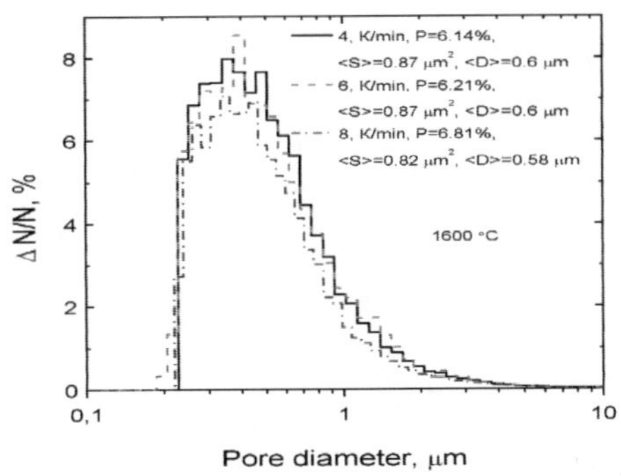

Figure 3. 2D pores size distributions of samples sintered at 1600 °C with different heating rates.

On the figure 3 it is shown 2D pore size distribution function of samples sintered with heating rates 4, 6 and 8 K/min at annealing temperature 1600 °C. On the same figure are shown average porosity of pellet, average areas of pores and their average diameter. It can be seen that change of heating rate does not cause change of pore diameter. Average characteristics of the pore ensemble determined only by the duration and temperature of isothermal annealing. 2D histograms of pores size distributions have the same form.

Final density of samples calculated using 2D pore size distribution and measured density by usual methods are shown in the Table 1. The average density of pellets calculated by pore size distribution function is close to measured density within the deviation 1%. This coincidence within the empirical error indicates on good accuracy of the method of distribution functions in comparison with other methods of determining the density of the samples.

Table 1. Density of the samples before and after sintering.

	Initial density	Final density		
Heating rate (K/min)	volume measurement (% TD)	volume measurement (%TD)	hydrostatic weighing in water (%TD)	pore size distribution (%TD)
4	49.5	94.2	94.7	93.9
6	49.6	94.2	94.7	93.8
8	49.5	93.2	94.8	93.2

4. Pore size distribution during sintering at different annealing temperatures

Series of long-term isothermal annealing of UO_2 pellets carried out at temperatures 1300 °C, 1400 °C for 40 hours and at temperature 1500 °C during 8 hours. Sintering at 1500 °C is less because sintering curve have reached the plateau. The sintering curves of these samples are shown on figure 4. It is seen that greatest density is achieved by tablet sintered at highest sintering temperature. The densities of sintered pellets obtained by dilatometer measurements are represented in table 2.

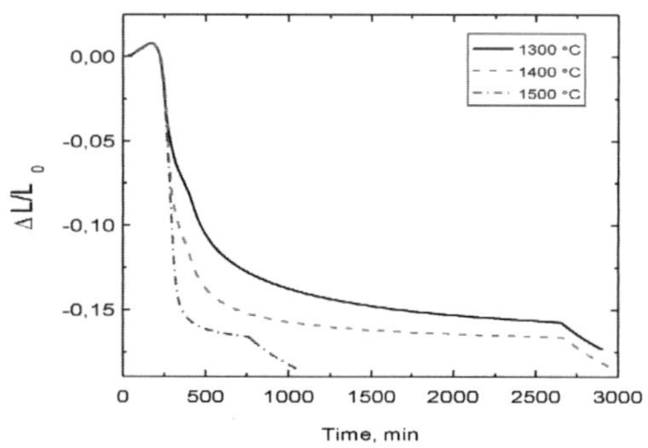

Figure 4. Sintered curves of UO_2 samples

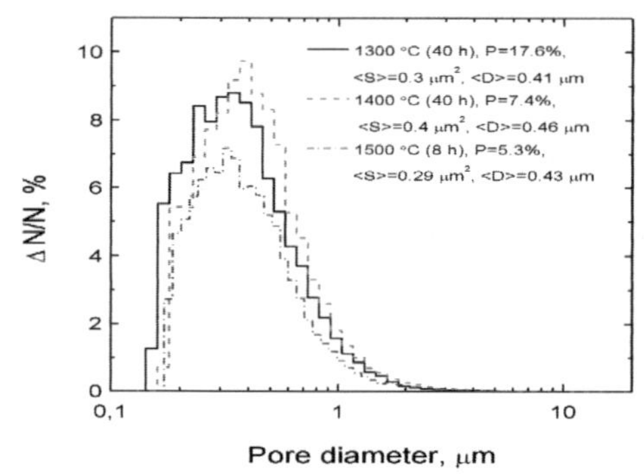

Figure 5. 2D pore size distribution of UO_2 samples sintered at different temperatures

2D pore size distribution functions of sintered samples with different sintering temperatures are shown on figure 5. On the same figure are shown average porosity of pellet, average areas of pores and their average diameter. The figure 5 shows that the average characteristics of the distribution - the middle area and pore diameter are constant during sintering. Maximum of pore size distribution function as seen from figure 5 practically does not change while sintering temperature increased.

Table 2. Final parameters of samples

Annealing temperature (°C)	Final density	
	volume measurement (%TD)	based on pore size distribution (%TD)
1300	85.4	82.4
1400	89	92.6
1500	88.9	94.7

The total amount of pores with all sizes is reduced by increasing the sintering temperature. Only small pores dissolved completely. Thus, as a result of long annealing, function of pore size distribution is narrowed, and the amount of pore size is reduced.

5. Pore size distribution during sintering in atmospheres with different oxygen content

It is known that the beginning of the sintering temperature depends on oxygen contenf of atmosphere in a furnance. Sintering in atmospheres with extremely low oxygen partial pressure is shifted the beginning of the sintering to the higher temperatures. Moreover, the deviation from stoichiometric composition of UO_{2+x} depends on oxygen content in atmosphere. The sintered curves of UO_2 samples annealed during 40 h (sample №1) and for 20 h (the other samples) at 1300 °C are shown on figure 6. Samples with conventional numbers № 2, 3 and 4 contained 180 ppm Al $(OH)_3$ and sample № 1 was made without alloying additions. Figure 6 shows the value of oxygen potential in the atmosphere for each sample. For each value of the oxygen potential is conform deviation (x) from the stoichiometric composition. So, sintering of green pellet of uranium dioxide in atmosphere with ΔG_{O_2} = -440, -395, -365 kJ/mol will result hyperstoichiometric composition of UO_{2+x} and deviation from stoichiometry: x = 0.0000, 0.0002 and 0.0007 respectively [8].

Figure 6 shows that the alloying pellet sintered more efficiently in atmosphere with the highest oxygen potential. Final densities of the sintered samples in Figure 6 are presented in Table 3. Difference of the final density for the samples № 1 and 2 is associated with their difference in green density.

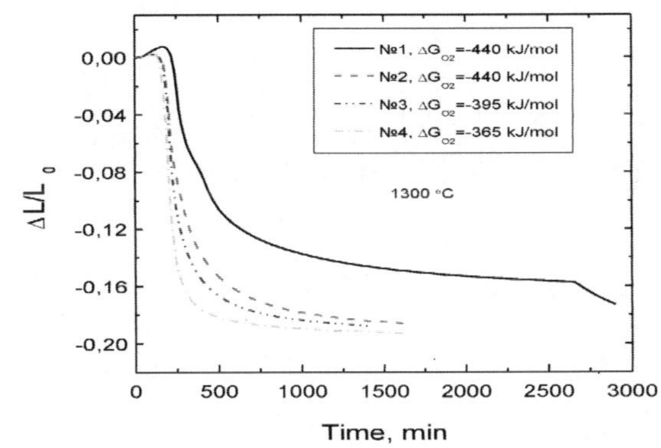

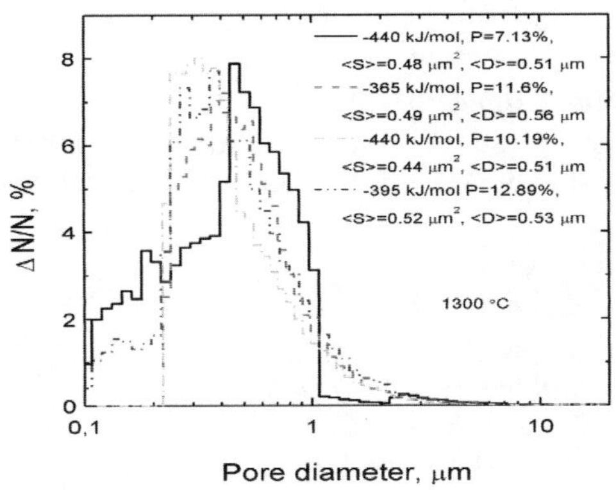

Figure 6. Sintering curves of unalloyed and alloyed uranium dioxide samples sintered at 1300 °C in atmospheres with different oxygen potential

Figure 7. Pore size distribution in UO_2 samples sintered at 1300 °C in atmospheres with different oxygen potential

Figure 7 shows the pore size distribution function of sintered samples. It can be seen that the greatest density is achieved through the complete dissolution of submicron pores with diameters D < 0.3 μm. Even a slight increase in the partial pressure of oxygen increases the rate of dissolution of the pores. Note that the average pore area and its equivalent diameter are practically independent of the sintering atmosphere.

Table 3. Samples parameters

№	Oxygen potential (kJ /mole)	Final density		
		hydrostatic weighing (%TD)	volume measurement (%TD)	based on pore size distribution (%TD)
1	-440	91.1	88.7	92.9
2	-440	90.5	87.2	89.8
3	-395	91.8	89.9	87.1
4	-365	92.5	90.1	88.4

6. Conclusions

Analysis of all obtained distributions functions sintered UO_2 showed that the characteristic pore size is about 0.1<D<2 μm. During sintering average pore diameter and their average area was almost unchanged. It is found that the pore size distribution of the sintered UO_2 pellets wasn't depended on heating rate but it is determined by sintering temperature and duration of sintering process and also by oxygen potential of atmosphere. During the heating, pore size distribution function redused. This is due to complete dissolution of submicron pores (characteristic dimensions less than 0.3 micron). The total quantity of pore size decreased. It is shown that the δ-function approximation of size distribution function is incorrect. Dissolution kinetics of isolated pore does not coincide with densification kinetics of fuel pellet.

Thus, it is necessary to take into account the evolution of the pore ensemble for the correct description of the sintering process.

Acknowledgements

This work was performed within the framework of the Center of Nuclear Systems and Materials supported by MEPhI Academic Excellence Project (contract № 02.a03.21.0005, 27.08.2013).

References

[1] Coble R L 1959 *Kinetics of High Temperature Processes.* (N.Y., Wiley)
[2] Coble R L 1961 *J. Appl. Phys.* **32** (5) 787
[3] Coble R L 1961 *J. Appl. Phys.* **32** (5) 793
[4] Pines B Ya 1946 *J. Techn. Phys.* **16** 137
[5] Pines B Ya *1954 Usp. Fiz. Nauk (Advances in Physical Sciences)* **4** 501
[6] Frenkel' Ya I 1946 *J. Techn. Phys.* **16** 29
[7] Johnson D L 1969 *J. Appl. Phys.* **40** 192
[8] Baranov V G, Deviatko Y N, Tenishev A V and Khomiakov O V 2014 *Nuclear Physics and Engeneering* **5** (2) 101

Superelastic alloy Ti-22%Nb-6%Zr: X-ray study of deformation features

Yu Perlovich, M Isaenkova, S Chekanov, V Fesenko and O Krymskaya

National Research Nuclear University MEPhI (Moscow Engineering Physics Institute), Kashirskoe highway 31, Moscow, 115409, Russia

E-mail: yuperl@mail.ru

Abstract. In this paper features of deformation of superelastic Ti-22%Nb-6%Zr alloy were studied. Using X-ray method of generalized pole figures for study orientation dependence of martensite transformation in superelastic Ti-22%Nb-6%Zr alloy it was shown that: under compression the strain hardening due $\beta \rightarrow \alpha''$ transformation is significantly greater in grains with rolling plane {100} than in grains with a rolling plane {111}; the lattice microdeformation at β- and α'' - phase grains under loading is redistributed so that it extremely assist the sample macrodeformation (most of all); most intense $\beta \rightarrow \alpha''$ transformation occurs in grains, where $\Delta d / d_{av} > 0$, most prone to increasing of its lattice volume. Diagrams of the cross-correlation substructural parameters built on the basis of generalized pole figure indicate that in loaded sample the deviation of interplanar space from the weight average level promotes the activation of martensite transformation at which the X-ray line broadening is minimal.

1. Introduction

Superelastic alloys are now popular objects of studies both because of interesting scientific aspects and owing to perspective practical applications. The mechanism of superelasticity consists in continuation of the elastic deformation by ~0.2%, connected with interaction of neighboring atoms, by reversible elastic deformation by 2÷3% and more due to martensitic transformations (MT). The given paper is devoted to X-ray study of the superelastic alloy Ti-22%Nb-6%Zr (at. %) (TNZ) using methods of texture analysis. Several detailed papers on this alloy were published recently in different countries [1-4]. In particular, it was shown that superelasticity of TNZ alloy is conditioned by $\beta \leftrightarrow \alpha''$ MT, where β-phase has BCC crystalline lattice and α''-phase – orthorhombic lattice. It was found as well that $\beta \rightarrow \alpha''$ MT realizes with an increase of the elementary cell volume [4]. But the approach of authors was directed mostly to integral description of material and they did not split it into separate texture components, playing different roles in behavior of the alloy by MT. Since TNZ alloy is used usually in the rolled state, its samples have always the developed crystallographic texture and we need to take into account significant regular substructure inhomogeneity, accompanying formation of this texture. In the course of preliminary experiments it was shown that in TNZ alloy MTs and connected with them superelasticity take place at the room temperature.

The foil of superplastic alloy Ti-22%Nb-6%Zr (at. %) was rolled down to thickness of ~0.2 mm. Then the foil was deformed in different manners in order to observe reversible MTs: (1) the foil was located in the special set for the X-ray diffractometer, where it could be stretched under X-ray irradiation; (2) the foil was bent by winding on arbor, so that at one its side the surface layer was extended, whereas at another side the surface layer was compressed. In order to promote MT, before

IOP Conf. Series: Materials Science and Engineering **130** (2016) 012034 doi:10.1088/1757-899X/130/1/012034

deformation of the foils they were annealed at the recrystallization temperature of 700°C [3], since in perfect, coarse grains martensitic needles move easier. Then winded annealed foils were straightened partially or completely for the sake of convenient X-ray measurements, especially – texture measurements. By straightening in annealed foils MT $\beta \rightarrow \alpha''$ begins, so that both stages of superelasticity can be observed.

X-ray studies of foil samples were conducted by automated texture diffractometers DRON-3 and D8 DISCOVER of Russian and German (BRUKER) manufacture, respectively. Both diffractometers were used by registration of X-ray line profiles, DRON-3 – by measurement of direct texture pole figures (DPF) [5], D8 DISCOVER – by measurement of generalized pole figures (GPF) [6].

2. Experimental results

2.1. Stretching of foil

The foil sample was stretched in the special set to diffractometer up to ε=2.5% by continuous registration of the diffraction spectrum using a position-sensitive detector. Under tensile deformation X-ray lines of the initial β-phase shift to higher Bragg angles (Figure 1-*a*,*b*) and their profiles become wider (Figure 1-*c*), but by removal of loading both positions of lines and their angular widths return to their initial values. When taking into account, that the foil is subjected to tension in the rolling plane, it is understandable, that along perpendicular to rolling plane (ND, that is normal direction) material of the foil experiences compression, – namely therefore its diffraction lines shift to higher angles. Since the almost complete recovery of X-ray line parameters occur after deformation degree of 2.5%, it becomes evident, that we deal with superelasticity.

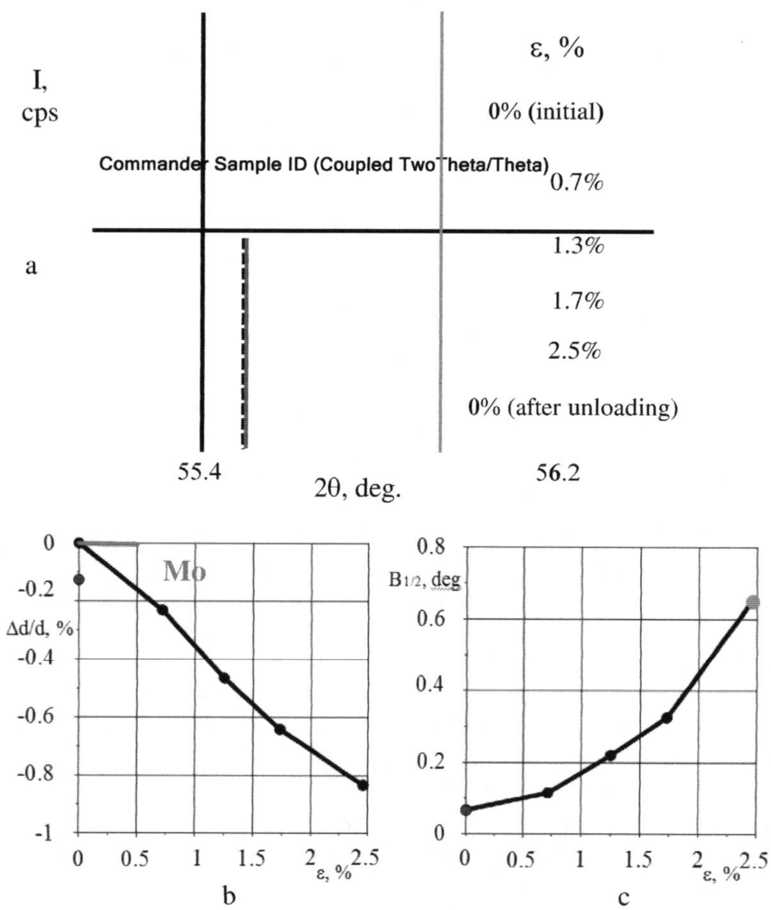

Figure 1. Change of line profile (a), interplanar spacing (b) and dispertion of crystal lattice (c) during increasing sample deformation by tension. Green color is used for sign of data on sample with the maximal deformation, blue – for unloaded sample.

Behavior of the foil from TNZ alloy under tension within the limits of elastic deformation was compared with that of the molybdenum foil (Figure 1). The return of X-ray line parameters, i.e. peak position and half-width of the profile β, to their initial values after unloading of samples is the general property of metal materials, though concrete trajectories of β recovery vary in details. Mechanisms of this effect are different in cases of Mo and TNZ alloy: broadening of X-ray lines due to usual elastic deformation is connected with inhomogeneous changes of interplanar spacing in grains of β-phase, whereas broadening of X-ray lines in the course of superelastic deformation is conditioned by spreading of martensitic needles, subdividing β–grains. Recovery of X-ray lines half-width by unloading of sample in cases both of elastic deformation and superelasticity occurs by redistribution of regions with different interplanar spacings or different degrees of β↔α″ MT. Finally, the half-width decreases down to its initial value independently on the trajectory of this return, if only irreversible plastic deformation would not prevent it.

2.2. Bending of foil

The samples of foil 17x17 mm in size were winded on the arbor 11 mm in diameter around rolling direction (RD) or transverse direction (TD), annealed at 700°C, straightened and stuck to the plain holder in the predeformed state. At that the deformation degree of the surface layer was equal to ~1.5%. For both sides of straightened foil samples partial DPF {002} with angular radius 70° were measured (Figure 2). In order to see visually changes in DPF due to bending of foils, subtraction pole figures DPF Δ were constructed as well. Here $\Delta(\psi, \varphi) = P_{def}(\psi, \varphi) - P_{in}(\psi, \varphi)$, P_{in} and P_{def} – pole densities in points with angular coordinates (ψ, φ) of DPF {002} for initial rolled foil after annealing and for straightened foil, annealed after winding. Since shown DPF {002} are normalized [5], the real decrease of pole density in the center of DPF, corresponding to grains with rolling plane {001}, is accompanied with the supposed its increase in points, where pole densities do not change, including those, corresponding to grains with rolling plane {111}. This explains features of the observed distribution in DPF Δ(ψ, φ) (Fig. 2). It is evident, that as a result of MT β→α″ the volume fraction of β-grains in the texture component with rolling plane {001} decreases, whereas the volume fraction of β-grains in the texture component with rolling plane {111} remains the same or changes insignificantly.

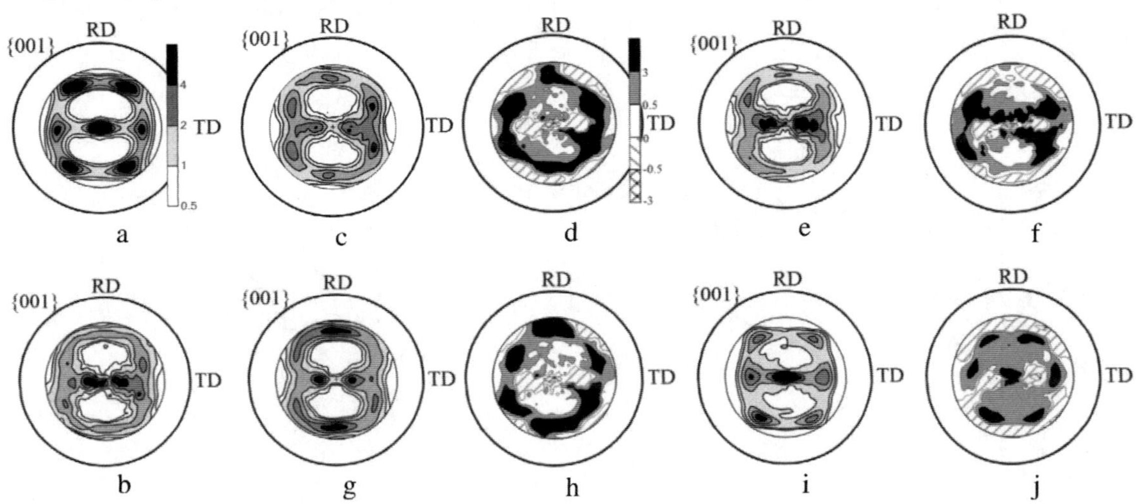

Figure 2. Change of texture (PF {001}) of foil from Ti-22%Nb-6%Zr alloy as a result of its bending: a – cold rolling; b – annealing at temperature 700°C during 1 h; c, d – compression along RD (c – PF, d – differential PF); e, f – compression along TD (e – PF, f – differential PF); g, h – tension along RD (g – PF, h — differential PF); i, j – tension along TD (i – PF, j— differential PF).

On the basis of experimental partial DPF{011}, {002} and {112} by use of ODF method [7] with the help of software LABOTEX inverse pole figures (IPF) were obtained for ND, RD and TD of the foil. IPF for ND, shown in Figure 3, confirm data of DPF in Figure 2.

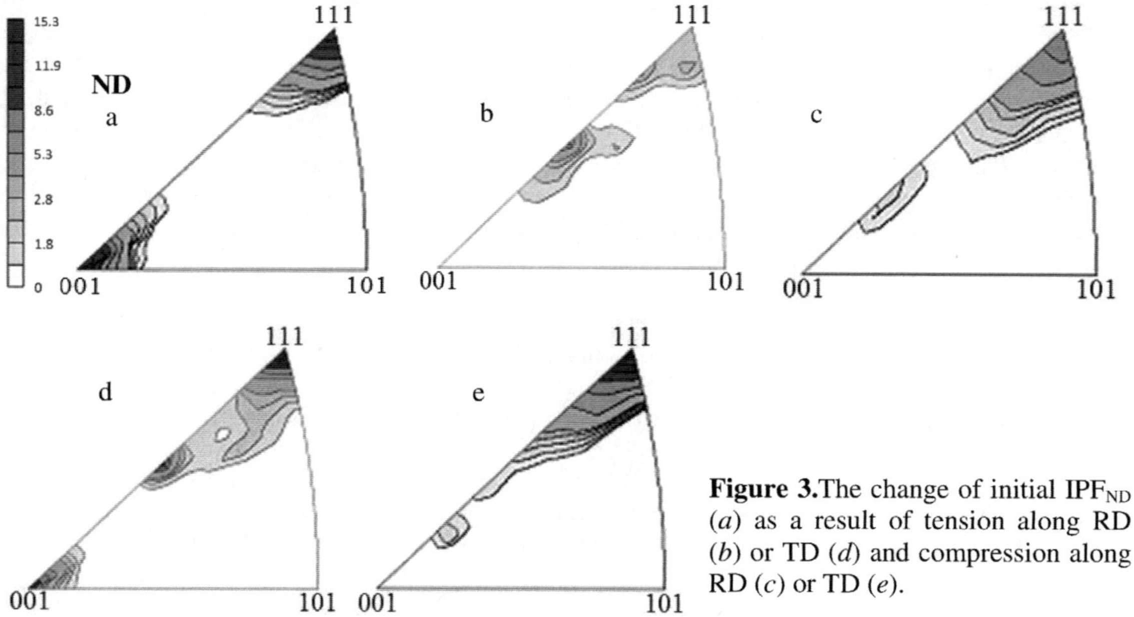

Figure 3. The change of initial IPF_{ND} (*a*) as a result of tension along RD (*b*) or TD (*d*) and compression along RD (*c*) or TD (*e*).

One more experiment consists in gradual straightening of the winded annealed foil by use of the series of arbors with increasing radii; at that, by each radius, corresponding to the definite deformation degree of the surface layer by compression up to 2%, X-ray lines (002) and (222) were registered. As a result of compression along RD the lattice parameter *a*, measured along ND, increases as well as angular widths of both X-ray lines. When the increase of the lattice parameter stops due to development of dislocation plastic deformation, the loading was gradually removed and parameter *a* returns to the initial value (Figres 4 and 5).

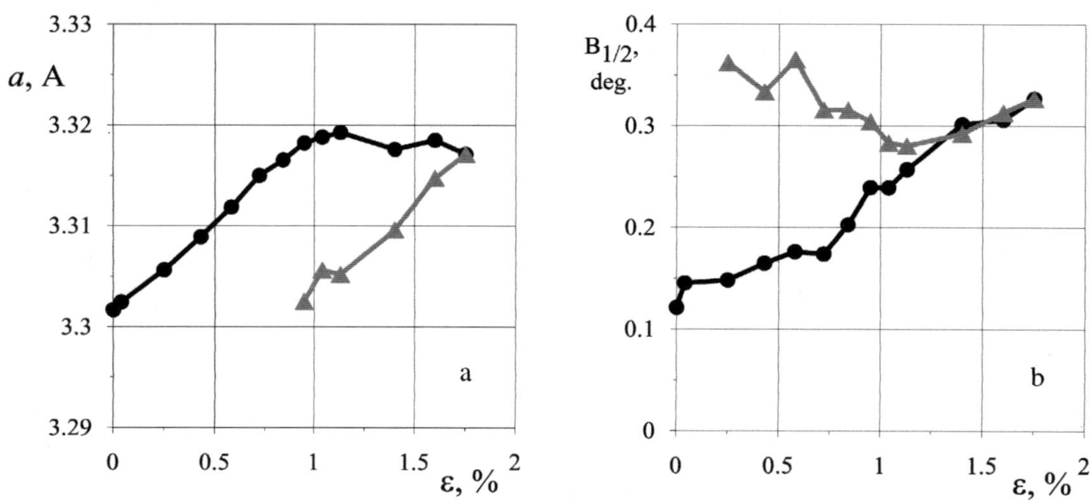

Figure 4. Changes of crystal lattice parameter (a) calculated by using of X-ray reflection (200) and half-width of this reflection (b). Black color indicates the process of loading, red – unloading.

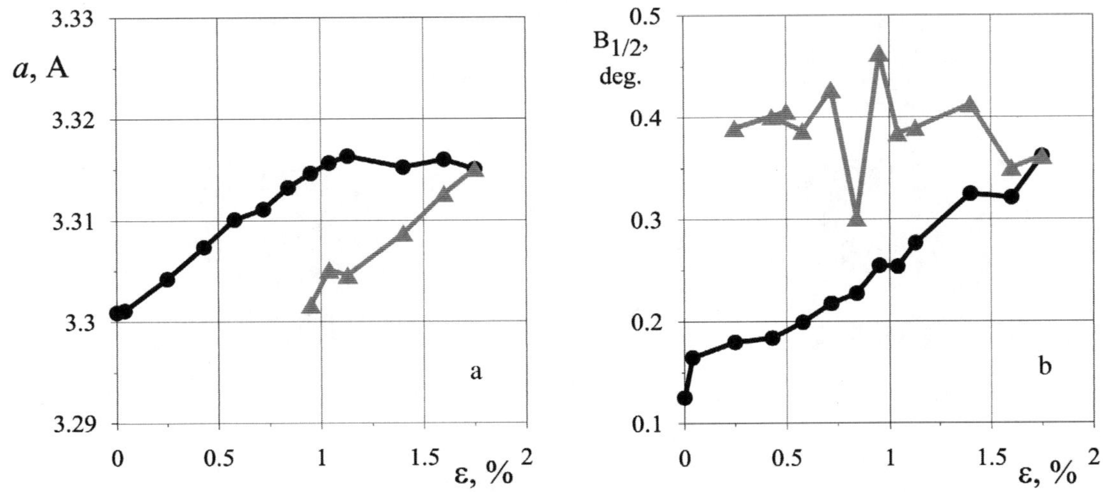

Figure 5. Changes of crystal lattice parameter (a) calculated by using of X-ray reflection (222) and half-width of this reflection (b). Circle points correspond process of loading, and triangle points – unloading.

The most noteworthy feature of shown curves is theelastic recovery of the lattice parameter *a*, whereas the angular half-width of X-ray lines do not show the similar tendency. By unloading of sample the curve $B_{1/2}(\varepsilon)$ for grains with rolling plane {001} coincides with the curve of loading within the interval $\Delta\varepsilon=1.7\div1.3$, while for grains with rolling plane {111} it deflects from the loading curve starting with the very beginning of unloading. This difference indicates, that in latter grains MT was absent.

3. Application of Generalized Pole Figures

A series of data, confirming the non-uniform character of $\beta\leftrightarrow\alpha''$ MT in the foil of TNZ alloy, was obtained by the comparatively new X-ray method of Generalized Pole Figures (GPF) [6]. The method involves successive registration of the same X-ray line profile for grains of all orientations in the course of texture measurement using a position sensitive detector. Correction of measured data by use of standard samples and computer treatment of obtained results allowed to construct so called GPF {hkl}, where some measured diffraction parameter or calculated substructure parameter is plotted, so that points (ψ, φ) correspond to reflecting grains with the same coordinates of axes {hkl}.

In particular, in Figure 6 GPF β_{002} and GPF $(\Delta d/d_{mid})_{002}$ are presented, respectively, for foils, winded along RD. Winded samples were annealed, straightened and fixed on the flat holder. GPF were measured both from compressed and stretched sides of the foil by deformation degree of the surface layer $\varepsilon\approx1.5\%$. Presented GPF β_{002} for the compressed side (Figure6-a) directly confirms, that grains with rolling plane {001}, localized near the center of PF, are characterized with higher strain hardening, i.e. with higher X-ray line broadening β_{002}, than grains with rolling plane {111}, localized at the angular distance of 55° from the center. The reason of this effect is the following. According to the main regularity of texture formation, in rolled BCC-metals grains with rolling plane {001} are larger, than grains with rolling plane {111} [7]. This principle remains true both after recovery and primary recrystallization. Therefore formation of martensitic needles develops in grains with rolling plane {001} more freely, than in grains with rolling plane {111}, without obstacles from grain boundaries.

Since MT $\beta\rightarrow\alpha''$ occurs with an increase of interplanar spacings [4], stretching of the surface layer promotes this MT, while compression of the surface layer hinders MT. Therefore MT $\beta\rightarrow\alpha''$ develops more actively on the stretched side of the bend foil, than on its compressed side. As a result, it is difficult to reveal in GPF regions of preferred passing of MT β→α″, though one can suppose, that in grains with rolling plane {001} MT would be most active both under compression and tension.

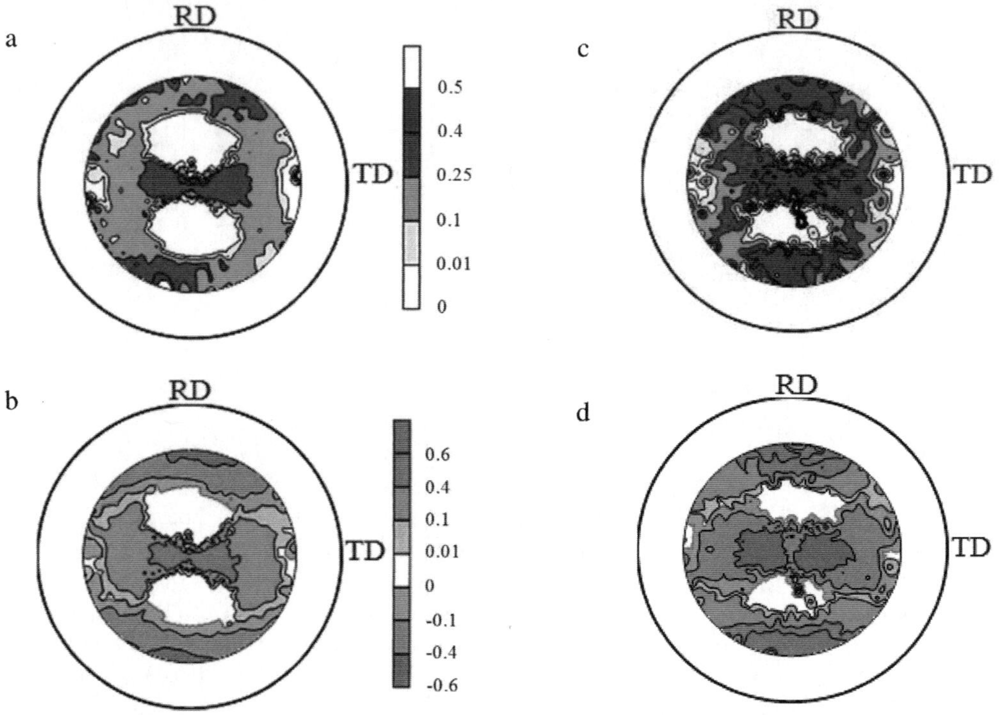

Figure 6. GPF β_{200} (*a, c*) and GPF $\Delta d/d_{200}$ (*b, d*) for the foil sides compressed (*a, b*) or stretched along RD (*c, d*).

As for GPF MT β→α″, it ought to be noted, that in conditions of compression and tension studied material has a number of resources, realization of which would promote operation of external loading. A stereographic projection of the loaded foil allows to see, how these resources are realized. When, for example, in GPF$(\Delta d/d_{av})_{002}$ the value $(\Delta d/d_{av})(\psi, \varphi) > 0$, it signifies, that in β-grains with axes <001>, having coordinates (ψ, φ), the interplanar spacing d under applied loading proves to be higher, than its weighted average level, responsible for macrostress. Points (ψ, φ) near the center of GPF correspond to axes <001>, directed close to the normal to rolling plane (ND), whereas points in the vicinity of rolling direction (RD) correspond to axes <001>, close to RD. Since elastic compression of metal material along some direction is accompanied with its elastic tension in the perpendicular plane and vice versa, the distribution of values $(\Delta d/d_{av})_{002}$ in the corresponding GPF under uniaxial loading.

As it is seen in Figure 6-b, elastic deformation of β-grains under compression along RD redistributes in such a manner, that along direction of compression in these grains axes <001> are situated with values $\Delta d/d_{av} < 0$, while across this direction – axes <001> with values $\Delta d/d_{av} > 0$. Therefore along GPF diameter, perpendicular to direction of compression, the zone, where $\Delta d/d_{av} > 0$, is situated.

Along direction of tension in β-grains axes <001> with values $\Delta d/d_{av} > 0$ take up position, while across this direction – axes <001> with values $\Delta d/d_{av} < 0$ (Fig. 6-d). Therefore along GPF diameter, perpendicular to direction of tension, the zone, where $\Delta d/d_{av} < 0$, is situated.

Grains of martensitic phase α″ develops in such a manner, that along direction of tension their axes <111> with values $\Delta d/d_{av} > 0$ are situated, while across this direction in GPF $(\Delta d/d_{av})_{\alpha''}$ – axes <111> with values $\Delta d/d_{av} < 0$ take position.

4. Conclusions

(1) Superelasticity of Ti-22%Nb-6%Zr alloy is conditioned by passing of elastic deformation of the β-phase lattice into its reversible deformation by $\beta \rightarrow \alpha''$ MT, beginning from stresses, which are smaller, than the yield point of β-phase.

(2) An increase of interplanar spacings in the course of $\beta \rightarrow \alpha''$ MT promotes growth of β-phase elastic deformation by stretching of sample, at least, up to ε=2.5% and impedes elastic deformation of β-phase by compression of the surface layer of sample down to deformation above ε=1%.

(3) By bending of the foil of TNZ alloy, which, as all metal materials with BCC crystalline lattice, has the two-component rolling texture, when the deformation degree of the surface layer is close to ε=1.5%, MT $\beta \rightarrow \alpha''$ is localized on the compressed side in less fragmented grains with rolling plane {001}, whereas in grains with rolling plane {111} MT is practically absent.

(4) On the stretched side of the bent foil MT $\beta \rightarrow \alpha''$ covers the more significant part of volume and its selectivity is not evident.

(5) The reversible widening of X-ray lines for samples of superelastic TNZ alloy is conditioned by fragmentation of β-phase grains in the course of MT $\beta \rightarrow \alpha''$ due to spreading of martensitic needles, whereas their irreversible widening is connected with dislocation plastic deformation; at that, by unloading of sample the reversible part of line widening rehabilitates itself first of all even in the case, when MT $\beta \rightarrow \alpha''$ is followed by dislocation plastic deformation, since taking down of loading initiates at once MT $\alpha'' \rightarrow \beta$ and to significant degree removes fragmentation of β-grains, connected with formation of martensitic needles.

(6) By unloading of the bent foil at its compressed side grains with rolling plane {001} reveal partial narrowing of X-ray line (002) profile, while grains with rolling plane {111} do not show the similar narrowing of line (222), confirming the fact of reversible MT $\beta \leftrightarrow \alpha''$ in the first group of grains and its absence in the second group.

(7) It was shown by use of the X-ray method of Generalized Pole Figures, that:
- by compression in grains with rolling plane {001} strain hardening due to MT $\beta \rightarrow \alpha''$ is significantly higher, than in grains with rolling plane {111};
- in grains of β- and α″-phases microdeformation of crystalline lattice in the course of loading redistributes in such a manner, that maximally promotes macrodeformation of the sample;
- the most intensive MT $\beta \rightarrow \alpha''$ takes place in grains, showing the greatest tendency to increase the volume of crystalline lattice, i.e. where $\Delta d/d_{av}>0$.

Acknowledgements
This work was performed within the framework of the Center of Nuclear Systems and Materials supported by MEPhI Academic Excellence Project (contract № 02.a03.21.0005, 27.08.2013).

References
[1] Ma J, Karaman I, Maier H J and Chumlyakov Y I 2010 *Acta Materialia* **58** 2216–24
[2] Tobe H, Kim H Yand Miyazaki S 2009 *Materials Transactions* **50** (12) 2721–25
[3] Kim J I, Kim H Y, Inamura T, Hosoda H andMiyazaki S 2006 *Materials Transactions* **47** (3) 505
[4] Kim H Y, Ikehara Y, Kim J I, Hosoda H and Miyazaki S 2006 *Acta Materialia* **54** 2419–29.
[5] Isaenkova M G and Perlovich Yu A 2014 *Regularities of formation of crystallographic texture and substructure inhomogeneities during plastic deformation and heat treatment* (Moscow, NRNU MEPhI) p 528
[6] Bunge H J, Perlovich Yu and Isaenkova M 1997 *Textures and Microstructures* **29** 241
[7] Perlovich Yu A 1983 *Structure, texture and mechanical properties of deformed Mo-based alloys* Trefilov V I (ed) (Kiev: Naukova dumka) chapter 5 p 146

IOP Conf. Series: Materials Science and Engineering **130** (2016) 012035 doi:10.1088/1757-899X/130/1/012035

Two-ply anode X-ray tube for computed tomography scanner

D Ignatyev[1,2], M Taubin[1,2], D Chesnokov[1], V Malyshev[1] and A Yaskolko[1,2]

[1] Federal State Unitary Enterprise "Science Research Institute "LUCH", 24 Zheleznodorozhnaya st, Podolsk, Moscow region, 142100, Russia
[2] National Research Nuclear University MEPhI (Moscow Engineering Physics Institute), Kashirskoe highway 31, Moscow, 115409, Russia

E-mail: taubin@sialuch.ru

Abstract. This report presents a method of the formation of tungsten layer on the graphite surface. The described method can be used to create the anode of powerful x-ray tubes for medical purposes, in particular, a computer tomograph (CT). The thermal properties of the graphite base and the deposited tungsten coating, as well as the strength of the resulting coating were studied. Thermal fields in the CT-anode with a power of 100 kW were calculated.

1. Introduction

One of the major factors determining the progress of the development of medical X-ray equipment is increasing the power of X-ray sources. This is especially important for the X-ray tube of CT. It is known that the X-ray tube is a thermal device in which more than 90% of input energy is converted into heat. Also, the focal track in tubes with a rotating anode for CT scanners is exposed to considerable temperature fluctuations (up to 1000°C per 1 mm) during a single shot. These fluctuations are repeated during the rotation of the anode. Rapidly changing stresses appear as a result of temperature changes that are accompanied by the appearance of micro-cracks on the anode surface, metal melting and evaporation. This is accompanied by degradation of the anode surface, which reduces the X-ray intensity by the absorption of radiation by protruding surface irregularities.

One of the possible effective methods of heat removal and, as a consequence, reduction of thermal and mechanical stresses is the selection of anode material and the technology of its manufacture.

The anode material must provide:

1. A thermal stability of the target made of refractory material, as the local overheating of the focal spot of the anode reaches 2500°C.
2. A high thermal capacity of the anode disk. The maximum value of the thermal capacity anode is an important parameter for the X-ray tube. The higher the value of the thermal capacity of the anode base material, the more the number of switching the tube without its overheating during the operation of CT is allowed. In addition, the high heat capacity makes it possible to minimize the size of the anode disc.
3. A high heat exchange coefficient between its components and the use of materials with a high coefficient of thermal conductivity.
4. A high emissivity of the anode surface.
5. A high atomic number of the material, because it determines the integrated intensity of radiation;
6. A coherence of thermal expansion of anode materials and the focal track material working in conditions of thermal cyclic loads.

7. A mechanical strength of the anode disc with the lowest possible moment of inertia.
8. A purity of the X-ray (no diffusion escape of the substrate material on the surface of the anode).
9. The joint of the target material with the substrate must be durable, continuous, and able to provide effective heat dissipation from the focal track of the combined anode.

Tungsten satisfies all of these requirements. However, a disadvantage of tungsten is that it is very fragile, therefore uncomfortable to handle. In addition, the high specific weight also imposes limitations. Therefore, in modern designs of anodes, tungsten and tungsten-rhenium alloys are used as a thin target layer formed on the anode substrate having a high heat capacity.

This work is dedicated to the optimization of the method of forming a tungsten layer on the surface of the graphite substrate having high operating properties (a high specific heat, a low density and a high emissivity). The method used in this paper can be applied to create anodes of powerful x-ray tubes for medical purposes, in particular for computed tomography and angiography.

2. Experimental data

The formation of a tungsten layer on the graphite surface was performed by CVD (chemical vapor deposition). The main advantage of crystallization from the gas phase is that the deposition takes place at the atomic level. This allows getting a refractory material with a density close to the theoretical value. The coating may be applied to any form of the product. The starting material for the coating is tungsten hexafluoride.

The reaction of tungsten hexafluoride with hydrogen is described by the overall equation:

$$WF_6(g)+3H_2(g)=W(solid)+6HF(g).$$

The anode is a product that works in conditions of thermal cycling. For successful operation of the anode, equal or similar coefficients of linear thermal expansion of the constituent materials are required. For that, the thermal linear expansion coefficients of a deposited tungsten coating and graphite in various directions of its compression were identified by dilatometry (Figure 1). Graphite can have two directions of the compaction: the press direction is coincident with the axis of symmetry of the anode ($\alpha \|$) and the press direction perpendicular to the axis of symmetry of the anode ($\alpha \perp$).

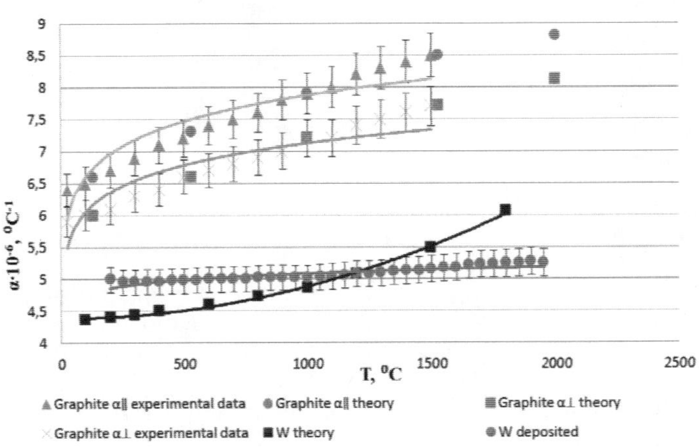

Figure 1. Coefficients of linear thermal expansion of the anode materials.

The obtained data shows that it is advisable to use graphite $\alpha\perp$ in the direction perpendicular to the pressing axis. The difference in the thermal linear expansion coefficients of materials in this case is no more than $2{\cdot}10^{-6}$ K^{-1}.

To estimate the influence of the difference in the linear expansion coefficients of materials by Lame equations, the internal mechanical stresses occurring at the tungsten-graphite boundary were calculated. The temperature dependence of the thermal linear expansion factor of materials can be determined using the analytical expression $\gamma = \varphi \cdot T^{0,67}$, where φ depends on the material properties.

Arising radial and hoop stresses can be written in the following form:

$$\sigma_r = -\frac{1}{R_H^2 - R_B^2}\left[P_H \cdot R_H^2 \cdot \left(1 - \frac{R_B^2}{R^2}\right) + P_B \cdot R_B^2 \cdot \left(\frac{R_H^2}{R^2} - 1\right)\right]$$

$$\sigma_t = -\frac{1}{R_H^2 - R_B^2}\left[P_H \cdot R_H^2 \cdot \left(1 - \frac{R_B^2}{R^2}\right) - P_B \cdot R_B^2 \cdot \left(\frac{R_H^2}{R^2} - 1\right)\right]$$

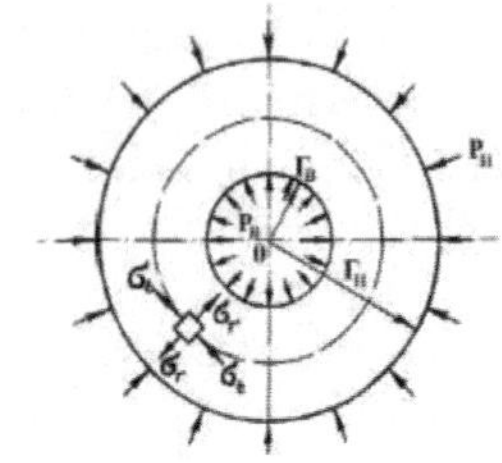

where σ_r and σ_t - radial and circumferential stresses, R - linear coordinates in the anode body, R_B and R_H - inner and outer radii, and P_B , P_H - pressure acting from the cylinder axis (inside) and to the cylinder axis (from the outside).

The internal stresses arising at the tungsten-graphite boundary with the use of graphite in the direction parallel to the pressing axis is 0.33 GPa and 0.18 GPa in the direction perpendicular to the pressing axis.

For products operating at high temperature, the efficiency of heat dissipation is an important parameter. The heat dissipation in vacuum is possible with radiation from the surface and thermal conductivity. From the analysis of the results of our studies [5] and the literature data, it is shown that the thermal conductivity of polycrystalline tungsten decreases linearly from 104 to 90 W/(m·K) with increasing the temperature from 1200 to 2400°C.

Based on our results obtained, the degree of blackness increases monotonically in the temperature range of 1200-2200 °C and its value is higher by 10-15% than that of the reference data.

In this work, the heat capacity was measured and it was showed that the specific heat of polycrystalline tungsten varied in the range of 130-160 J/(kg·K) at the temperature of 1500 to 2100 °C.

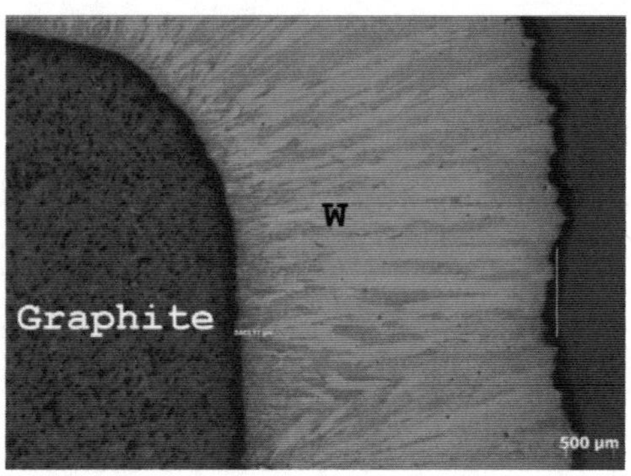

Figure 2. Structure of the deposited W coating.

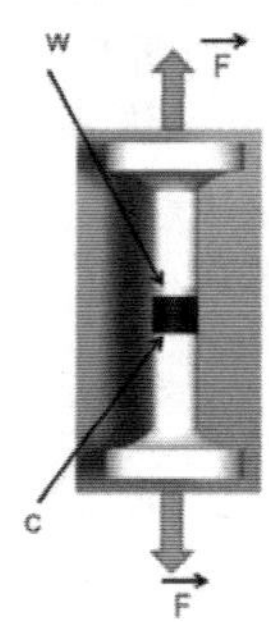

σ border > σ graphite

Figure 3. Qualitative determination of the adhesion of the tungsten coating for a small sample.

To analyse the state of the boundary after the deposition of tungsten, the graphite sample was cut into several parts for metallographic analysis.

Figure 2 shows that the surface relief of the tungsten coating is 40 μm. The quality of adhesion of the applied layer to the substrate is satisfactory. To determine the quality of the adhesion of the tungsten boundary, standard graphite samples were tested using a tensile testing machine (Figure 3). It is shown that failure occurs along the body of graphite. This shows qualitatively that the strength of adhesion of the tungsten-graphite boundary is higher than the graphite tensile strength on stretching. Open pores, cavities and cracks on the tungsten-graphite boundary are not detected. The result is formation of a coating with a columnar structure, a high density, and the presence of a certain growth texture. The initial layers of the deposited tungsten consist of small grains (≤1 mm), which indicates the presence of a plurality of nucleation.

Taking into account the results obtained by the methods of mathematical modelling, the anode temperature fields were calculated. The maximum temperature at the tungsten-graphite boundary at 100 kW for 1 s is 1450°C, the maximum temperature on the surface of the target under the same conditions is 1800°C.

To determine the growth rate of the carbide layer (Figure 4) at the tungsten-graphite interface from temperature, a series of annealing was carried out. The carbide layer thickness was 10 μm after annealing at 1400°C for 5 h and 100 μm at 1700°C for 1 h (Figure 5).

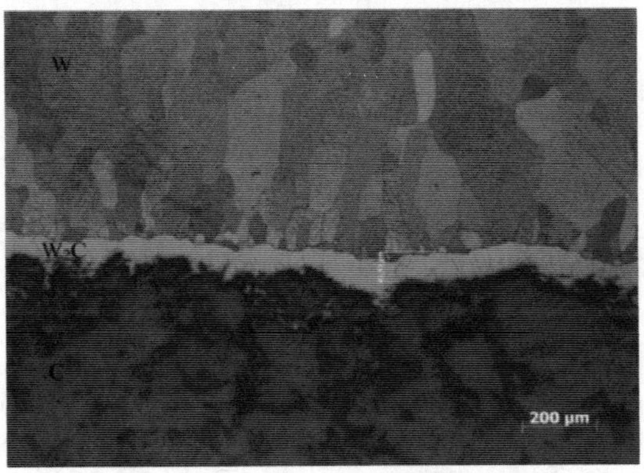

Figure 4. Structure of the tungsten-graphite boundary after annealing at 1700°C for 1 h.

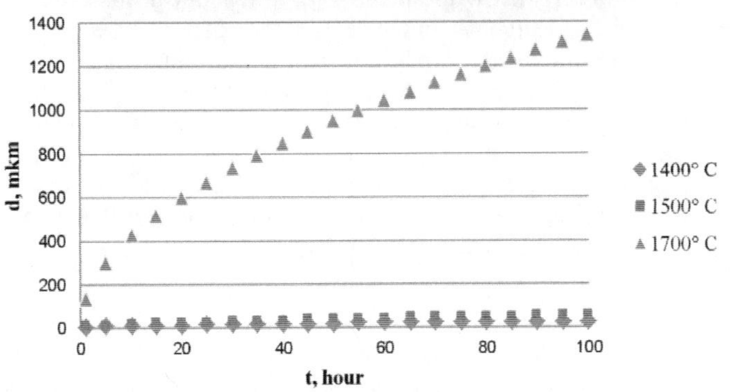

Figure 5. Dependence between the calculated thickness of the carbide layer and the time of annealing.

3. Conclusions

A calculated-experimental research of the effectiveness of the CVD method for forming a target tungsten layer on the graphite anode has been carried out. It has been shown that the given method can be regarded as an advanced one in the manufacture of X-ray tube anodes working in conditions of high-intensive, alternating mechanical and thermal loads.

Acknowledgements

This work was performed within the framework of the Center of Nuclear Systems and Materials supported by MEPhI Academic Excellence Project (contract № 02.a03.21.0005, 27.08.2013).

References

[1] Taubin M, Platonov V and Yaskolko A 2009 *Medical Technology* **1** (in Russian)
[2] Espe V 1962 *Technology of vacuum materials (Moscow: ln. German)* p 631
[3] Alekseev S, Taubin M and Yaskolko A 2013 *Proc. 2nd Biotechnology World Congress* (Dubai) 18
[4] Evstyuhin A, Rusakov A and Gavrilov E 1984 *Sat Refractory metals, alloys and compounds with a monocrystalline structure* (Moscow: Nauka*)* p.259 (in Russian)
[5] Taubin M, Platonov V and Yaskolko A 2007 *Tcvetnie metally* **11** 29 (in Russian)

IOP Conf. Series: Materials Science and Engineering **130** (2016) 012036 doi:10.1088/1757-899X/130/1/012036

Imitators of plutonium and americium in a mixed uranium-plutonium nitride fuel

S N Nikitin, D P Shornikov, B A Tarasov, V G Baranov and M A Burlakova

National Research Nuclear University MEPhI (Moscow Engineering Physics Institute) Kashirskoye Highway 31, Moscow, 115409, Russia

E-mail: d.p.shornikov@mail.ru

Abstract. Uranium nitride and mix uranium nitride (U-Pu)N is most popular nuclear fuel for Russian Fast Breeder Reactor. The works in hot cells associated with the radiation exposure of personnel and methodological difficulties. To know the main physical-chemical properties of uranium-plutonium nitride it necessary research to hot cells. In this paper, based on an assessment of physicochemical and thermodynamic properties of selected simulators Pu and Am. Analogues of Pu is are Ce and Y, and analogues Am - Dy. The technique of obtaining a model nitride fuel based on lanthanides nitrides and UN. Hydrogenation-dehydrogenation-nitration method of derived powders nitrides uranium, cerium, yttrium and dysprosium, held their mixing, pressing and sintering, the samples obtained model nitride fuel with plutonium and americium imitation. According to the results of structural studies have shown that all the samples are solid solution nitrides rare earth (REE) elements in UN.

1. Introduction

At this time discusses the use of mixed uranium-plutonium nitride (MUPN) as a fuel for fast reactors and addition to fuel minor actinides (Am, Cm, Cf). The key problem of this research is to expand knowledge of the changes of occurring in nitride fuel with the introduction of plutonium and minor actinides (MA), as well as during reactor irradiation. When the main volume of specified data is only available during reactor irradiation and subsequent studies in hot cells, which in turn is associated with a number of technological and methodological difficulties. The study of mixed uranium-plutonium fuel containing MA, also faced with the need to work in heavy protective box that greatly complicates and expensive experiments. Preliminary results (phase composition, thermal properties, etc.) can be obtained based on surrogate-materials in which radioactive elements (Pu, Am, etc.) are replaced by elements-imitators. This method is widely used for research oxide fuel [1, 2]. Therefore, the aim of this work is the rationale for the choice of imitators nitride plutonium and americium nitride.

2. Method of creation model nitride fuel

Metallic uranium initially hydrogenated at a temperature of 250-270 ° C to obtain a powder with a high active surface area that reacts with nitrogen at a faster rate than the compact metal [3]. Then uranium hydride dehydrogenation is conducted at a temperature (430-450) ° C resulting in the evolution of hydrogen gas, the destruction of the compact metal and fragmentation it into powder in its reiteration "hydrogenation - dehydrogenation" several times leads to reduction of the specific surface of the powder. Empirically it found that 3 cycles "hydrogenation - dehydrogenation" is enough to produce a powder grain size of about 20 microns. The reaction of uranium hydride and a powder of uranium to nitrogen is carried out in the temperature range 500-800 ° C for 2 hours. The resulting U_2N_3 decomposed in a vacuum at a temperature of about 1100 ° C. Uranium nitride has a relatively low oxygen content (100-200) ppm and the carbon (less than 100 ppm), which usually depends entirely upon the purity of the metal used. It has also been found possible hydrogenation not in hydrogen, but in argon-hydrogen mixture, in this case, due to the reduced content of hydrogen in the mixture of 9% and hence lower the partial pressure of hydrogen is necessary to raise the temperature to the hydrogenation reaction (500-600) ° C, in turn, in the dehydrogenation as necessary to reduce the temperature to 100 °C, and the process carried out at a diffusion pump for pumping within (30-40) min. At the same time, marked advantages of this method: a nitride obtained in the form of larger chips rather than powder with a characteristic gray-steel surface, the purity of the obtained product is slightly higher purity powder of size 20 microns. After receiving it mononitride uranium milled in an agate mortar, the powder particle size is about 50 microns. Nitrides of cerium and dysprosium are obtained in a similar way. Then, powders of nitrides of uranium and lanthanides is mixed in a vortex mixer was filled in a mold and compressed with a force of 200 MPa. The obtained compact was unloaded from the box into a molybdenum boat and sintered in a furnace at 1600 °C for 5 hours to a density of (80-95)% of the theoretical.

The microstructure nitride modeling performed by the scanning electron microscope JEOL 6610 LV, definition of uniformity distribution plutonium-imitators was determined by EDX microanalysis consoles.

In addition to scanning electron microscopy as part of this work carried out X-ray phase analysis of powders modeling nuclear fuel. The analysis was carried out in the X-ray camera RKD (X-ray Debay camera) diameter of 57.3 mm, recording the interference pattern was performed on a cylindrical film while using normal (balanced) and asymmetrical methods of charging the film. Radiation used copper $K\alpha$. Despite the presence of more modern methods of X-ray analysis, we used the powder method with the registration of the interference maxima in the film on the grounds that the method makes it possible to equip the samples with little or no risk of oxidation, as well as working with a very small sample weight.

3. Selecting simulators Pu and Am in nitride nuclear fuel

According to some literature data, such as replacement of the lanthanides may be considered. This is due to the fact that both the lanthanide and actinide series in the filling of the electrons is not the outer but the inner shell *5f*, and according to the theory Seaborg lanthanides are physico-chemical analogues actinides.

One of the key parameters that determine many properties of the compound, and is the type of lattice parameter. All the nitrides of rare earth metals, including scandium and yttrium, crystallize in the face-centered cubic lattice of the NaCl type and correspond to the composition MeN. The phases of other compounds in the rare earth metal oxide is not observed [4]. Also for lanthanide nitrides well-known pattern, is to reduce the lattice parameter with increasing atomic number of the element, the so-called effect of "lanthanide contraction."

The lattice parameters of lanthanide nitrides generally much larger lattice parameters actinide nitrides. There are four lanthanide nitrides, which have values of the lattice parameter, next to the nitrides of actinides - Ce, Dy, Gd, Y, so in the future we will stop its attention on these elements.

Dissolution nitrides of lanthanides nitride uranium will lead to an increase in the lattice parameter is thus the closest analogue is the material, which causes an increase in uranium nitride lattice parameter close to this nitride plutonium. Analysis of the relationship indicates that dissolution of lanthanide nitrides leads to a strong increase in the lattice constant than the dissolution of plutonium nitride. At the same time, it is clear that at low concentrations of plutonium nitride (up to 20 mol.%), The closest analogue may be considered material nitride cerium. In addition, we note that according to [4-6] dissolution of yttrium nitride uranium nitride virtually no increase in the lattice parameter of the latter, due to the proximity of the values of the lattice parameters YN and UN. In this respect yttrium nitride can be considered very close substitute nitride plutonium.

IOP Conf. Series: Materials Science and Engineering **130** (2016) 012036 doi:10.1088/1757-899X/130/1/012036

Analysis [7,8] showed that in the range of operating temperatures (the sintering temperatures) (1700-2000 K) the partial pressure of nitrogen and the metal nitride plutonium fairly close to the pressure of yttrium nitride, lanthanum nitride curves and cerium are somewhat lower, but when the temperature rises more than 2500 K, the curves are the same.

Thus, on the basis of this analysis it is expedient to choose as the simulator nitride plutonium when tested at low temperatures and high temperatures - nitrides of yttrium and cerium, respectively. The simulator can select americium nitride dysprosium, all the above materials are available and can be synthesized in the laboratory.

4. Results

Uranium nitride is a powder with a particle size of from 0.1 to 10 microns. After compacting and sintering powder of uranium mononitride obtained tablets. The tablet has few pores, cracks are absent. As a result of electron microprobe and X-ray phase analysis of the oxide phase is not detected. Figure 1 shows the microstructure of tablets Mononitride uranium with addition of 5 wt. % Yttrium. Figure 1 (a) visible white phase separation, according to microanalysis phase defined as yttrium oxide (Y_2O_3). Its occurrence is associated with short-term stay of the sintered sample mints in the air. Subsequently mints treatment technology has been changed and the oxide phase of the samples is not detected. Figure 1 (b) clearly visible large pores in which the particles were yttrium nitride, which during sintering dissolved uranium nitride.

(a)

(b)

Figure 1. (a) The microstructure of tablets UN-5 wt. % YN, contaminated yttrium oxide; (b) the microstructure of the sintered tablet UN-5 wt. % YN, without yttrium oxide.

A similar situation was observed for the mixture of uranium nitride - nitride, cerium, in one of the samples were in the air, traces of the oxide phase. This fact is clearly demonstrated in Figure 2 (a). Figure 2 (b) shows the microstructure of a sample in which no traces of the oxide phase.

The worst situation is with pills containing nitrate, dysprosium, whose microstructure is shown in Figure 3. Clearly visible white discharge oxide phase oxidation occurs even during a short stay in the air while carrying samples of argon box under a microscope. Due to the lack of homogenization of a large number of dysprosium nitride is insoluble in uranium nitride. Subsequently, dysprosium oxides to nitride oxide.

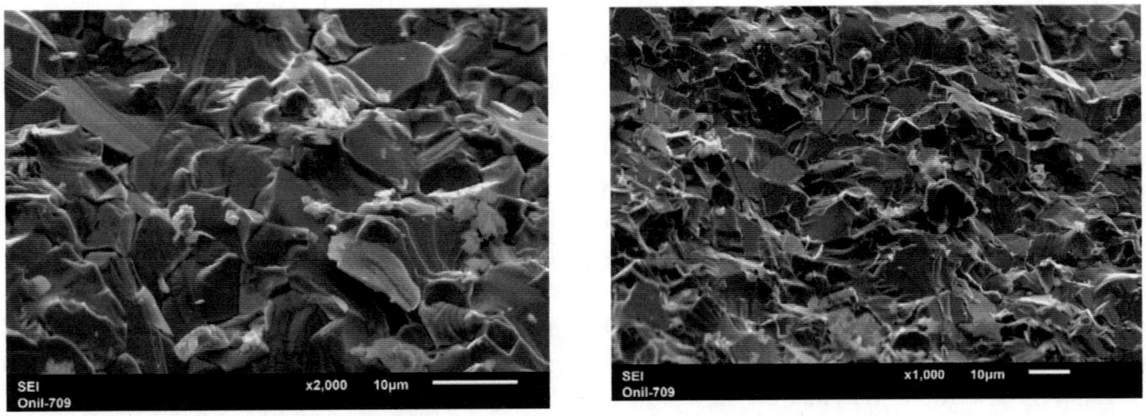

(a) (b)

Figure 2. (a) The microstructure of tablets UN-10 wt. % CeN; (b) the microstructure of the sintered tablet UN-20 wt. % CeN.

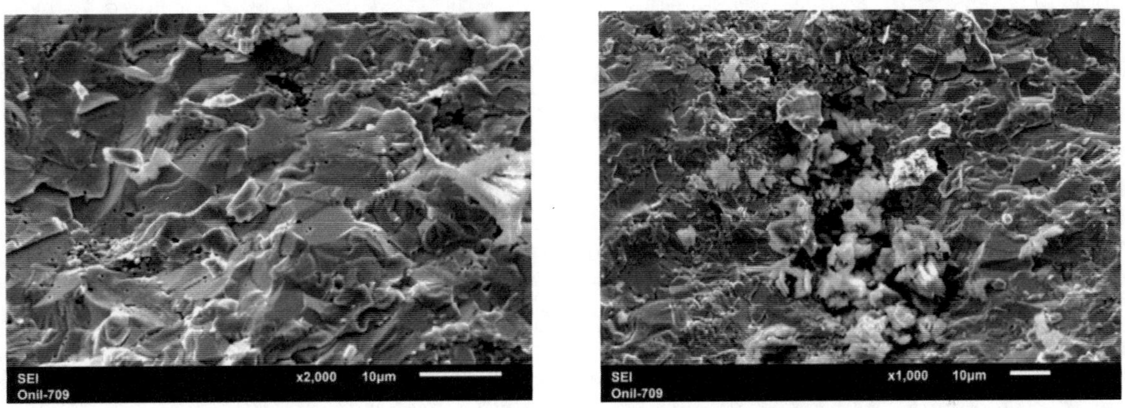

(a) (b)

Figure 3. (a) and (b) the microstructure of tablets UN-10 wt. % DyN.

Table 1. Results of measurement and calculation of the lattice parameter.

Composition	Calculated a (pm)	Referential a (pm)
UN – 5 % YN	488.4	488.5
UN – 10 % YN	489.9	488.5
UN – 10 % CeN	490.5	490.8
UN – 20 % CeN	491.7	491.9

According to the analysis we can conclude a good agreement between the results obtained in this study with the literature data. Also, comparison of the data in table 1 suggests that in the synthesis of nitride powder and sintering was noticeable evaporation simulators plutonium nitride (yttrium and cerium).

5. Conclusion

The results of the work can be done preliminary conclusions:

1. On the basis of an assessment of physicochemical and thermodynamic properties of selected lanthanides, actinides are the closest analogues. Analogues of plutonium are cerium and yttrium, americium and analogues - dysprosium. The introduction of these elements in mononitride uranium produces fuel Snoop model for the study of a number of its properties;
2. A method of obtaining a model Snoop-based fuels and lanthanide nitrides Mononitride uranium.
3. The method of hydrogenation-dehydrogenation, nitration obtained nitride powders of uranium, cerium, yttrium and dysprosium, conducted their mixing, pressing and sintering, obtained samples of the model nitride fuel with plutonium and americium imitation of a given composition; According to the results of structural studies have shown that all the samples are solid solution of nitride of REE in uranium nitride. It is found that the sintered pellets Snoop model fuel are heavily oxidized in air, as opposed to tablets Mononitride uranium.

Acknowledgements

This work was performed within the framework of the Center of Nuclear Systems and Materials supported by MEPhI Academic Excellence Project (contract № 02.a03.21.0005, 27.08.2013).

References

[1] Bibilashvili J K, Baranov V G and Godin Y G 2002 *Problems of Atomic Science and Technology. Series: Materials and New Materials* **59** 55
[2] Laverov N. P. et al. 2006 *Geology of Ore Deposits* **48** **5** 335
[3] Brauer G 1985 *Guidelines for inorganic synthesis* 4 (Moscow Mir)
[4] Samsonov G V 1964 *Refractory compounds of rare-earth metals with nonmetals* (Moscow Metallurgy)
[5] Yasufumi S, Yasuo A, Takashi I and Toshihiko O 1990 *J. Nuc. Sci. and Techn.* **28** 689
[6] Tennery V J, Bomar E S 1971 *J. Amer. Ceram. Soc.* **54** 247.
[7] Benedict U 1979 *Thermodynamics of nuclear materials* **1** 453
[8] Wicks C E and Block F E 1961 *Thermodynamic properties of 65 elements: their oxides, halides, carbides, and nitrides* (Bureau of Mines, Albany, OR (USA). Albany Metallurgy Research Center) **BM-BULL-605**

IOP Conf. Series: Materials Science and Engineering **130** (2016) 012037 doi:10.1088/1757-899X/130/1/012037

The interaction between nitride uranium and stainless steel

D P Shornikov, S N Nikitin, B A Tarasov, V G Baranov and M S Yurlova

National Research Nuclear University MEPhI (Moscow Engineering Physics Institute) Kashirskoye Highway 31, Moscow, 115409, Russia

E-mail: d.p.shornikov@mail.ru

Abstract. Uranium nitride is most popular nuclear fuel for Fast Breeder Reactor New Generation. In-pile experiments at reactor BOR-60 was shown an interaction between nitride fuel and stainless steel in the range of 8-11% burn up (HA). In order to investigate this interaction has been done diffusion tests of 200 h and has been shown that the reaction occurs in the temperature range 1000-1100 ° C. UN interacted with steel in case of high pollution oxygen (1000-2000 ppm). Also has been shown to increase interaction UN with EP-823 steel in the presence of cesium. In this case the interaction layer had a thickness about 2-3 μm. Has been shown minimal interaction with new ODS steel EP-450. The interaction layer had a thickness less then 2 μm. Did not reveal the influence of tellurium and iodine increased interaction. It was show compatibility at 1000 °C between UN and EP-450 ODS steel, chrome steel, alloying aluminium and silicium.

1. Introduction

Currently, one of the directions of development of nuclear power is a widespread introduction of fast breeder reactors. At the same time as the fuel is considered to use a nitride mixed uranium-plutonium fuel, which has high thermophysical and radiation properties [1]. As construction materials (fuel cladding and fuel assemblies) are treated corrosion-resistant steel (CRS) different classes of EP-823, EP-450/450 ODS, EK-164, etc. The most important aspect of the nuclear fuel is its compatibility with structural materials in a range of operating temperatures characteristic for fast neutron reactors, ie 600-1000 °C. For a long time it was believed that the interaction of nitride fuel and steel shell, up to 1000 °C is absent, and the main cause of damage to the fuel element is associated with swelling of the fuel, offensive of thermomechanical contact with the cladding and damage the latter [2-4]. However, the test reactor in recent years of fuel assemblies with nitride fuel in the experiment Bora-Bora, and their subsequent postirradiation research in "hot" cells using the methods of electro-probe microanalysis showed the presence of the interaction with the fuel cladding.

2. Experimental method

To obtain preliminary data on the interaction of the fuel-cladding was used method manufacturing composite ie co-pressing and sintering powders Mononitride uranium and corrosion-resistant steel. UN powder obtained by the hydrogenation-dehydrogenation-nitriding has a particle size of 10-20 micrometers, and the powder of corrosion-resistant steel, obtained by mechanical grinding of ingot and subsequent milling with a ball mill into powder particles had a size range of 20-30 micrometers. Subsequently, the powders UN and corrosion-resistant steel are mixed in a ratio of 20:80 or 30:70, respectively. Sample compacts in the form of plates and sintered at 1000 °C in vacuum furnace for 5 hours. Due to the large volume of content steel powder provided its sintering, ie the sample is a dispersion composition, wherein as the dispersed phase acts mononitride uranium. The significant disadvantages of this method is the impossibility of introducing chemically active elements (cesium, iodine and tellurium), since through the pass-through porosity said elements vaporized in the annealing process.

In this regard, was developed a method to test the compatibility of the powder of nuclear fuel with cladding material, consisting in the fact that the preliminary from the cladding-material is made crucible with a lid, wherein the inner surface of the crucible is polished to roughness values of Ra = 0.16-0.125 mm, and the lid is made in the form of a truncated cone, after which the crucible filled with the powder of the test fuel, then added simulators chemically active fission products and sealing of the crucible is carried out in inert gas atmosphere, followed by annealing in the temperature range 600-1000 °C.

After a series of diffusion annealing conducted metallographic preparation of the samples to determine the magnitude of the interaction layer. Initially, all the samples (crucibles) were cut into two halves along the cut portion the generatrix part of the diameter of the crucible. Cuts on the diamond saw Buehler IsoMet, as well as the coolant used isopropyl alcohol (IPA), which reduces oxidation of uranium nitride. Metallographic preparation of thin sections was performed according to standard procedures, a grinding carried out on sandpaper with fineness 1200 and then polished on a cloth alternately with diamond paste of size 10, 3 and 1 micrometers. To prevent oxidation, in all cases using IPA.

Investigation of the microstructure of diffusion couples UN / SS carried by the scanning electron microscope JEOL 6610 LV. As a further method for determining the thicker layer of interaction used method of measuring microhardness. The measurement was performed by pressing a diamond tetrahedral pyramid with an apex angle of 136°, and then determines the size of the diagonal and perform calculations microhardness. The purpose of this work is determined by the values of the microhardness near the border "nitride - CRS", as well as an array of CRS. Any deviation microhardness of the border array clearly indicates diffusion processes or the formation of intermetallic compounds. Measurements were performed at a load of 100 g, loading time was 10 s.

3. Experimental results

3.1. Features of interaction between the UN and the CRS

Analysis of the sintered powders UN and CRS, implemented by the X-ray phase analysis, is highly sensitive and allows to detect interactions at the initial stage. According to the results presented above, we can conclude that there is no interaction up to 1000 °C and only at exposure 200 hours were found traces of interaction products. According to test results, may be noted the good compatibility of the steel EP-823 and EP-450 ODS, also need to pay attention to chromium steels, alloyed with aluminum. Table 1 shows the results of X-ray phase analysis of powder compacts. From the data in the table it can be seen that during the annealing at 1000 °C compacts powders mononitride uranium and CRS begin to appear traces of oxide and nitride phases (Fe_3N, Fe_2O_3, (Fe, Mo) $_3O_4$) and ternary phase U_2CrN_3, indicating that beginning of the reaction. With regard to the diffusion couples with steel EP-450 it is possible to note the presence of the interaction, as evidenced by the presence of the triple phase U_2CrN_3. Otherwise, there is a presence of mononitride uranium ferrite and uranium dioxide. It is also considered the interaction with uranium nitride dispersion-strengthened steel EP-450 ODS.

Analysis of the diffraction pattern shows a lower level of interaction, the presence of triple-phase have not been identified, apparently this could be explained by the presence of particles of yttrium oxide that is effective diffusion barrier. Steel EP-823 also interacts with uranium nitride to form iron nitride, positive aspects may include less, compared with the EP450, peak height, what indicating a very weak interaction. However, in this case was found the traces of oxide and iron nitride.

Table 1. Results of X-ray phase analysis of diffusion couples «UN-CRS» manufactured from a mixture of powders.

Labelling	Sample status	Detected phases	Extras. phases
UN-EP450	Sintered	UN, α-Fe, UO_2, U_2CrN_3	
UN-EP450 ODS	Sintered	UN, α-Fe, Y_2O_3, UO_2	tracks: $(Fe,Mo)_3O_4$
UN-EP823	Sintered	UN, α-Fe, UO_2	tracks Fe_3N and Fe_2O_3
UN-H18N10T	Sintered	UN, α-Fe, UO_2, U_2CrN_3, γ-(Fe,Ni)	
UN-(Fe, Cr, Al)	Sintered	UN, α-Fe, UO_2, U_2CrN_3	tracks U_2N_3

Austenitic steels X18H10T also interacts with nitride fuel to form a ternary phase. The powder compacts UN- (Fe, Cr, Al) traces ternary phase, but its amount is small. Also found a number of one and a half nitride, the appearance of which can be attributed to exchange reactions between the aluminum nitride. A similar situation exists for steels alloyed with silicium, but in this case no phase one and a half nitride. This was followed by an analysis of the microstructure and MXSA samples.

3.2. The influence of oxygen on the interaction between the UN and CRS

One reason for the strengthening of interaction between the UN and the steel can be increased content of oxygen in mononitrides uranium. To verify this fact made UN powders with different oxygen content: № 1 - Powder UN, containing ~ 100-200 ppm oxygen; №2 - Powder UN, containing ~ 1000-2000 ppm oxygen. Each of these powders was charged in a steel crucible EP-450, EP-823, as well as chrome steel crucibles with additions of silicium and aluminum.

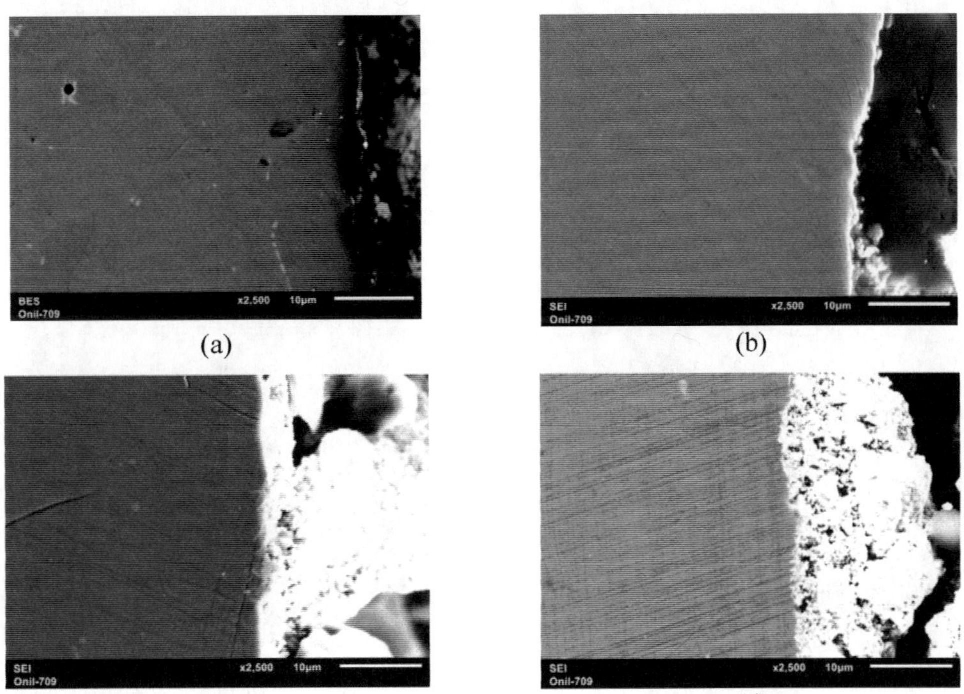

Figure 1. (a) The microstructure of the the interaction layer diffusion couples UN-EP450, the oxygen content of 100 ppm; (b) diffusion couples UN-EP450, the oxygen content of 1000 ppm; (c) UN (100)-EP823; (d) UN (1000)-EP823.

Metallographic analysis of diffusion couples UN (100 ppm) with steel EP-450 allow note the very small width of the interaction zone (not more than 2 micrometers), as well as the emergence of low porosity. This fact is illustrated in figure 1 (a, b). A similar situation exists for the steel EP-823, on the border observed porosity what, apparently, due to the Kirkendall effect. Also not found significant difference for diffusion couples nitride-EP823 with different oxygen content (100 and 1000 ppm), the width of the interaction zone differs slightly and varies in the range of 2-5 micrometers. This fact can be explained by the low partial pressure of oxygen (over the oxide phase) as shown in figure 1 (c, d).

3.3. Influence of chemically active fission products on interaction between the UN and the CRS

Accelerate the interaction may also presence of corrosive fission products (cesium, iodine and tellurium). In this paper, added a chemically-active products in an amount equivalent to 10.2% of HA burnout (Cs 12.017, I 0.899, Te 1.661 g / kg of irradiated fuel), which is due to the fact that this value is the threshold at which there is damage to the cladding. These elements were introduced into the box with an inert atmosphere, due to the high activity of cesium and tellurium toxicity. The uranium nitride powder poured alternately cesium, iodine and tellurium, mixed in an agate mortar and pressed into the crucible of the investigated steel. In this work, mainly conducted corrosion tests have become EP823 in the presence of cesium, iodine and tellurium, due to its high potential as a fuel rod cladding material. Photograph of the microstructure shown in figure 2 a and b (the oxygen content was 1000 ppm). From the analysis of the microstructure, it follows that the interaction layer has a width of about 5-9 micrometers, which is somewhat higher than in the case of a pure nitride or in case of couples diffusion test without CAPD. Also, with CAFP were tested the samples chromium steel (oxygen content was 1000 ppm), doped with 2 wt. % of aluminum and silicium, which is shown in figure 2 c, and d.

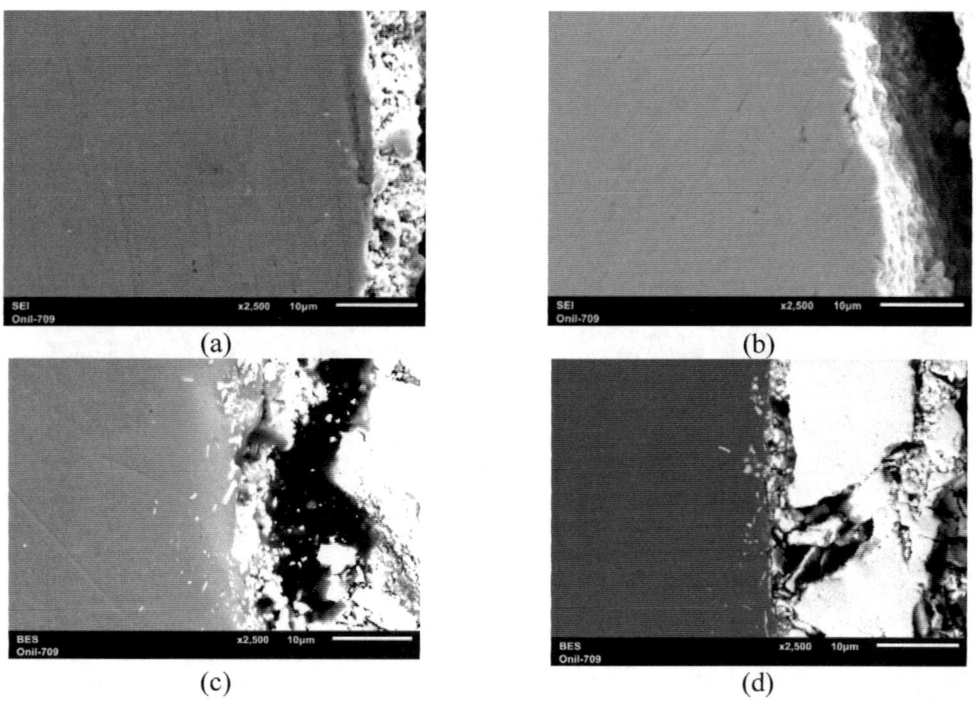

Figure 2. (a) The microstructure of the the interaction layer diffusion couples UN-EP823 with oxygen content 1000 ppm; (b) diffusion couples UN-EP823; (c) diffusion couples UN with FeCr-2% Al; (d) diffusion couples UN with FeCr-2% Si.

In this case, the interaction is almost undetectable in some cases there is a weak interaction, wherein the the interaction layer thickness was not greater than 1 micrometers. Note that such a value is very low, even with the error and resolution scanning microscope one can speak about a significant reduction in the interaction layer for steels alloyed with aluminum and silicium, even in the presence of CAFP.

Based on the concepts of a parabolic law of growth of the layer, the parabolic growth constants. The constants for similar compositions UN (100) + EP450, UN (100,1000) + EP823, UN (1000) + 2% FeCr-Al, of the order of 10-17 m^2/s. For the composition of UN (1000) + EP823 + CAFP there is an increase the kinetics of the interaction with the UN on the order of CRS. The activation energy of the interaction is not possible due to lack of cooperation with UN CRS at 600 °C.

3.4. Evaluation microhardness of diffusion couples

Studies microhardness of diffusion couples UN-CRS allows us to conclude that the microhardness of areas CRS near the border and in the array varies. Basically microhardness boundaries is 10-15% lower compared to the array, which can be explained by the weak interaction CRS with nitride fuel microhardness decrease due to the appearance of porosity and intergranular corrosion. Also, from the obtained data it is clear that such samples UN (1000 ppm O2) + FeCr-2% Al and UN (1000 ppm O2) + FeCr-2% Si difference between microhardness array and the boundary is within measurement error, which may be indicative the lack of interaction. Strong distinction between an array of micro-hardness and abroad observed for samples UN (1000) -EP823 and UN (1000) -EP823-Cs. Microhardness of distance samples UN (1000) + EP823 and UN-EP823-Cs is shown in figure 3.

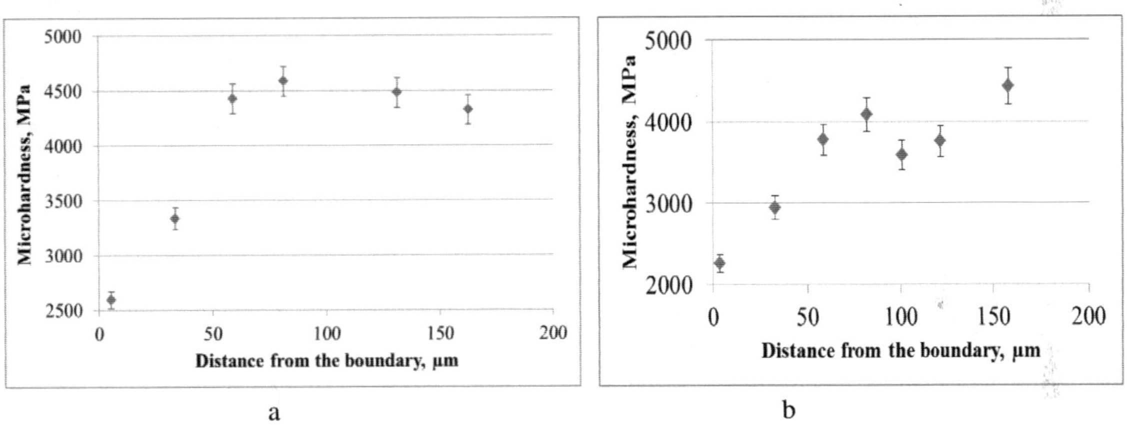

Figure 3. (a) Microhardness diffusion couples UN (1000) - EP 823; (b) diffusion couples UN-EP823-Cs.

From these dependencies can be seen that the graphs of microhardness similar, which suggests about the similarity of corrosion damage to the contact area in the presence of high oxygen content and in the presence of cesium.

4. Conclusions

Developed two methods of diffusion test compatibility nitride fuel and CRS. One of which is annealed and pressed powders UN CRS followed by X-ray phase analysis. It is shown that the time of occurrence of nitrides or triple phase clearly indicates the beginning of the interaction. Another technique is based on the study of the filling of the crucible steel powder with mononitride uranium, the process allows you to enter chemically-active fission products, as well as uniquely determine the value of the interaction layer.

According to the results of diffusion tests for 200 hours showed no interaction at a temperature of 600 ° C nitride fuel with the CRS. At the same time shows the appearance of the interaction layer thickness of 2-3 micrometers in the temperature range 1000-1100 ° C for 200 hours.

Noted increased interaction and thus increase the interaction layer to 5-7 micrometers in the case of tests Mononitride with increased oxygen content (1000-2000 ppm). During the diffusion test steels EP-450 and 823 and 18Cr10NiTi shown appearance iron nitride Fe3N phase and ternary phase U2CrN3, the presence of which is uniquely enabled to determine the beginning of interaction. It showed good compatibility of the steel EP-450 ODS and chromium steel, alloy aluminum and silicium at 1000 ° C. Analysis of the boundaries of the "nitride - steel", as well as microhardness revealed no occurrence of the interaction layer.

Shown increased interaction UN and steel EP823 in the presence of cesium, in this case, the interaction layer was 15 micrometers. There was no impact of tellurium and iodine to enhance interaction.

Acknowledgements
This work was performed within the framework of the Center of Nuclear Systems and Materials supported by MEPhI Academic Excellence Project (contract № 02.a03.21.0005, 27.08.2013).

References

[1] Streit M and Ingold F 2005 *J. Eur. Cer. Soc.* **25**(12) 2687-92
[2] Zaboudko L et al. 2001 *ANL Transmuter Fuel Development Workshop* 19
[3] Arai Y et al. Development Status of Metallic, Dispersion and Non-oxide Advanced and Alternative Fuels for Power and Research Reactors IAEA, 2003 VIENNA
[4] Bauer A A et al. 1977 *In: Proc. of Intern. Meeting on Advanced LMFBR Fuels* (Tucson) 299
[5] Nebeisen J C et al. 1998 *Trans. Amer. Nucl. Soc.* **11** 104

Basic principles of creating a new generation of high-temperature brazing filler alloys

B A Kalin and A N Suchkov

National Research Nuclear University MEPhI (Moscow Engineering Physics Institute) Kashirskoe Highway 31, 115409 Moscow, Russia

E-mail: BAKalin@mephi.ru

Abstract. The development of new materials is based on the formation of a structural-phase state providing the desired properties by selecting the base and the complex of alloying elements. The development of amorphous filler alloys for a high-temperature brazing has its own features that are due to the limited life cycle and the production method of brazing filler alloys. The work presents a cycle of analytical and experimental materials science investigations including justification of the composition of a new amorphous filler alloy for brazing the products from zirconium alloys at the temperature of no more than 800 °C and at the unbrazing temperature of permanent joints of more than 1200 °C. The experimental alloys have been used for manufacture of amorphous ribbons by rapid quenching, of which the certification has been made by X-ray investigations and a differential-thermal analysis. These ribbons were used to obtain permanent joints from the spacer grid cells (made from the alloy Zr-1% Nb) of fuel assemblies of the thermal nuclear reactor VVER-440. The brazed samples in the form of a pair of cells have been exposed to corrosion tests in autoclaves in superheated water at a temperature of 350 °C, a pressure of 160 MPa and duration of up to 6,000 h. They have been also exposed to destructive tests using a tensile machine. The experimental results obtained have made it possible to propose and patent a brazing filler alloy of the following composition: Zr–5.5Fe–(2.5–3.5)Be–1Nb–(5–8)Cu–2Sn–0.4Cr–(0.5–1.0)Ge. Its melting point is 780 °C and the recommended brazing temperature is 800°C.

1. Introduction

The development of new materials and the improvement of the existing ones is a complex process, of which the purpose or the main task is the directed formation of an element composition, a phase state and a structure in the volume of a material and (or) in its surface layers, i.e. formation of a predetermined structural-phase state (SPhS). At that, an important task is to ensure the stability of this state in the operating conditions of the material (the product) [1].

The development of materials for a specified product is based on an analysis of the operating conditions of the product in the construction under development, an analysis of a single-item and combined action of structural components of the product, determination of a manufacturing technology of individual parts and the construction in general, knowledge of the achievements in the corresponding field of materials science.

An analysis of all these factors ends by formulation of characteristics of the developed material and drawing up requirements to the main and auxiliary properties of the developed material.

The development of such functional materials as high-temperature brazing filler alloys has its own features that are due to the limited life cycle and the production method. Solders obtained by rapid quenching of the melt are called rapidly quenched filler alloys (RQFA). At present, rapidly quenched filler alloys in the form of ribbons with a thickness of 20-80 μm have become widely used in mechanical engineering. Such filler alloys are obtained by ultrarapid quenching of the melt on a rotating disk-refrigerator with a cooling rate of 10^4-10^6 K/s. The high cooling rate allows obtaining brazing filler alloys at room temperature with the structure of a supercooled liquid and in the state of a supersaturated solid solution with a uniform distribution of the alloy components over the entire volume, which determines the unique properties (wettability, capillary activity, etc.) of the filler alloy in the process of melting and interaction of the melt with the brazed materials [2].

Features of the application of brazing filler alloys determine the requirements for the structural-phase state of alloys to ensure an amorphous or nanocrystalline state, and chemical and microstructural homogeneity as they determine narrow temperature ranges of melting, a high diffusion activity of atoms, an increased capillary and adhesion activity and an increased flowing of the melt into the gap for providing a high strength and corrosion resistance of the brazed joints, especially for creating permanent joints of dissimilar materials, dispersion-strengthened alloys and steels, beryllium, refractory metals, solid alloys, ceramics, graphite and others [1, 3].

On the example of practical application of RQFAs in the energy-intensive techniques, for example, for the manufacture of the first wall and the divertor of ITER, rocket engine nozzles, the fuel elements of nuclear reactors, it is shown that profound studies of the structural-phase state of developed RQFAs at every stage of their life cycle: developments, manufacture and application, are necessary to ensure the operability of important units. The step of producing alloys at fast (10^4-10^6 K/s) quenching rates is crucial for obtaining a homogeneous distribution of components in the alloys with complex composition. Not only the brazing temperature, but also the brazing time requires deep justification in the process of forming a permanent joint [4-7].

In this context, the aim of this report is to present the fundamental materials science investigations that made it possible to obtain specified structural-phase states of both the RQFAs themselves and in the zones of permanent joints of metals and dissimilar materials on the example of development of a new amorphous filler metal based on zirconium alloys.

2. Basic requirements to the properties of the brazing filler alloy

The requirements are determined by the appointment of a brazing filler alloy. For example, the fuel elements and other details of fuel assemblies of the thermal neutron reactor active zone must meet the following basic requirements for a high-temperature brazing of details made from zirconium alloys [8]:

- the brazing filler alloy must be compatible with brazed materials, i.e. not to induce a significant chemical erosion and not to form chemical compounds;
- it must have a low melting point and a small brazing time to prevent the formation of intermetallic compounds in the brazed seam and changes of the structural-phase state of brazed materials; with respect to this work, the brazing temperature should be below the temperature range of the existence of a two-phase area in zirconium alloys, i.e. the melting point of the alloy should be below 800°C and, in this case, the brazed alloy should be in a single-phase area during the brazing;
- the brazed joint should have an unbrazing temperature no less than 1200 °C (this is a stability condition of the fuel assembly at a complete loss of the coolant, accidents of the LOCA type - Loss of Coolant Accident);
- should have a minimum thermal-neutron capture cross-section and an absence of the formation of relatively long-lived radioactive nuclides;
- must provide a corrosion resistance of the brazed product and the brazed seam for the entire operation period (3-6 years);
- should provide a predetermined resistance of the brazed seam to radiation creep;
- must provide a production of a brazing filler alloy in the amorphous state.

3. Principles of the choice of the composition of zirconium brazing filler metals

The developed alloy should have such a composition and a structural-phase state that will provide the required level of properties and characteristics. Studying the experience of the development of brazing filler metals based on zirconium showed that the following compositions and brazing temperatures were tested in experiments: (80–91)Zr–(5.5–15)Fe–1.5Ge–(1–3)Cr–2.5Be (the brazing temperature is 960–1030 °C), Zr–12.5Cu–2.5Be (brazing at the temperature of 1130 °C), Zr–5Be (the brazing temperature is 965 °C) and others [8]. The main disadvantage of the developed brazing filler metals is their high brazing temperature resulting in a significant oxidation of the brazed seams during tests in reactor conditions.

When selecting the alloy composition, a classic scheme of synthesis containing the following main stages was developed:

- choice of the alloy *base*, of which the state corresponds to the basic requirements to its properties;
- choice of *alloying elements* and identification of harmful impurities;
- choice of *a complex* and a composition of alloying elements with regard to all the requirements to the properties of the alloy;
- *obtaining* (smelting) of an alloy;
- obtaining of an amorphous ribbon by rapid quenching of the melt on the surface of a rapidly rotating disk-refrigerator;
- study of the structural-phase state and the temperatures of phase transformations of the resulting ribbon.

3.1. The choice of an alloy base and alloying elements

In materials science of brazing filler alloys for a high-temperature brazing, in choosing the alloy base it is oriented on known requirements to avoid changes of the state of brazed materials and formation of brittle intermetallic compounds in the brazed seam [9, 10]. In this case, the base element of the brazed composition should be selected as the base of a filler alloy and the filler alloy may include alloying elements of the brazed material. In addition, the experience of the development of filler alloys for brazing of zirconium was taken into account. Therefore, the composition Zr-Nb was selected as the base for the future brazing filler alloy.

Niobium has a number of positive properties, including the formation of a solid solution with β-phase of zirconium. Niobium dissolves in α-phase of zirconium at the monotectoid temperature in the amount of up to 1.5 wt.%, it has a small capture cross section of thermal neutrons, stabilizes the corrosion resistance of unalloyed zirconium, eliminates the harmful effects of small amounts of such impurities as carbon, aluminum and titanium, and effectively decreases the quantity of hydrogen absorbed by the zirconium alloy [11].

When developing the brazing filler alloy composition, two conjugated scientific and technical problems were solved in the first place. The first problem is the choice of the filler alloy composition with a minimum melting point, i.e. with a temperature less than 800 °C. The second one is improvement of the composition content to obtain an amorphous state by rapid quenching of the melt.

3.1.1. The choice of a composition with the minimum melting point

There are no alloys with the melting point less than 800 °C among the known two- or three-component systems based on zirconium. Therefore, the development of the composition of alloying elements was made with taking into account the compositions of existing commercial alloys (Nb, Sn, and Fe) and in the direction of creating a multi-component system with a melting point below 800 °C. Analyzing the composition of known filler alloys mentioned above, it can be seen that Fe, Ge, Cr, Be and Cu were used as alloying elements. In general, a preliminary chemical and physical analysis of potential alloying elements shows that Sn and rare-earth metals are of interest in addition to the above-mentioned elements [8].

Small quantities of iron insoluble in α-zirconium and present in the form of fine-dispersed crystals of intermetallic compounds increase the strength of a zirconium alloy. It should be noted that niobium and iron slightly lower the melting point of zirconium [11].

Beryllium greatly lowers the melting point of zirconium and forms a eutectic at a content of 5 wt. %. It has a low capture cross section of thermal neutrons. The data on the corrosion resistance of brazed joints obtained using brazing filler alloys of the system Zr-Be are ambiguous, but it is possible to increase their corrosion resistance by additional alloying elements [11].

Copper was chosen as a depressant, since it most strongly lowers the melting point of zirconium at a content of up to 8 wt.% and decreases the liquidus temperature of zirconium from 1860 to 995 °C at increasing the content up to 21 wt.%. The eutectic reaction Lq ⇔ $CuZr_2$ + βZr occurs in the binary alloy Zr-Cu at the temperature of 995 °C [11].

To neutralize the possible influence of copper on the deterioration of the alloy's corrosion resistance and additionally strengthen the alloy, it was alloyed by Sn (2%) and Cr (0.4%).

The quantitative composition of the selected alloying elements was clarified in the process of the analysis of binary and ternary phase diagrams of the alloy systems.

The performed physicochemical analysis of alloys of the systems Zr-Be, Zr-Fe, Zr-Nb, Be-Fe, Zr-Cu, Zr-Be-Fe, Zr-Be-Nb, Fe-Nb-Zr and Ze-Fe-Nb-Be made it possible to choose optimal compositions and produce a pilot batch of alloys for the investigation of phase transformations and temperatures of liquidus and solidus. It was taken into account that the system Zr-Be has a eutectic Lq ↔ Zr + $ZrBe_2$ at the temperature of 965 °C, of which the composition is 95 wt.% (65.2 at.%) Zr and 5 wt.% (34.8 at.%) Be, the beryllium concentration of about 5 wt.% being optimal.

The system of Zr-Fe alloys has a eutectic Lq ↔ β-Zr + Fe_2Zr at 934 °C. The liquid composition at the eutectic temperature is 16 wt.% Fe, Fe decreasing the polymorphic transformation temperature in zirconium to 735±5°C. The eutectoid concentration is 3.5 at.% Fe.

The system of Be-Fe alloys has two eutectic reactions at 1165 °C (the eutectic liquid composition is 8 wt.% Be) and 1225 °C (the eutectic liquid composition is 77 wt.% Be).

The following reactions are present in the system of Cu-Zr alloys in the copper content range of 0-20 wt.%: a) eutectic: Lq ↔ β-Zr + $CuZr_2$ at 995 °C (at the zirconium content in phases of 72: 94.7: 66.67 at.%) and b) eutectoid: β-Zr ↔ $CuZr_2$ + α-Zr at 882±10 °C (at the zirconium content in phases of 97.7: 66.7: ~ 99.7 at.%).

In the system of Zr-Be-Fe alloys, the alloys quenched from the temperature of 800 °C are still in a liquid-solid state and contain beryllium and iron in the following proportions:

Fe:Be = 4:1 (Be+Fe) ≥ 3%, i.e. Be ≥ 0.6% and Fe ≥ 2.4%;
Fe:Be = 3:2 (Be+Fe) ≥ 2%, i.e. Be ≥ 0.8% and Fe ≥ 1.2%;
Fe:Be = 2:3 (Be+Fe) ≥ 1%, i.e. Be ≥ 0.6% and Fe ≥ 0.4%;
Fe:Be = 1:4 (Be+Fe) ≥ 1%, i.e. Be ≥ 0.8% and Fe ≥ 0.2%.

From the performed analysis it follows that the eutectic composition of the alloy is approximately the following one: > 5 wt.% Fe and > 2.5 wt.% Be.

In the system of Zr-Nb-Be alloys at the temperature of about 855 °C, there is a eutectic equilibrium $Lq^7 \leftrightarrow \alpha^7_{Zr} + \beta^7_{Zr} + ZrBe_2$ at a composition of the point Lq^7 ~ 5 wt.% Be and ~ 1 wt.% Nb. Thus, there is one more confirmation of the concentration of ~ 1% Nb.

In the system of Zr-Fe-Nb alloys, the following reaction Lq ↔ β-Zr + Zr_2Fe + T (~ 5 at.% Fe in the liquid composition) taking place at the temperature of 850°C has the lowest temperature of all the possible eutectic reactions.

An analysis of alloys of the system Ze-Fe-Nb-Be showed that the brazing filler alloy composition having the lowest melting point is located in the range of (3-6)% Fe, (2-5)% Be, and (1–3) % Nb. On this basis a filler alloy of the following composition Zr-5.5% Fe-1% Nb-2.5 % Be was selected.

Samples of the alloy were smelted for a complex analytical research and tests of its properties. The samples were used to obtained foils by a spinning technique and then these foils were applied to braze the spacing grid cells of the VVER-440 fuel assemblies using the following regime: the brazing temperature was 850 °C and the holding time at the temperature of 580 °C was 6 h.

The temperature melting ranges of the smelted ingots were determined at the heating rate of 20 K/min by a differential thermal analysis using the installation VDTA-8M3. Tensile tests of the brazed samples showed that the joints have a satisfactory strength. The destruction of the samples took place mainly at the base metal. However, autoclave corrosion tests showed that the brazed joints had an insufficient corrosion resistance. This means that the brazing temperature of 850 °C is overrated and it is necessary to decrease it. To do this, 8 wt.% of Cu was added. To increase the corrosion resistance and additionally strengthen the alloy, 2% of Sn and 0.4% of Cr were additionally added into the alloy. Thus, an alloy of the following composition Zr-5.5Fe-2.5Be-1Nb-8Cu-2Sn-04Cr (in wt.%) was smelted. After the spinning technique, this alloy was used to braze cells made from the alloy E110 for corrosion tests.

The brazed joints withstood corrosion tests in a distilled steam-water mixture in an autoclave at the temperature of 350 °C and the pressure of 16.5 MPa for 6,000 h. A further increase of the time resulted in formation of elements of a noticeable corrosion in the zone of fillets of the brazed seam, which indicated the necessity to further decrease the melting point of the alloy and, therefore, the brazing temperature up to 780 °C. A possible reduction of the copper content of up to 5 wt.% was justified. However, decreasing the content of copper is accompanied by increasing the melting point of the alloy (820-830 °C at 5%). To decrease the melting point of the brazing filler alloy, the content of Be was increased up to 3.5 wt.%. The resulting alloy **Zr- 5.5 Fe-3.5Be-1Nb-5Cu-2Sn-0.4Cr (wt.%)** has a start melting temperature of ~ 780 °C.

3.1.2. The refinement of the composition for obtaining an amorphous state by rapid quenching of the melt

The amorphous state is most likely for the elements with polymorphic transformations that are trend to form various phases and have a high degree of anisotropy of the interatomic bond [1]. From this viewpoint, zirconium is the most suitable metal. All the amorphizing metal melts are like in having a quick growth in the intensity of *the covalent component* of forces of the interatomic interaction at decreasing the temperature. The growth in the covalent interaction intensity under supercooling of the melt leads to the binding of components into associative complexes, which, on the one hand, makes the processes of mass transfer more difficult and inhibits crystallization and, on the other hand, creates an ordered state in the liquid phase that has a smaller amount of Gibbs energy in relation to that of the crystal. Thus, the transfer of metal melts into the glassy state at a rapid cooling is completely controlled by the covalent component of the chemical bond of atoms.

The interval of compositions in which this is realized is in the concentration area of a dominance of associative complexes with low *entropy of formation*. For example, the tendency of the Cu-Zr melts to glass formation in wide concentration ranges is predetermined by the fact that all the kinds of associative groups present in them have large negative entropies of formation. It is seen from Table 1 that presents the enthalpy ($-\Delta H$, J/mol) and the entropy ($-\Delta S$, J/(mol×K)) of the formation reactions of associations and data on the concentration ranges of the alloy amorphization [1].

It is the entropy term of the atom association reactions that determines to a decisive extent the values of the dynamic contributions to thermodynamic (configurational entropy and heat capacity) and kinematic (viscosity, the activation energy of crystallization, crystallization rate) properties of the melt that determine its transition to the amorphous state, bypassing the crystallization under decreasing the temperature.

This allows selecting the formation entropy of associative complexes as the main thermodynamic value to estimate the tendency of alloys to amorphization and forecast the compositions of alloys that are the most prone to amorphization.

Table 1. Thermodynamic parameters of association reactions ($-\Delta H$, J/mol and $-\Delta S$, J/(mol×K)) and data of the concentration ranges of amorphization for a number of alloys

System of elements	Parameter	Parameters for associations		Amorphization interval
		CuZr	Cu_2Zr	
Cu–Zr	$-\Delta H$	54800	73100	$0.20 < x < 0.70$
	$-\Delta S$	40.5	40.3	

At the same time, it is quite convincingly established that the intensity of the interatomic (interparticle) interaction in metal melts, in contrast to the oxide ones, has only an indirect impact on their glass-forming ability.

Beryllium at a concentration of 3.5 wt.% and more significantly lowers the melting point of zirconium and is a strong amorphizer and a strong depressant.

To improve the brazed seam homogeneity, decrease the formation possibility of intermetallic compounds in the brazed zone and thus increase the strength and the corrosion resistance, an addition of germanium at a concentration of 0.5-1.25 wt.% was justified. At that, germanium slightly lowers the melting point of the alloy. Detailed investigations of the microstructure of brazed joints by a scanning electron microscope using 6 alloys containing Ge from 0.5 to 1.25 wt.% at various brazing temperatures ranging from 780 to 860 °C to analyze the distribution of beryllides showed that wedging-out and decreasing the quantity of beryllides until their practically complete disappearance were observed at the concentrations of germanium in the filler metal of 0.75 and 1.0 wt.% and in regulation of the gap between the brazed details. The investigations performed make it possible to conclude that alloying the brazing filler metal by germanium in an amount of 0.5-1.0 wt.% should be considered the most optimal from the viewpoint of the brazed seam structure.

Thus, the following composition of a brazing filler metal, Zr–5.5Fe–(2.5–3.5)Be–1Nb–(5–8)Cu–2Sn–0.4Cr–(0.5–1.0)Ge, for brazing of zirconium alloys of the system Zr-Nb is proposed and patented. Its melting point is 780°C and the recommended brazing temperature is 800°C.

4. The analytical and instrumental maintenance of the development of alloys

Stages of selecting the alloy composition were accompanied by the manufacture of ingots, obtaining a rapidly quenched ribbon of the filler alloy, investigation of the ribbon state, carrying out a pilot brazing of Zr-1%Nb plates or brazing the spacing grid cells of the VVER-440 fuel assemblies, studies of the structural-phase state in the razed seam and performing of corrosion tests and destruction tests of the brazed samples under tensile loads.

Ingots of alloys were obtained using arc and induction melting furnaces. Experiments on improving the technology to manufacture amorphous ribbon-type brazing filler alloys were carried out on the installation Crystal-702 for ultrarapid quenching of the melt at a rate of up to 10^6 K/s. The experiments performed made it possible to determine the optimum parameters of the process of rapid solidification of the melt for the developed zirconium-based alloy and obtain an amorphous ribbon with a width of 10 mm and a thickness of 40-45 μm. Then, a complex technological problem of obtaining ultrathin AMA-filler alloys (i.e. brazing filler alloys based on amorphous metal alloys) with a thickness of 20-25 μm was successfully solved. The thickness of the filler alloy influences the brazed seam thickness and, accordingly, the structural-phase state in the seam zone. The geometry, the structure and the element composition of the resulting ribbon were investigated. The chemical composition of the ingots and rapidly quenched ribbons was determined by a micro X-ray spectrum analysis using the installation "CamScan-4DV".

The starting melting point of an amorphous ribbon of the brazing filler alloy was determined by a differential thermal analysis. An X-ray pattern of the ribbon was obtained using the installation DRON-3 that showed the presence of an amorphous structure.

Brazing of the elements of zirconium spacing grids was performed in vacuum furnaces. Then, the brazed joints were exposed to corrosion tests in a distilled steam-water mixture in laboratory autoclaves with a volume of 1 dm^3 at the temperature of 350 °C and the pressure of 16.5 MPa for 6,000 h.

The distribution of elements in the area of brazed seams was investigated using the scanning electron microscope "CamScan-4DV" equipped with a wave crystal diffraction spectrometer WDX-3PC.

The unbrazing temperature of the brazed joints was determined in accordance with the previously developed technique [8]. The failure of the brazed joint was shown to start at the temperature above 1400 °C.

The strength of the brazed joints was determined on pairwise brazed cells using the tensile testing machine R-05 with registration of the failure load.

5. Conclusion

The following composition of a brazing filler alloy with a melting temperature of 780 °C and a recommended brazing temperature of 800 °C has been proposed and patented as a result of a physicochemical analysis and an experimental study of a series of zirconium-based alloys: Zr-5.5 Fe-(2.5-3.5)Be–1Nb-(5-8)Cu-2Sn-0.4Cr–(0.5-1.0)Ge.

The developed brazing filler alloy provides the formation of a permanent corrosion-resistant joint of the Zr-1%Nb alloy with a working temperature higher than 1200 °C.

Acknowledgements

This work was performed within the framework of the Center of Nuclear Systems and Materials supported by MEPhI Academic Excellence Project (contract № 02.a03.21.0005, 27.08.2013).

References

[1] Kalin B A 2012 *Fizicheskoe materialovedenie* vol 5 (Moscow, MEPHI) p 700
[2] Rabinkin A 1989 *Welding Journal* B **10** 39
[3] Rabinkin A, Liebermann H and Pounds S 1999 *Scripta Metallugica et Materialia* **25** 399
[4] Kalin B A, Suchkov A N, Fedotov V T, Sevryukov O N, Ivannikov A A, Polyansky A A, Mazul I V, Makhankov A N and Gervash A A 2012 *Fusion Science and Technology* **61** 147
[5] Suchkov A N, Kalin B A, Fedotov V T, Sevryukov O N, Ivannikov A A, Polyansky A A, Mazul I V, Makhankov A N, Gervash A A, Dzhumaev P S, Yakushin V L and Pol'skii V I 2012 *Fusion Science and Technology* **61** (2) 147
[6] Kalin B A, Sevryukov O N, Fedotov V T, Suchkov A N and Fedotov I V 2013 *Proc. 7th Asia Pacific IIW Int. Congress* 172
[7] Suchkov A N, Kalin B A, Fedotov V T, Sevryukov O N, Morokhov P V, Ananiyn V M, Ivannikov A A, Polyansky A A, Mazul I V, Makhankov A N and Gervash A A 2014 *Fusion Science and Technology* **65** 212
[8] Kalin B A, Мамедова T T, Fedotov V T, Sevryukov O N, Isaenkova M G, Perlovich Yu A, Timoshin S N, Sinelnikov L P and Kuznetsov V R 2007 *Svarochnoe proizvodstvo* **10** 10
[9] Suchkov A N, Kalin B A, Fedotov V T, Sevryukov O N, Morokhov P V, Ivannikov A A and Grigoryev A E 2014 *Advances in Science and Technology* **88** 156
[10] Suchkov A N, Kalin B A, Fedotov V T, Sevryukov O N and Ivannikov A A 2014 *Tsvetnye metally* **12** 32
[11] Kalin B A 2012 *Fizicheskoe materialovedenie* vol 6 (Moscow, MEPHI) p 736 (in Russian)

The multicomponent doping of surface layers of materials under the influence of ion beams with a broad energy spectrum

B A Kalin, N V Volkov, R A Valikov, A S Yashin and T V Yakutkina

National Research Nuclear University MEPhI (Moscow Engineering Physics Institute) Kashirskoe highway 31, Moscow,115409, Russia

E-mail: NVVolkov@mail.ru

Abstract. The paper discusses the various factors that influence the efficiency of ion mixing. It was found that in the base of penetration of atoms multilayer films in polycrystalline substrate is the process of energy transfer from ions and primary knocked-on atom (PKA) of films to subsequent displacement cascade. At the same time the penetration of implanted atoms to great depths determined by the density of defects, radiation-stimulated migration of interstitial atoms and their physico-chemical interaction with the atoms of the matrix, which can be described by the model of an isotropic mixing. It is shown that doping atoms of the multilayer films, possibly the formation of gradient layers, which are determined by radiation traces in the substrate implanted atoms and their migration under irradiation by the ion beam with a broad energy spectrum.

1. Introduction

The use of new implant techniques to create a gradiently structurally-phase state of construction and functional materials with improved operating and electro-physical properties, is an important problem of materials science.

Previously shown [1,2] that irradiation of systems "film-substrate" by ion beam with a broad energy spectrum, along with purely kinematic interaction between the ion implanted atoms films and the substrate material contains radiation-induced processes to penetrate deep into the implanted atoms in the substrate the depth is much greater than the projected range R_p.

Kinematic interaction in the energy range up to 100 keV is determined by the energy exchange $<E>_{if}$ between ions Ar^+, primary knock-on atom (PKA) films and the substrate material $<E>_{fs}$, which can be accounted for by the reduced mass μ [1,3]:

$$\mu_{if} = 2 \cdot \frac{M_{Ar} \cdot M_f}{(M_{Ar} + M_f)^2}, \quad (1)$$

$$\mu_{fs} = 2 \cdot \frac{M_f \cdot M_s}{(M_f + M_s)^2}, \quad (2)$$

where μ_{if}, μ_{fs} – reduced masses of the interaction of atoms "ion-film" and "film-substrate", respectively; M_{Ar} и M_f – atomic mass of ions Ar^+ and film; M_s – atomic mass of substrate material; $<E'>$ – average transmitted power $<E'> = \mu E_i$; E_i – the energy of the incident particle Ar^+.

The effectiveness of the implementation of atoms at great depths can be estimated as a function that depends on the intensity of the ion beam, its energy and irradiation time [4]. In this regard, it is important to determine the optimum exposure conditions that ensure penetration of the implanted atoms to greater depths in their maximum possible concentration, especially in case of doping the polycrystalline substrates.

This paper presents the results of studying laws of formation of a given structural-phase state of the depth of the surface layers in a multilayer film and polycrystalline Ti (Zr) substrates at various modes of irradiation beams of ions He^+ and Ar^+ ions with a broad energy spectrum with an average energy of 10 keV.

2. Materials and tools

Were selected for study polycrystalline samples of Zr and Ti. The doping atoms of Si, Al, Fe, Mo, Zr and W formed as atoms of one kind or more atoms at the same time of the multilayer films in the following combinations: SiAl; AlFe; AlMo; TiSiAl-Me; MoFeAl-Me; MoFeTiAl-Me; MoFeTiSiAl-Me; MoFeAlTiSi-Me, where Me - a matrix material, i.e. Zr or Ti.

Samples were standard treatment consisting in mechanical grinding and polishing, electropolishing in the respective electrolytes. Some samples undergo additional purification by ion-beam treatment and the vacuum annealing ($p=3\times10^{-4}$ Pa) at a temperature of $0.7T_m$ for 1 hour.

The multicomponent doping carried out by ion mixing of the film when irradiated beams of ions He^+ and Ar^+ ions with a broad energy spectrum on the "VOKAL" [5], the average energy of the ions in the beam $\langle E \rangle = 10$ keV, the ion current density $j = 25$ $mA{\cdot}cm^{-2}$, dose $F \approx (1\text{-}5) \times 10^{18}$ cm^{-2}, the temperature of the samples under irradiation $T \leq 500$ °C, residual gas pressure $p \approx 1 \times 10^{-6}$ Pa. For more information about the distribution of interstitial atoms in the depth C(x) and polycrystalline Ti- and Zr-matrix method was used to measure the precision spectrometer SA-2000 (c. LECO) with a depth resolution of ± 5 nm.

3. Experimental results

3.1. Surface roughness

The experiments to study the dependence of change of surface roughness R_a of the dose of irradiation beams of ions He^+, Ar^+ and $He^+ + Ar^+$ ($C_{He+} : C_{Ar+}$= 1: 1, 1: 2, 1: 5, 1:10) with a broad energy allowed to identify the most optimal conditions exposure (Figure 1) R_a<0.6 micrometers.

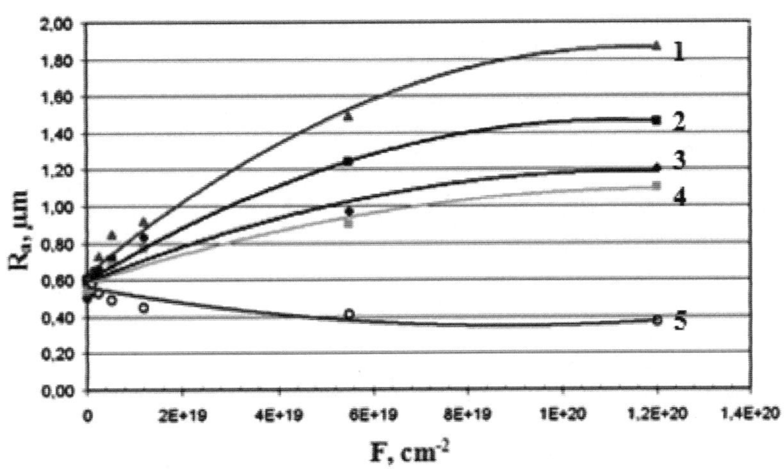

Figure 1. Dependence of surface roughness on exposure dose: 1 – Ar^+, 2 – He^+, 3 – $He^+ + Ar^+$ ($C_{He+} : C_{Ar+}$ = 1: 1), 4 – $He^+ + Ar^+$ ($C_{He+} : C_{Ar+}$ = 1: 5) 5 – $He^+ + Ar^+$ ($C_{He+} : C_{Ar+}$ = 1: 10).

From which it follows that the two-component ion beam irradiation of He^+ + Ar^+ with the content in the beam of He^+ ions at 10% (Figure 1, curve 5) creates favorable conditions for spraying the surface of the alloy in the "ion polishing". In this mode, a multi-component alloying matrix allows you to minimize distortions in the distribution of interstitial atoms, arising from the different rates of dispersion of individual crystallites, which is especially important when irradiated with high doses.

3.2. Single-component implementation

Figure 2 shows the typical distribution of atoms of Si, Al, Fe and Mo in polycrystalline zirconia after irradiation with Ar^+ ions dose of 1.3×10^{18} cm^{-2}. As seen from the figure Mo and Fe atoms penetrate the greatest depth of 400-420 and 250-280 nm, respectively, and the atoms of Si and Al are arranged mainly close to the surface of <200 nm.The maxima of distributions of atoms Si, Fe and Mo are located at the surface at a depth of 20-40 nm, and the Al atoms at a depth of 60-80 nm. Distribution C(x) of Si atoms and Mo has the form of a monotonically decreasing dependence, and Al atoms distribution corresponds to the dome depending on (asymmetric).Form distributions of embedded Fe atoms is more complex relationship and has two segments monotonic decrease in the concentration - 40-160, and 180-280 nm. Their appearance may be due to the processes of radiation exposure while stirring (first section) and radiation-stimulated diffusion under irradiation (second section) [3,6].

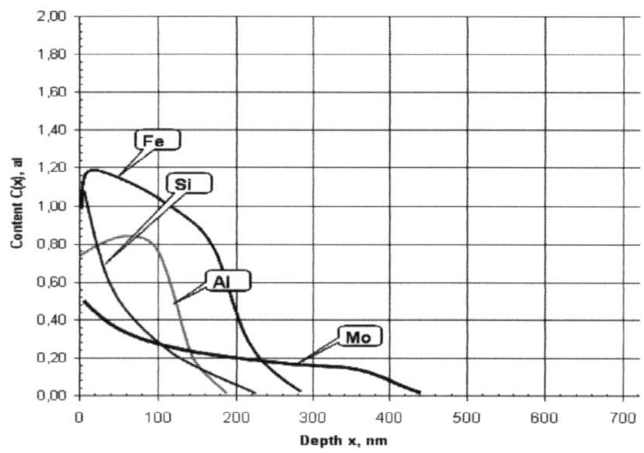

Figure 2. Distribution interstitial atoms in zirconium of monolayer films by irradiating a beam of Ar^+ ions with an average energy <E> = 10 keV, dose F = 1.3×10^{18} cm^{-2}, irradiation temperature T<100 °C.

3.3. Multi-component implementation

Figure 3 shows the dependence of the distribution of atoms Al, Ti and Fe from which it is clear that the function C(x) are characterized by several maxima of distributions: Al - 0-120 nm, Ti - 10-30 and 80-120 nm, Fe - 40-60 , 200-220 and 380-420 nm. There is a deep penetration of Fe atoms (to x_m = 500 nm).

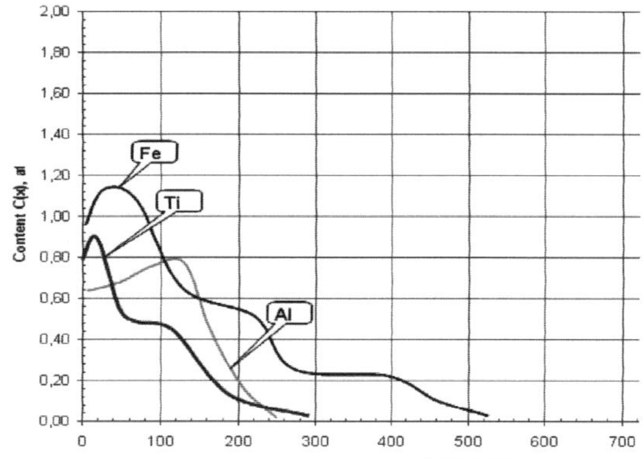

Figure 3. Distribution atoms Fe, Ti, Al in a polycrystalline sample Zr100 after irradiation with Ar^+, dose F = 2.1×10^{18} cm^{-2}, temperatureT≤ 100-150 °C, three-layer films FeTiAl- Me.

Changes in the mutual arrangement of the layers in a multilayer film MoFeTiSiAl-Me (Figures 4 and 5) is only one Al-layer (MoFeAlTiSi-Me) significantly changes the form of the functions C(x). On the distribution of C(x) (Figure 5) atoms Al, Ti, Fe concentrations appear secondary maxima located at great depths (Al - 80-100 nm, Ti - 100-120 nm, Fe - 160-180 nm). This is accompanied by the penetration of Si atoms and Mo to make greater depth (Si - 200-220 nm and Mo - 600-620 nm) compared with the distributions given in Figure 3 for a three-doping. Analysis of the distribution C(x) of interstitial atoms represented in Figure 5, reveals three specific areas: 1 - the area of radiation exposure, 2 - radiation-stimulated migration of interstitial atoms, 3 - area of diffusion processes in the fields of internal stress.

For example, for the atoms Al, Si, Ti, Fe concentration maxima located in the first region (depth of 20-30, 0-10, 20-40, 40-60 nm, respectively), further extends the area of radiation-stimulated migration of physico-chemical interaction between interstitial atoms and the atoms of the substrate (for Al, Si, Ti, Fe, it extends to the 100-120, 120-140, 160-180, 220-240 nm, respectively), the diffusion region extends to depths of Al - 220-240 nm , Si - 220 nm, Ti - 300-320 nm, Fe - 360-380 nm. Distribution molybdenum atoms C(x) has a maximum at a depth of 80-120 nm, and extends to a depth of 600 nm with a monotonic decrease of the largest concentration.

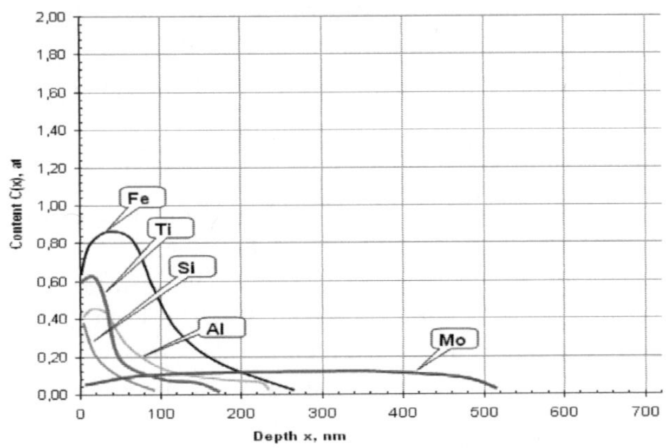

Figure 4. Distribution atoms Si, Al, Ti, Fe, Mo in a polycrystalline sample Zr-100 after irradiation ionsHe^+ + Ar^+(C_{He+} : C_{Ar+}= 1: 10), dose F =5,5 × 10^{18} cm^{-2}, temperatureT≤100-150°C, five-layer films MoFeTiSiAl – M.

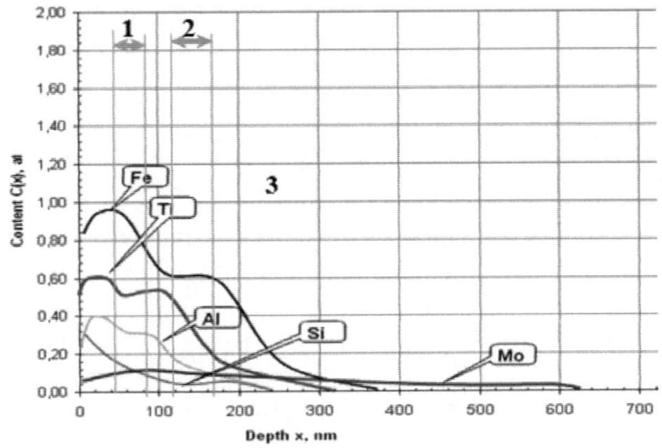

Figure 5. Distribution Si, Al, Ti, Fe, Mo atoms in Zr polycrystalline substrate (Zr-100) after the exposure to the ion beam He^+ + Ar^+ (C_{He+} : C_{Ar+} = 1:10), dose F = 5.5 × 10^{18} cm^{-2}, temperature T≤ 100 – 150 °C, five-layer film MoFeAlTiSi - Me: 1-region of radiation exposure, 2-radiation-stimulated migration of interstitial atoms, 3-area diffusion processes in the fields of internal stress.

4. Discussion

Earlier experiments [7,8] showed that for more effective implementation of the atoms of the films on the one hand it is necessary to reduce speed of dispersion films V_{Spf} =dx_f/dt, on the other hand seek to increase the speed of implementation of their atoms V_{implf}= dx_m/dtin matrix. As part of the cascade theory of sputtering and isotropic mixing it is possible to show that these quantities are related as follows [7-9]:

$$V_{Spf} = \frac{A}{\pi^2 \rho N_A s_o} \times \frac{\Delta x j F_q}{U_o} \quad (3)$$

$$V_{implf} \cong \sqrt{\frac{BjF_q}{4t}}, \quad (4)$$

where B - gram mole, ρ - the proportion of the film material, N_A - Avogadro's number, s_o - irradiated area, U_o - the binding energy of the atoms on the surface, - the coefficient of ion mixing, F_q - the energy in the substrate material, j - current density ion beam, t - time of exposure.
From these relations we can draw the following conclusions:

- speed reduction sputtering films possible by reducing energy F_q and increasing the energy of the atoms U_o;
- implantation rate increase may be accomplished by increasing the energy F_q and the coefficient of ion stirring B.

It is also one of the essential factors to increase the value of B can also be an optimum relative positioning of the layers of the films themselves [6-11]. For example, the outer layer of the multilayer films is expedient to spray with heavier atoms, and an inner, adjacent to the surface of the matrix - the lighter.

In this case, the conditions are favorable for intensification of atomic mixing and spraying rate of the film is reduced.

To explain the observed laws of change of the maximum depth of penetration of the atoms of the film substrates used in different model of an isotropic mixing proposed by Sigmund Peter and Andreas Grass Marty [3]. In the case of irradiation of the ion beam with a broad energy spectrum introduced these features. The amount of energy in the implanted layer F_q calculated from the condition that the thickness of the layer is equal to twice the projected range of ions in a suitable substrate - $2R_p$. Kinematic parameters μo was replaced by 1- μ*, where $\mu^* = 0{,}5\mu_{fs}$. The value of (1- μ*) is responsible for maintaining the kinetic energy of interstitial atoms.

At the same time we have discussed various physical and chemical mechanisms of interaction between the implanted atoms and the atoms of the substrate that allowed to enter into the expression for the calculation of the relative depth migration (x_m/R_p electronegativity of the atoms of the multilayer film and the substrate in the form of a power function $\Delta\chi = \Sigma(A_i \times \chi^{mi})$,and the expression for the relative depth of migration of interstitial atoms can be written as:

$$\frac{x_m}{R_p} = C_o \frac{M_i M_f}{(M_i + M_f)^2} \times \left[0.202 \frac{F_q(x)\Phi}{N_o E_d (1-\mu_{if})} \right]^{0.3} \times \Phi_{fs}\left(\left| \Delta\chi \right|^n \right), \quad (5)$$

гдеx$_m$ – depth migration of interstitial atoms; R_p – projected range of interstitial atoms; F_q – energy release in implantable layer; Φ – exposure dose; N_o – density target material; E_d – energy atomic displacements; μ_{if} –linkage parameter, C_o – coefficient; $\Phi_{fs}\left(\left|\Delta\chi\right|^n \right) = \sum_{k=0}^{k} A_k \chi_k^{n_k}$ – function describing the physical and chemical interaction (A_k – coefficients depending on the film material; χ_o, χ_I–electronegativity substrate and film material, respectively, tl.Pauling). For single-layer film k=3, a two-layer film – k = 4-5, a three-layer film - k = 6, five-layer film – k = 7-8 (Al-Fe-Mo-Y-Zr on Be, Fe, Cu, Zr и Ti alloys).

In multilayer systems, the distribution of interstitial atoms may issue contribute diffusion and radiation-induced processes. For example, in the "three-layer film - polycrystalline Zr-substrate" found that the mutual arrangement of layers significantly affects the amount of implanted atoms.The optimal relative positioning film layers allows substantially increase the concentration of interstitials in the matrix, e.g., by irradiation system «MoFeAl-Zr» atoms content in the matrix of the outer layer (Mo) is higher than the Al upon irradiation system «AlFeMo-Zr», i.e. the maximum number of interstitial atoms in the surface layer of the film substrate was observed for the case where the outer layer has a higher atomic weight with respect to the atomic weight of the ions Ar^+, and an inner layer adjacent to the substrate surface smallest atomic mass.

Experimental data on the introduction of atoms from one-, two-, three- and five-layer films in polycrystalline zirconia substrate and its alloys under irradiation with a beam of Ar + ions and He^+ + Ar^+ (C_{He+} : C_{Ar+}= 1:10) shows that in the range of doses irradiation (1-100) × 10^{18} ion·cm^{-2} doping the surface layers is advantageously carried out at relatively low temperatures of <100-250 °C.Wherein assessment of the effectiveness of the ion mixing in can be carried out according to the formula [6,7] $B = \langle x_m \rangle 2/\Phi * F_q$, where $\langle x_m \rangle$- the maximum depth of penetration of the atoms, Φ - dose, F_q - the energy in the substrate material.From the expression that improving the efficiency of ion mixing is to increase the penetration depth of the dopant atoms with decreasing radiation dose Φ (ie, the thickness of the sprayed layer) and reducing energy F_q in the substrate where $F_q \approx \langle E \rangle / R_p$,, $\langle E \rangle$- the average energy of the ions in the beam, R_p - projected range of ions in the matrix. This condition managed to fulfill in experiments by irradiating an ion beam with a broad energy spectrum, and the use of the beam containing ions He^+ + Ar^+, the coefficient of efficiency of ion mixing in reaching a value B = (5-20)×10^{-3}nm^{5}eV^{-1}.

In the case of irradiation of five-layer film Mo, Fe, Al, Ti, Si beam of Ar + ions and He^+ + Ar^+ (C_{He+} : C_{Ar+}= 1:10) repositioning the Al-layer relative to the other layers leads to both a deeper penetration of Fe atoms and the Mo and the appearance of secondary maxima distribution atoms Fe, Al, Ti. These changes can be explained by taking into account the processes of ion mixing, radiation-stimulated migration and physico-chemical interactions of atoms introduced both among themselves and with the atoms of the matrix, ie, condition Darken-Guri-Sud (DGS) [1,5,12].

Analysis of changes in dependencies C(x) for a variety of one, two, three and five-layer films leads to several conclusions:

- with increasing depth of penetration into the matrix of implanted atoms concentration gradient grad $C(x)$ is minimal;
- less deep penetration of the interstitial atoms in a matrix corresponding to the maximum value of grad $C(x)$;
- doping allows to create multicomponent fibers at greater depths with the value grad $C(x)$ similar as in the one-component alloying grad $C(x) \approx$5-10 at.% μm^{-1};

As follows from the data presented, the introduction of single-layer film the maximum penetration depth xm, e.g., in the Zr-matrix is observed for the Mo atoms (440 nm), Minimal - Ti (150 nm), and the concentration gradient is maximal for the atoms of Si and Ti (20 ± 3 at.% μm^{-1}), the minimum - Fe and Mo.Alloying Zr-matrix from bilayer films shows that the maximum depth of penetration is observed for the Fe atoms (460 nm) and Mo (420 nm), and the minimum - Al (75 nm) film AlTi-Me. The concentration gradient for a maximum of Si atoms and Al (10±2 at.%·μm^{-1}) of the film SiAl-Me, AlTi-Me, AlFe-Me and the minimum for the Mo atoms (1.0±0.1 at.% μm^{-1}).

Doping of Zr-matrix of three-layer film offers insight to the maximum depth of atoms of Fe (520 nm) film FeTiAl and Mo (460 nm) film MoAlSi-Me. The minimum depth of penetration is observed in atoms Al (220 nm) film MoAlSi-Me. The concentration gradient maximum atoms Si (12 ± 1 at.% μm^{-1}) film MoAlSi-Me, and the minimum of the atoms Mo (3 ± 1 at.% μm^{-1}), the film MoFeAl-Me for atoms Si, Ti, Fe (11 ± 1 at.%·μm^{-1}), and the minimum for Mo atoms (1.0 ± 0.1 at.%·μm^{-1}). Simultaneous introduction of a five-layer film enables to introduce a maximum depth of atoms of Mo (620 nm) film MoFeAlTiSi-Me, a minimum depth to penetrate atoms Si (100 nm). The concentration gradient with a maximum (12 ± 1 at.%·μm^{-1}).

5. Conclusion

The experiments showed that irradiation of the "film-substrate" change of value relative depth of penetration of the interstitial atoms (x_m/R_p) is satisfactorily described by the model of an isotropic mixing and physico-chemical interaction with its environment. In the case of the introduction of the matrix atoms of the multi-layer film (two, three and five -ply) physico-chemical interaction between the implanted atoms and the atoms of the matrix can be accounted for by the value of the electronegativity of elements, presented in the form of a power series.

It was revealed that multicomponent doping can create gradient layers at greater depths with the value of grad $C(x) \approx$ 5-10 at.% μm close, as in the one-component doping.

Due to the presence of multiple peaks of concentration in multicomponent doping in the matrix formed by several sections (layers) of a high concentration gradient of the interstitial atoms, which are formed as a result of redistribution of flows of displaced atoms of the multilayer film and substrate.

Analysis of the data suggests that the basis of the penetration of the atoms in the oxide film at the ion-doped multicomponent stirring under irradiation with ion beams of He^+ and Ar^+ with a broad energy spectrum lies several processes including:

- transfer of energy from the primary ions and sputtered atoms to subsequent displacement cascade;
- radiation-induced migration and physico-chemical interaction between the implanted atoms and the atoms of the substrate;

The effectiveness of penetrate of the implanted atoms to great depths B is defined energy release in implantable layer F_q, which is suitable to reduce the largest. One of the possible ways to implement the optimal conditions of doping by ion mixing the irradiation of the ion beam with a broad energy spectrum, and improve the efficiency of penetration into the substrate implanted atoms can be an increase in energy of the ions in the beam with a simultaneous decrease in the reduced atomic mass of the beam, which can be realized while irradiating ions of several mass For example, light and heavy ions (He^++Ar^+) to determine their ratio ($C_{He+} : C_{Ar+}$) and the optimal arrangement of the layers in the multilayer film, i.e. outer layer - heavy atoms, the lower layer the lightest atoms.

Acknowledgements
This work was performed within the framework of the Center of Nuclear Systems and Materials supported by MEPhI Academic Excellence Project (contract № 02.a03.21.0005, 27.08.2013).

References

[1] Komarov F F 1990 *Ion implantation in metals* (Moscow: Metallurgy) p 261
[2] Kalin B A and Volkov N V 2003 *Proc. of 12th Int. Conf. on Physics and Chemistry of Inorganic materials* (Tomsk: TPU) 231
[3] Sigmund P and Gras-Marti A 1980 *J. Nucl. Instrum. Methods Phys. Res.* **168** 389
[4] Kalin B A and Volkov N V 2006 *Proc. 8th Int. Conf. on Modification of Materials with Particle Beams and Plasma Flows* (Tomsk: TPU) 127
[5] Volkov N V and Kalin B A 1991 *Coll. Material science questions of nuclear technology* (Moscow: Energoatomizdat) 64
[6] Sood D K 1978 *Physics Lett.* **68** 469
[7] Volkov N V 2011 *J. of Surface Investigation. X-ray, Synchrotron and Neutron Techniques* **3** 40
[8] Kalin B A and Volkov N V 2007 *Proc. Sci. session of the MEPhI-2007*(Moscow: MEPhI) **9** 44
[9] Kalin B A, Volkov N V and Yakushin V L 2007 *Issues of atomic science and technology. Ser.The physics of radiation damages and radiation materials* **90** 164
[10] Kalin B A and Volkov N V 2003 *J. of Surface Investigation. X-ray, Synchrotron and Neutron Techniques* **5** 43
[11] Didenko A N and Sharkeev Yu P 1983 *J. of Surface Investigation. X-ray, Synchrotron and Neutron Techniques* **3** 120
[12] Darken L S and Gurry R W 1953 *Physical Chemestry of Metals* (Inc. New York Toronto London, Mc. Granw - Hill Book Company) p 582

Influence of the pulsed plasma treatment on the corrosion resistance of the low-alloy steel plated by Ni-based alloy

P Dzhumaev, V Yakushin, B Kalin, V Polsky and M Yurlova

National Research Nuclear University MEPhI (Moscow Engineering Physics Institute) Kashirskoe highway 31, Moscow, 115409, Russia

E-mail: psdzhumaev@mephi.ru

Abstract. This paper presents investigation results of the influence of high temperature pulsed plasma flows (HTPPF) treatment on the corrosion resistance of low-alloy steel 0.2C-Cr-Mn-Ni-Mo cladded by the rapidly quenched nickel-based alloy. A technique that allows obtaining a defect-free clad layer with a good adhesion to the substrate was developed. It is shown that the preliminary treatment of steel samples by nitrogen plasma flows significantly increases their corrosion resistance in the conditions of intergranular corrosion test in a water solution of sulfuric acid. A change of the corrosion mechanism of the clad layer from intergranular to uniform corrosion was observed as a result of sub-microcrystalline structure formation and homogeneous distribution of alloying elements in the plasma treated surface layer thus leading to the significant increase of the corrosion resistance.

1. Introduction

Reliability, durability and performance of parts and components of various mechanisms are largely determined by their surface properties. In particular, modification of only the surface layers of products by pulsed laser, electron, ion or plasma sources is often sufficient and cost-effective for improving their corrosion and wear resistance [1-5]. Therefore, an important task is to develop new methods of obtaining the targeted microstructure and properties of the surface layers of materials.

Carbon and low-alloy steels are widely used as a structural material. However, they tend to intensive corrosion damage due to various corrosive environments, which makes the use of these steels in parts and components of systems economically unprofitable. Therefore, at this time, an urgent task is to improve the corrosion resistance of these materials for use in various branches of industry. A technique of improving the corrosion resistance of the low-alloy steels by their pre-cladding with high quenched solder alloys and subsequent treatment by high-temperature pulsed plasma flows (HTPPF) is presented in this work.

2. Experimental procedure

Low-alloy steel 0.2C-Cr-Mn-Ni-Mo was used as the substrate material for cladding. Samples were made in the form of disks with a diameter of 20 mm and a thickness of 5 mm. The surface of the disk samples was mechanically polished to a roughness Ra=0.4. Thin tapes of a thickness about 35-50 μm were used for cladding of the samples. To obtain the sub-microcrystalline structure tapes were produced by rapid quenching from the melt of the nickel-based alloy (Ni (base) - 18% Cr -10% Si).

The chosen method of cladding was the following. A thin foil of the fine-crystalline nickel-based alloy was welded to the disk substrates by spot welding along the disk periphery (figure 1). On a next step, samples with a welded tape were heated to a temperature of 1150 °C in a vacuum furnace at a residual pressure not higher than $6.5\cdot10^{-3}$ Pa and subsequently cooled at the different regimes: 1 - "sawtooth" cycle (heating to 1150 ° C → Cooling); 2 - heating with the exposure (Δt) for 1-5 min at the maximum heating temperature of 1150 °C followed by cooling.

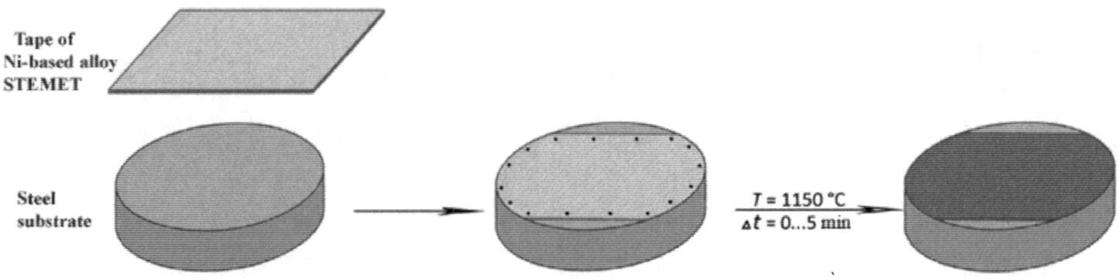

Figure 1. Method of nickel based alloy cladding of steel.

Samples were treated by means of the Z-pinch type plasma installation "Desna-M" [6] by high-temperature pulsed nitrogen plasma flows in various regimes. The energy density (Q) of plasma flows ranged from 22 to 31 J/cm^2 and their power density from $1.1\cdot10^6$ to $1.5\cdot10^6$ W/cm^2. The number of pulses was $N = 2$ for each cycle of irradiation. The pulse duration was 20 μs.

Intergranular corrosion tests were performed using the method based on a Russian state standard GOST 6032-2003. The samples were exposed for 8 h in a boiling aqueous solution containing 120 g/dm^3 of $CuSO_4$ and 120 cm^3/dm^3 of H_2SO_4. The tests were carried out in a flask with a reflux condenser. Copper shavings were located at the bottom of the flask to create a redox potential of +0.1 V. The samples were placed into the flask so as to avoid any contact with each other.

The nature and level of corrosion interaction were determined by ZEISS EVO 50 XVP (Zeiss, Germany) scanning electron microscope (SEM) using cross-section samples and by mass loss of the sample.

3. Results and discussion

Microstructure studies of plated samples showed (figure 2, a) that the surface microstructure was fairly homogeneous. Regardless of the cladding regimes, three distinct zones with different phase contrast were observed in all the samples: equiaxed grains with a mean size of about 20 μm (1), an intergranular layer along the grain boundaries (2) and faceted inclusions (3) of various shapes with sizes varied from 2 to 10 μm. SEM study of the microstructure of plated samples cross-sections (figure 2, b) showed the formation of a clad layer with relatively uniform thickness (~ 35-40 microns) on the entire surface of the sample. During the cladding process diffusion bonding of cladding alloy with the substrate occurs thus leading to the good adhesion of the clad layer to the substrate without the formation of microcracks and pores.

The results of the X-ray microanalysis (Table 1) showed that the equiaxed grains and the intergranular layer are nickel based solid solution containing different amount of chromium, iron and silicon. The atomic concentration of silicon in the intergranular layer is higher and chromium concentration is lower than in the grains. The fine-dispersed inclusions are carbosilicides of chromium and nickel Cr_3Ni_2SiC. It should be noted that the formation of precipitates is observed both in the bulk of grains and along their boundaries. The quantity of such precipitates increased more than twice with the increase of duration of cladding process (the exposure time at the maximum temperature varied from 1 to 5 min). At that, carbosilicide precipitates were predominantly located along the grain boundaries. Quantitative analysis of the SEM micrographs showed that the most homogeneous structure with a relatively low density of formed precipitates and equiaxed grains of similar sizes were observed at the exposure time during cladding of 1 and 3 min.

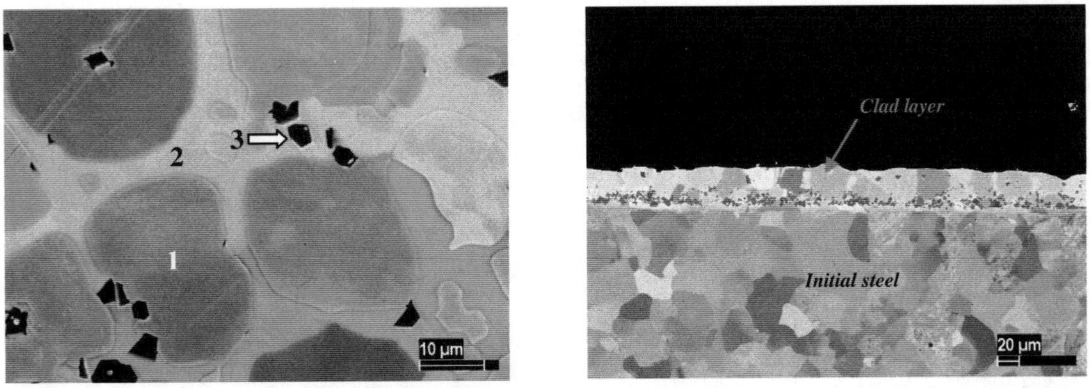

(a) (b)

Figure 2. The surface (a) and cross-section (b) microstructure of steel cladded by nickel-based alloy (Δt =1 min).

Table 1 – Elemental composition of steel 0.2C-Cr-Mn-Ni-Mo samples cladded by STEMET1305.

Area of the element analysis (according to Fig. 2, b)	Main elements, wt. %				
	C	Si	Cr	Fe	Ni
Initial STEMET 1305 tape	-	10	18	-	Bal.
1 – grain	< 0.5	6.67	16.23	9.73	Bal.
2 – intergranular layer	< 0.5	15.23	3.43	3.30	Bal.
3 – precipitates	5.28	8.64	47.63	1.33	Bal.

The treatment of cladded samples by nitrogen plasma results in melting of their surface layer and significantly smoothers the surface relief. At that, traces of the original microstructure of clad samples are practically not observed (figure 3, a). At higher magnification a submicrocrystalline ordered cellular structure with a typical cell size about of 100-200 nm (figure 3, b) is observed in the clad samples for all the regimes of cladding. However, intensive cracking of the surface of samples cladded at the maximum temperature for Δt = 5 min took place (figure 4, a). The cracking is due to the temperature gradient, different coefficients of thermal expansion of various phases formed during the cladding and subsequent radiation heating and rapid heating and subsequent cooling of the surface layer during plasma treatment.

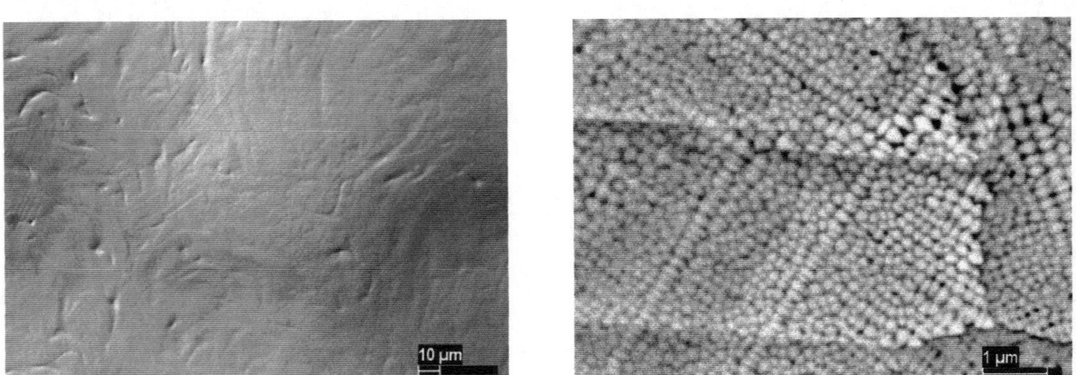

(a) (b)

Figure 3. The surface microstructure of clad (Δt =1 min) steel samples after the treatment by nitrogen pulsed plasma flows (Q = 22 J/cm^2).

SEM investigations of finely polished treated samples have shown (figure 4, b) that the origin and the development of cracks occur mainly in the grain boundary interlayer (2), i.e. in areas in which in the initial (untreated by HTPPF) state there is an increased content of silicon, as well as along the brittle precipitates of chromium and nickel carbosilicides (3), while the equiaxed grains themselves (1) are not subjected to cracking. The surface cracking at Δt = 1 and 3 min is not observed because of decreased number of brittle phases.

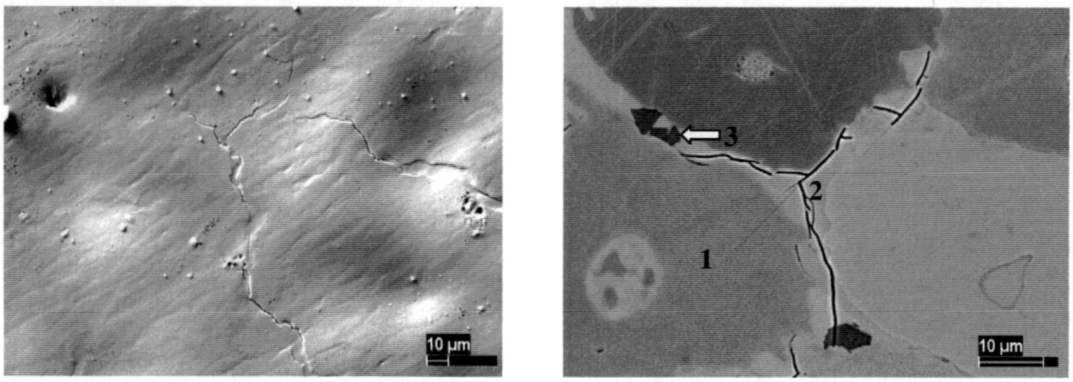

(a) (b)

Figure 4. The surface microstructure of clad (Δt = 5 min) steel samples after the treatment by nitrogen pulsed plasma flows (Q = 22 J/cm^2).

Microstructure analysis of the clad samples treated by HTPPF showed that the degree of decreasing the surface relief increased with decreasing the holding time of samples at the maximum temperature. The most smooth relief is observed for samples cladded using the regime of Δt = 1 min. The energy density of plasma flows and the total energy release also influence the smoothing of the initial surface relief and the uniform mixing of elements after plasma treatment. Studies of clad samples (Δt = 1 min) treated by plasma flows using the regime of 2 pulses in a cycle with a sequential increase of the flow energy density ($Q_1 = 22 + Q_2 = 25 + Q_3 = 28 + Q_4 = 31$ J/cm^2) showed that the most homogeneous and practically defect-free surface layers are formed after sequential treatment by 6 pulses ($Q_1 + Q_2 + Q_3$, $N = 2$). Whereas after 4 irradiation cycles ($Q_1 + Q_2 + Q_3 + Q_4$, $N = 2$), there is a formation of local cracks and micro pores in the surface of samples which can significantly decrease the corrosion resistance.

Table 2 – Elemental composition of samples of steel 0.2C-Cr-Mn-Ni-Mo cladded by STEMET1305 after HTPPF treatment.

HTPPF treatment regime	Main elements, wt. %				
	C	Si	Cr	Fe	Ni
Q_1=22 J/cm^2	< 0.5	8.8 ± 1.7	13.4 ± 3.3	8.8 ± 1.6	Bal.
Q_1=22 J/cm^2 + Q_2=25 J/cm^2+ Q_3=28 J/cm^2	< 0.5	8.2 ± 0.3	13.6 ± 0.5	10.3 ± 0.40	Bal.

Microstructure studies of the samples after corrosion tests have shown that the initial samples of steel are exposed to significant corrosion damage that is quite uniform over the sample surface. Treatment of the sample surface by HTPP flows decreases the corrosion rate of samples by up to 3-10 times depending on the treatment conditions. However, corrosion interaction with the initial material matrix begins to intensively occur after destruction of a thin (~ 10 µm) modified layer. More detailed results of the effect of plasma treatment on the corrosion resistance of steels were reported earlier.

IOP Conf. Series: Materials Science and Engineering **130** (2016) 012040 doi:10.1088/1757-899X/130/1/012040

Corrosion tests were carried out for two types of clad samples (without any treatment): those obtained using the "sawtooth" regime and samples held at the maximum heating temperature for 1-3 min. In all cases there was a significant corrosion damage of the surface layer (figure 5, a), the corrosion of the cladded layer occured predominantly along the intergranular layer (figure 5, b and c). In addition, delamination of the clad layer and formation of long cracks in it (figure 5, c) took place due to penetration of an aggressive medium to the substrate through the clad layer.

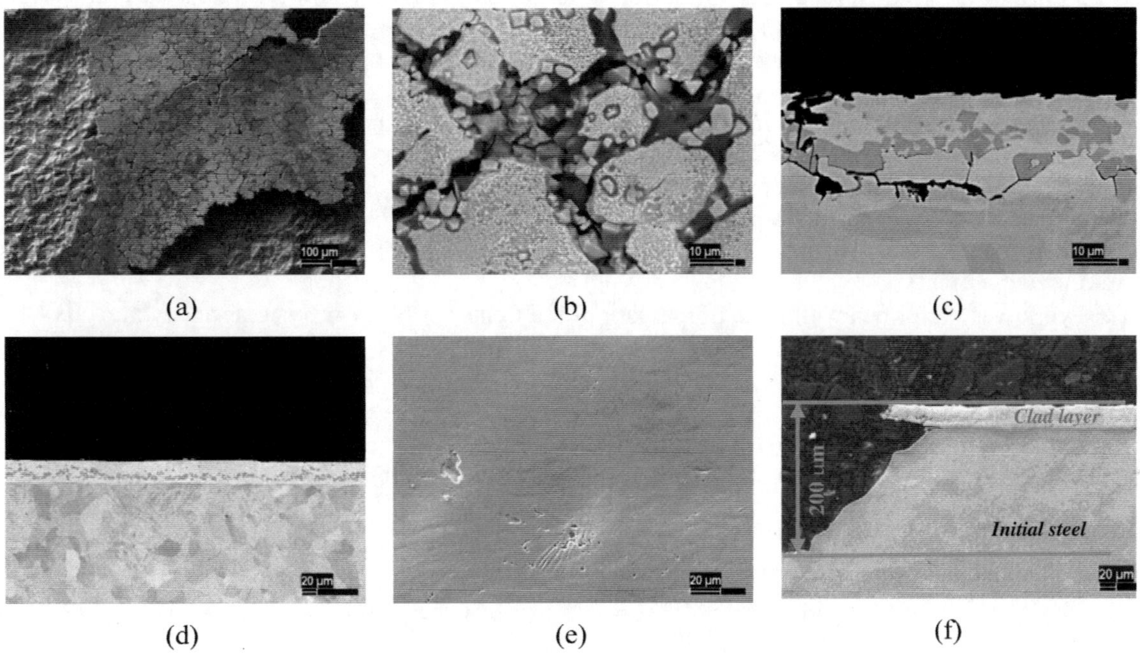

Figure 5. Microstructures of surface (a, b and e) and cross-sections (c, d and f) of clad samples after corrosion tests: without (a-c) and after the HTPPF treatment (d-f).

Studies of samples cladded and treated by plasma flows with $Q_1 = 22 + Q_2 = 25 + Q_3 = 28$ J/cm^2 after corrosion tests showed that, in general, only edge areas of the samples were exposed to damage, i.e. the interface between the clad layer and the substrate. At that, the modified layer preserves undamaged on the surface of samples (figure 5, d) and only local areas of corrosion damage (figure 5, e) were observed. The estimated difference in levels between the modified clad surface and the corroded area of the initial steel at the sample edge is ~ 200 μm (figure 5, f) which corresponds to 25 μm/h corrosion rate for initial steel and therefore clearly shows the significant corrosion protection effect of the plasma treated clad layer.

These results allow us to make the following assumptions about the mechanism of the corrosion resistance increase of the plated and treated by HTPPF samples. Corrosion of the cladded samples occurs mainly as a result of damage of the clad layer due to selective dissolution of intergranular layer. Penetration of the aggressive medium to the substrate leads to the destruction and exfoliation of the cladding layer. Corrosion of plasma modified cladding layer occurs by a mechanism different from the corrosion destruction of untreated samples. As shown above, plasma treatment leads to the formation of a homogeneous nanostructured layer with uniform distribution of the elements and this layer contains ~ 13.5 wt. % Cr (table 2). Creating such a layer prevents intergranular corrosion. Also, as is known, alloying nickel alloys with chromium (typically 10-25%) provides high corrosion resistance due to the formation of the protective oxide film, and thus prevents destruction of the material in contact with a corrosive medium.

IOP Conf. Series: Materials Science and Engineering **130** (2016) 012040 doi:10.1088/1757-899X/130/1/012040

4. Conclusions

A technique of cladding of the samples of low-alloy steel 0.2C-Cr-Mn-Ni-Mo by rapidly quenched tape of STEMET 1305 nickel-based alloy has been developed. It allows obtaining of a 35-40 μm thick clad layer which is sufficiently uniform in microstructure and the elemental composition and has a good adhesion to the substrate.

It was found that the treatment of the cladded samples by pulsed plasma flows leads to the smoothing of the surface relief and to the formation of sub-microcrystalline cellular structure with a mean size of cell about 150-200 nm and uniform distribution of alloying elements.

It was found that the cladding of low-alloy steel samples by nickel-based alloy STEMET 1305 followed by pulsed plasma flows treatment significantly reduces the corrosion rate compared to untreated samples. The optimal mode of cladding and plasma treatment for improving the corrosion resistance of low-alloy steel was chosen. At the same time there is a change of corrosion mechanism from intergranular to uniform corrosion that caused by microstructure and phase composition changes in the surface layers after plasma processing.

Acknowledgements

This work was performed within the framework of the Center of Nuclear Systems and Materials supported by MEPhI Academic Excellence Project (contract № 02.a03.21.0005, 27.08.2013).

References

[1] Wei Cong, Zhengjun Yao and Xiaolin Zhu 2010 *Wear* **268 (5–6)** 790
[2] Yilbas B S, Ihsan-ul-Haq Toor, Jahanzaib Malik and F Patel 2014 *Optics and Lasers in Engineering* **54** 8
[3] Pacquentin W et al 2015 *Applied Surface Science* **356** 561
[4] Abdolahi B et al 2011 *Applied Surface Science* **257** 9921
[5] Sari A H et al 2015 *Vacuum* **115** 39
[6] V A Gribkov, F I Grigoriev, B A Kalin and V L Yakushin 2001 *Prospective Beam Irradiation Technologies of Materials Processing* (Moscow: Krugliy god, In Russian)

IOP Conf. Series: Materials Science and Engineering **130** (2016) 012041 doi:10.1088/1757-899X/130/1/012041

Influence of the chemical composition of rapidly quenched amorphous alloys (Ni, Fe, Cr)-B-Si on its crystallization process

G Elmanov, P Dzhumaev, E Ivanitskaya, V Skrytnyi and A Ruslanov

National Research Nuclear University MEPhI (Moscow Engineering Physics Institute) Kashirskoe highway 31, Moscow,115409, Russia

E-mail: GNElmanov@mephi.ru

Abstract. This paper presents results of research of the structure and phase transformations during the multistage crystallization of the metallic glasses with the compositions $Ni_{71,5}Cr_{6,8}Fe_{2,7}B_{11,9}Si_{7,1}$ and $Ni_{63,4}Cr_{7,4}Fe_{4,3}Mn_{0,8}B_{15,6}Si_{8,5}$ labeled as AWS BNi-2 according to American Welding Society. Differential scanning calorimetry (DSC), X-ray diffraction analysis (XRD), scanning electron microscopy (SEM) and energy-dispersive X-ray microanalysis (EDX) were used as experimental research methods. The influence of the alloys chemical composition (boron, manganese and iron) on the temperatures and the exothermic heat effects of phase transformations, as well as on the phase composition of alloys at three stages of crystallization was analyzed. We present a thermodynamic explanation of the observed heat effects. It has been shown that manganese has the main influence on the phase transformations temperatures and heat effects in these two alloys. It is also assumed that at the final crystallization stage simultaneously with the formation of phases Ni_3B and β_1-Ni_3Si should occur the nucleation of borides of CrB type with high Cr and low Si content.

1. Introduction

Alloys based on the ternary system Ni-Si-B (AWS BNi-2) in crystalline and amorphous states are widely used as a brazing alloys and corrosion-resistant coatings. It is known [1-4], that crystallization of amorphous BNi-2 alloys proceeds in three stages: the first stage of crystallization (450–500°C) is associated with the formation of solid solution crystals based on nickel (α-Ni). The second stage (500–550°C) is associated with the formation of a metastable τ-phase with $M_{23}B_6$-type structure. At the final stage (550–650°C), the formation of Ni_3B and Ni_3Si [2, 4] or $Ni_{31}Si_{12}$ [1] as a result of the decomposition of τ-phase was reported. It should be noted that other authors observed another phases in BNi-2 alloys [3]: Ni_4B_3, CrB, $B_2Fe_{15}Si_3$, $CrSi_2$ and $Ni_{45}Si_2B$ [5]; Ni_2B, Ni_3Si_2 and $Fe_3(Si,B)$. Data obtained for the phase composition of such alloys produced by crystallization from the amorphous state, as well as data on the temperatures and heat effects of the phase transformations are slightly different due to different chemical compositions of the investigated BNi-2 alloys.

Therefore, a purpose of this work is to explore the influence of the chemical composition of the alloys on their phase composition at all stages of crystallization, as well as on the temperatures and the exothermic heat effects of phase transformations.

2. Materials and Experimental Methods

The amorphous alloys in the form of tapes with a thickness of 40–50 μm with two similar to BNi-2 compositions $Ni_{71,5}Cr_{6,8}Fe_{2,7}B_{11,9}Si_{7,1}$ (STEMET® 1301A) and $Ni_{63,4}Cr_{7,4}Fe_{4,3}Mn_{0,8}B_{15,6}Si_{8,5}$ (STEMET® 1301) (hereinafter referred to as 1301A and 1301) were produced by the rapid solidification techniques [6]. The main difference between alloys is the content of boron, manganese and iron. Crystallization process of the amorphous alloys was carried out in a vacuum furnace with heating rate of 20°C /min.

Differential scanning calorimetric (DSC) analysis of alloys was performed using the Netzsch STA409CD calorimeter. The samples of the amorphous alloys weighing approximately 20 mg were heated in the scanning calorimeter at a heating rate of 20°C/min under a purified argon atmosphere.

X-ray diffraction phase analysis (XRD) was carried out by means of DRON-3M and D8 DISCOVER (Bruker) at CuK_{α}-radiation using crystallographic database ICDD PDF-2 and software Bruker AXS DIFFRAC.EVA. Determination of the chemical composition of the phases formed during crystallization was carried out using the scanning electron microscope (SEM) EVO 50 XVP (Zeiss) with an energy-dispersive X-ray microanalysis spectrometer (EDX) INCA x-act (Oxford Instruments).

3. Results and Discussion

The results of differential scanning calorimetry of 1301A and 1301 alloys are presented on figure 1 (curves 1 and 2 respectively). This figure also shows the DSC curve 3 for alloy of similar composition ($Ni_{64,4}Fe_4Cr_{4,9}Mn_2B_{16,2}C_{0,5}Si_8$), presented in [1]. It can be seen that curves differ significantly despite the similarity of the compositions of these three alloys. Primary crystallization of 1301 compared to 1301A alloy begins at a slightly higher temperature, while the temperature of formation of τ-phase is below it. Thus, the first and second peaks on the DSC curve of the alloy 1301 are partially overlapped. For the third alloy ($Ni_{64,8}Cr_{4,9}Fe_{4,0}$ Mn_2 $B_{16,2}Si_8C_{0,5}$) this overlap is considerably more visible. Overlapping of the first and second crystallization peaks means that the formation of α-Ni and τ-phase crystals occurs substantially simultaneously.

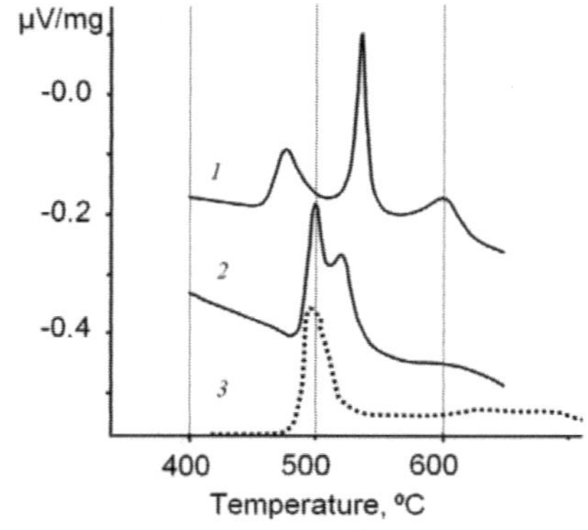

Figure 1. DSC curves of alloys: 1 - 1301A; 2 - 1301; 3 - $Ni_{64,8}Cr_{4,9}Fe_{4,0}Mn_2B_{16,2}Si_8C_{0,5}$ [1].

Table 1 presents data on the temperatures of the three most high heat peaks (T_1, T_2, T_3) on the DSC curves for different stages of crystallization of alloys 1301 and 1301A in comparison with the published data for other amorphous alloys with similar chemical composition. Because the available DSC data were obtained at the various heating rates of the samples, the values were given for the one heating rate (20 °C/min). For this purpose the data on the crystallization temperature dependence of the heating rate for the amorphous alloy $Ni_{65}Cr_{6,6}Fe_3B_{17,5}Si_{7,6}C_{0,3}$ for all three stages of the crystallization process were used [5].

Analysis of these data shows that at the beginning of the primary crystallization generally not boron content, but the amount of manganese and possibly iron affects temperature T_1. Indeed, alloys No.2 and No.3, which contain a large amount of boron, iron and manganese as compared with other samples have a maximum value of this temperature (close to 500°C). While sample No.6, which has the high boron content, but do not contain manganese and iron, has a primary crystallization temperature around 480 °C that is typical for the majority of alloys of this type. It also can be seen that alloys with manganese and possibly with increased iron content have the lower values of temperature T_2 and the higher temperature T_3.

Table 1. Temperatures of highs for the heat peaks on the DSC-curves for three stages of crystallization.

№№	Alloy	Temperatures, °C			Reference
		T_1	T_2	T_3	
1	$Ni_{71,5}Cr_{6,8}Fe_{2,7}B_{11,9}Si_{7,1}$	478.4	535.5	597.1	our data
2	$Ni_{63,4}Cr_{7,4}Fe_{4,3}\mathbf{Mn_{0,8}B_{15,6}}Si_{8,5}$	499.7	519.1	610	our data
3	$Ni_{64,4}Cr_{4,9}Fe_{4}\mathbf{Mn_{2}B_{16,2}}Si_{8}C_{0,5}$	497.5	509	629	[1]
4	$Ni_{69,1}Cr_{6,6}Fe_{2,7}B_{13,7}Si_{7,9}$	482	517	583	[4]
5	$Ni_{68.7}Cr_{6.6}Fe_{2.65}B_{14}Si_{7.8}C_{0.25}$	472	520	582	[3]
6	$Ni_{65}Cr_{6,6}Fe_{3}B_{17,5}Si_{7,6}C_{0,3}$	480	525	600	[5]

Note that the relative and total heat effects for three crystallization stages of 1301 and 1301A alloys are approximately the same (table 2). As shown below the relative amount of phases in 1301A and 1301 alloys after the first and second stages are significantly different (the amount of α-Ni in 1301A is approximately 1.4–1.6 times higher, and amount of τ-phase is 1.1–1.3 times lower than in 1301). Thus, formation of the same amount of α-Ni phase in the alloy 1301A is accompanied by a less heat effect than in the alloy 1301, and a larger heat effect for τ-phase formation, respectively.

Table 2. Relative and total (per mole of alloy) heat effects of three crystallization stages

Process / Alloy	Δ ***H*** (conventional units)	
	1301	**1301A**
1 A →α-Ni + A′	0.31	0.30
2 A′ → τ ($M_{23}B_6$)	0.45	0.51
3 τ → Ni_3B + Ni_3Si	0.24	0.23
Σ	1.00	1.06

[a] A and A′ – amorphous phases of different composition

These results of DSC can be explained in terms of thermodynamics. The process of crystallization of the metal glasses is always exothermic and reduces its free energy ΔG. At the first crystallization stage a metastable equilibrium between the amorphous phase and the α-Ni crystal phase is established. The compositions of these phases are determined by the common tangent to the curves of Gibbs free energy dependence on boron concentration (figure 2).

It is known that manganese lowers the melting point of BNi-2 alloy (to 20°C at 2% of Mn) [1]. Therefore, the ΔG curve of the amorphous phase (supercooled liquid) should move down with the addition of manganese. Thus, increase of the manganese content in the amorphous alloy should lead to decrease of the thermodynamic stimulus of α-Ni phase nucleation and to the growth of the onset crystallization temperature.

Increase of the boron content in the amorphous alloy also reduces the thermodynamic factor. However, the crystallization temperature remains nearly unchanged. It can be explained by the fact that not only the thermodynamic, but also kinetic factor affects the nucleation rate. Kinetic factor is determined by the rate of the boron atoms transition through the boundary between crystalline and amorphous phase and increases with the boron concentration increase. Thermodynamic and kinetic

IOP Conf. Series: Materials Science and Engineering **130** (2016) 012041 doi:10.1088/1757-899X/130/1/012041

factors compensate each other with the change of the boron concentration, causing the temperature remains unchanged.

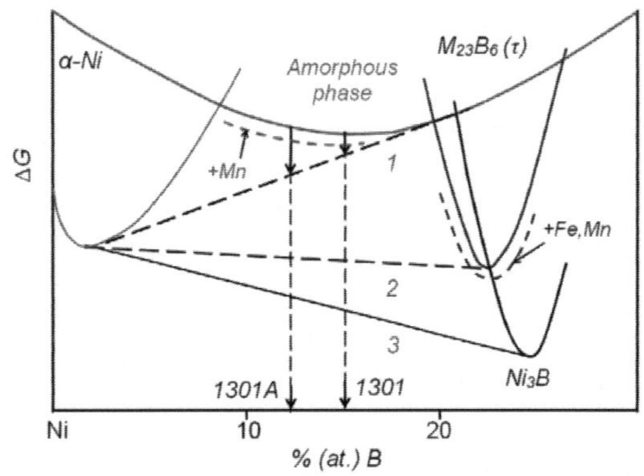

Figure 2. Hypothetical diagram of Gibbs free energy dependences on boron concentration for different phases and phase equilibria in the Ni – B binary system:
1 – metastable equilibrium of α-Ni and A;
2 – metastable equilibrium of α-Ni and τ-phase ;
3 – stable equilibrium of α-Ni and Ni_3B.

It should be noted that increase of the both boron and manganese concentration should reduce the heat effect of the first crystallization stage (per mole of α-Ni phase), because it reduces the difference of enthalpies of the crystalline and initial amorphous phase.

The estimation of the relative volume of the crystalline phase in the alloy 1301A after the first stage of crystallization was carried out. The XRD data from figure 3 and the formula $V_{cr} = I_{cr}/(I_{cr} + \alpha I_a)$, where I_{cr} and I_a – integrated intensities of reflections from the crystalline and amorphous phases, and α - constant (the Huang parameter), were used for this purpose. The values of α for similar alloys ($Fe_{73.5}Cu_1Nb_3Si_{13.5}B_9$ and $Fe_{73.9}B_{13.2}Si_{10.9}C_2$) are known: α ≈ 1.05 and 0.77 [7]. The estimated volume and mole fractions of crystalline phase calculated using these parameters are (49-57) vol. % and (44-52) mol. %. Considering that the solubility of boron in α-Ni is negligible, the boron content in amorphous phase at the metastable equilibrium can be estimated as 21-25 at.%, that is close to τ-phase ($M_{23}B_6$) composition. On the second stage of crystallization, the formation of τ-phase occurs by the polymorphic transformation without significant change of α-Ni phase amount (Figure 4). In the alloy $Ni_{64,4}Cr_{4,9}Fe_4Mn_2B_{16,2}Si_8C_{0,5}$ with the composition close to 1301 alloy significantly greater amount of τ-phase have been formed (~65 vol. %) [1], which is in a good agreement with our data (figure 4).

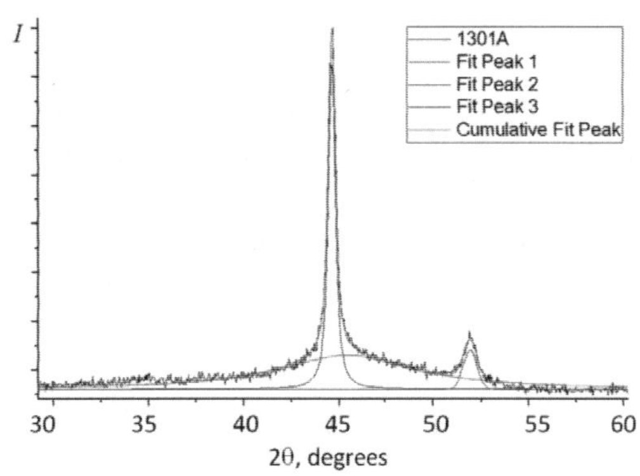

Figure 3. XRD pattern of the 1301A alloy after the first stage of crystallization (annealing at 500 °C, 1 min) with separated peaks of crystalline and amorphous phases.

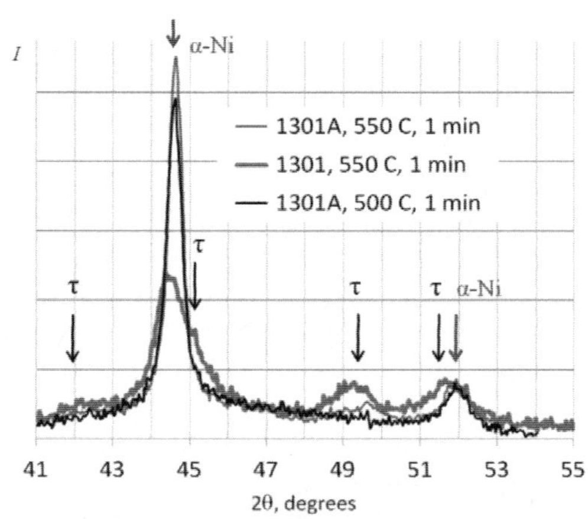

Figure 4. XRD pattern of 1301A and 1301 alloys after the first and second stages of crystallization.

It is also known that in the Ni-B system metastable τ-phase can be fixed by addition of Fe and probably, Cr, into its composition [1]. In the absence of Fe this phase does not form. Note that the formation of compounds of type $M_{23}(B,C)_6$ is typical for such metals as Fe, Cr, Mn, but not Ni. Thus, these metals contribute to the stabilization of τ-phase by decrease of its free energy. This leads to the increase of temperature of τ-phase formation (T_2) and to increase of its decomposition temperature T_3. Indeed, alloy 1301A completely decomposes at a temperature 600°C, while the alloy 1301 still contains τ-phase at temperatures up to 650°C (figure 5). Annealing at 900°C led to the decomposition of this phase. The presence of two cubic phases (α-Ni and Ni_3Si) with fcc-structure and different lattice parameters at the end of the crystallization process of alloys 1301 and 1301A was observed. Ni_3Si-based phase can hardly be discriminated from α-Ni by XRD since it has the same structure, but the presence of the silicide phase is well observed in XRD pattern of alloy 1301. According to EDX analysis of alloy 1301 annealed at 900°C, in addition to phases mentioned above, formation of chromium boride of CrB type with low Si content can also be observed (figure 6). Application of the EDX method in scanning electron microscope for alloys annealed at low temperatures is limited because of the small size of crystals (30-100 nm). XRD method did not detect the presence of CrB phase in the crystallized amorphous and massive cast alloys, although this phase was detected by EDX analysis [4, 8, 9].

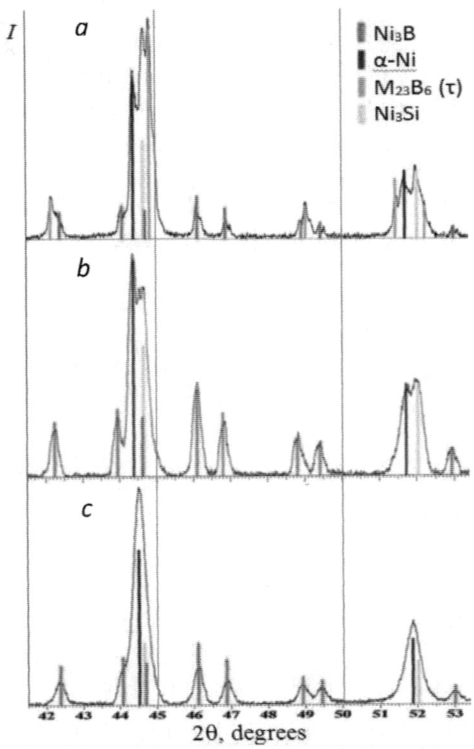

Figure 5. XRD phase analysis of alloys 1301 and 1301A after annealings: *a* –1301, 650°, 1h ; *b* –1301, 900°, 1h; *c* – 1301A, 600°C, 1 h.

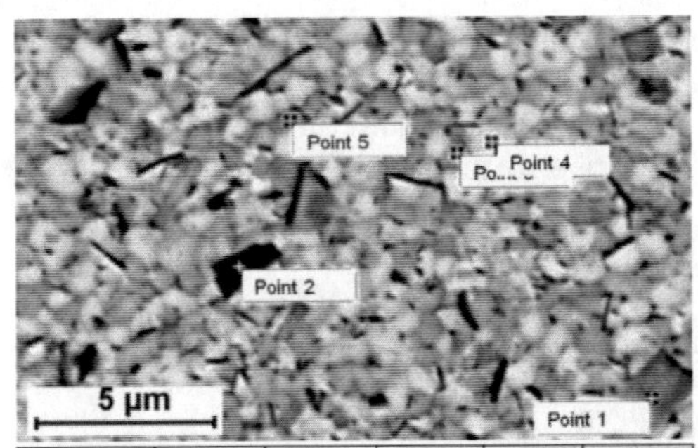

Point	B	Si	Cr	Fe	Ni
1	45.1	1.1	50.2	1.2	2.4
2	49.2	3.3	27.0	1.2	19.3
3	0.0	21.0	3.2	3.1	72.7
4	26.2	0.2	2.3	3.5	67.8
5	0.0	9.7	1.9	2.7	83.4

Figure 6. Surface microstructure and EDX results (in at.%) of alloy 1301, annealed at temperatures of 900°C for 1 hour.

Assuming that Ni_3B does not contain silicon and Ni_3Si does not contain boron, the concentration of silicon in α-Ni and Ni_3Si corresponds to the Ni-Si diagram, and iron is not redistributed between the phases, the phase compositions of alloys were estimated from material balance equations. These estimations were carried out for two cases: considering of formation CrB and without it (table 3). The obtained estimations of phase composition do not contradict to the rigorous thermodynamic calculations for alloy BNi-2 at 650 °C performed using the program ThermoCalc [4]. We can conclude that even a small increase of the boron content in the alloy leads to a mentioned increase of the amount of Ni_3Si and reduce of α-Ni. Note that the amounts of α-Ni and Ni_3Si phases in the crystallized alloy 1301 are comparable as shown in figure 1 *a*, *b*. The observed high amount of α-Ni phase in the alloy 1301 and low amount of Ni_3Si phase in the alloy 1301A after the complete crystallization cannot be explained without assumption of the formation of phase CrB.

Table 3. The phase composition of alloys, calculated according to different models.

Phase amount,	1301A		1301A		BNi-2
% (wt.)	with CrB	without CrB	with CrB	without CrB	ThermoCalc [4]
α-Ni	16.2	33.3	39.6	66.2	43.9
Ni_3B	50.0	31.8	43.8	18.9	29.1
Ni_3Si	33.8	25.2	16.6	6.4	18.8
CrB	–	9.7	–	8.5	8.2

3. Conclusion

The influence of the alloys chemical composition (boron, manganese and iron) on the temperatures and the exothermic heat effects of phase transformations, as well as on the phase composition of alloys at three stages of crystallization were analyzed.

It is shown that the onset crystallization temperature is practically independent of the boron content in alloys, but it increases with the increase of manganese content. The manganese and iron stabilizes a metastable τ-phase, resulting in decrease of the onset temperature of the second crystallization stage (τ-phase formation) and increase of the onset temperature of the third stage (decomposition of τ-phase). The thermodynamic explanation of these phenomena is proposed.

It is assumed that at the final crystallization stage simultaneously with the formation of Ni_3B and Ni_3Si phases the nucleation of borides of CrB type with high Cr and low Si content should occur.

Acknowledgements
This work was performed within the framework of the Center of Nuclear Systems and Materials supported by MEPhI Academic Excellence Project (contract № 02.a03.21.0005, 27.08.2013).

References

[1] Kurakova N V, Umnov P P, Molokanov V V, Sviridova T A and Kovneristyiy Yu K 2007 *Perspektivnyie materialyi* **4** 66–72 (in Russian)
[2] Elmanov G N, Ivanitskaya E A and Zharikov E S 2015 *Non-ferrous Metals* **1** 36–41
[3] Poloczek T, Griner S and Nowosielski R 2007 *Archives of Materials Science and Engineering* **28** 353–356
[4] Ruiz-Vargas J, Siredey-Schwallera N, Geya N, Bocherb P and Hazottea A 2013 *Journal of Materials Processing Technology* **213** 20–29
[5] Raju S, Arun Kumar N S, Jeyaganesh B, Mohandas E, Kamachi and Mudali U 2007 *Journal of Alloys and Compounds* **440** (1/2) 173–177
[6] Fedotov V T, Suchkov A N, Kalin B A, Sevryukov O N and Ivannikov A A 2014 Tsvetnye metally **12** 32–37 (in Russian)
[7] Abrosimova G, Afonikova N, Estrin Y, Kobelev, N, Kolyvanov E 2010 *Physics of the Solid State* **52** (3) 445–45
[8] Ivannikov A A, Suchkov A N, Tukhbatov V A Ustyuzhaninov A L and Bogachev I A 2014 *Tsvetnye metally* **12** 27–31 (in Russian)
[9] Yakushin V L, Dzhumaev P S, Kalin B A et al 2011 *Inorganic Materials: Applied Research* **2** (5) 503–511

IOP Conf. Series: Materials Science and Engineering **130** (2016) 012042 doi:10.1088/1757-899X/130/1/012042

Investigation of microstructure and thermal stability of pulsed plasma processed chromium ferritic-martensitic steels

O Emelyanova, P Dzhumaev, V Yakushin and V Polsky

National Research Nuclear University MEPhI (Moscow Engineering Physics Institute), 31, Kashirskoye shosse, Moscow, 115409, Russia

E-mail: OVEmelyanova@mephi.ru

Abstract. This paper presents results of the microstructural evolution and thermal stability of the promising Russian ferritic-martensitic steels (EP 823, EP 900, EK 181 and ChS 139) for the nuclear and fusion application after surface modification by high temperature pulsed plasma flows (HTPPF) treatment. Investigations of microstructure, topography and elemental content changes associated with irradiation by nitrogen plasma with energy density 19-28 J/cm^2 and pulse duration 20 μs were carried out. Changes in microstructure and elemental content occurring in the modified surface layer were characterized by means of scanning electron microscopy (SEM) and X-ray microanalysis (EDS and WDS). It was shown that independently of initial microstructure and phase composition, HTPPF treatment of ferritic-martensitic steels leads to formation of ultrafine homogeneous structure in the near surface layers with typical grain size ~100 nm. Results of microstructure investigations after annealing during 1 hour demonstrates significant thermal stability of nanostructure formed by HTPPF treatment.

1. Introduction

12% chromium ferritic-martensitic steels are planned to be used as structural materials for novel fission (Generation IV) and fusion reactors. Development of these new facilities demands to increase the operation temperature range of the materials up to 650 °C. Operating 12% chromium steels at the elevated temperature leads to new requirements both to the bulk properties and to the surface which will ultimately determine the applicability of the materials under corrosion/erosion effects of the environment. Hence, the optimization of surface structure has become a feasible and effective method of improving the overall properties of structural materials. Relatively novel methods such as ultrasonic treatment [1], electron and ion beams [2,3], pulsed laser and plasma irradiation [4,5] have been admitted as a promising methods for surface modification over the last few years. Such treatment results in the evolution of dislocation structure of the material, grain refinement and increase of the grain misorientation angles, etc., and eventually allow to create nanocrystalline materials. Ultrafine grains and large density of grain boundaries of nanomaterials possess unique properties in particularly increased strength/hardness, enhanced diffusivity, improved toughness/ductility, corrosion and erosion resistance. Numerous papers have been focused on the investigation of the nanostructuring methods and unique properties of produced materials [1-5], though, their thermal stability for high temperature applications such as nuclear and fusion industry is far from being clarified.

2. Materials and methods

The samples for this study are 12% Cr steels EP823, EP900, ChS139, EK-181 (Table 1) in the heat treated condition (quenched from 1100 C, followed by tempering at 720 C for 3 hours, provided by Bochvar institute, Moscow) and EK-181 nanostructured by severe plastic deformation by torsion under high pressure (further HPT, samples provided by Ufa State Aviation Technical University). HPT nanostructuring technique was performed in Bridgman anvil cell by torsion for 10 turns under pressure $P \sim 6$ GPa at a temperature of 400 °C. Processing of samples by HTPPF conducted in experimental z-pinch type pulsed plasma device "Desna-M" [4]. "Desna-M" is designed to produce and study gas plasma flows with energy of flow up to ~ 50 kJ and energy of ions up to 2 keV. HTPPF action with power densities above the critical value for irradiated material leads to melting and subsequent high-speed (up to 10^6 K / s) cooling of the surface layer [4]. This process, in turn, leads to the formation of gradient microstructure of the material surface: a modified surface layer with enhanced physical-mechanical and physico-chemical properties without pronounced interface between modified layer and bulk material. The microstructural investigation and the elemental composition analysis of the samples were carried out by scanning (EVO 50 XVP SEM) and transmission electron microscopy (Technai G2 TEM) and X-ray microanalysis using energy-dispersive spectrometer (EDS) Inca x-act and wave-dispersive spectrometer (WDS) Inca Wave 500 (Oxford Instruments, UK).

Steel	Concentration of main alloying elements (% wt.)													
	C	Si	Mn	Cr	Ni	Mo	Nb	Ti	W	V	B	Ta	Ce	N	Zr
EP823	0.14-0.18	1.2	0.6-0.7	11.0-12.0	0.7	0.8	0.3	-	0.3-0.6	0.3-0.4	0.006	-	-	-	-
EP900	0.15	1.14	0.7	11.5	0.7	0.7	0.3	-	0.75	0.3	0.0045	-	0.03	0.15	-
ChS139	0.19-0.25	0.1-1.0	0.5-0.8	10.0-12.5	0.5-0.8	0.4-1.1	0.2-0.4*	0.03-0.3	0.5-2.0	0.2-0.4	0.002-0.006	0.2-0.4*	0.001-0.10	0.02-0.15	0.05-0.2
EK181	0.10-0.21	0.1-0.8	0.5-0.8	10.0-12.5	<0.1	<0.01	<0.01	0.03-0.3	0.8-2.5	0.2-1.0	0.003-0.008	0.05-0.2	0.001-0.10	0.02-0.15	0.05-0.2

3. Results and discussion

Figures 1 *a*, *d* show typical electron micrographs of the microstructure of 12% chromium steels in the heat treated state. Electron microscopic studies and the results of elemental analysis showed that in the initial state steel EP823 and EP900 have two-phase microstructures of the lath martensite, ferrite and finely dispersed precipitates of different types (Figures 1 *a*, *b*). In particular, in EP823 steel complex carbides based on niobium were observed. EP900 steel, which is additionally doped with nitrogen, as well as large carbides type (Fe, Nb, Cr) C were observed fine nitrides (Figure 1 *b*) on the basis of Fe and Cr. At the same time nitrogen has not been observed in the matrix of EP900 steel.

Steel CHS139 in heat treated state demonstrates structure of lath martensite with carbide precipitates of $M_{23}C_6$ type, mostly located along the grain boundaries of the martensitic plates and a size of about 150-200 nm. The transverse dimension of the martensite plates varies in the range from 0.1 to 1 μm. Also, there are relatively large (about 1 μm) niobium carbide precipitates in the CHS139 steel.

Steel EK181 has similar microstructure with ChS139 steel (Figure 1 *c*), except of the presence of minor amount of ferrite phase. Carbide precipitates in this steel has less volume fraction and sizes which might be due to reduced carbon content and absence of a strong carbide-stabilizing elements such as molybdenum and niobium in EK181 steel in comparison with ChS139 steel. Also in the structure of the EK181 steel large (up to 5 μm) precipitates of zirconium nitride were observed.

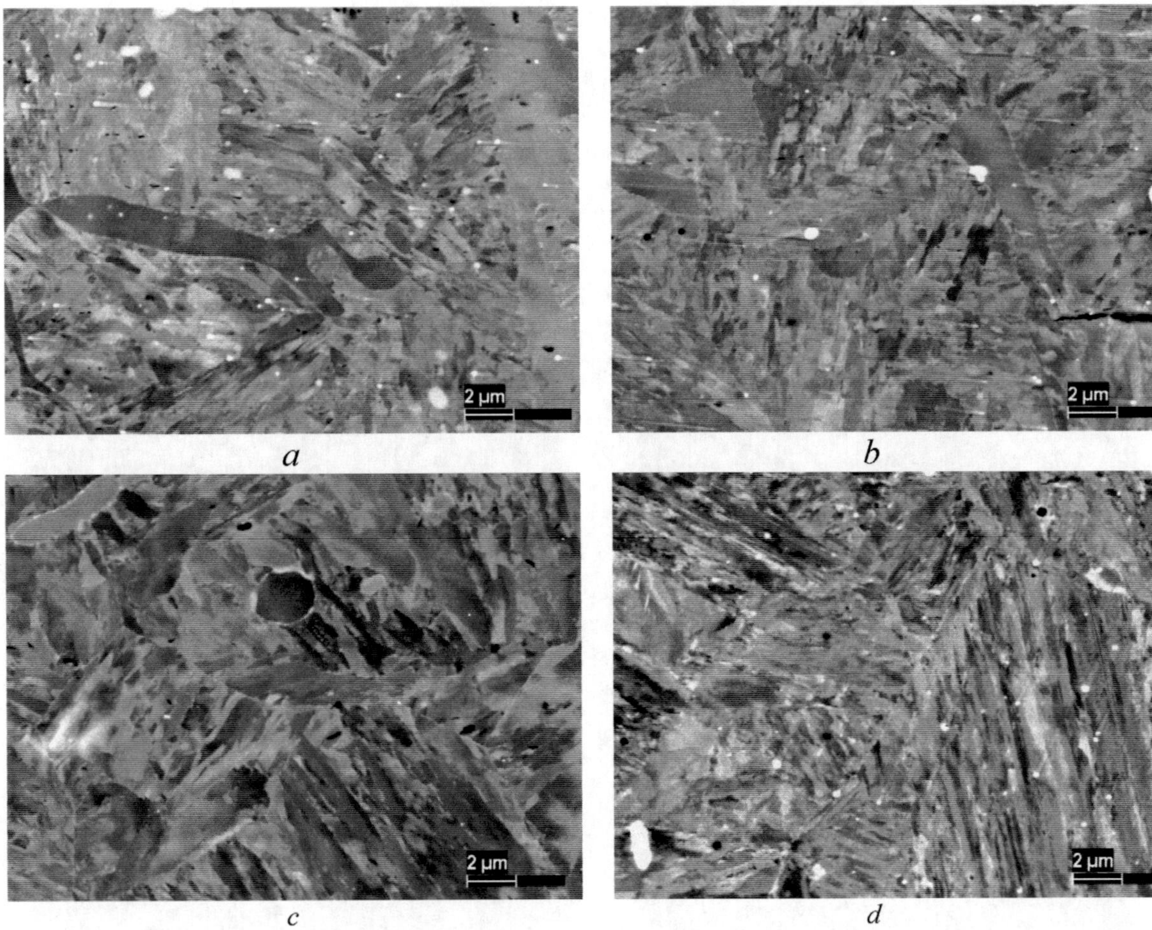

Figure 1. Microstructure of fuel cladding tubes made of EP823 (*a*), EP900 (*b*), EK-181 (*c*), ChS139 (*d*) in initial state.

Figure 2 shows electron micrographs of the sample surface after the processing by nitrogen HTPPF. It can be seen that wave-like structure of the solidified melt was formed on the surface of EK-181 steel after treatment by pulsed plasma flows (Figure 2, *a*). Under the high magnification (Figure 2, *b*), the surface cellular structure with an average size 80-150 nm of sub-grains is observed. Importantly, precipitates typical for untreated EK-181 steel are completely dissolved during HTPPF processing which indicates the transition of alloying elements to the solid solution and corresponding homogenization of the material elemental content. Dissolution of carbides occurs throughout the thickness of the modified layer. Analysis of cross section SEM micrographs shows that modified layer has columnar microstructure with transverse grain size about 2 µm (Figure 2, *c*). However precision TEM cross-section investigations reveal formation of columnar ultrafine subgrains in near surface layer (Figure2, *d*). Combined analysis of SEM and TEM cross-sectional data leads to conclusion that HTPPF treatment results in formation of two-dimensional nano-sized structure with transverse subgrain size ~80-150 nm and longitudinal size ~1-1,5 µm. Also it was found that processing by nitrogen plasma leads to additional doping of steel surface by nitrogen up to 4 % wt. (depending on the plasma processing regime).

It was shown that for the same plasma processing conditions there are no significant differences in microstructure formed of steel samples EP823, EP900, ChS139 and EK181: the phase composition and grain/subgrain size are independent of the initial state of the material, the distribution of alloying elements between matrix and carbide particles in the initial material which is due to close the

IOP Conf. Series: Materials Science and Engineering **130** (2016) 012042 doi:10.1088/1757-899X/130/1/012042

elemental composition and the initial structure of the investigated steels, and therefore they are almost identical thermal properties.

This leads to the same values of characteristics of the processes occurring during high speed quenching surface layer, particularly the cooling rate. As for HTPPF treatment, formation of cellular microstructure is the result of the cellular solidification of a melt, and can be described by constitutional supercooling model [6]. Due to intense heat transfer inside the bulk material the cooling rate reaches very high level (more than 10^6 K / s), thus leading to realization of the extreme conditions of this phenomenon.

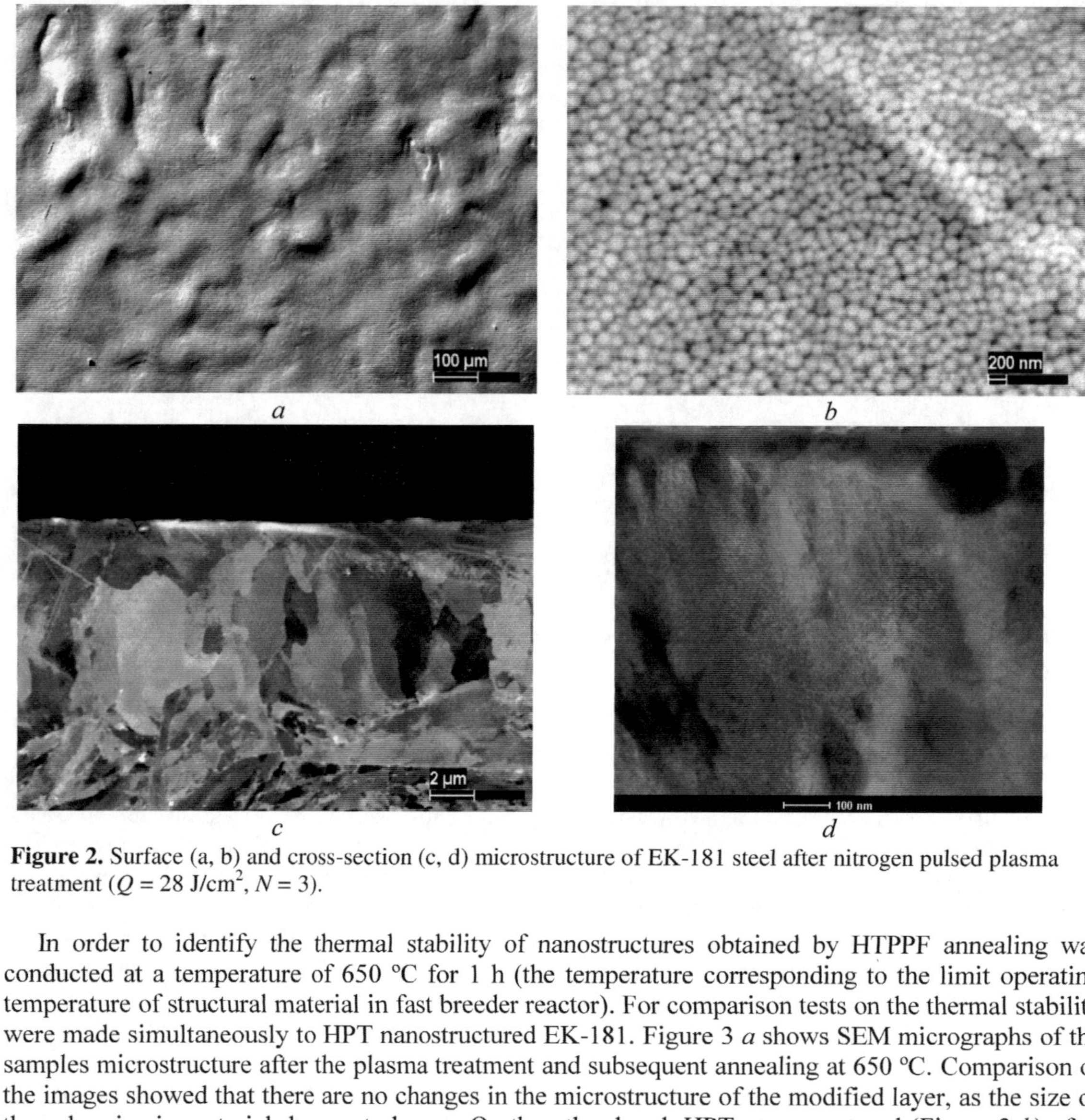

Figure 2. Surface (a, b) and cross-section (c, d) microstructure of EK-181 steel after nitrogen pulsed plasma treatment (Q = 28 J/cm^2, N = 3).

In order to identify the thermal stability of nanostructures obtained by HTPPF annealing was conducted at a temperature of 650 °C for 1 h (the temperature corresponding to the limit operating temperature of structural material in fast breeder reactor). For comparison tests on the thermal stability were made simultaneously to HPT nanostructured EK-181. Figure 3 *a* shows SEM micrographs of the samples microstructure after the plasma treatment and subsequent annealing at 650 °C. Comparison of the images showed that there are no changes in the microstructure of the modified layer, as the size of the subgrains in material does not change. On the other hand, HPT nanostructured (Figure 3 *b*) after annealing showed a significant increase in the grain size from 170 nm to the initial state to 1 micron after annealing due to past processes of recrystallization.

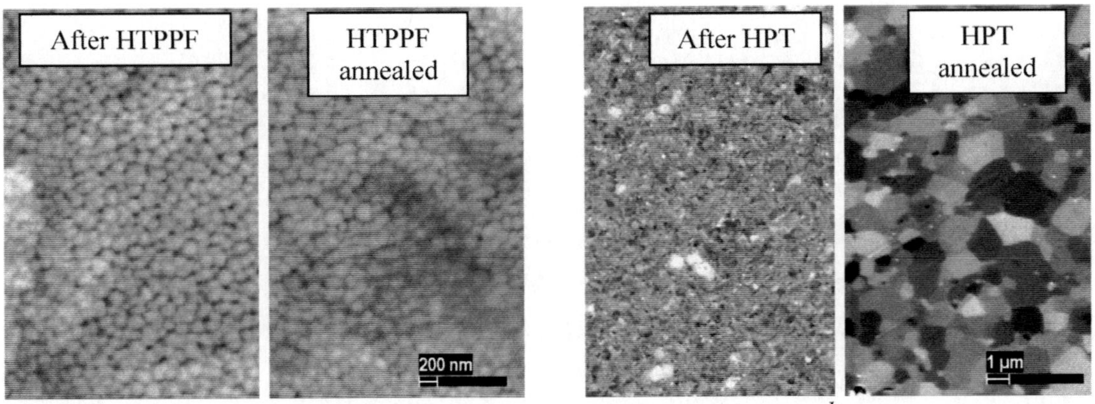

a *b*

Figure 3. Change of microstructure of EK-181 in HTTPF (a) and HPT (b) as-processed state and annealed at 650°C for 1 h (right part of image).

4. Summary

It was shown that surface modification of 12 % chromium ferritic-martensitic steels EP823, EP900, and EK181 CHS139 by pulsed nitrogen plasma flows leads to formation of two-dimensional (2D) surface nanostructure with a main grain size 80-150 nm. It was revealed that the characteristics of formed nanostructure such as thickness of the modified layer and grain size are independent on the initial steel microstructure (phase composition and / or grain size). It was found that the modified layer manifests significant thermal stability and annealing of the samples at the temperature up to 650 °C, 1 hour, has almost no effect on its microsctructure.

Acknowledgements

This work was performed within the framework of the Center of Nuclear Systems and Materials supported by MEPhI Academic Excellence Project (contract № 02.a03.21.0005, 27.08.2013).
Authors would like to thank Dr. Maria Leontieva-Smirnova from the Bochvar institute (Moscow, Russia) for EP823, EP900, ChS139, EK-181 heat treated steel samples provided for the investigation; Prof. Ruslan Valiev and Dr. Nariman Enikeev from Ufa State Aviation Technical University (Ufa, Russia) for high pressure torsion samples of EK-181 steel.

References

[1] Panin A V , Melnikova E A , Perevalova O B , Pochivalov Yu I , Leontyeva-Smirnova M V , Chernov V M and Ivanov Yu F 2009 *Phys. Mesomech.* **12** 150
[2] Fangjun Xu, Guangze Tang, Guangwei Guo, Xinxin Ma and Ozur G 2010 *Nucl. Instrum. Methods Phys. Res. B* **268** 2395
[3] Uglov V V, Remnev G E, Kuleshov A K, Astashinski V M and Saltymakov M S 2010 *Surf. Coat. Tech.* **204** 1952
[4] Kwok C T, Leong K I, Cheng F T and Man H C 2003 *Mat. Sci . Eng. A* **357** 94
[5] Kalin B A, Yakushin V L, Vasiliev V I and Tserevitinov S S 1997 *Surf. Coat. Tech.* **96** 110
[6] Glicksman M E 2011 *Principles of Solidification* (Springer-Verlag New York) pp 213–235

IOP Conf. Series: Materials Science and Engineering **130** (2016) 012043 doi:10.1088/1757-899X/130/1/012043

Influence of technology of nanopowder production on the microstructure of the sintered by spark-plasma material on the example of aluminum oxide

A G Zholnin[1], A P Melekhov[1], R S Hafizov[1], S A Vakulenko[1], N A Rubinkovskiy[1], A V Samokhin[2], N V Alekseev[2] and E G Grigoryev[1]

[1]National Research Nuclear University MEPhI (Moscow Engineering Physics Institute), Kashirskoe highway 31, Moscow, 115409, Russia
[2]Federal State Institution of Science Institute of Metallurgy and Materials Science, RAS, Leninsky Avenue 49, Moscow, 119991, Russia

E-mail: azholnin@list.ru

Abstract. A comparative study of the results of spark-plasma sintering (SPS) of two-types of aluminum oxide nanopowders, obtained by the method of conductor explosion and plasma synthesis. When the parameters of both powders are similar (spherical form of the particles, size, phase composition) as well as SPS modes the properties of the resulting compacts are significantly different both in mechanical properties and microstructure. The reason of differences in the properties of the obtained compacts is in technological impurities in powders, obtained by different methods. Artificial addition of impurities, contained in the nanopowder, obtained by electro explosion of conductor, into the powder, made by synthesis in plasma and not containing these impurities, allowed to reveal their effect on the formation of the microstructure and properties of the sintered by SPS method sample.

1.Introduction

Production of bulk materials by spark-plasma powders sintering having a nanostructure, allows to control formation of the microstructure of the sintered material and to achieve unique strength and special properties due to the short duration of the process [1-5]. At the same time a significant influence on the properties of the sintered material has a technology of obtaining initial nanopowder. For example, the presence of even very insignificant amounts of certain impurities can dramatically affect the properties of the sintered material and change the microstructure. The aim of this work is to study the causes of the formation of dissimilar structures of sintered by the method of spark-plasma using identical modes from nanopowders of similar properties but obtained in different ways aluminum oxides. The main focus of the work was placed on differences in the microstructure. All other properties were left behind the scenes.

2.Experimental equipment and methods

Two kinds of aluminum oxide nanopowder were used as an initial material for SPS:

1. The powder Al_2O_3, obtained by the explosion of aluminum wire in an oxygen-containing gas produced by "Advanced Powder Technologies", Tomsk. According to the certificate the powder

contains kappa, theta, and alpha phases, the traces of metal aluminum are present. Specific surface is 35-40 m^2/g. The arithmetic average size of particles is 36 nm, average size by surface – 46 nm, average mass size – 54 nm. Hereinafter – "powder 1" and sintered compacts "compacts 1".

2. The powder Al_2O_3, obtained by oxidation of the disperse aluminum in air plasma of electric arc discharge [6, 7]. Delta phase dominates in the initial phase composition. Specific surface is 35,8 m^2/g close to the specific surface of the powder 1. Hereinafter – "powder 2" and sintered compacts "compacts 2".

The form of particles in both kinds of powder is spherical what is proved by REM.

Spark-plasma sintering (SPS) of both powders was carried out in vacuum using LABOX-625 Sinter Land device in the graphite matrix with an internal diameter of 10 mm at a pressure of 50 MPa. Spraying of aluminum oxide was 0.4 g, which provided a height of the tablet slightly higher than 1 mm. Temperature was set and monitored using pyrometer. The heating rate was 100°C/min, maximum heating temperature – 1400°C, holding at max temperature – 10 min. This mode was obtained empirically, as a mode, providing high homogeneity at maximum average micro hardness on the surface of the sintered compacts.

Properties of compressed tablets were studied at different stages of sintering at normal temperature using a raster electron microscopy (REM TESCAN model VEGA). REM was carried out on tablets hear with spraying on it of platinum – palladium coating of thickness of a few atomic layers.

Semi qualitative analysis of micro impurities and compacts was carried out on time-of-flight mass spectrometer analyzer with lazed excitation [3-Nukleonika]. Both initial powders, pressed using a hydraulic press and microscopic sections of compacts, obtained from these powders by SPS were studied.

3. Experimental results

Microstructures of cross shears of tablets, obtained by the SPS of both types of powder are shown in Figures 1 and 2.

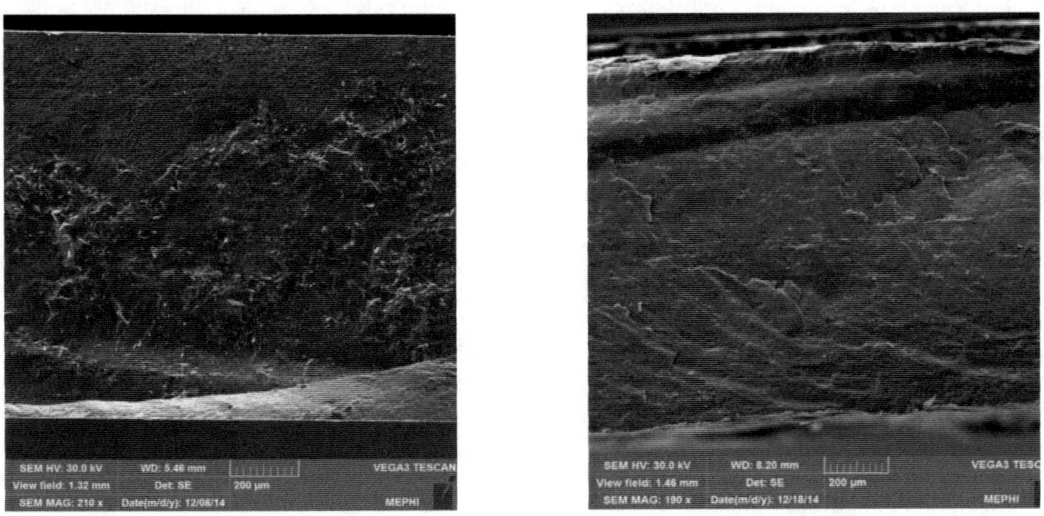

Figure 1. Microstructure of the fracture of compact 1 (left) and compact 2 (right).

It can be seen on the pictures that the compacts, obtained from the powder 1, in addition to the small grain areas have large grain areas of 300-500 microns. Such areas are absent in compacts obtained from powder 2, the microstructure is almost homogenous across the surface of shear. Deviations from the plane by the shear are caused by the destruction of the tablets (a blow with a sharp rod on the center of tablets).

Of particular interest are large grain areas. At high magnification, it can be noted that they are not simply composed of large grains but have a bimodal structure, consisting of a mixture of super-large (up to 40 microns) and ultrafine (0.5-1 microns) grains. (Figure 2)

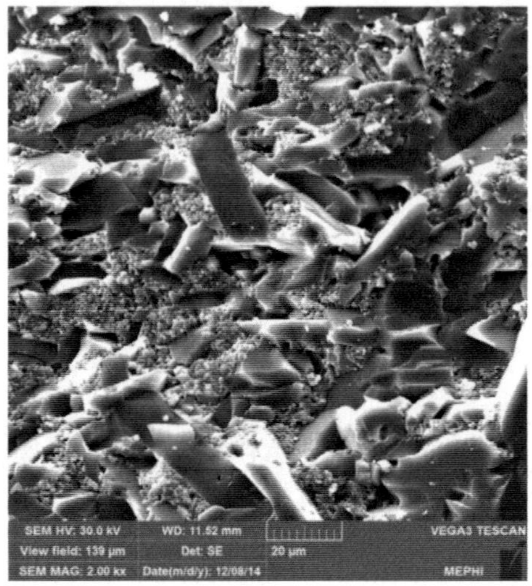

Figure 2. Bimodal structure observed on the shear of compacts obtained from powder 1.

This inhomogeneity of the structure, observed in compacts from powder 1, could be attributed to the poor quality of nanopowder and close the topic. However, mechanical test and measurements of micro hardness showed a significant excess of these characteristics not only over compacts, obtained from powder 2, but (at least in micro hardness) over literature data on corundum. In areas with bimodal structure micro hardness reached 35 MPa, instead of 20-24 MPa, specific to corundum or leucosapphire. This has stimulated the study of the causes of the observed anomaly.

Theoretically, at similar external parameters of particles of both powders they can vary in phase composition and by the presence of impurities. Aluminum oxide has a few low temperature modifications the ratio of which can be influenced by its production method. However, at temperature higher than 1200-1500°C all aluminum modifications proceed to a stable state α-Al_2O_3. Sintering and formation of grains occur at temperatures which are higher than temperatures of this phase transition. Therefore, differences in the initial phase composition of the initial powders cannot influence the formation of the microstructure of the compact and properties. A more significant impact on the formation of the microstructure has to be expected from the presence of technological impurities.

Composition and approximate concentration of the impurities in four studied samples are given in the Table 1. Both powder samples and their compacts contain a lot of iron and silicon impurities (identification ambiguity is possible due to the undecidability of peaks Fe^{+2} and Si^{+1}) and impurities of carbon and magnesium (peaks Mg^{+2} and C^{+1} are undecidable). These peaks, due to the almost equal m/z, are difficult to clearly define. M, z – mass and charge of the ion.

The powder 1 and its compact contain a significant amount of carbon (possibly magnesium). Powder 2 and compact 2 also contain a significant amount of carbon (possibly magnesium) and copper, which is absent in the powder 1 and compact 1. These elements are not considered further because their presence in detected quantities does not influence the homogeneity of the microstructure.

Iron and silicon can be possible sources of inhomogeneity of the structure during the sintering (not distinguished in mass-spectrometer), the total number of which is in many times higher than in powder 2 and compact 2. Titanium can also be the source of inhomogeneity, which is absent in powder 2.

Titanium was found only in powder 1, moreover its presence has random nature, i.e. it appears not in all measurements (Figure 3).

Table 1 Composition of impurities in the powders and compacts obtained therefrom.

Sample	Fe, ppm	Cu, ppm	Ti, ppm	Mg, ppm	C, ppm	Cr, ppm	V, ppm	F, ppm	Se, ppm	Sum, ppm
Powder 1	2900(Si)	>10	2900	>10	>10					5800
Compact 1	4000(Si)	>10	<10	>10	>10					4100
Powder 2	760(Si)	630	<10	720(C)	720(Mg)	60	50	20	50	2300
Compact 2	740(Si)	6700	<10	>10	120(Mg)					7600

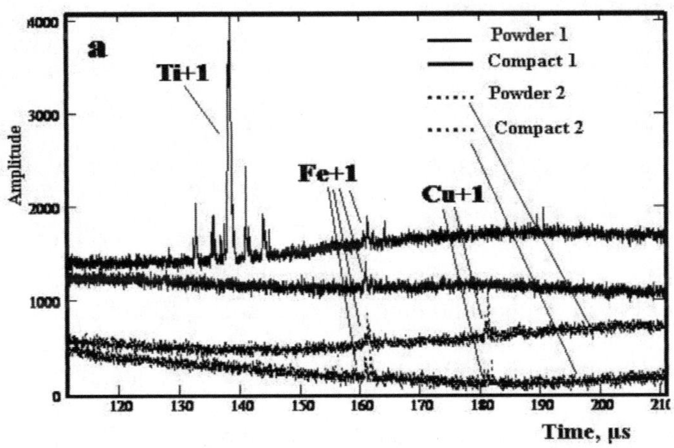

Figure 3. Fragments of mass spectrum for four independent measurements of both powders and their compacts. The multiplicity of the peaks for titanium for sample "Powder 1" is caused by the presence of five stable isotopes in its structure.

Thus, it was found out:

- In compact 1, in contrast to compact 2, the areas with bimodal structure having increased micro hardness are formed in the volume of compact.
- In the powder 1 and compact 1 the content of iron (silicon) is in several times higher than their content in the powder 2 and compact 2.
- In individual measurements for powder 1 the presence of titanium in its structure was revealed, which is absent in other measurements of the same sample, titanium is absent in the compact obtained from this powder. In the compact from powder 2 the presence of titanium was not found.

These findings suggest the existence between them of the casual relationships. Let's start from the titanium.

3.1. The correlation between the doped titanium and bimodal structures

The random nature of the presence of titanium powder in powder 1 and its absence in compact indicates the presence in compact structure of relatively rare inclusions containing titanium. If the

IOP Conf. Series: Materials Science and Engineering **130** (2016) 012043 doi:10.1088/1757-899X/130/1/012043

micro-object in the impact zone of the laser beam, titanium is present in the mass spectrum. In the process of SPS the inclusion dissolves in the surrounding material, the titanium concentration falls below the detection limit (10 ppm). REM confirms the presence of foreign formations in compact 1 in the form of relatively large 10-2 microns) hollow spheres (Figure 4). Such objects are absent in compacts 2. Similar object in the nanopowder of aluminum oxide, obtained by blasting from metallic aluminum, are noted in operation. [8]

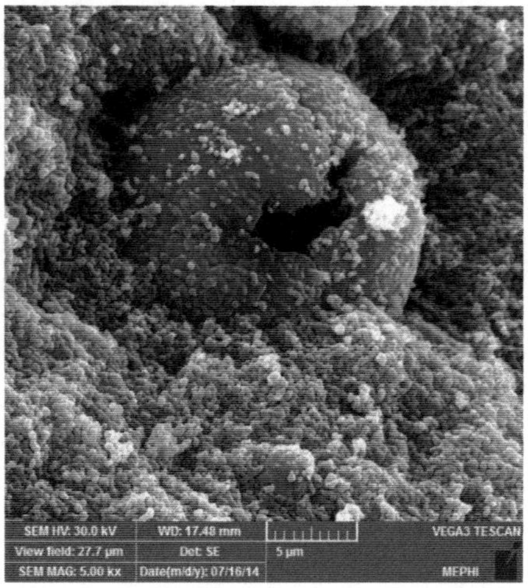

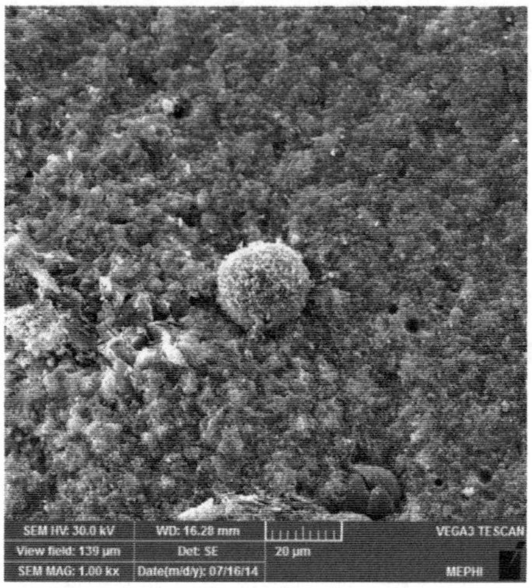

Figure 4. More examples of spherical objects in the compacts of powder 1 after free sintering in the dilatometer. Heating to 1460°C without holding (left), heating to 1380°C without holding.

The presence of micro analyzer in REM allowed to confirm the presence of increased content of titanium in this objects.

However, the addition of titanium oxide nanopowder into the powder 2, in the amount close in weight to sphere-like structures in powder 1 and compacts 1, didn't have a significant effect on the microstructure of obtained by the SPS method from this mixture compacts 2. These studies exclude the role of the titanium impurity in the powder 1 in formation of bimodal areas with increased micro hardness.

3.2. The correlation between the impurities of iron and silicon and bimodal structures

According to the Table 1, powder 1 in comparison to powder 2 has a multiple excess of iron and silicon, as used mass-spectrometric method gives only rough estimates of concentrations of impurities. The influence of these elements on the microstructure of compacts 2 was studied separately through the adding of them of their oxides to the powder 2 and subsequent sintering of the resulting mixtures. Determining the value of additives, the studies were based on data from mass-spectrometer, shown in Table 1.

Mixing of carbonyl iron powder into powder 2 and consequent SPS of the mixture didn't influence the structure of the compact. In contrast, the addition of silicon oxide into the nanopowder 2 after the sintering resulted in a formation of structure, similar to the bimodal with increase micro hardness, but with smaller size of large grains, uniformly distributed over the entire volume of the compact (Figure 5). This experiment confirms the special role of silicon oxide.

In the powder 1 the impurity is present initially. The fact that bimodal during the sintering bimodal areas have local nature can be caused due to an inhomogeneous distribution of silicon oxide in the powder, for example its agglomeration. The anomalous areas a formed where there is a cluster of silicon oxides agglomerates.

IOP Conf. Series: Materials Science and Engineering **130** (2016) 012043 doi:10.1088/1757-899X/130/1/012043

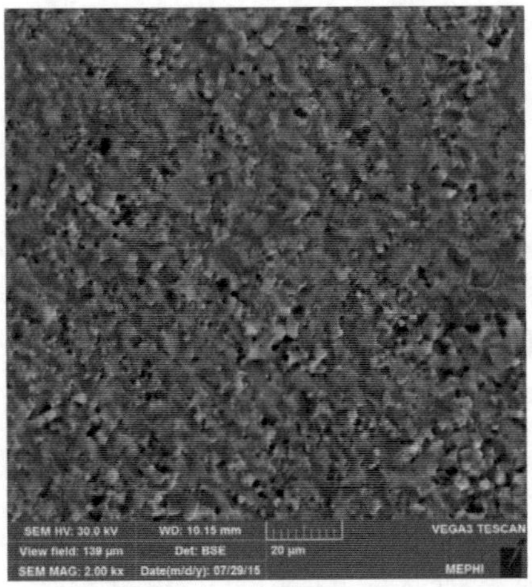

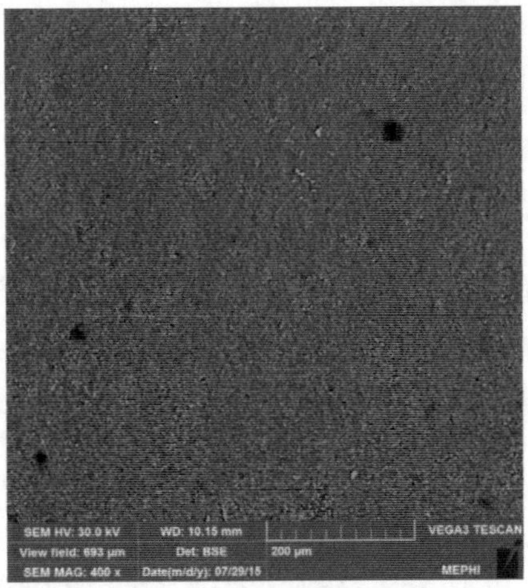

Figure 5. Compact 2 with silicon oxide impurity.

To test this hypothesis, powder 1 was subjected to the intense ultrasonic dispersion in alcohol, followed by drying in the air with continuous stirring. Though the bimodal structures changed the configuration, but were evenly distributed throughout the sample. This confirms the presence of the impurity of silicon oxide in the powder 1 in the agglomerated state.

4. Conclusions

The study showed a significant effect of technological impurities in nanopowders on the microstructure of the materials obtained in spark-plasma sintering.

Considering the identified technological impurities, the greatest influence on the formation of microstructure during SPS showed an impurity of silicon, as a natural contaminant present in the powder 1.

Acknowledgements

This work was performed within the framework of the Center of Nuclear Systems and Materials supported by MEPhI Academic Excellence Project (contract № 02.a03.21.0005, 27.08.2013).

References

[1] Kisliy P S, Gevorkyan E S, Shkuropatenko V A and Gutsalenko Y G 2010 *Superhard materials* **68** 24
[2] Rodriguez-Suarez T, Dhaz L A, Torrecillas R and others 2009 *Sci. and Technol.* **69** 2467–73
[3] DongTao, Jiang W Z, Dustin M, Hulbert Z, Umberto Anselmi-Tamburini et al. 2008 *The American Ceramic Soc.* **91** 151
[4] Wang S W, Chen L D, Hirai T 2000 *Mat. Res. Soc.* **15** (4) 982
[5] Dyatlova Y G, Ploshchanskiy Y S, Boikov S Y and others 2011 *Proc. of the int. conf. on nanomaterials*
[6] Samokhin A V, Alekseev N V, Tsvetkov Y V 2006 *High energy chemistry* **40** (2) 120
[7] Sirotinkin V P, Shamrai V F, Samokhina A V and others 2012 *Inorganic materials* **48** (4) 409
[8] Bukaemsky A A, Beloshapko A G and Puzyr A P 2000 *Physics of combustions and explosions* **36** (5) 119

Development of rapidly quenched nickel-based non-boron filler metals for brazing corrosion resistant steels

A Ivannikov, B Kalin, A Suchkov, M Penyaz and M Yurlova

National Research Nuclear University MEPhI (Moscow Engineering Physics Institute) Kashirskoe highway 31, Moscow, 115409, Russia

E-mail: ivannikov7@rambler.ru

Abstract. Corrosion-resistant steels are stably applied in modern rocket and nuclear technology. Creating of permanent joints of these steels is a difficult task that can be solved by means of welding or brazing. Recently, the use rapidly quenched boron-containing filler metals is perspective. However, the use of such alloys leads to the formation of brittle borides in brazing zone, which degrades the corrosion resistance and mechanical properties of the compounds. Therefore, the development of non-boron alloys for brazing stainless steels is important task. The study of binary systems Ni-Be and Ni-Si revealed the perspective of replacing boron in Ni-based filler metals by beryllium, so there was the objective of studying of phase equilibrium in the system Ni-Be-Si. The alloys of the Ni-Si-Be with different contents of Si and Be are considered in this paper. The presence of two low-melting components is revealed during of their studying by methods of metallography analysis and DTA. Microhardness is measured and X-ray diffraction analysis is conducted for a number of alloys of Ni-Si-Be. The compositions are developed on the basis of these data. Rapidly quenched brazing alloys can be prepared from these compositions, and they are suitable for high temperature brazing of steels.

1. Introduction

Nowadays the structure having lower performance because of the presence of heterogeneities, which can lead to the degradation and destruction of the joint structures during operation, is forming in the brazed joints while using filler metals containing boron [1]. Therefore, an important task is to develop non-boron filler metal.

Currently the system Ni-Si-Be is poorly studied. The action of beryllium in nickel alloys in many ways similar to the action of boron; but in contrast to boron, it is highly soluble in nickel, forming an eutectic with the beryllium content of 23.8 at.%, melting at 1150 °C and consisting of a solid solution Ni (α) and low ductile intermetallic BeNi (β) [2]. The intermediate phases, worsening properties of the compound will not form in iron during brazing because of high solubility of Be, as is the case with boron. Therefore, nickel alloy containing Be, are promising for brazing steels on a par with containing boron alloys. Studies of various alloys containing beryllium, showed that they have a high specific strength and corrosion resistance [3].

Analyzing all the advantages and disadvantages of the impact of beryllium on the properties of the brazed joint, can be concluded that the study of nickel alloys containing this element is extremely important field and can solve a number of problems that arise when working with boron filler metals, and therefore the purpose of research is to develop filler metals based on Ni-Si-Be system for brazing stainless steels.

2. Experiment and methods

The microstructure of the samples and elemental composition alloys of Ni-Si-Be, which were pre-annealed for 6 hours at 950 °C in a vacuum of 10^{-4} Torr, was investigated by the scanning electron microscope JEOL JSM-6610LV. The study of alloys containing beryllium by techniques of electron microprobe microanalysis is difficult, because Be – can`t be detected by EDX.

Differential thermal analysis, which was performed on the thermal analyzer SDTQ600, was used to determine the intervals of melting alloys Ni-Si-Be. This thermal analyzer allows record the change in mass of the sample (thermal gravimetric analysis) and processes, accompanied by the release or absorption of heat simultaneous.

X-ray diffractometer Bruker D8 Discover with international X-ray data base ICDD PDF-2 was used to determine the quality of the phase composition of samples.

Determination of Vickers microhardness was performed on digital micro durometer models Future-Tech FM-800. The rate of load 50 m/s, the load time of 20 seconds, the accuracy in accordance with GOST 9450 (government state). The measurement was carried out by indentation in the test material the Vickers diamond tip with a square base, four-sided pyramid, providing a mechanical and geometric similarity of fingerprints for the deepening of the indenter under load.

3. Results and Discussion

3.1 Study of the structure alloys Ni-Si-Be

Eutectic alloys in the Ni-Si-Be system, which have low melting points and better fluidity as compared to non-eutectic alloys, are most interesting alloys for the creation of filler metals.

We can assume by analyzing the binary systems Ni-Si and Ni-Be that most fusible alloys (eutectic) are close to the composition of Ni- (4..6)Si- (3..5)Be. To study the effect of alloying beryllium and silicon on the structural-phase state of nickel alloys were selected the compositions containing different amounts of silicon and beryllium (Table 1).

Table 1. Nickel alloys with various contents of Si and Be.

1	Ni-4Si-(2; 3; 3.5; 4)Be
2	Ni-5Si-(2; 3; 3.5; 4)Be
3	Ni-6Si-(2; 3; 3.5; 4; 5)Be
4	Ni-7Si-(4; 5)Be
5	Ni-8Si-4Be

These alloys are selected in such way as to capture supposed area amounted to approximately the ternary eutectic

To investigate influence of the Be and Si amount on the structural-phase state of nickel alloys, the microstructure of the samples was studied and carried out differential thermal analysis (DTA). The effect of silicon and beryllium content on the structure of nickel alloy was studied based on this data. Figure 1 shows the results of metallographic analysis.

Let us consider the alloys Ni-4Si-2Be, Ni-4Si-4Be, Ni-6Si-4Be, Ni-6Si-5Be for study of the effect of the beryllium and silicon amount on the temperature and melting range. Alloy with the lowest amount beryllium Ni-4Si 2Be has the structure of a solid solution with a small amount of the eutectic. The forming of second phases in the form of needles into the solid solution is observed due to the reduction the solubility of Be and Si in nickel, according to the contrast and shape of these emissions it is the phase with a high content beryllium that is beryllide.

IOP Conf. Series: Materials Science and Engineering **130** (2016) 012044 doi:10.1088/1757-899X/130/1/012044

Further increase of the amount of beryllium up to 4 wt. % leads to formation of a new structural component which has a finely dispersed eutectic structure. Its amount increases with the increase of beryllium, and enhanced thermal effect associated with the crystallization / melting of the structural component.

The thermal effect is in the temperature range 885-985°C on the thermograms by increasing the silicon content up to 5 wt. % that indicates the appearance of a new structural component having a low transformation temperature. Thus, as in the case of the alloy Ni-4Si-4Be, new fine eutectic can be observed in the structure. The solidus temperature of this alloy series is practically unchanged and is 1115-1118°C (heating), 1107-1112°C (cooling).

The effect of increasing the melting range is observed in alloys with 5 wt. % silicon by increasing the amount of beryllium from 3.5 to 4%. This is related to a decrease of the amount of binary eutectic (1107-1112°C) and the appearance of the ternary eutectic (880-936°C), while on the border of the ternary eutectic crystallized solid solution, which gives the increase of the melting range.

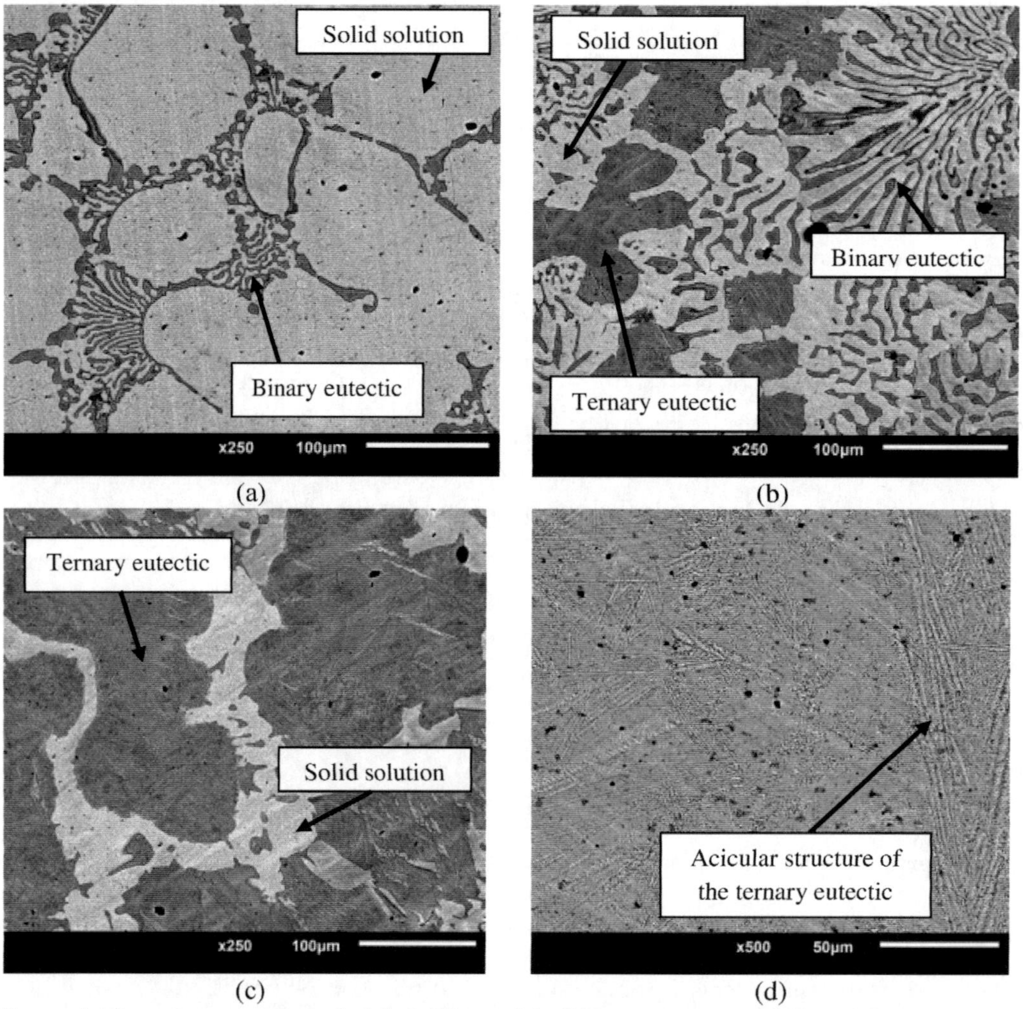

Figure 1. The microstructure of nickel alloys with different contents of Si and Be, (a) – Ni-4Si-2Be; (b) – Ni-4Si-4Be; (c) – Ni-6Si-4Be; (d) – Ni-6Si-5Be.

Figure 2 shows the results of differential thermal analysis for alloys with constant silicon content of 6 wt. % and variable content of beryllium (all DTA curves are offset for clarity). The microstructure of the alloy Ni-6Si-5Be most interesting as it is fully eutectic and has a needle structure. Analyzing the thermogram it can be compared observed eutectic structure with low-temperature peak in the range of consumable 933-944°C. Supposedly the composition is close to the composition of the low-temperature ternary eutectic. However, large melting range of the solid solution, which is apparently present in the alloy, is still observed on the thermogram.

It may be noted that the same pattern repeats for the alloys with a silicon content of 6 wt. % as for the previous alloys containing 5 wt. % silicon. Specifically, the amount of beryllium has no effect on the solidus temperature. Melting range decreases with increasing of beryllium (3 wt. % and 3.5 wt. % Be), and an eutectic structure becomes larger. And when it reaches content of 4 wt. % of beryllium range begins to increase again, which is associated with the formation of low-temperature and over-eutectic solid solution on its border.

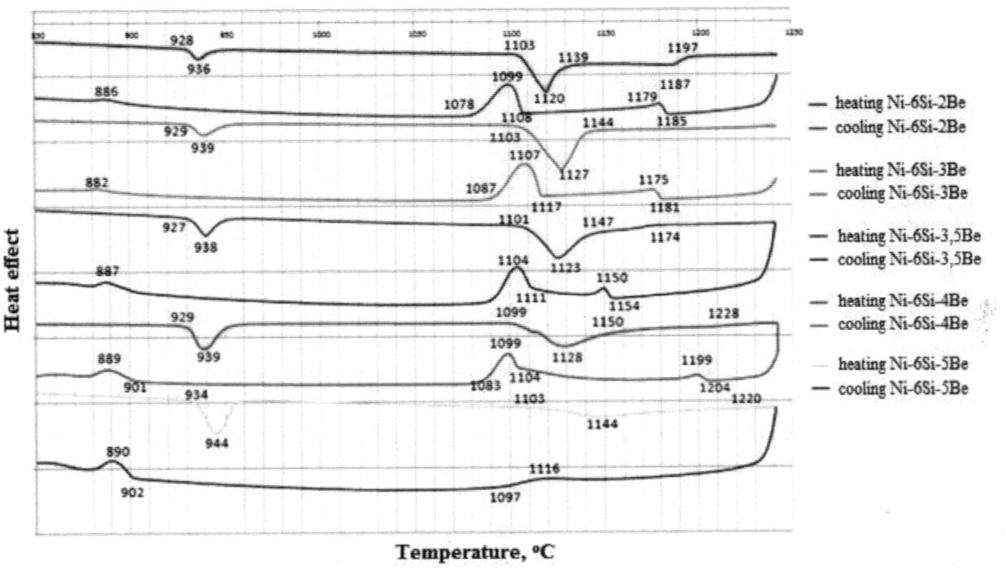

Figure 2. Differential thermal analysis for alloys Ni-6Si-2Be, Ni-6Si-3Be, Ni-6Si-3,5Be, Ni-6Si-4Be, Ni-6Si-5Be.

It was further considered the change of temperature and melting range of alloys with the same amount of beryllium and various amount of silicon. Thermogram was made for alloys with constant content of beryllium and varying content of silicon, and it was found that silicon lowers the temperature of solidus. In alloys with a constant content of beryllium observed effect of the amount of silicon on the eutectic structure.

With the increase of silicon up to 6 wt. % the eutectic structure becomes fine and uniform that is the proportion of the ternary eutectic increases with an increase of silicon. With a high silicon content (7-8 wt. %) the structure of alloys becomes needle, and there are observed silicides (white allocation).

The peak corresponding ternary eutectic are clearly defined in the alloy Ni-6Si-5Be due to the high amount of Be and Si, and practically no peaks corresponding binary eutectic (little thermal effect), which were pronounced in the previous alloys containing less of silicon and beryllium.

It was concluded about the influence of Be and Si to nickel alloys on the basis of the thermograms and the microstructures. Increasing the amount of Be and Si in the alloy is accompanied by an increase of the amount of low-temperature ternary eutectic in structure (appearance of the eutectic changes - it becomes fine, and there is an eutectic with larger plates). Also, melting range varies strongly enough with increasing the silicon content - it becomes smaller and decreases the solidus temperature at 10 -15 °C that is the peaks shifted to the left. The narrowest melting range of all considered alloys characterized for the beryllium content of 3-3.5 wt. % independently of the silicon content. Also, thermal effect of melting ternary eutectic enhanced, when the silicon content of 6 wt. %.

After analyzing the results of studies it can be concluded that a high amount of silicon in the alloy uniquely effect on its properties. On the one hand adding more silicon which lowers a melting point – are increased the melting range of the solid solution. On the other hand there is the low-temperature ternary eutectic.

3.2 X-ray diffraction analysis

In addition for the qualitative study of alloys Ni-Si-Be was carried out X-ray analysis of the alloy Ni-4Si-2Be, which has the structure: dominated solid solution and the double eutectic, and Ni-6Si-5Be, which has the structure: ternary eutectic.

According to the results of X-ray analysis of the sample Ni-4Si-2Be present: a solid solution based on Ni, BeNi, Ni_3Si. The phases presenting in the sample Ni-6Si-5Be are identical with the phases of the sample-Ni-4Si 2Be.

It can be concluded that by increasing the content of silicon and beryllium in the alloy, phases involved in the formation of structures remain unchanged.

Analyzing the experimental data, we can conclude that alloys compositions close to: Ni-5Si-3Be and Ni-6Si-5Be, - are interesting for further research. The latter has a low melting point due to the ternary eutectic (brazing temperature of the brazing alloy so 1070°C).

3.3 Measurement of microhardness nickel-based alloys with various contents of silicon and beryllium

In order to evaluate the mechanical properties of the developed filler metals and determine the form in which they will be used (in the form rapidly quenched ribbon or powder) microhardness of the alloys was measured. The microhardness of near eutectic nickel alloys containing boron was measured for comparison. Table 2 shows the measurement results.

If the value of microhardness is too high, then the alloy will not apply some methods of getting the status of a rapidly quenched. Alloy with high microhardness - not technological, and because of low ductility it can easily evolve microcracks, so that it is can show low performance.

According to the obtained values it can be concluded that an increase the silicon and beryllium content in the alloy increases the value of microhardness. The alloys, which is present as a solid solution and eutectic is high magnitude of the spread of values of microhardness. Alloys, having mainly the solid solution in structure, have lower values of microhardness.

Compared with the boron alloys it should be considered that the ribbon received from the alloy Ni-3.5Fe-7.5Cr-4.5Si-2.6B is amorphous [4, 5] due to the presence of silicon and boron, in contrast to ribbon produced from alloys system Ni-Si-Be with the same cooling rate.

Table 2. The microhardness alloys of the Ni-Si-B and Ni-Fe-Cr-Si-B systems.

Alloy	Average micro-hardness, (GPa)	The mean square deviation (GPa)	Alloy	Average micro-hardness, (GPa)	The mean square deviation (GPa)
Ni-4Si-2Be	3.0	0.5	Ni-6Si-3Be	4.1	1.2
Ni-4Si-3Be	3.7	0.4	Ni-6Si-3.5Be	5.7	1.5
Ni-4Si-3.5Be	3.7	0.6	Ni-6Si-4Be	6.1	1.1
Ni-4Si-4Be	4.0	1.0	Ni-6Si-5Be	6.4	0.9
Ni-5Si-2Be	3.4	0.7	Ni-7Si-4Be	6.6	1.1
Ni-5Si-3Be	4.6	0.9	Ni-7Si-5Be	6.5	0.8
Ni-5Si-3.5Be	4.6	0.7	Ni-8Si-4Be	6.4	1.5
Ni-5Si-4Be	4.8	1.9	Ni-3.5Fe-7.5Cr-4.5Si-1.7B	6.0	1.5
Ni-6Si-2Be	4.2	1.2	Ni-3.5Fe-7.5Cr-4.5Si-2.6B	6.8	1.6

Therefore the microhardness for 1301A alloy does not effect on the flexibility of the final ribbon, but for alloys with beryllium effect: there are limiting the amount of silicon and beryllium in the Ni-Si-Be, which you can get flexible processing ribbon. On the basis of the data we concluded that to obtain high-quality ribbon the amount of silicon shall not exceed (4.5-5.0) wt. % and the amount of beryllium (3.0-3.5) wt. %.

4. Conclusions

From the analysis of the research of Ni-Si-Be discovered that the system Ni-Si-Be is promising to create of a rapidly quenched filler metal without boron. In the system Ni-Si-Be has been established the presence of two eutectic with melting temperatures ~1100°C and ~915°C. With regard to such factors as the melting range, the brazing temperature and microhardness alloys of Ni-5Si-3Be and Ni-6Si-5Be are the most interesting for further studies, the compositions of which are mounted near two eutectics respectively.

Composition of Ni-6Si-5Be is promising because it could be get the filler metal with brazing temperature from 1070°C, but only in the form of powder, therefore, such a composition requires additional alloying increases the ductility of the alloy. In contrast, the alloy composition of Ni-5Si-3Be can be obtained tech flexible ribbon but the temperature of brazing of this filler metal is little bit higher than 1130°C.

Acknowledgements

This work was performed within the framework of the Center of Nuclear Systems and Materials supported by MEPhI Academic Excellence Project (contract № 02.a03.21.0005, 27.08.2013).

References

[1] Steuer S and Singer R F 2013 *The Minerals, Metals & Materials Society and ASM Int.* 2226-32
[2] *Phase diagrams of binary systems* (electronic resource: http://www.himikatus.ru/ free)
[3] Kenneth A 2009 *Walsh Beryllium Chemistry and Processing* ASM International 400 p
[4] Elmanov G N, Ivanitskaya E A and Zharikov E S 2015 *Non-ferrous Metals* **1** 36-41
[5] Elmanov G N, Ivanitskaya E A, Sevryukov O N, Skrytnyi V I and Suchkov A N 2014 *Inorganic Materials: Applied Research* **5** (3) 237–244

IOP Conf. Series: Materials Science and Engineering **130** (2016) 012045 doi:10.1088/1757-899X/130/1/012045

The use of Ni-Cr-Si-Be filler metals for brazing of stainless steels

A Ivannikov, V Fedotov, A Suchkov, M Penyaz, I. Fedotov and B Tarasov

National Research Nuclear University MEPhI (Moscow Engineering Physics Institute), Kashirskoe highway 31, Moscow, 115409, Russia

E-mail: ivannikov777@gmail.com

Abstract. Nanocrystalline ribbon filler metal-alloys of system Ni-Cr-Si-Be are produced by the rapidly quenching of the melt method. By these filler metals carried out hight temperature vacuum brazing of austenitic steels (12Kh18N10T and Kh18N8G2) and austenitic-ferritic class EI-811 (12Kh21N5T). The basic laws of structure-phase state foundation of brazed joints are determined, features of the interaction of the molten filler metal to the brazed materials are identified, the optimal temperature and time parameters of the brazing process are determined.

1. Introduction

Nowadays, brazing alloys containing boron are widely used for connecting stainless steels. One of the major problems in the application of boron filler metals is formation of heterogeneities in the brazed seam and heat affected zone, [1] which can negatively effect on the performance characteristics of the brazed joints and even cause destruction, as they adversely affect its fatigue strength and corrosion resistance [2]. Reducing the mass percentage of boron in the composition of the filler metal does not completely solve the problem. Therefore, new brazing alloys that do not contain boron on the basis of Ni-Cr-Si-Be for brazing stainless steels austenitic and austenitic-ferritic class are developing now.

The requirements for strength properties and technology for preparing compounds of corrosion resistant steels are constantly increasing with the development of aerospace technology, the creation of advanced missile, space and aircraft design and manufacture of cellular or lattice panels and plate heat exchangers. Increasing the strength of connections can be achieved through the technological solutions or the use of new materials.

The brazing has received a significant development among the methods of creating a permanently connection, which widely used in mechanical engineering. The use of brazing allows provides the compounds that can withstand high temperatures, mechanical stress and the effect of the active media for a long time without significant deterioration of their properties [3]. Therefore, the actual problem is to improve the mechanical properties of brazed joints by optimizing of structures and the development of new filler metals, improving their properties and technological modes of brazing.

Nowadays rapidly quenched filler metals is widely apply. They are used in powder form or in the form of a ribbon with thickness of 20 ... 80 μm. This is a special category of filler metals, which represent a new class of advanced materials [4]. Such alloys produced ultrarapid quenching technique that gives certain advantages in the process of melting and interaction of the melt with brazed materials. It improves the quality of brazing, reduces the amount of defects and the formation of intermetallics in the brazed joints [5].

The high strength of the brazed joint provides a diffusion brazing. Objective of the diffusion brazing, as an independent process for producing brazed joints, is to perform the crystallization process so as to provide the most equilibrium structure of compounds and thereby increase the temperature of desoldering of the seam, improve the strength, plasticity, electric conductivity, and to prevent the formation of low-strength and fragile cast structure of intermetallic compounds in the seams. The solidification of the joint occurs with this brazing during the isothermal holding and the structure of the seam is more homogeneous [6, 7].

2. Experiment and methods

The temperature characteristics and microstructure of brazing alloys of Ni-Si-Be and Ni-(Cr)-Si-Be were analyzed and two brazing alloys Ni-7Cr-5Si-3Be and Ni- 6Si-5Be wt. % have been selected to study by its temperature and mechanical properties. Figure 1 shows the differential thermal analysis (DTA) for these alloys (graphs are shifted for clarity).

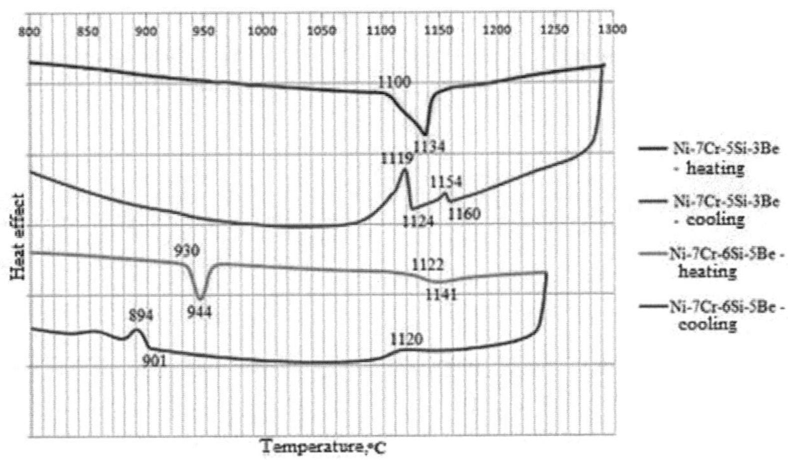

Figure 1. DTA of alloys Ni-7Cr-5Si-3Be and Ni-7Cr-6Si-5Be, wt. %.

The most perspective alloys in the system Ni-Cr-Si-B for brazing stainless steels are Ni-7Cr-5Si-3Be and Ni-7Cr-6Si-5Be. The joint brazed by filler metal Ni-7Cr-5Si-3Be has high strength and corrosion resistance due to chromium and relatively low melting point. As the alloy is ductile enough, a flexible tech ribbon was obtained from it. Alloy composition Ni-7Cr-6Si-5Be has a low melting point due to low melting eutectic (brazing temperature of such filler metal proceeding from the DTA may begin with 1070°C). However, this alloy has very high strength and low ductility, so the result the brittle ribbon was obtained by spinning. The filler metal was used for brazing in the form of pieces of ribbon.

Cylindrical (12Kh18N10T) and rectangular patterns (12Kh18N10T and EI-811 (12Kh21N5T, austenitic-ferritic)) were used for brazing. Cylindrical samples have a diameter of 10 mm and a height of 70 mm (GOST 28830-90). Rectangular samples have a long (wide) to (20 ... 30) mm and a height (3 ... 5) mm. The ribbon of filler metal was attached on the surface of sample by point capacitor welding. Filler metal thickness of 50 μm was used in a single layer and in the case of brazing the cylindrical samples it was placed directly into the brazing gap, and in the case of rectangular samples – near the brazing gap, so that the filler metal contact with a gap. Brazing of rectangular samples was carried out in clamps, and brazing of cylindrical samples was carried out in a special conductor.

The wedges were made specially to investigate the dependence of the joint overgrowing during the brazing process from the width of the gap. The gap between the rectangular steel plates was provided by tungsten wire thickness of 100 μm.

IOP Conf. Series: Materials Science and Engineering **130** (2016) 012045 doi:10.1088/1757-899X/130/1/012045

Structural studies of the brazed joints were carried out using metallographic microscope METAM PB-21-1 for the observation of the microstructure in the reflected and polarized light. It increases from x50 to x1000.

Determination the Vickers microhardness of brazed joint was carried out on the digital micro durometer Future-Tech FM-800, which has the rate of load 50 m / s, the load time of 20 seconds, the accuracy in accordance with GOST 9450.

Mechanical properties of brazed cylindrical samples at tensile were determined using two-column universal testing machine QUASAR 50,which has the maximum force is 50 kN.

3. Results and Discussion

It is necessary to determine the optimal mode of brazing to investigate the mechanical properties of the brazed joint, since the structure of the brazing joint, and therefore the mechanical properties, are depended strongly on the gap width and the holding time at brazing temperature.

The gap width corresponding to that time brazing, at which the fusible elements diffuse fully into the base material and a solid solution forms, was determined, because the eutectic has a low ductility and its presence in the joint is undesirably. Figure 2 shows the structure of the seam, which obtained after brazing of Kh18N8G2 (austenitic) and EI-811 steels by filler metal Ni-7Cr-5Si-3Be wt. %, with a different holding time.

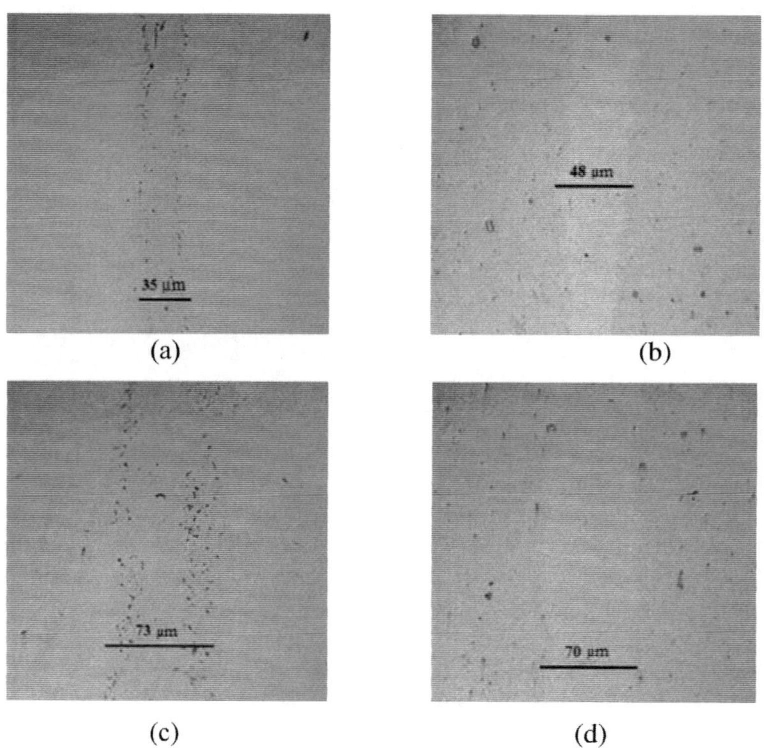

Figure 2. Structure of the brazed joint and the magnitude of the gap, at which the overgrowing of joint is complete, $T_{brazing}$ = 1150 °C (filler metal Ni-7Cr-5Si-3Be, wt.%), (a) – Kh18N8G2, $t_{brazing}$=30мин; (b) – EI-811, $t_{brazing}$=30мин; (c) – Kh18N8G2, $t_{brazing}$=1час; (d) – EI-811, $t_{brazing}$=1час

According to the structure of brazed joints it can be seen that overgrowing of joint occurs for the steel in the gap size of 30-50 µm during brazing at the holding time of 30 minutes and the holding time at 60 minutes - 70 µm. Therefore, heating to 1150 °C for 40 minutes at the gap size 45-50 µm were selected for brazing cylindrical samples for mechanical testing. The black particles are observed in the structure of joint and heat affected zone for Kh18N8G2 steel. Apparently, this is related to a high content of impurities in this steel (the steel are used in the technical purity).

After the optimal modes of brazing were installed, the plates were brazed with this filler metal considering gap size, brazing temperature and holding time. The values of microhardness of brazed joints were obtained. The microhardness of brazing joint obtained by the filler metal Ni-3.5Fe-7.5Cr-4.5Si-2.6B (1301A) wt. % were measured to compare. The results are shown in Table 1 (error ± 120 MPa). Alloy STEMET 1301A is studied in detail in the paper [8].

Table 1. Microhardness of samples brazed with different modes, in MPa.

Sample	Steel	Joint		Brazing mode
		fillet	solid solution	
Ni-7Cr-5Si-3Be/Kh18N8G2	1900	2800	2200	1150 °C, 1 hour
Ni-7Cr-5Si-3Be/EI-811	2700	2600	2000	1150 °C, 1 hour
Ni-7Cr-5Si-3Be/X18H8Г2	1500	2700	2000	1150 °C, 30 min
Ni-7Cr-5Si-3Be/ EI-811	3100	2800	2200	1150 °C, 30 min
1301A/ Kh18N8G2	1600	8700	3000	1150 °C, 15 min

The best mechanical characteristics of brazed joint will be achieved if the steel and the joint with loads will behave as a single entity. Therefore, the hardness of them should not vary greatly. In the case of using filler metal containing beryllium microhardness of the steel and the joint (fillet as well) are equal within the measurement error. The microhardness value in fillet area is extremely high compared with the microhardness value in the joint, when using boron filler metals. As known, the strength of fillets contributes significantly to the strength of the brazed joints [9].

The cylindrical samples were brazed at 1150 °C and holding time of 40 minutes for tensile test. The specimens brazed three filler metals: Ni-7Cr-5Si-3Be, Ni-7Cr-6Si-5Be and Ni-3,5Fe-7,5Cr-4,5Si-2,6B (1301A) were tested.

During tensile test of samples brazed with filler metal Ni-7Cr-6Si-5Be the destruction was not along the seam, it occurred on steel; therefore, the connection can withstand loads greater than steel. This result says about the prospects of this filler metal.

The compounds, which brazed with filler metal Ni-7Cr-5Si-3Be, showed better mechanical properties than the compounds, which brazed with boron filler metal 1301A. Averaging the value of tensile strength, which can withstand the connection brazed with filler metal Ni-7Cr-5Si-3Be, the value obtained (530 ± 50) MPa, and for 1301A - (420 ± 30) MPa.

Analyzing the experimental data obtained as a result of tensile tests, it can be concluded that the connection Kh18N10T / Ni-5Si-7Cr-3Be / Kh18N10T shows better strength values than the connection brazing with boron filler metal (Ni-3,5Fe-7, 5Cr-4,5Si-2,6B), which speaks about the prospects of filler metals of Ni-Si-Cr-Be.

If we choose the optimal mode of brazing and clearance, in which the eutectic in the joint will disappear completely, it is possible to get stronger compounds in combination with a low temperature brazing.

The experiment on spreadability was carried out with the filler metal Ni-7Cr-6Si-5Be, wt.% on substrate of steel 12Kh18N10T at 1100°C for 1 hour. A good interaction with the steel was received (Figure 3a). The diffusion zone is 230 μm. Therefore, this filler metal is more promising for brazing, because most of stainless steels is better brazing at temperatures not exceeding 1120°C as the loss of strength and grain growth are occurred above this temperature, which leads to embrittlement of the material.

Beryllium - an element that has a greater affinity with oxygen, and there is a possibility of formation of oxides, so additional studied of the effect of calcium, as a deoxidizer, on the spreadability

IOP Conf. Series: Materials Science and Engineering **130** (2016) 012045 doi:10.1088/1757-899X/130/1/012045

of filler metal Ni-5Si-7Cr-3Be was carried out. It was added 0.1 wt.% Ca. Figure 3b shows the results of an experiment on spreading filler metal with and without calcium on steel 12Kh18N10T at the temperature of 1150 °C and holding time 30 minutes. Calcium increases the spreading about three times.

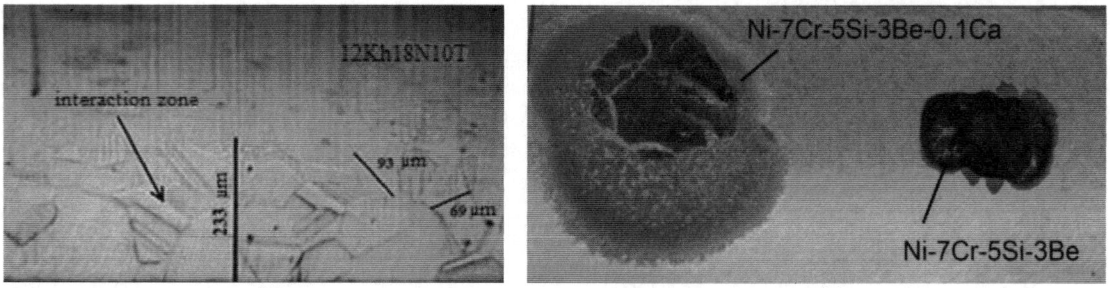

(a) (b)

Figure 3. A study of physico-chemical interaction of the filler metal with the substrate of steel 12Kh18N10T, (a) - filler metal Ni-7Cr-6Si-5Be, (b) - Ni-5Si-7Cr-3Be, wt.%.

4. Conclusions

The wedge patterns for different modes to determine the optimal mode of brazing steels Kh18N8G2 and EI-811 were brazed. As a result the mode was selected: 40 minutes at the temperature of 1150 °C and at the gap size of 50 µm.

The brazing of cylindrical samples made of steel Kh18N10T by filler metals Ni-7Cr-5Si-3Be and Ni-7Cr-6Si-5Be was carried out under this mode. Tensile strength in uniaxial tension of these samples exceeded 500 MPa.

It is shown that the filler metal Ni-7Cr-6Si-5B interacts well with the steel at the temperature of 1100°C (diffusion zone is 230 µm). The modification by calcium of the original filler metal was carried out and the spreadability was increased is about three times.

Acknowledgements

This work was performed within the framework of the Center of Nuclear Systems and Materials supported by MEPhI Academic Excellence Project (contract № 02.a03.21.0005, 27.08.2013).

References

[1] Steuer S and Singer R F 2013 *The Minerals, Metals & Materials Society and ASM Int.* 2226–32
[2] Maksimova S V 2007*Adhesion of melts and brazing of materials* p 70 (in Russian)
[3] Horunov V F 2008 *Basics of brazing of thin-walled structures made of high-alloy steels* (Kiev: Naukova Dumka Publishing House) p 238 (in Russian)
[4] Kalin B A, Pleshcheev A N, Fedotov V P, Sevryukov O N and Goltsev V Yu 2001 *Welding Production* **8** 38 (in Russian)
[5] Kalin B A, Fedotov V T, Sevryukov O N, Grigoriev A E et al 1996 Welding production **1** 15 (in Russian)
[6] 1984 *Reference on brazing Ed. Petrunina O E* **2** (Moscow, Mechanical engineering.) p 400
[7] Lasko S V, Lasko N F 1988 *Brazing of metals. 4th ed.* (Moscow, Mechanical engineering) p 376
[8] Elmanov G N, Ivanitskaya E A and Zharikov E S 2015 *Non-ferrous metals* **1** 36
[9] Novosadov V S 2014 Welding Production **10** (in Russian)

Study of phase transitions in NbN ultrathin films under composite ion beam irradiation

K Prikhodko[1, 2], B Gurovich[1] and M Dement'eva[1]

[1]National Research Centre "Kurchatov Institute", Kurchatov sq.1, Moscow, 123182, Russia
[2]National Research Nuclear University MEPhI (Moscow Engineering Physics Institute), Kashirskoe highway, 31, Moscow, 115409, Russia

E-mail: prihodko_ke@nrcki.ru

Abstract. This work demonstrates implementation of Selective Displacement of Atoms (SDA) technique to change the crystal structure and atomic composition of thin superconductive film of NbN under low dose composite ion beam irradiation. All structure investigations were performed using High Resolution Transmission Electron Microscopy (HRTEM) technique by the analysis of Fourier transformation of bright field HRTEM images. It was found that composite ion beam irradiation induces the formation of niobium oxynitrides phases.

1. Introduction

Last years in the area of radio electronics superconducting microelectronics comes to the forefront. The phenomenon of superconductivity is widely used for creation of electronic devices large capacity memory, high-speed digital switches, the highly sensitive radio receiving equipment.

The advantages of the phenomenon of superconductivity in comparison with the most common semiconductor should include lower energy dissipation caused by disappearance of resistance at cryogenic temperatures, as well as faster performance digital devices. This allows the use of superconducting elements as the basic units of electric circuits with a frequency range up to hundreds of GHz.

In accordance with active progress in the physics of thin films, which began after the creation of the microscopic theory of superconductivity Bardeen-Cooper-Schrieffer and development of technological approaches that can generate low-temperature materials, has opened up new prospects for the development of integrated Cryoelectronics.

Niobium nitride (NbN) is one of the most promising materials in the field of low-temperature superconducting electronics to build electronic devices such as bolometers (THz HEB), single-photon counters superconductor (Superconducting Single-Photon Detector, SSPD), quantum devices [1]. This advantage over other materials caused by high strength characteristics, resistance to thermal cycling. Of all the refractory metals NbN has the highest critical temperature of the superconducting transition ~ 14 K.

Today there is the problem of creating technological approaches for the production of nanoscale cryoelectronics elements and improve their quality of basic parameters. In our institute developing many methods of radiation-induced modification of atomic structure, electrical, optical and magnetic properties of materials under irradiation with accelerated particles. The basis of these methods are the mechanisms of selective removal of atoms (SRA), a selective association of atoms (SAA) and selective displacement of atoms (SDA) [2, 3]. These approaches allows to create a composite structure with a locally changing the chemical composition and properties of thin film materials by a controlled way.

2. Experiment

Ultrathin films of niobium nitride (NbN) were grown on a mono-crystalline silicon substrate covered with a 0.15 μm amorphous oxide SiO_2 layer by cathode sputtering method using a niobium target controlled gaseous mixture of nitrogen and krypton at an operating vacuum of ~ 10^{-5} Torr. A necessary condition for the growth of film is heated substrate up to temperatures of ~ 800°C [4]. The thickness of the starting film was 5 nm. The critical temperature of the superconducting transition ~12 K [4,5].

Samples were irradiated by composite ion beams extracted from the high-frequency plasma discharge and consisting of oxygen ions and protons [3] with energies (0.1-4) keV in a dose range (1-2) d.p.a. for nitrogen. The ratio of oxygen to hydrogen ion composition in the composite ion beam was monitored by measuring the partial pressures of the residual H_2O component and hydrogen supplied [6]. In our case the beam composition was $c = 1.2 \cdot 10^{-3}$. The dose rate of nitrogen atoms displacement was performed using SRIM code [7] and in our case was $2.25 \cdot 10^{-2}$ d.p.a./s.

To studies the microstructure and phase composition of the initial and the irradiated samples were performed using High Resolution Transmission Electron Microscopy (HRTEM) technique at Titan 80-300 electron microscope at an accelerating voltage of 200 kV. Cross section samples NbN/SiO_2/Si were manufactured at the facility FIB Helios Nanolab 650.

The phase composition of individual grains for initial and irradiated films was determined by analyzing the corresponding diffraction patterns. However, because that films were characterized by an ultra-small grain size (~ 5 nm), the analysis of the pictures of conventional electron selected area micro-diffraction pattern was ineffective due to a strong blurring of the reciprocal lattice points.

In this paper for obtaining diffraction patterns were used bright-field high-resolution images of a cross section of films, obtained at magnifications more than x500000. For phase identification the region of interest in the HRTEM image was selected and construction performed by the Fourier transform of the art, for example, in Digital Micrograph program.

Computer calculated Fourier transform of HRTEM image is a complete analogue of the physical picture of micro-diffraction pattern, because the intensity of reflections on an experimental micro-diffraction pattern is proportional to the square of the Fourier transform of the electron density of the study phase.

The procedure for phase identification by pair reflex technique includes the following. In the diffraction pattern defines a set of non-parallel reciprocal lattice vectors $H_{h1,k1,l1}$, $H_{h2,k2,l2}$,…,etc., with indexes (h_1,k_1,l_1), (h_2,k_2,l_2), ..., etc., by measuring the distance from the center of the diffraction pattern to the corresponding reflexes. Interplanar distances for the reflex data d_1, d_2, d_3, ..., etc. calculated as the inverse of the length of the reciprocal lattice vectors: $d_{hkl} = 1/H_{hkl}$. We also defined a set of angles between different reciprocal lattice vectors α_1,, α_2,… α_i.

For phase identification was carried out comparing the experimental values of the interplanar distances and angles between the vectors of the reciprocal lattice with published data for a large number of chemical compounds containing niobium, nitrogen and oxygen in various combinations. In this paper, the identification phase was carried out using X-ray spectra database ICSD PDF-4.

For the phase analysis the correspondence between measured and database interplanar distances were compared. Then we calculated all the possible angles between the crystallographic planes $(h_1k_1l_1)$ with interplanar distance d_1 and $(h_2k_2l_2)$ with interplanar distance d_2. During angles calculations we took into account the symmetry elements in each family set for a given syngony. If there was a match between both the lattice parameters and all angles for the experimental and calculated diffraction pattern from the database, then concluded the successful identification of this phase was found.

Such an approach allows us to determine the phase composition of the films with nano-sized grains in the initial state and also after composite ion beam irradiation to the different doses.

3. Results and discussion

The objective of research was an experimental study of the evolution of structural-phase state of thin superconducting films NbN, irradiated by composite ion beams by using a high-resolution bright field TEM images. It is well known [4], the irradiation of thin films of superconducting niobium nitride by composite ion beams causes a reduction in the superconducting transition temperature and increasing resistance of the film in a wide temperature range from 4.2 K to 300 K.

Typical HRTEM image of a cross section of the starting niobium nitride film on a substrate of amorphous silicon oxide is shown in Figure 1. As seen in Figure 1, after the deposition the niobium nitride film has a polycrystalline structure with a grain size of (4-5) nm. If different grains of polycrystalline films are differently oriented to the electron beam, a contrast with the image of the atomic planes for them is significantly different. The most successful and sophisticated are the images of the grains, the orientation of which corresponds to a slight deviation angles between crystallographic zone axis of the grain and incident electron beam direction. According to the pair reflexes technique we must have as big numbers of pairs of different reflections as possible. To get these conditions we selected grains oriented with crystal zones along the electron beam for analysis. An example of such grain is shown by isolated square in Figure 1.

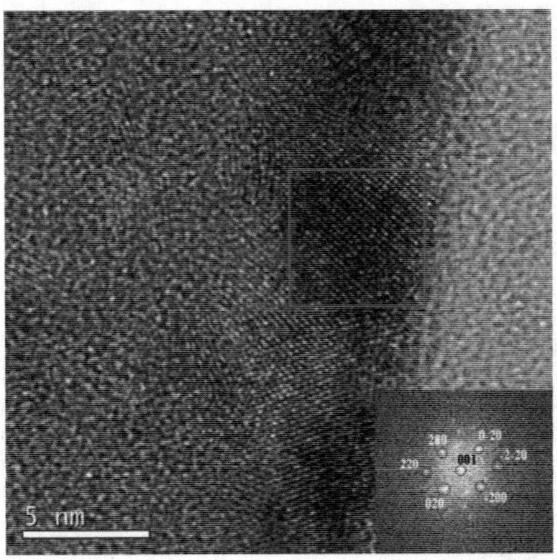

Figure 1. Bright-field HRTEM image of the cross section of the film NbN, synthesized on a substrate of SiO_2, diffraction (inset).

An analysis of the diffraction pattern in Figure 1 shows that grain starting material corresponds to the phase NbN cubic system (Fm-3m) with the lattice parameter a = 0.4394 nm.

By increasing the dose to ~1 d.p.a. for nitrogen observed change in the phase composition of the grains, as shown in Figure 2. The analysis of pictures of high resolution to grains in Figure 2 showed the formation of the monoclinic phase NbNO (P21/c(14)) with unit cell parameters a=0.4977 nm, b=0.50217 nm, c=0.52053 nm, $\alpha = \gamma = 90$ °, $\beta = 100$ °.

A further increase of the radiation dose up to ~(1.5-2) d.p.a. for nitrogen showed the formation of individual grains of $NbN_{0,64}O_{1,36}$ phase monoclinic (P21 / c (14)) with unit cell parameters a=0.49808 nm, b=0.50250 nm, c=0.52097 nm, $\alpha = \gamma = 90$ °, $\beta = 100°$. Together with grains of $NbN_{0,64}O_{1,36}$ phase under dose of ~(1.5-2) d.p.a. we also observed above mentioned NbNO phase grains (Figure 3).

It should be noted that closeness of the unit cell parameters of niobium oxynitride phases NbNO and $NbN_{0,64}O_{1,36}$, allowing to talk about their semi-identical atomic structure. At the same time the ratio of oxygen atom numbers to the nitrogen atom numbers in the latter phase is higher than the first. The increase in the proportion of oxygen atoms with increasing radiation dose reflects the general tendency of flow selective displacement of atoms (SDA) process.

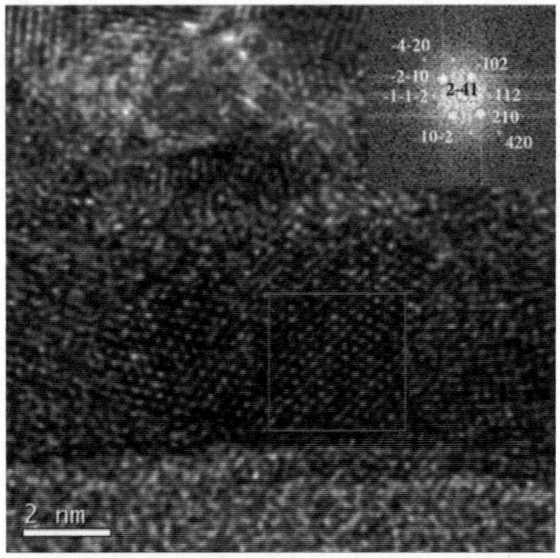

Figure 2. Bright-field HRTEM image of a cross section of the irradiated NbN film at a dose of ~1 d.p.a. for nitrogen, diffraction pattern (inset).

Figure 3. Bright-field HRTEM image of a cross section of the irradiated NbN film after composite irradiation to a dose of ~(1.5-2) d.p.a. for nitrogen, diffraction pattern (inset).

It should also be noted that delivery of oxygen atoms from the composite beam during irradiation remains approximately constant, while in terms of phase transformations of initial NbN to niobium oxynitrides it is not possible to implement the "continuous" phase transformation of the structure since there are only a fixed number of niobium oxynitride phases “available”.

For this reason, it is expected that in the region of existence a certain phase of a niobium oxynitride, for example NbNO, within a certain range of irradiation doses the concentration of oxygen in it will grow, and besides part of the oxygen atoms will be in the lattice interstitial positions. The formation of the next phase will take place only when there is a phase for it with the respective intermediate atomic composition, for example, $NbN_{0,64}O_{1,36}$ in this case. If the crystalline phase with a suitable atomic structure does not exist, it should be expected to increase the distortion of the crystal lattice with increasing radiation dose due to the increase of the concentration of interstitial oxygen atoms, up to a total loss of crystallinity, and form an amorphous film.

4. Acknowledgments

The authors thank Stolyarov V.L. and Olshansky E.D. for the production of the original films of niobium nitride, Komarov D.A. for carrying out experiments on ion irradiation, and Kutuzov L.V. for making the cross-sectional samples. The work was supported by the Ministry of Education and

Science of the Russian Federation - Agreement on Subsidies № 14.607.21.0005 (project code RFMEFI60714X0005) under the Federal Program "Research and development on priority directions of scientific-technological complex for 2014-2020."

5. Conclusions

By methods of high resolution transmission electron microscopy were studied changes in the phase composition of ultra-thin films of superconducting niobium nitride as a result of the process of selective displacement of atoms under the influence of composite ion beam irradiation. It was shown that the composite ion beam irradiation of initial NbN film to doses (1-2) d.p.a. for nitrogen initiated the formation of niobium oxynitride phases NbNO and $NbN_{0,64}O_{1,36}$ of monoclinic crystal system while maintaining the polycrystalline structure of the film.

Acknowledgements

This work was performed within the framework of the Center of Nuclear Systems and Materials supported by MEPhI Academic Excellence Project (contract № 02.a03.21.0005, 27.08.2013).

References

[1] Guillet B, Arthursson O, Mechin at all. 2008 *Low Temp Phys.* **151** 570
[2] Gurovich B A and Prikhodko K E 2009 *Physics–Uspekhi* **52** 2 165
[3] Gurovich B A, Prikhodko K E, Kuleshova E A, Maslakov K I and Komarov D A 2013 *Journal of Experimental and Theoretical Physics* **116** 916
[4] Gurovich B A, Tarkhov M A, Prikhodko K E, Kuleshova E A, Komarov D A, Stolyarov V L, Olshanskii E D, Goncharov B V, Goncharova D A, Kutuzov L V and Domantovskii A G 2014 *Nanotechnologies in Russia* **9** (7-8) 386
[5] Gurovich B A, Prikhodko K E, Tarkhov M A, Kuleshova E A, Komarov D A, Stolyarov V L, Olshanskii E D, Goncharov B V, Goncharova D A, Kutuzov L V, Domantovskii A G, Lavrukhina Z V and Dement'eva M M 2015 *Nanotechnologies in Russia* **10** (7-8) 530
[6] Prikhodko K E, Gurovich B A, Komarov D A, Goncharova D A, Kutuzov L V 2014 *Nuclear Instruments and Methods in Physics Research B* **326** 273
[7] Ziegler J F, Ziegler M D and Biersack J P 2010 *Nuclear Instruments and Methods in Physics Research B* **268** 1818

IOP Conf. Series: Materials Science and Engineering **130** (2016) 012047 doi:10.1088/1757-899X/130/1/012047

Peculiarities of the structural-phase state of a brazing zone of dissimilar metal joints

O N Sevryukov, E V Guseva and A A Polyansky

National Research Nuclear University MEPhI (Moscow Engineering Physics Institute), 31 Kashirskoe highway, Moscow, 115409, Russia

E-mail: sevr54@mail.ru

Annotation. The present work is devoted to the study of the structural-phase state of the brazed joints of dissimilar materials obtained by rapidly quenched filler metals before and after tests in the working conditions. The constructive elements of the International Thermonuclear Experimental Reactor (ITER) working chamber were selected: mock-ups of the first wall (a beryllium cladding joined with the heat-removing bronze by high-temperature brazing) and the divertor (a tungsten cladding with built-up copper, brazing of copper with bronze). An analysis of the investigated samples and the data on the structural-phase state of the brazed joints and its evolution as a result of a cyclic thermal impact was carried out by metallography and electron microscopy.

1. Introduction

At present, materials with various physical and chemical properties (for example, ceramics and metal, alloys and metals with very different coefficients of thermal expansion, fusible and refractory metals, etc.) are used in many constructive elements. Brazed constructions often work under extreme conditions, which make them more demanding to thermal-, wear- and corrosion-resistance, mechanical strength, and radiation resistance. Damages can occur both at the unsatisfactory quality of the brazed seam in the initial state and at its non-compliance to the acceptance criteria of the existing standards, as well as in the event of defects in the course of operation. Amorphous and nanostructured brazing filler metals, possessing high characteristics of diffusion and capillary activity and a high homogeneity of the chemical composition, make it possible to solve the complex tasks of creating original and workable constructions from various materials. In particular, the cladding of the divertor (tungsten) and the first wall (beryllium) of the International Project ITER (International Thermonuclear Experimental Reactor) assumes brazing of the above-mentioned elements with bronze as this part of the construction is in direct contact with plasma [1]. As the right choice of the brazing filler metal, the knowledge of regularities of the structure- and phase-formation during the brazing, as well as an understanding of the processes occurring during the operation of the brazed joint are the necessary conditions to ensure a long-term operation of the product, an object of the present work was to study the evolution of the structural-phase state of the brazed seam during its operation by methods of the modern metallography.

2. Experimental part

Brazing filler metals based on copper (STEMET 1101 and 1108) were used in the work. Their compositions and characteristics are shown in Table 1. They were in the form of both rapidly-quenched amorphous and nanostructured ribbons, as well in the form of powders obtained by grinding of a preliminarily embrittled amorphous ribbon. Ribbon-type brazing filler metals were prepared at the installation Crystal-702 that makes it possible to produce such alloys in the form of a ribbon (with a thickness of 25 - 100 μm and a width of 1.5 - 50 mm) by an ultrafast solidification of a planar jet of the melt on a rotating copper disc. The achievable cooling rate of the melt at the installation was regulated in the range of 10^4 - 10^6 °C/sec. The maximum quantity of the ribbon obtained for one technological cycle was 0.8-1.0 kg for copper alloys.

Table 1. The composition and characteristics of the brazing filler metals.

Brand of the filler metal	Nominal composition in (wt. %)	$T_{melting}$ (°C)	$T_{brazing}$, (°C)	Brazing conditions	Brazed materials
STEMET 1101	8.8 Ni 7.0 P 3.5 Sn Cu- the rest	615 – 660	650 – 670	Gas flame, electrocontact, and induction brazing.	Copper and its alloys. Beryllium - copper alloys.
STEMET 1108	12 Sn 5 In 2 Ni 1 P Cu - the rest	750 – 870	870 – 950	Gas flame, induction, electrocontact, and furnace brazing. Brazing in a vacuum or inert atmosphere.	Copper, copper alloys and steels in any combination. Beryllium - copper alloys, and tungsten.

3. Objects of the investigation

1. Sample 1. An element of the ITER first wall mock-up

The first wall is a multi-layer construction, of which the water-cooled bronze structure is covered with brazed beryllium tiles with a size of 16x16 mm and a thickness of 8 mm. The operation regime of the ITER reactor is thermocycling. Thermal stresses occur on the "beryllium-bronze" border during the heating cycle. Therefore, thermocycling experiments are the main workability criterion of the beryllium coating. To braze the beryllium tiles, a rapidly quenched amorphous ribbon-type non-silver filler metal STEMET 1101 is used [2]. The brazing of the mock-up is carried out either by an induction method or by heating with an electron beam. To preserve the desired mechanical properties of the temperature-sensitive bronze in each of the above-mentioned methods, it is necessary to minimize the time of heating / cooling of the CuCrZr alloy in the temperature range of 450–700–450 °C (figure 1).

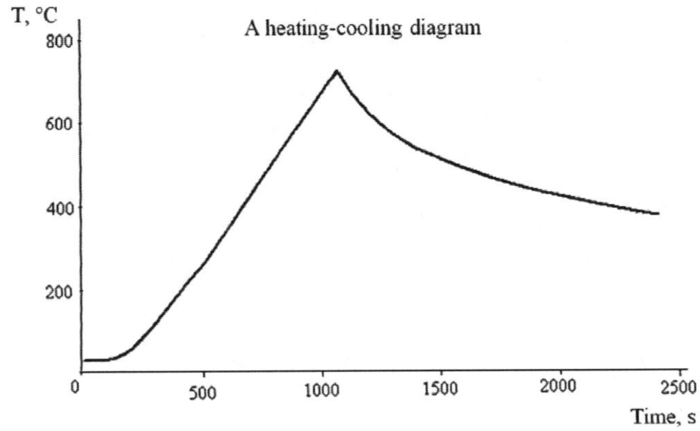

Figure 1. A diagram of brazing the mock-ups with a beryllium coating.

Mock-ups are installed on the stand Cepheus-M and covered with a water-cooled mask.

The mock-ups are supplied by cooling water and the beryllium surface, seen from under the masks, is exposed from above to a cyclic heat flow that is simulated by an electron beam.

The main parameters of the experiment expected in the ITER reactor are shown in Table 2.

Table 2. The main parameters expected in the ITER reactor.

Parameter	Value
The absorbed heat power density	0.5 – 5.9 MW/m^2
The water pressure at the inlet into the mock-up	2 ±5% MPa
The water temperature at the inlet into the mock-up	70 ± 5 °C
The water consumption through the mock-up	0.42±5% kg/s
The water speed	2 ±5% m/s
The duration of a pulse / pause during the thermocycling experiments	15/15 s
The heat loading area	75 cm^2 (the entire surface of the coating)

A fragment of the ITER first wall mock-up is shown in Figure 2.

Figure 2. An element of the ITER first wall mock-up.

2. Sample 2. An element of the ITER divertor

The divertor consists of two main parts, a support construction made from stainless steel and a water-cooled part facing to the plasma that is coated by erosion- and heat-resistant tungsten.

Table 3. Working conditions of the protective elements of the ITER divertor from tungsten (the average heat flow is 0.3 MW/m^2).

Action type	Vertical target	Dome	Liner
The number of cycles (without a replacement)		~3000	
The maximum heat flow, MW/m^2	5	3	0,2–1
The maximum damage degree, dpa	~0.6	~ 0.7	~0.1–0.2
The maximum generation of helium, appm	< 0.35	< 0.4	< 0.1
The working temperature, °C	200–600	500–1200	500–1200
The energy at the plasma disruption, MJ/m^2	1–5	10	1
The number (*N*) / duration of plasma disruptions, ms	500/1…2	500/1…2	500/1…2

Figure 3 shows a fragment of the ITER divertor. The tungsten is coated by a copper layer with a thickness of 3 mm; the copper is brazed to the bronze that is welded to a heat-removing steel base. For brazing, a filler metal STEMET 1108 was used [2].

Working conditions of the protective elements of the divertor made from tungsten (the average heat flow is 0.3 MW/m^2) are presented in Table 3.

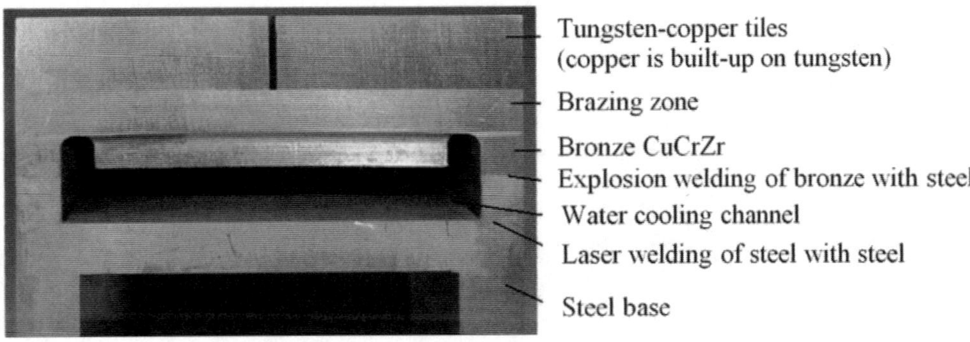

Figure 3. An element of the ITER divertor

4. Metallographic analysis

A metallographic analysis was carried out at the starting stage of the experiment to obtain primary information about the microstructure of the seam and the heat-affected zone, to confirm or refute the presence of structural-phase changes and their qualitative estimation, and to analyze the presence defects in the brazed seam. The images were obtained by a metallographic microscope METAM PB-21 using the Video Test "Structure"-5.2 software package. To reveal the grain boundaries, chemical etching (at the etchant of $FeCl_3 + HCl + H_2O$) was carried.

4.1. Sample 1. The joint of beryllium with bronze

According to the present investigation, it can be said that the brazed joint of beryllium with bronze has a high quality: pores, cracks and faulty brazed areas are absent; the joint width is the same in all areas; the joint structure is fine and presumably of eutectic composition. The structural-phase changes in the sample after operation in relation to the initial one were not detected (Figure 4).

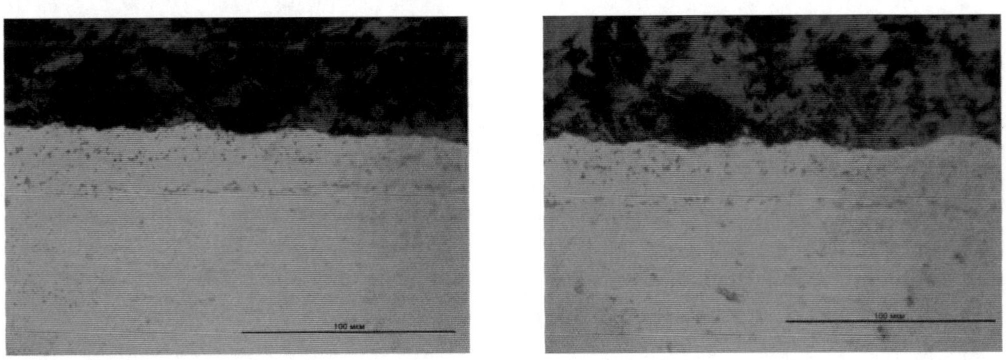

a) b)

Figure 4. Microstructure of the brazed seam and the heat-affected zone of the joint of beryllium with bronze (the marker is 100 μm): a) the sample before the tests and after the chemical etching; b) the sample after the tests and after the chemical etching.

IOP Conf. Series: Materials Science and Engineering **130** (2016) 012047 doi:10.1088/1757-899X/130/1/012047

4.2. Sample 2. The joint of copper with bronze

It is necessary to note a high content of defects (faulty brazed areas) of up to100 μm in length in the brazed seam of both the samples. They are mainly observed at the edges of the brazed joint. The quantity of defects and their size is much less in the central zone of the brazed seam (Figure 5).

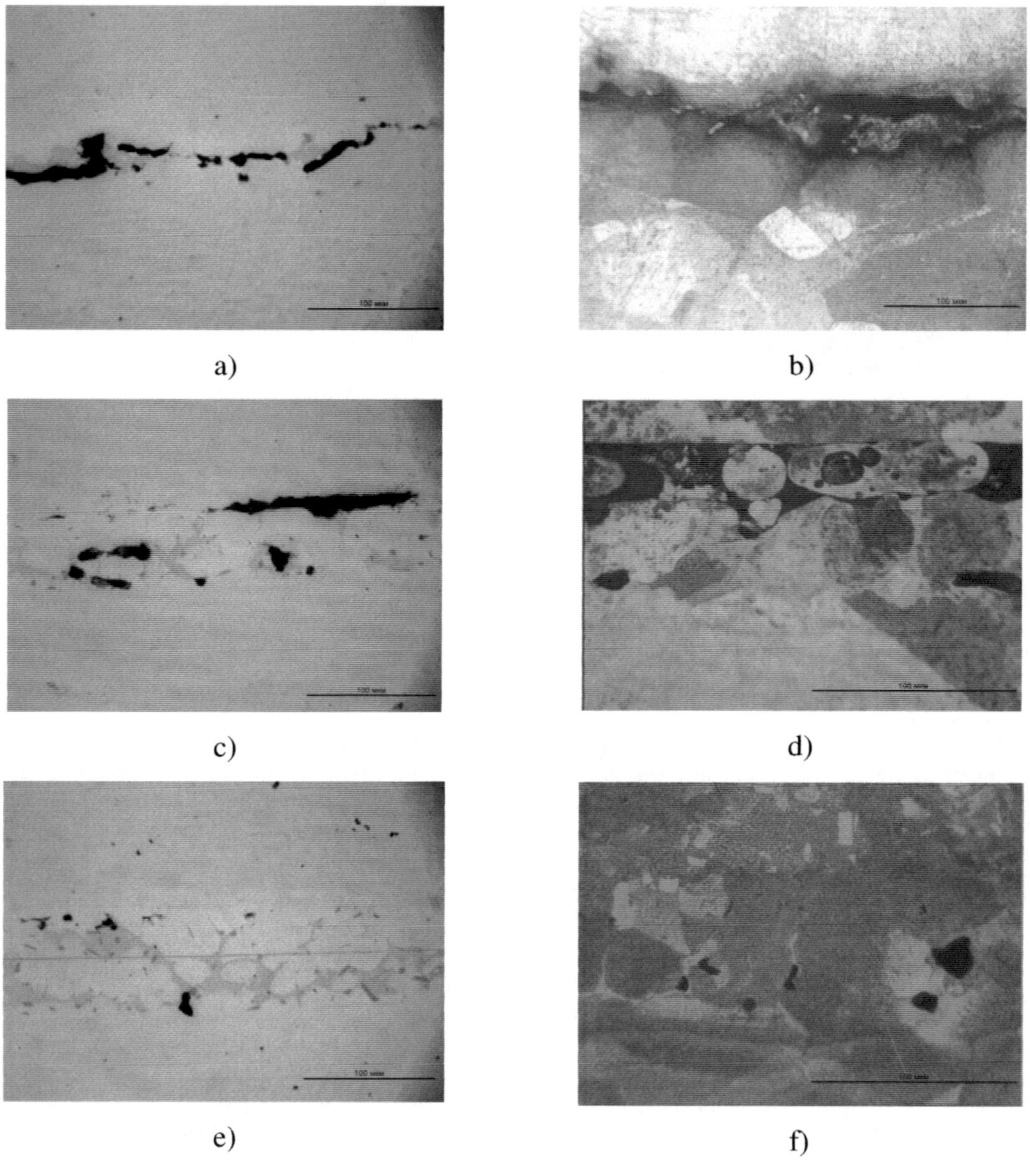

Figure 5. Microstructure of the brazed seam and the heat-affected zone of the joint of copper with bronze (the marker is 100 μm): a) the sample before the tests and before the chemical etching (edge); b) the sample before the tests and after the chemical etching (edge); c) the sample after the tests and before the chemical etching (edge); d) the sample after the tests and after the chemical etching (edge); e) the sample after the tests and before the chemical etching (center); f) the sample after the tests and after the chemical etching (center).

The presence of structural changes, such as an increase in the width of the diffusion zone and formation of grains of a new phase, was revealed in the sample exposed to thermal cycling tests.

5. Electron-microscopic investigations, an analysis of state diagrams

Images of the brazed seam microstructure for each of the prepared sections of the brazed joints were obtained at various magnifications by an electron microscope Carl Zeiss EVO-50XVP with EDS Inca x-act in the regime of reflected and secondary electrons. To estimate the distribution of elements, a chemical analysis of the brazing filler metal was carried out after operation of the product; maps of the distribution of elements were made as well as [3].

5.1. An element of the ITER first wall mock-up. Sample 1

The seam microstructure of the sample before the tests (Figure 6) consists of three phases that are equally distributed throughout the whole volume of the seam. Its width varies in a small range and is of the order of 40 ± 2 μm. The beryllium contains implanted particles of the abrasive material, the relief is strong enough, and a step is clearly visible at the junction of the brazing filler metal and beryllium. Large pores and cracks are absent, small pores (less than 10 μm^2) are extremely rare. The sample does not contain significant changes after operation (Figure 7), the microstructure is also three-phase, and only a slight increase of the dispersion is visible owing to the thermal treatment. Additional detailed investigations were not carried out due to the lack of structural-phase changes. An analysis of the phase diagrams made it possible to determine phases that are most likely present in the brazed seam structure (Figure 8).

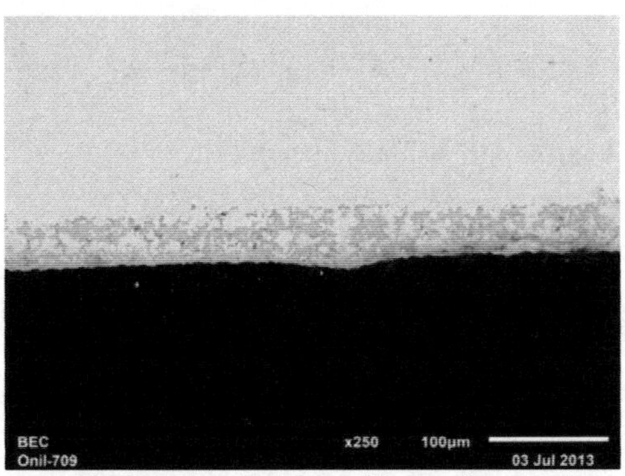

Figure 6. Brazing of beryllium (bottom) with bronze (top). The sample before the operation.

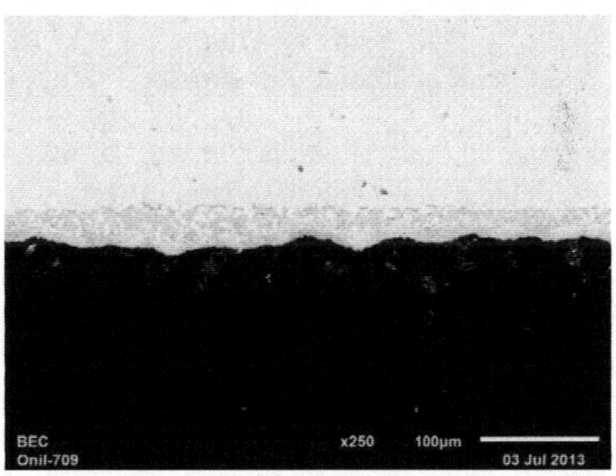

Figure 7. Brazing of beryllium (bottom) with bronze (top). The sample after the operation.

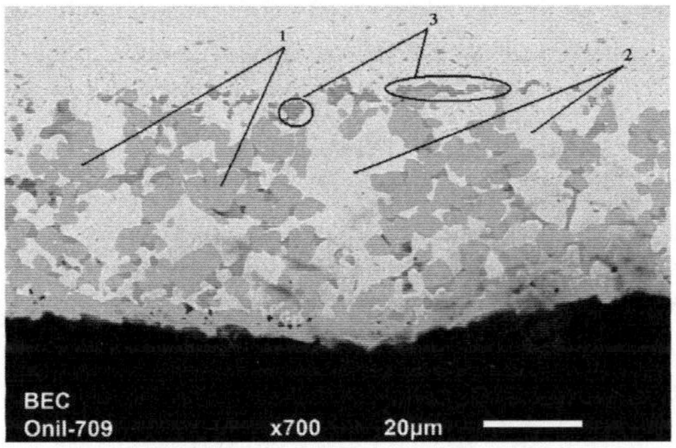

Figure 8. Brazing of beryllium (bottom) with bronze (top). The sample before the operation. 1) α-solid solution based on copper; 2) eutectic of ε and η phases; 3) phase with a maximum content of phosphorus - (37–39) % Mn, (34–36) % Ni, ≈ 24% P).

5.2. An element of the ITER divertor. Sample 2

There is a great number of small pores in the brazed seam both before and after its operation (Figures 9 and 10) which are rounded and elongated along its line, as well as larger pores and faulty brazed areas that reach a length of 40 μm (the area is of the order of 400 μm^2) are also present. The seam width of the initial sample varies from 20 to 50 μm, its boundaries are slightly visible due to the fact that the materials of the brazing filler metals and the joined elements have an equal metal base

IOP Conf. Series: Materials Science and Engineering **130** (2016) 012047 doi:10.1088/1757-899X/130/1/012047

(copper), and the seam is non-uniform in phase composition. Its structure is coarse two-phase. The first phase, heavier, is presumably of the eutectic composition with a grain size of 100–700 μm^2 and it is mainly located at the central zone of the seam. The second one is a copper-based solid solution. The brazed seam of the sample after the tests has a more ordered microstructure and a relatively equal width. The dispersion of its structure is significantly increased and there is also a great number of pores and cracks. When a larger magnification is used, it is noticeable that the eutectic phase is distributed in the form of a grid. It is necessary to point out the presence of chemical compounds (phosphides) absent in the initial sample. They are distributed chaotically, have a very various form and are found in the form of both globules (≈ 10 μm^2) and thin elongated needles (1×50 μm).

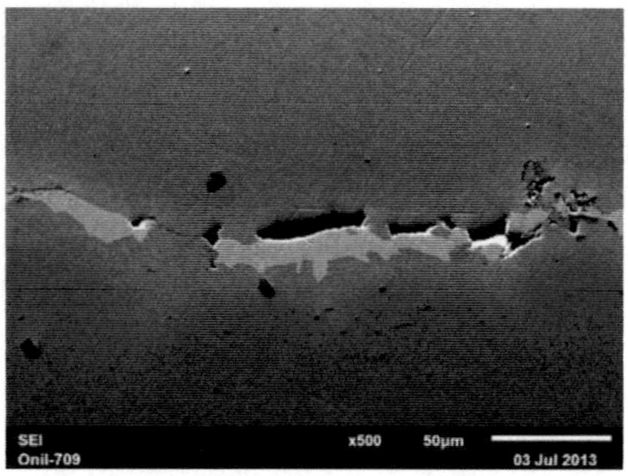

Figure 9. Brazing of copper with bronze. The sample before the operation.

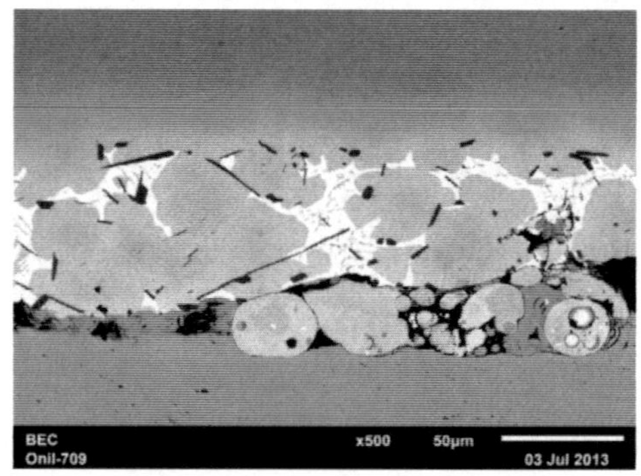

Figure 10. Brazing of copper with bronze. The sample after the operation.

As shown by metallographic analysis, elongated tangential cracks, formed most likely in the process of crystallization of the brazing filler metal, are present in the seams of the edge zones of both the samples. However, the electron image shows that these cracks end at the grain boundaries and do not penetrate into the base material. There is even no growth of the structure imperfection in the product after its operation, which indicates the ability of the brazed joint to withstand the planned loads.

6. Conclusion

Based on the experimental results obtained, the following conclusions can be made:

1. As a result of the tests of details exposed to thermal loads, the brazed joints can experience structural-phase transformations and preserve the initial microstructure without any changes.

2. The seam of the joint of beryllium with bronze, obtained by a copper brazing filler metal STEMET 1101 and possessing a high quality has undergone the minimum structural-phase changes. A small increase of the structure dispersion should be interpreted as the result of the heat treatment without a phase recrystallization.

3. It is necessary to point out the low quality of the initial brazed joint obtained using the brazing filler metal STEMET 1108 (brazing of copper with bronze). Numerous defects in the form of faulty brazed areas, pores and cracks are present in the brazed seam, which to some extent is quite strange as the brazed materials have similar physicochemical properties and, as well as the brazing filler metal, have the same metal base, i.e. copper. The differences in the microstructure of the samples before and after the tests (an increase in the number and the size of defects, the seam width and the dispersion, the allocation of intermetallics and a eutectoid grid) indicate an intensive diffusion of elements of the brazing filler metal and the brazed materials. The formation of chemical compounds of the Mn-Ni-P system having a high brittleness is a negative factor, which in turn may negatively affect the mechanical properties of the brazed joint and lead to failure of the constructive element.

Acknowledgements
This work was performed within the framework of the Center of Nuclear Systems and Materials supported by MEPhI Academic Excellence Project (contract № 02.a03.21.0005, 27.08.2013).

References
[1] Kalin B A, Suchkov A N, Fedotov V T and Ivannikov A A et al 2012 *Fusion Science and Technology* **61** (2) 147
[2] Kalin B A, Fedotov V T, Suchkov A N et al 2014 *Fusion Science and Technology* **65** 212
[1] Elmanov G N, Ivanitskaya E A and Zharikov E S 2015 *Non-ferrous metals* **1** 36

IOP Conf. Series: Materials Science and Engineering **130** (2016) 012048 doi:10.1088/1757-899X/130/1/012048

Plasma methods of obtainment of multifunctional composite materials, dispersion-hardened by nanoparticles

O N Sizonenko[1], E G Grigoryev[2], A D Zaichenko[1], N S Pristash[1], A S Torpakov[1], Ye V Lipyan[1], V A Tregub[1], A G Zholnin[2], A V Yudin[2] and A A Kovalenko[1]

[1]Institute of Pulse Processes and Technologies of NAS of Ukraine
Oktyabrski ave. 43-a, Mykolaiv, 54018, Ukraine
[2]National Research Nuclear University MEPhI(Moscow Engineering Physics Institute), Kashirskoe highway 31, Moscow, 115409, Russia

E-mail: sizonenko43@rambler.ru, eggrigoryev@mephi.ru

Abstract. The new approach in developed plasma methods consists in that dispersion-hardening additives (TiC, TiB_2 in particular) are not mechanically added to powder mixture as additional component, as in conventional methods, but are instead synthesized during high voltage electric discharges (HVED) in disperse system "hydrocarbon liquid – powder"; preservation of ultrafine structure is ensured due to use of spark plasma sintering (SPS) as a consolidation method. HVED in disperse system "hydrocarbon liquid – powder" due to impact of plasma discharge channel, electromagnetic fields, shock waves mechanical impact, hydro flows and volume microcavitation leads to synthesis of nanocarbon, metal powders dispersion and synthesis of micro- (from 10^{-6} to 10^{-7} m) and nanosized (from 10^{-7} to 10^{-9} m) composite powders of hardening phases. SPS is the passage of pulsed current (superposition of direct and alternating current) through powder with the simultaneous mechanical compressing. The formation of plasma is initiated in gaseous phase that fills gaps between particles. SPS allows targeted control of grain growth rate and thus allows obtainment of multifunctional composite materials dispersion hardened by nanoparticles. Processes of HVED synthesis of micro- and nanosized powders of new compositions from elemental metal powders and their mixtures with the subsequent application of high-speed SPS of obtained powders create conditions for increase of strength (by 10 – 20 %), hardness and wear-resistance (by 30 – 60 %) of obtained materials.

1. Introduction

Nuclear energetics is one of the most dynamic sectors of the world economy. Each new generation of nuclear plants becomes more efficient and safe, but demands the use of materials that have high strength, wear-resistance and are capable of working in aggressive environments. Thus, development of construction materials for nuclear plants is important scientific and technical task.

Creation of heterogeneous structure in material, which is a plastic matrix with hard inclusions, is a necessary condition to ensure materials strength and wear-resistance. Heterogeneous materials include metal-matrix composites of Fe – Ti – C system, obtained with methods of powder metallurgy. These materials have high strength, hardness, their wear resistance is close to tungsten-containing alloys and their modification by boron carbide leads to increase in corrosion and radiation resistance [1].

Present methods for obtainment of highly durable and wear resistant materials in powder metallurgy need a new complex approach, which would unite the processes of powders preparation with methods for their effective consolidation.

It is known that high voltage electric discharge (HVED) in disperse system "hydrocarbon liquid – powder" is used for dispersion, activation and synthesis of bimodal micro- (from 10^{-6} to 10^{-7} m) and nanosized (from 10^{-7} to 10^{-9} m) composite powders. HVED in disperse system "hydrocarbon liquid – powder" leads to a synthesis of nanocarbon due to pyrolysis of liquid, and the impact of plasma channel, electromagnetic fields, mechanical impact of shock waves, hydro flows and volume microcavitation leads to dispersion of metal powders and formation of their carbides [2, 3]. SPS of composite powders allows to control the targeted grains growth rate and thus to form microheterogeneous structure of multifunctional composite materials with high physical-mechanical and performance properties.

New approach in developed plasma methods consists in that dispersion-hardening additives are not mechanically added in powder mixture as additional component during powders mixing, like in conventional methods, but are instead synthesized during powders HVED treatment because of reactive synthesis under the impact of microplasma discharges, and SPS consolidation ensures preservation of ultrafine structure.

The aim of present work is an analysis of possibilities of HVED and SPS plasma methods with means of obtaining multifunctional composite materials, dispersion-hardened by nanoparticles, based on Fe – Ti – C system.

2. Object and methods of studies

HVED treatment of powders of initial mass composition 75 % Fe – 25 % Ti, 50 % Fe – 50 % Ti, 25 % Fe – 75 % Ti and 75 % Fe – 20 % Ti – 5 % B_4C was performed on experimental stand, which is described in detail in paper [2]. Galvanic contact of point-electrode with layer of processed powder was ensured during treatment. Specific treatment energy of micropowders of different mass composition was chosen in accordance to condition of full carbidization of titanium to obtain blend of Fe – Ti – C system with hardening TiC phase content from 25 to 75 % (see table 1) [4, 5].

Table 1. Powder mixtures treatment regimes.

Number of regime	Mixture composition	Specific treatment energy (MJ/kg)
1	75 % Fe – 25 % Ti	6.25
2	50 % Fe – 50 % Ti	12.5
3	25 % Fe – 75 % Ti	18.75
4	75 % Fe – 20 % Ti – 5% B_4C	6.25

Computer granulometric and X-ray diffraction analysis were used to study the impact of HVED on dispersity, form and phase composition of powder mixtures [6].

Specific electrical resistance of powder mixtures before and after HVED treatment was determined on experimental stand based on MCP "BR 2820" LCR METER device, which is described in detail in paper [2].

Consolidation of powder mixtures was performed by method of spark-plasma sintering on experimental complex "Gefest – 10", which allows to consolidate powders at mechanical load in vacuum with superposition of direct and alternating currents with amplitude of 1.1 kA and frequency of alternating component of 10 kHz [7, 8].

IOP Conf. Series: Materials Science and Engineering **130** (2016) 012048 doi:10.1088/1757-899X/130/1/012048

Hardness of consolidated specimens was measured according to [9]. Wear resistance was studied on friction machine СМЦ – 2 (SMC – 2) with roller – block scheme. Diamond circle 1A1 with AC4 80/63 grain was used as counter-body. Mechanical properties of obtained specimens were determined on P-5 (R-5) research machine by TRS method of determining strength limit [10].

Physical-mechanical properties of obtained specimens were compared with properties of P6M5 (R6M5) steel.

3. Results of experimental studies

Initial Fe and Ti powders had similar granulometric composition. Mean diameter of Fe particles was ~ 59 μm, mean diameter of Ti particles was ~63 μm and mean diameter of boron carbide particles was ~ 5 μm. Size distributions of initial mixtures particles are shown on figure 1.

HVED treatment of powder mixtures leads to more than 3 times decrease of mixtures particles mean diameters (to ~ 20 μm) (see figure 2).

Comminution efficiency of Fe – Ti system powder mixtures does not depend on their composition despite the fact that strength of pure Ti is much higher than of Fe. Most likely this is due to embrittlement of Ti because of reaction with H, which is synthesized during pyrolysis of hydrocarbon liquid.

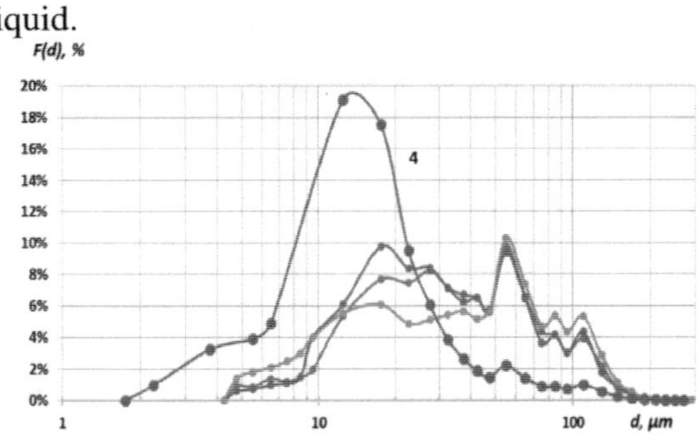

Figure 1. Size distribution of initial powder mixtures particles: 1 – 75 % Fe – 25 % Ti; 2 – 50 % Fe – 50 % Ti; 3 – 25 % Fe – 75 % Ti; 4 – 75 % Fe – 20 % Ti – 5 % B_4C.

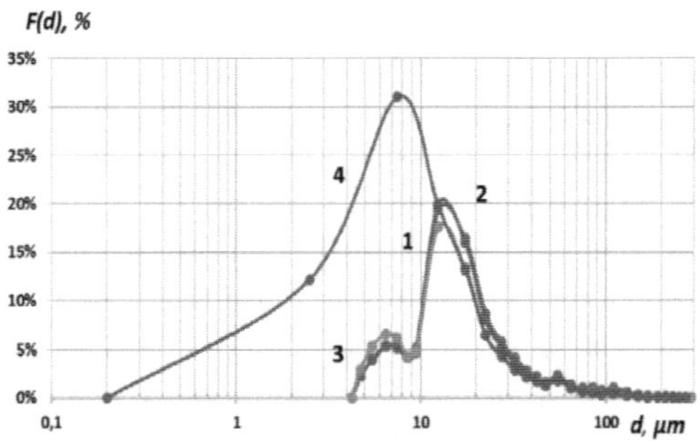

Figure 2. Size distribution of powder mixtures particles after HVED treatment: 1 – 75 % Fe – 25 % Ti; 2 – 50 % Fe – 50 % Ti; 3 – 25 % Fe – 75 % Ti; 4 – 75 % Fe – 20 % Ti – 5 % B_4C.

Initial mean diameter of powder mixture that contains boron carbide was ~ 22 μm (see figure 1, line 4) and after HVED treatment it decreases to ~ 17 μm, while quantity of particles with diameter more than 30 μm is less than 5 % (see figure 2, line 4 and figure 3).

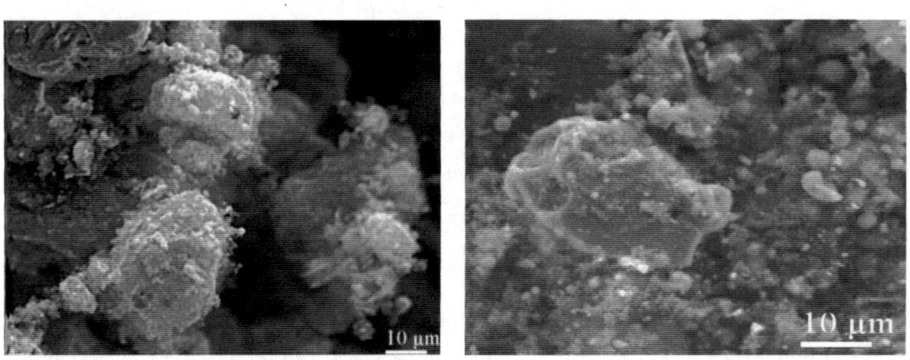

a *b*

Figure 3. Electron microphotographs of powder mixtures after HVED treatment: *a* – 75 % Fe – 25 % Ti; *b* – 75 % Fe – 20 % Ti – 5 % B_4C.

X-ray diffraction analysis of powder mixtures after HVED treatment showed and confirmed the presence of hydrides, carbides and some amount of residual Ti in mixtures composition. Presence of residual Ti indicates that part of nanocarbon reacted with Fe forming Fe_3C. Destruction of boron carbide during HVED treatment in considered regimes have not happened (see figure 4).

Formation of Fe_3C, Fe_2Ti and TiC with size in range from 100 to 600 nm, which are dispersion-hardening phases, occurs in all considered treatment regimes (see figures 3 and 4) [11].

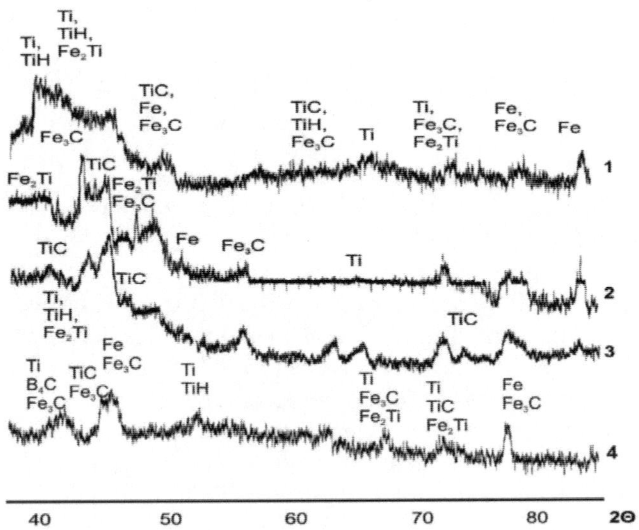

Figure 4. Phase composition of mixtures after HVED treatment: 1 – 75 % Fe – 25 % Ti; 2 – 50 % Fe – 50 % Ti; 3 – 25 % Fe – 75 % Ti; 4 – 75 % Fe – 20 % Ti – 5 % B_4C.

Composition of powder mixtures has significant impact on their specific electrical resistance, which increases in proportion to Fe content from ~ 20 Ω·m (for 25 % Fe – 75 % Ti) up to ~ 610 Ω·m (for Fe – 25 % Ti) even while specific electrical resistance of pure Fe powder is significantly lower than of Ti. This can be explained by the presence of oxide films on surface of powder particles, which greatly reduce their conductivity. Addition of 5 % B_4C, which is a semiconductor, leads to a ~ 2 times increase of specific electrical resistance (up to ~ 1100 Ω·m). Change of powders dispersity and phase composition after HVED treatment leads to decrease of specific electrical resistance by ~ 4 orders of magnitude for 75 % Fe – 25 % Ti (to ~ $1{,}6\cdot10^{-2}$ Ω·m) and 50 % Fe – 50 % Ti (to ~ $8\cdot10^{-2}$ Ω·m) powder mixtures, and to decrease by ~ 2 orders of magnitude (to ~ 0,12 Ω·m) for 25 % Fe – 75 % Ti powder mixture. Specific electrical resistance of mixture with addition of boron carbide decreases by ~ 5 orders of magnitude (to $1{,}6\cdot10^{-2}$ Ω·m) (see figure 5, a). Change of specific electrical resistance occurs due to increase of contact surface and destruction of oxide films, which is confirmed by results of X-ray diffraction analysis.

Mechanical load on f powder leads to decrease of specific electrical resistance of blend – as the pressure was increased to 10 MPa specific electrical resistance of the mixtures decreases ~ 40 times (see figure 5, a), and at pressure of 60 MPa it decreases by 3 – 4 orders of magnitude.

Mechanical load on blend during SPS leads to its shrinkage, which occurs abruptly when pressure increases up to 10 MPa and totals from 20 to 30% due to the removal of the arch effect (stable large voids) in the filling (see figure 5, b), and then shrinkage increases monotonically and at a pressure of 60 MPa is from 35 to 45%, due to plastic deformation of the particles. Increase of mechanical pressure over 60 MPa is inappropriate because it is limited by strength of matrix.

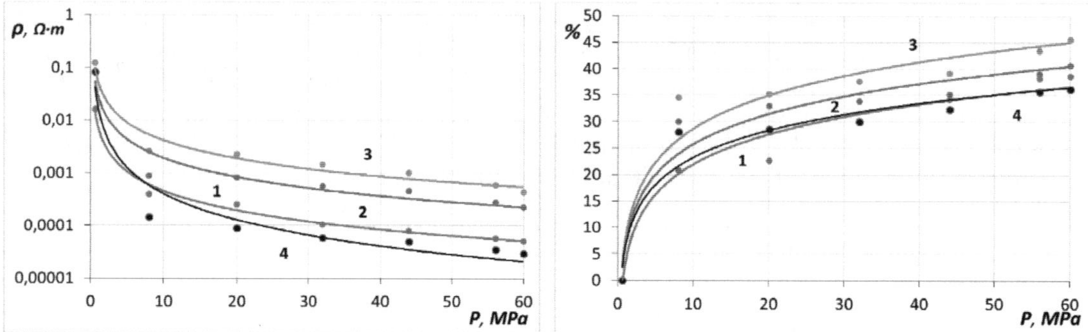

a *b*

Figure 5. Change of specific electrical resistance (*a*) and shrinkage (*b*) of HVED processed powders under mechanical pressure: 1 – 75 % Fe – 25 % Ti; 2 – 50 % Fe – 50 % Ti; 3 – 25 % Fe – 75 % Ti; 4 – 75 % Fe – 20 % Ti – 5 % B_4C.

Analysis of dispersity, phase composition and specific electrical resistance of powder mixtures after HVED treatment allows choosing the regime of their SPS consolidation. Fe – Ti system phase diagram [12] shows that in area of Ti concentration near 53 – 77 % there is a low-temperature eutectic zone with melting temperature of 1085 °C. If the mixture is heated over given temperature the melt will consist of FeTi intermetallic and Ti. If nanocarbon, which could not react during HVED, is present in system, then, given the high affinity of carbon to titanium and its solubility in iron, favorable conditions for active interaction of components with synthesis of Fe and Ti carbides are created during sintering.

SPS regime was chosen in accordance to phase diagram – 1100 °C temperature with subsequent isothermal holding during 3 minutes, which allows obtaining materials with high density and wear-resistance (see figure 6).

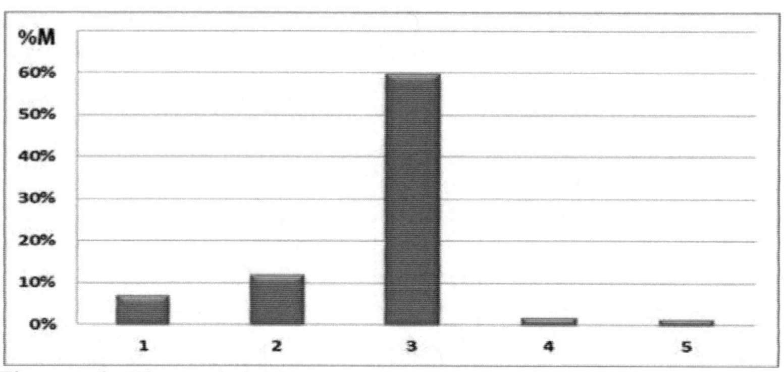

Figure 6. The weight loss of the samples in contact with the diamond wheel 1A1 of AC4 80/63 grain: 1 – sample, consolidate from HVED treated mixture № 1 (see table 1); 2 – from mixture № 2; 3 – from mixture № 3; 4 – from mixture № 4; 5 – P6M5 (R6M5) steel.

Samples of initial mass composition 25 % Fe – 75 % Ti show the highest mechanical characteristics after SPS among all considered mixtures of Fe – Ti – C system after HVED treatment. Their bending strength was 800 MP, hardness was 48 HRC, and weight loss of the samples in contact with the diamond wheel 1A1 of AC4 80/63 grain was ~ 7 % of sample mass for 1 km of way. Increase

IOP Conf. Series: Materials Science and Engineering **130** (2016) 012048 doi:10.1088/1757-899X/130/1/012048

of carbide content in system composition leads to their concentration across grain boundaries and thus to embrittlement of material (see figure 7).

Addition of 5% B_4C to mixture leads to its destruction during SPS and formation of Fe and Ti borides in consolidated samples, which significantly increase their mechanical properties: weight loss was ~ 1.8 % (which is comparable to wear resistance of tungsten-containing steel P6M5(R6M5) – weight loss of ~ 1.4% for 1 km), bending strength increased up to 1100 MPa.

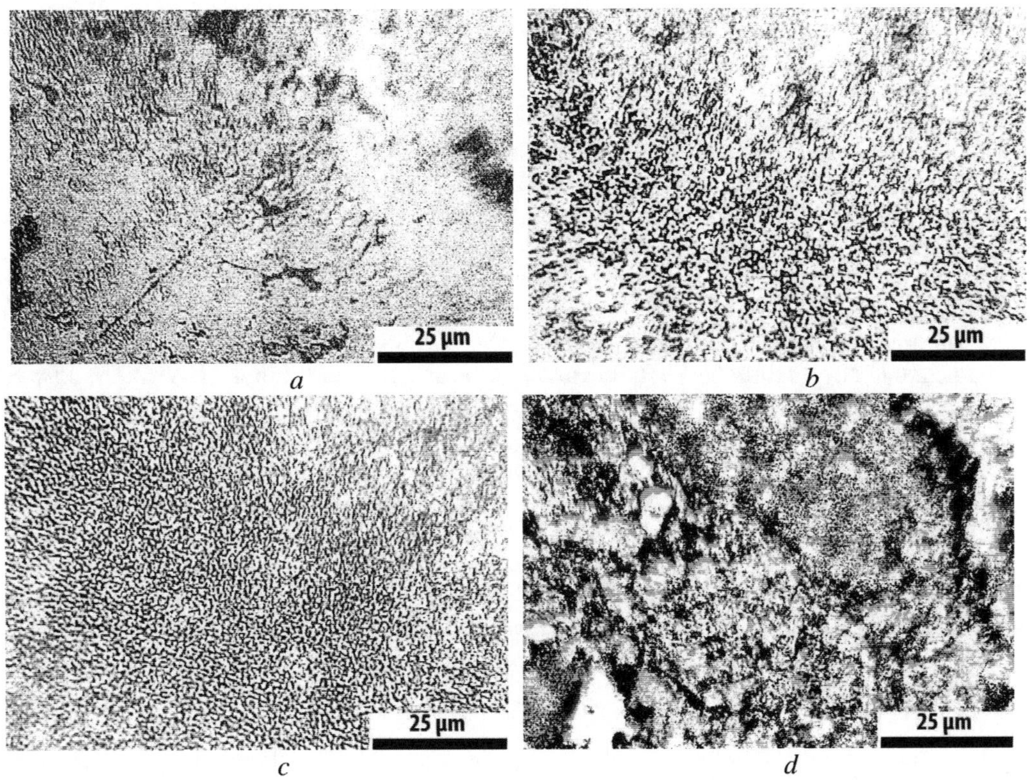

Figure 7. Microstructures of consolidated materials. Magnification ×250: *a* – sample, consolidate from HVED treated mixture № 1 (see table 1); *b* – from mixture № 2; *c* – from mixture № 3; *d* – from mixture № 4.

4. Conclusions

1. It is shown that plasma methods of HVED powders preparation for consolidation and their SPS allows obtaining multifunctional dispersion-hardened by nanostructured particles composite materials, based on Fe – Ti – C – B system.

2. Use of HVED in disperse system "powder – hydrocarbon liquid" allows performing dispersion of particles (mean diameter of Fe and Ti powders decreases ~ 3 times) and synthesis of TiC, Fe_2Ti and Fe_3C dispersion-hardening additives with size from 100 up to 600 nm in them.

3. Consolidation of blend after HVED treatment by SPS method allows obtaining material of Fe – Ti – C system with bending strength of 800 MPa and weight loss during contact with diamond circle of 7 % for 1 km.

4. Modification of Fe – Ti – C system by boron carbide leads to increase of bending strength up to 1100 MPa and to decrease of weight loss to 1.8 %, which is comparable to wear resistance of P6M5 (R6M5) tungsten-containing steel.

IOP Conf. Series: Materials Science and Engineering **130** (2016) 012048 doi:10.1088/1757-899X/130/1/012048

Acknowledgements
This work was performed within the framework of the Center of Nuclear Systems and Materials supported by MEPhI Academic Excellence Project (contract № 02.a03.21.0005, 27.08.2013).

References
[1] Risovanyj V D, Zaharov A V, Klochkov E R and Guseva T M 2011 *Bor v jadernoj tehnike* (Dimitrovgrad: GNC NIIAR) Russian
[2] Sizonenko O N, Pristash N S, Taftaj E I, Torpakov A S, Zajchenko A D, Lipyan E V and Zhdanov A A 2013 *Visnyk ukrayins'koho materialoznavchoho tovarystva* **6** 104 (in Russian)
[3] Sizonenko O N, Taftaj E I, Pristash N S, Zaichenko A D and Torpakov A S 2014 *Elektrychni kontakty ta elektrody* 129 (in Russian)
[4] Kuskova N I, Yushchishina A N, Malyushevskaya A P, Tsolin P L, Petrichenko L A and Smal'ko A A 2010 *Surface Engineering and Applied Electrochemistry* **46** 149
[5] Syzonenko O M, Taftai E I, Raichenko O I, Bahliuk H A, Torpakov A S, Lypian Y V and Zajchenko A D assignee 2012 Mar 26 *Ukraine patent UA 97890*
[6] Zaichenko A D 2015 *Udoskonalennya protsesiv elektrorozryadnoyi obrobky shykhty ta iskro-plazmovoho spikannya dlya otrymannya vysokoznosostiykykh karbidostaley* (Mykolaiv: Admiral Makarov National University of Shipbuilding) (inUkrainian)
[7] Sizonenko O N, Bagljuk G A, Rajchenko A I, Ivliev A I and Lipyan E V 2009 *Visnyk NTU KhPI* **39** 177 (in Russian)
[8] Sizonenko O N, Ivliev A I and Grigor'ev E G assignee 2015 Feb *Ukraine patent UA 107729*
[9] Euro-Asian Council for Standardization, Metrology and Certification 2001 *GOST 9013–59 (ISO 6508–86) Metals Method of measuring Rockwell hardness* (Moscow: Izdatel'stvo standartov) Russian
[10] Khaleghi E, Lin Y-S, Meyersb M A and Olevsky E A 2010 *Scr. Mater.* **63** 577
[11] Sizonenko O N, Baglyuk G A, Taftai E I, Zaichenko A D, Lipyan E V, Torpakov A S, Zhdanov A A and Pristash N S 2013 *Powder Metall. Met. Ceram.* **52** (491) 11
[12] Ljakishev N P, ed 1997 *Diagrammy sostojanija dvojnyh metallicheskih sistem* vol 2 (Moscow: Mashinostroenie) (in Russian)

IOP Conf. Series: Materials Science and Engineering **130** (2016) 012049 doi:10.1088/1757-899X/130/1/012049

Structural transformations in the α-tricalcium phosphate powders after mechanical activation and subsequent heat treatment

V F Shamray, A E Karpihin, A Y Fedotov, V S Komlev, S M Barinov and V P Sirotinkin

[1] Institution of Russain Academy of Science A.A. Baikov Institute of Metallurgy and Material Sciences of RAS, Leninskiy av. 49, Moscow, 119991, Russia

E-mail: shamray@imet.ac.ru, antigop90@rambler.ru

Abstract. X-ray diffraction analysis of the structural phase evolution in the α-TCP powder subjected to milling in the planetary mill in butanol, and followed by annealing at 1300 °C was performed. After milling, there was a systematic smoothing of the X-ray diffraction pattern of α-TCP. Significant changes in lattice constants and atomic parameters were not observed. The average particle size was reduced from 10 to 2 micrometers. The decrease of the size of coherently scattering domains CSD (250 to 80 nm) makes the main contribution to the X-ray lines broadening. The value of microstrain changes from $1*10^{-6}$ to $2*10^{-4}$. After annealing at 1300 °C for 2 hours the phase with apatite structure was found. The content of the apatite phase increases with increasing of time of mechanical activation. Possible mechanisms of deformation of α-TCP particles and structural transformations α-TCP → HA are considered.

1. Introduction

Tricalcium phosphate (α-TCP, α-$Ca_3(PO_4)_2$) represents a large family of phosphate-materials based on bioceramics [1]. On the phase equilibrium diagram CaO-P_2O_5 there are three polymorphs corresponding composition of $Ca_3(PO_4)_2$: α'-TCP, α-TCP and β-TCP. At room temperature β-TCP is stable; at t ~ 1125 °C it turns into α-TCP-phase which is stable in the temperature range 1125-1470 °C. α-TCP is usually prepared by heating β-TCP above α↔β transformations. The structure of α-TCP crystallizes in space group $P2_1/a$ with Z=24 (figure 1a). Due to a good solubility α-TCP is applying as a component of the cement used for filling bone defects including metal biomaterials with bone tissue. When interacting with water α-TCP forms a calcium-deficient hydroxyapatite $Ca_9(PO_4)_5(HPO_4)(OH)$ (Figure 1b) that enhances the biocompatibility of cements. It is well known that the dissolution of α-TCP powders in liquids during the manufacturing of cement is controlled primarily by surface reactions, which depend on the rate of dissolution of the powders. There are different ways to increase the dissolution rate, and the most common - increasing the contact surface of the particles with the liquid by milling. The factor stimulating dissolution of particles, is also a grinding of microblocks. In this regard, very relevant to research the changes of dispersion, phase composition, characteristics of the microblock, parameters of the crystal structure in α-TCP powder produced by milling and subsequent annealing typically used to optimize the structure of α-TCP powders. The purpose of this work is the study by X-ray diffraction analysis of structural changes in α-TCP powder, which was

IOP Conf. Series: Materials Science and Engineering **130** (2016) 012049 doi:10.1088/1757-899X/130/1/012049

treated in a planetary mill, and subsequently annealed at temperatures corresponding to its stability on the diagram $CaO-P_2O_5$.

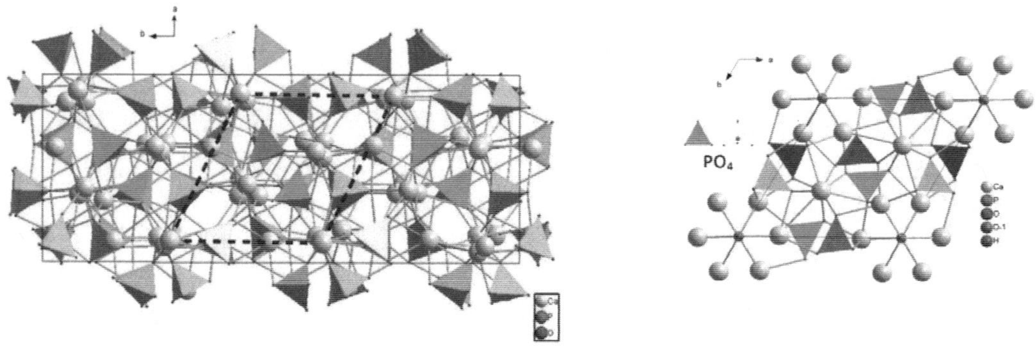

a b

Figure 1. Structure of α-TCP (a) and hydroxyapatite (HA) (b).

2. Experimental

Single-phase α-TCP powder was processed in a planetary mill for 15, 30, 60, 120 and 180 minutes in butanol at 60 °C and subsequent annealing at 1300 °C for 2 hours. Changes in the structure of α-TCP and microblock structure of powders were investigated using the X-ray diffractometer Rigaku Ultima IV in Cu-Kα radiation. The phase analysis is performed by means a software package PDXL. Refinement of the crystal structure parameters and microblock structure by the Rietveld method performed using software packages Jana 2006 and MAUD in the two-phase approach. As result, the volume fractions of α-TCP and phases formed during processing in a planetary mill was obtained.

Particle size determination by laser diffraction method was performed. The automatic unit Hydro 2000S and isopropanol alcohol as a dispersant were used.

1. Results and discussion.

In the X-ray pattern of the raw α-TCP powder, there are only reflections corresponding to α-TCP phase with a=1.2878, b=2.7271, c=1.5215 nm, β=126.19° (Figure 2).

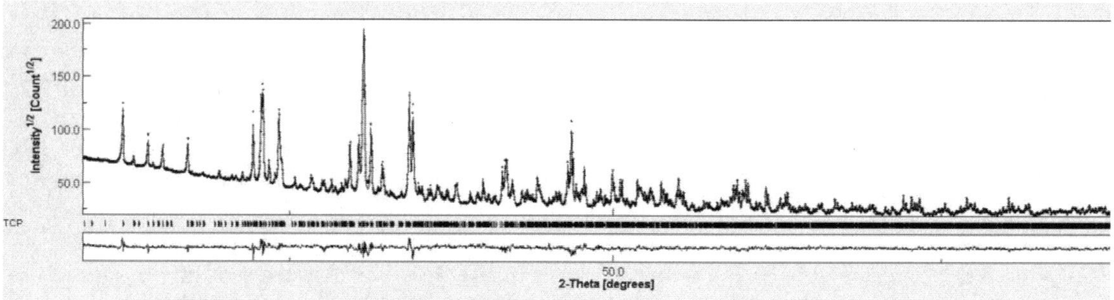

Figure 2. The XRD diffraction pattern of the raw α-TCP powder.

The best correspondence of the diffraction peaks on X-ray diffraction pattern and their intensities and 2Θ angles calculated during the refinement of the structure of α-TCP by the Rietveld method was observed with the card ICCD №70-364. Refinement the general parameters (background, shape parameters of lines, their asymmetry, texture March-Dollase parameters) atoms positions and thermal parameters was performed. As an initial atomic coordinates the values given in [2] were used. Refinement of Ca and P coordinates has not resulted in a significant change in their values. Coordinates of oxygen atoms were not refined. In general, the refining structure procedure completed

IOP Conf. Series: Materials Science and Engineering **130** (2016) 012049 doi:10.1088/1757-899X/130/1/012049

with R_{wp}= 6.25 (Table 1). The broadening of the lines of the XRD pattern is weak, and by *hkl* essentially does not depend. Subsequently, this sample was used as a standard for estimation of microblock values of powders after mechanical activation.

Table 1. The unit cell parameters α-TCP and the main characteristics of the structure refinement by the Rietveld method.

Parameters specification	Values
a, Å	12,8832
b, Å	27.2837
c, Å	15.2214
β, deg	126.1868
The absolute value of the maximum intensity, imp.	17449
R_{wp} (%)	6.25
R_p (%)	4.62
The number refined the parameters	67
Interval angles 2θ, deg	9-129
Increments for 2θ, deg	0.02
The number of data points	5001
The number of reflections	429

After processing of the powders in the planetary mill a noticeable reduction in grain size was observed (Figure 3). The mean particle diameter, equivalent specific surface area of the grain (D) after treatment for 180 minutes decreased 3.5 times. Such reduction in D and thus increasing their specific surface area leads to an appreciable reduction of solidification time cement [3].

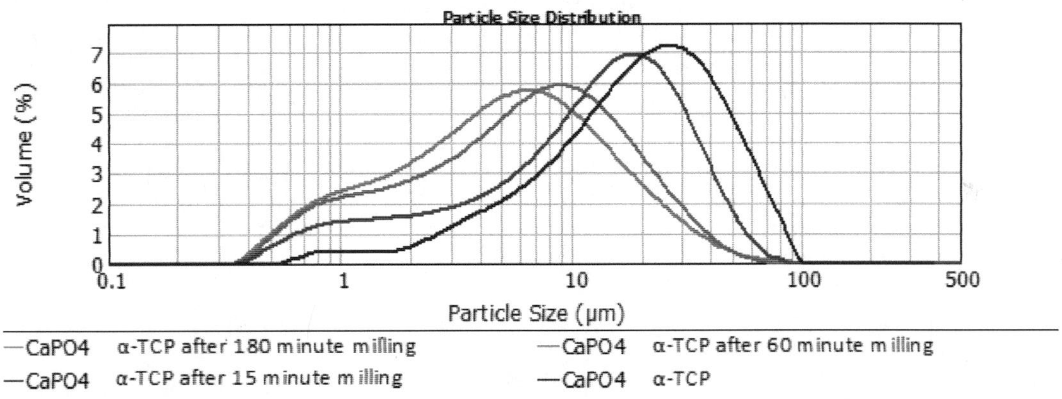

Figure 3. Dimensional distribution of α-TCP particles in size depending on milling time.

In the XRD patterns of powders after milling, the peaks of α-TCP phase are significantly broadened (Figure 4). The relatively weak reflections of $(CF_2)_n$ phase appear, apparently as a result of interaction powders with the material of the walls of the mill was observed. These results indicate that a significant change of the unit cell values and atomic parameters not happened. Shape anisotropy of microblocks and microstrain were not found. The values of the average size CSD and microstrain (Δ) after the processing of the powder for 60 minutes, where the effect of broadening of X-ray lines was maximal, were respectively: $d \approx 80$ nm; $\Delta \approx 2*10^{-4}$ (Table 2). The main contribution to the broadening of the X-ray peaks makes decrease of CSD. Just as in the case of hydroxyapatite, α-TCP plastic deformation perhaps describes as a processes of cluster fragmentation and cluster-boundary "sliding", the contribution of the plastic deformation by dislocation mechanism is negligible [4].

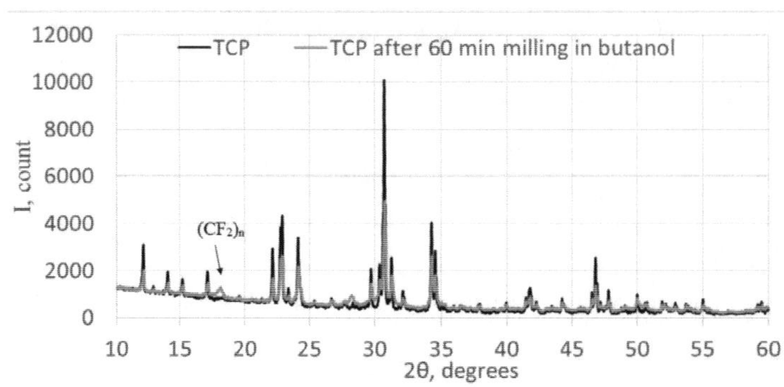

Figure 4. The broadening of the lines α-TCP after 60 min of milling in butanol

The XRD patterns of the samples after treatment in the mill and subsequent annealing at 1300 °C, as it was stated above, contains, as the main phase α-TCP. In addition, the reflexes apatite phase (a=0.9393, c=0.688 nm) observed. This is most clearly evident in the diffraction XRD pattern of the powder subjected to milling for 180 minutes. From the results of the quantitative phase analysis carried out with the plotting of the calibration curve, it was found that the contents of the apatite phase in powders after mechanical activation and annealing for 60, 120 and 180 min is 47, 58 and 75% respectively. The atomic parameters of apatite phase obtained in refining the structures correspond to the data of [2, 6].The diffraction spectra of powders were characterized by sharp peaks; microblocks and microstrains contributions in their width is minimized.

Table 2. Content of the phases, the size of CSD of α-TCP, microstrains of α-TCP (Δ) and the refinement factor (R_w) depending on the time of milling of α-TCP (t)

t, min	Content of the phase (%)		CSD (nm)	Δ	R_w (%)
0	αTCP 100		287	2,00E-04	6,25
15	αTCP 98,6	CF_2-1,4	172	6,56E-06	6,80
30	αTCP 97,8	CF_2-2,2	131	1,89E-05	7,20
60	αTCP 95,8	CF_2-4,2	79	2,72E-04	8,92
120	αTCP 95,3	CF_2-4,7	127	2,29E-04	6,10
180	αTCP 94,4	CF_2-5,6	123	2,19E-04	6,40
		After annealing			
60	αTCP 52,6	HA-47,4	222	3,15E-04	9,54
120	αTCP 42,2	HA-57,8	210	2,95E-04	6,54
180	αTCP 24,9	HA-75,1	202	2,34E-04	8,08

The presence in the α-TCP powder apatite phase hydroxyapatite (HA) is usually attributed to the presence of moisture in the air. Therefore, the synthesis under these conditions using phase diagram CaO-P_2O_5-H_2O, which includes phase regions comprising apatite [5]. The transition of α-TCP→HA facilitated by the similarity of the crystal structures, specifically, peculiarity of fragments structure and method of their packing.

The structure of α-TCP consist of calcium and phosphate ions arranged in the two types of columns along the *c* axis, one contains only the calcium, other calcium and PO_4 tetrahedron [2]. Each calcium column is surrounded by six calcium-phosphate columns (Figure 5a).

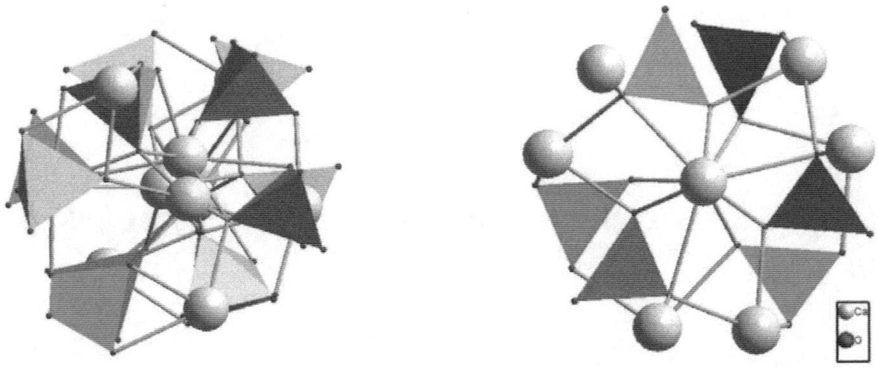

a b

Figure 5. Fragment cluster of structure α-TCP (a) and apatite (b)

However, if these columns in α-TCP are distorted; forming their calcium atoms are generally displaced in a *ab* plane with respect to the axis *c*, the apatite lattice arrangement of calcium atoms is more regular (figure 5b). In the lattice of α-TCP, it is possible to select the cell (figure 1a) in which the columns are packaged in a manner characteristic for HA. In the process of α-TCP in HA transition one of the main features of the cation-anion columns of α-TCP, calcium deficiency, may be compensated by displacement of cations Ca^{2+}.The calcium located at the vertices of this cell displaced in the *a,b* directions (Figure 1a).

3. Conclusion

As a result of milling α-TCP powders in planetary mill in butanol the average particle diameter *D* decreases in 3.5 times (10 to 2.5 micrometers). According to the literature this reduction of the particle size produces a substantial decrease of the setting time and accelerates the hardening of the cement without significantly affecting the final strength. Relatively weak decrease of microblocks dimension (CSD) to ~ 80 nm, presumably explained by the specific processes of deformation of the particles in the liquid medium and the material of the vessel. The broadening of the X-ray reflections is determined mainly by decreasing of CSD. The contribution of dislocation type of plastic deformation is small. Synthesized after mechanical activation and subsequent annealing at 1300 °C biphasic powders (HA + α-TCP) can be considered as two-component biomaterials with controlled solubility.

References

[1] Barinov S M and Komlev V S 2005 Bioceramics based on calcium phosphates (Moscow: Science) p 204
[2] Mathew M, Schroeder L W, Dickens B and Brown W E 1977 *Acta Cryst* **B33** 1325–33
[3] Ginebra M P, Driessens F C M and Planell J A 2004 *Biomaterials* **25** 3453–62
[4] Ievlev V M 2013 *Russian Chemical* **82 (2)** 131
[5] Raúl G Carrodeguas and Salvador De Aza June *Acta biomaterialia* **7** 3536–46
[6] Shamrai V F Sirotinkin V P Karpihin A E Kalita V I and Komlev D I 2014 *Crystallography* **59** 231

Properties of tungsten heavy alloys, prepared by spark-plasma sintering

N S Ermakova, M S Yurlova and E G Grigoryev

National Research Nuclear University MEPhI (Moscow Engineering Physics Institute) Kashirskoe highway 31, Moscow,115409, Russia

E-mail: yurmary@mephist.ru

Abstract. In this paper the effect of spark-plasma sintering parameters on the microstructure and mechanical properties of tungsten heavy alloys VNZHK-90 and VNZHK-93 are studied. The basic dependences of the density and strength characteristics from the sintering temperature and velocity are shown. Conclusions about the dependence of the method of preparation of the starting powders on the microstructure of alloys are formulated. It was found that by spark-plasma sintering technique it is possible to achieve the theoretical density of compact heavy alloys during solid-phase sintering. It is shown that the maximum level of alloys mechanical properties is observed when sintering temperature 1300 ° C. It is found that the change in the rate of heating does not contribute to change in the density of the samples. At the same time, on the mechanical properties the shrinkage rate has a significant effect.

1. Introduction

Heavy tungsten alloys (HTA) are widely used in various fields and have a number of valuable properties, chief among which is the ability to maintain a high hardness at elevated temperatures. The most widely used alloys VNZH (W-Ni-Fe), VNZHK (W-Ni-Fe-Co), VM (W-Cu), in which the tungsten content ranges from 90 to 98 weight percent. Nickel, iron and copper serve as a binder matrix which holds brittle tungsten grains together, thereby making alloys ductile and easily machined.

The heavy tungsten-based alloys are used as protection from radiation exposure, is widely used in the nuclear industry, in the missile technologies, as components of ammunition. HTA produced by powder metallurgy, in particular, a method of electro-plasma sintering. The properties of the alloy obtained by this method are very different, depending on the choice of the sintering conditions.

The aim of this work was to study the effect of laws modes spark-plasma sintering on the microstructure and mechanical properties of alloys VNZHK-90 and VNZHK-93.

2. Experimental results

2.1. Characteristics of sintered powders

As starting materials for the production of dense compacts using two kinds of powders VNZHK for which data is given in Table 1.

Spark plasma sintering, tungsten powders was conducted at the facility LABOX ™ Model 625 (SinterLand, Japan).

Table 1. Data on the source alloy powder VNZHK-90 and VNZHK-93.

Powder		VNZHK -90	VNZHK -93
A method for producing a powder		Mechanical mixing of metal powders W, Ni, Fe, Co	Joint reduction of oxides W, Ni, Fe, Co
The average grain size according to Fisher D_f, μm		4.5	1.6
Bulk density, g/cm^3		3.92	1.7
Composition, wt. %	W	other	other
	Ni	7.6	5.6
	Fe	1.7	1.4
	Co	0.97	0.73
	O_2	-	0.25

The microstructure of the powders obtained by a scanning electron microscope is shown in Figure 1. It is seen that alloy powders VNZHK 90 consist of mechanical mixtures of the individual particles. VNZHK-93 is a sponge-like structure and highlight the department-particles of different composition impossible. Spongy structure explains the low bulk density of the powder.

(a) (b)

Figure 1. The microstructure of the powders obtained by a scanning electron microscope: (a) – VNZHK-90, (b) – VNZHK-93.

2.2. Spark plasma sintering of alloys VNZHK

2.2.1. Changing the sintering temperature

Was researched the effect of sintering temperature on the microstructure and mechanical properties. The first series of samples were prepared as follows. Sintered samples were heated to various temperatures linearly from 900 to 1400 °C at a heating rate of 100 °C/min and holding at maximum temperature for 20 minutes. The applied external pressure to the powder was constant at 40 MPa. Parameters of the samples are shown in Table 2.

As seen from the table that the alloy VNZH-90 at a temperature of 850 to 900 °C manifested a marked increase in density. In the alloy VNZHK-93 it takes place only at 1100 °C.

Among the samples 1-1 – 6-1 stands out sample 6-1 (VNZHK-90 heated to 1400 °C), with a maximum density - 18.8 g/cm^3. This value is greater than the theoretical value for the mixture of powders (18.16 g/cm^3). This phenomenon is due to the fact that when the sintering temperature of 1400 °C bundle (Fe, Ni, Co) and softens during pressing occurs squeezing it from the sample.

With an increase in the sintering temperature at the same heating rate significant grain growth occurs and tungsten at 1300 °C the "separation" in pressing zones with large grains of tungsten (light areas in the figure) and a mixture of the binder with tungsten (dark areas). This phenomenon explains the overall increase in the density of alloys with increasing temperature. The microstructures of alloys annealed at 1300 °C at a rate of 100 °C/min are shown in Figure 2.

Table 2. Modes of sintering and the density values for samples obtained by SPS (the first series).

№	Alloy	Mode of sintering (Temperature °C - Heating rate, °C/min)	Density (g/cm^3)	Relative density (% TD)
1-1		900-100	15.2	84
2-1		1000-100	17.2	95
3-1	VNZHK-90	1100-100	15.3	84
4-1		1200-100	17.4	96
5-1		1300-100	16.9	93
6-1		1400-75	**18.8**	100
8-1		900-100	13.3	73
9-1		1000-100	13.0	72
10-1	VNZHK-93	1100-100	**17.3**	95
11-1		1200-100	17.2	95
12-1		1300-100	17.0	94

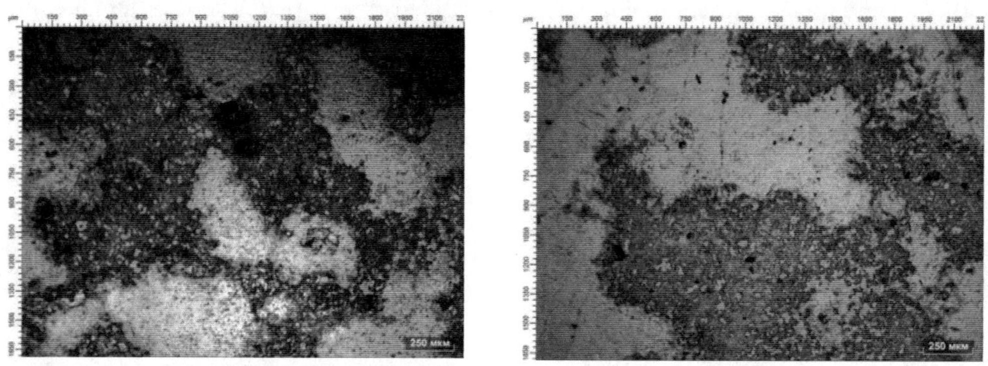

(a) (b)

Figure 2. The microstructure of samples obtained by SPS; T_{sint} = 1300 °C; heating rate 100 °C/min: *a* – VNZHK-90, *b* – VNZHK-93.

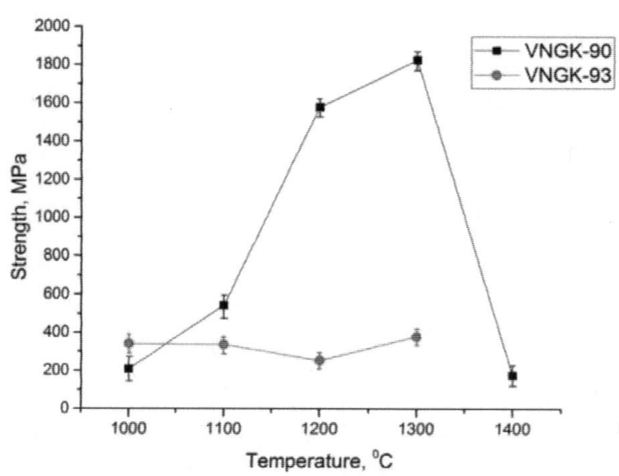

Figure 3. Dependence of tensile strength of the sintering temperature alloys.

For alloy VNZHK-93 at temperatures above 1100 °C increase in density is observed. This is due to the structure of the starting powders. Unlike alloy VNZHK-90, wherein the structure has been represented by individual particles, and as a result, shrinkage proceeded for a longer time, the powder VNZHK-90 had a spongy structure, so the active stage of the shrinkage proceeds at lower temperatures.

Strength tests were conducted. Their results are shown in Figure 3, a graphic image according to the tensile strength of the sintering temperature.

As might be expected, depending on strength agreement with the data on the density of the samples. For alloy VNZHK-90 an increase in the mechanical properties, which is caused by the overall increase in the density of the samples due to the shrinkage of the active phase. The point on the graph corresponds to a sintering temperature of 1400 °C is released in this series. Alloy, corresponding to it, has dramatically different from other low value of strength. This phenomenon is associated with the extrusion binder during sintering, as stated earlier. Extruding binder resulted in that the sample remained mainly tungsten grains, which are known, are fragile, so the sharp decline of strength at high temperatures naturally.

For the VNZHK-93 alloy residence permit significant changes in mechanical properties are observed. This is also consistent with the data on density, since shrinkage at temperatures in excess of 1100 °C does not occur and the mechanical properties values are in the same range.

2.2.2. Changing the heating rate

When manufacturing a second series of samples was fixed sintering temperature (1300 °C), while the process was performed at different heating rates of 10, 15, and 20 °C/min. The holding time and the applied voltage, remained constant at 20 minutes and 40 MPa, respectively. A second series of parameters of the samples are shown in Table 3.

Table 3. Modes of sintering and densities of the samples obtained by SPS.

№	Alloy	Mode of sintering (Temperature, °C - Heating rate, °C/min)	Density (g/cm^3)	Relative density (% TD)
1-2		1300-10	**16.9**	93
2-2	VNZHK-90	1300-15	16.8	93
3-2		1300-20	16.6	91
4-2		1300-10	**17.0**	93
5-2	VNZHK-93	1300-15	16.7	92
6-2		1300-20	16.9	93

As can be seen from Table 3, the maximum value of the density in the alloy VNZHK-90 and alloy VNZHK-93 achieved with a minimum heating rate of 10 °C/min. Dependence of density on the rate of sintering of the alloys is shown in Figure 4.

After analyzing the dependencies, it can be concluded that the density of the alloy sintering rate has no effect.

Figure 5 shows the dependence of microhardness for the alloys VNZHK at different heating rates and T_{sint} = 1300 °C.

As shown in the graphs, the light areas corresponding to the clusters of tungsten grains have significantly higher microhardness values, but with an increase heating rate to 100 °C/min are observed their decrease. This phenomenon can be explained by the fact that at low heating rates are formed tungsten grain size is much greater than at high temperatures.

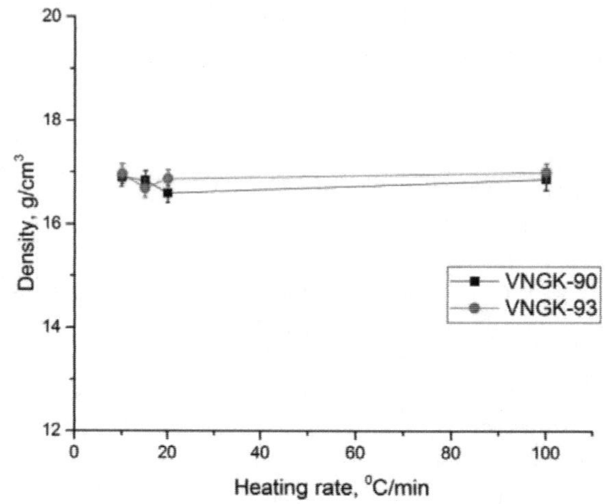

Figure 4. Density dependence on the heating rate of powders.

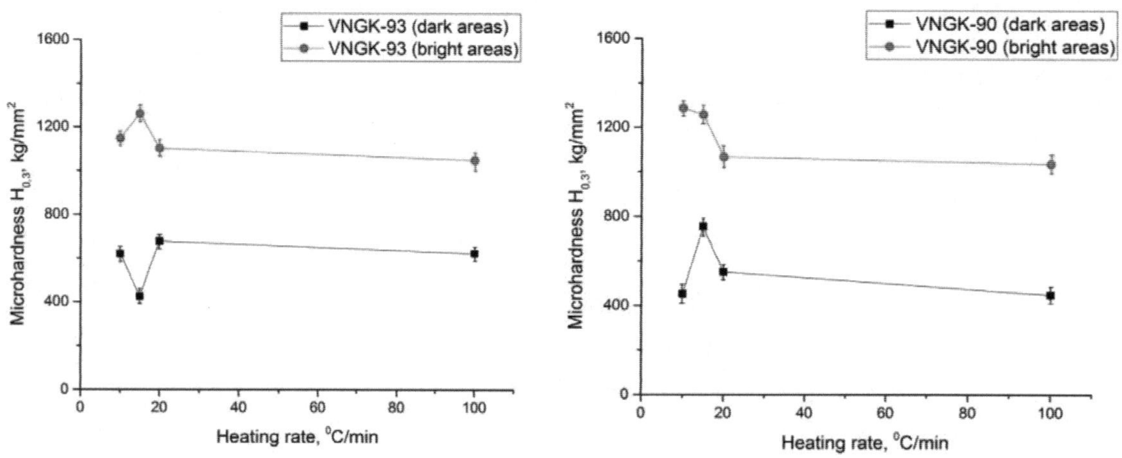

(a) (b)

Figure 5. Microhardness dependence of the rate of heating of the powder: *a* – VNZHK-90, *b* – VNZHK-93.

For dark areas that correspond to areas of Ni-Fe-Co ligaments, a tendency to decrease or increase the microhardness of indicators is not observed, because the contribution to the change in value is making only the grain growth of tungsten.

Figure 6 shows the dependence of the strength alloys VNZHK-90 and VNZHK-93 on the heating rate.

It should be noted that these data are correlated depending on the microhardness of alloys. As expected, with a decrease in the heating rate increases the strength characteristics as coarsens grain tungsten. Also, there is a dramatic difference in the values of tensile strength: for the alloy VNZHK-90 (powders obtained by mechanical mixing of the components), the tensile strength values are significantly higher than for the alloy VNZHK-93 (powder obtained by reduction of oxides of the joint). Thus, the mechanical characteristics of the alloys are significantly affected by a method of producing starting powders.

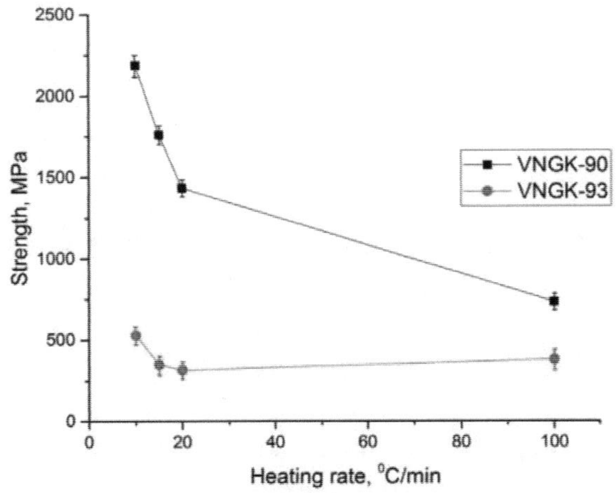

Figure 6. Tensile strength dependence of the rate of heating of the powder.

3. Conclusion

In this paper studied the effect modes spark-plasma sintering on the microstructure and mechanical properties of alloys VNZHK-90 and VNZHK-93. It is found that when spark-plasma sintering achieve the theoretical density of the compacts may heavy alloy by solid phase sintering, as opposed to free sintering at which the maximum density is attained in a melt binder. Was studied the effect of the method of producing powders of sintering kinetics and the final microstructure of the alloy. It is shown that in the case of making a joint reduced mixture of oxides, achieving theoretical density compacts is observed at temperatures up to 200 °C lower than when using a mechanical mixture of powders. It is shown that for the alloys maximum level of mechanical properties when spark-plasma sintering occurs at a sintering temperature of 1300 °C. It is found that the change in the rate of heating does not contribute to change in the density of the samples. At the same time sintering rate has a significant impact on the mechanical properties, due to the more uniform distribution of alloying elements in the more-length-heating. It is shown that the strength of the alloy depends on the morphology of the starting powders. Alloys made from mixtures of powders have high strength values than alloy powders which were obtained by reduction of oxides of the joint.

Acknowledgements

This work was performed within the framework of the Center of Nuclear Systems and Materials supported by MEPhI Academic Excellence Project (contract № 02.a03.21.0005, 27.08.2013).

References

[1] Chuvildeev V, Nokhrin A and Baranov G 2010 *Vestnik Nizhegorodski Univercity* **2** (1) 47 (in Russian)

[2] Povarova K, Alymov M, Gavrilin O, Drozdov A, Kachnov A, Korenovskii N and Bannykh I 2008 *Russian Metallurgy (Metally)* **6** 52

[3] Grigoryev E and Kalin B 2008 *Electropulse technology for creating materials from powders* (Moscow: NRNU MEPhI) p 152 (in Russian)

[4] Boldin M 2012 *Physical basis of the technology of spark-plasma sintering* (Nigniy Novgorod: NGTU) p 59 (in Russian)

[5] Bose A, Sims D and German R 1988 *Metallurgical Transactions March* **3** 487

Finish ion beam treatment of the longrange cylindrical products outer surface in automatic mode

B A Kalin[1], N V Volkov[1], R A Valikov[1], A S Yashin[1], V P Krivobokov[2], S N Yanin[2], O Kh Asainov[2] and Yu N Yurev[2]

[1]National Research Nuclear University MEPhI (Moscow Engineering Physics Institute), Kashirskoe highway 31, Moscow, 115409, Russia
[2] National Research Tomsk Polytechnic University, Lenin Avenue 30, Tomsk, 634050, Russia

E-mail: yashin_itf@mail.ru

Abstract. The results of using of ion-beam technologies methods for finish treatment of metal products are presented. The experiments were performed at the installation ILUR-03, which allows the operation of cleaning, polishing and surface layers doping of the material of unlimited length cylindrical samples by radial Ar^+ ions beam with energy up to 5 keV. The tubes from zirconium alloy E110 up to 500 mm length were used as samples for investigation. It is shown that selected automatic treatment modes reduce the surface roughness over the entire length of the samples and increase uniformity of the surface layer without observable effect on the bulk properties of material. Treatment promotes the formation of oxide films with improved defensive properties.

1. Introduction
Materials resistance to corrosion and wear largely depends on such characteristics of the surface layer of material, such as topography, elemental composition and the presence of impurities, grain size, etc. Wherein, state of the surface is often substantially different from the main volume material. For example, mechanical abrasive treatment, reducing roughness significantly deforms material barrier layer and the chemical etching contributes to the material saturation by difficult to remove impurities [1-3]. In connection with this the using of environmentally friendly and non-contact ion beam technologies for finishing of metal products is interesting.

2. Materials and tools
For the experiments carrying out was used the installation ILUR-03 [4], designed to clean and polish the outer of cylindrical samples surface with unlimited length by radial beam of Ar^+ ions with an energy of 5 keV. The design of installation also provides for local doping of the material product surface layerin the mode of ion mixing films, deposited built magnetrons. The appearance of the installation is shown in Figure 1.

IOP Conf. Series: Materials Science and Engineering **130** (2016) 012051 doi:10.1088/1757-899X/130/1/012051

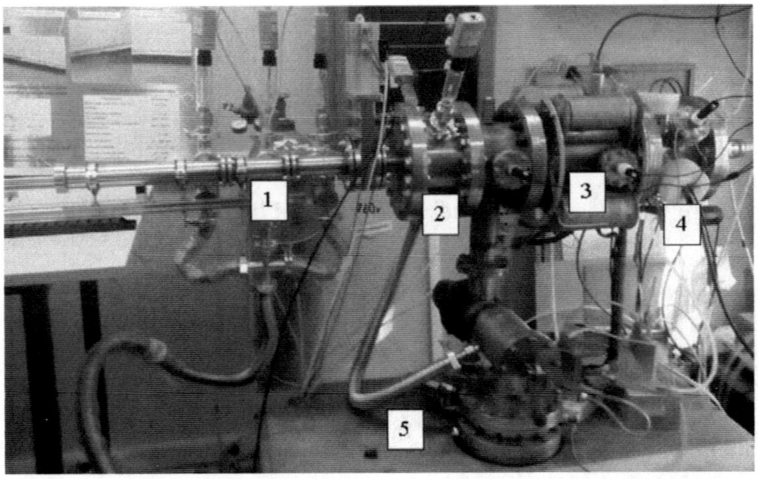

Figure1. Installation for ion-beam treatment of long range products ILUR-03:
1 – gateway chamber, 2 - preliminary pumping chamber, 3 – chamber with ion source, 4 – chamber with magnetrons, 5 – vacuum system desk.

Using an ion beam with a wide energy spectrum allows to increase the processing speed at a constant power load on the sample, as well as to increase the depth of the modified layer what can be seen in Figures 2 and 3, respectively.

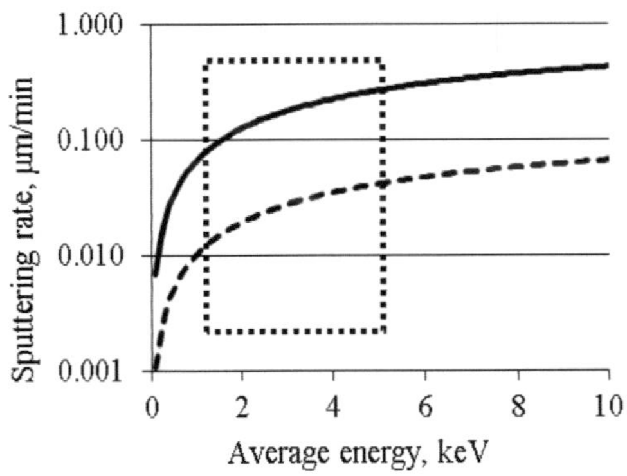

Figure 2. Dependence of sputtering rate of Zr sample on average energy of Ar^+ ion beam with narrow (dotted line) and wide (solid line) energy spectrum (calculationbySRIM-2012): area of ILUR-03 ion beam is shown by dotted rectangle.

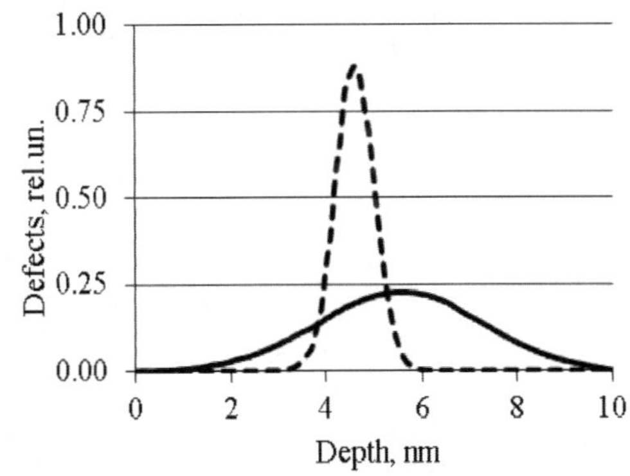

Figure 3. Distribution of defects generated by Ar^+ beam by depth of Zr sample for the ion beam with narrow (3.0-4.0 keV – dotted line) and wide (1.0-5.0 keV – solid line) energy spectrum (calculation by SRIM-2012).

Formed as samples were used fragments of staff claddings pressurized water reactors of E110 alloy (Zr-1% Nb, an outer diameter of 9.15 mm, wall thickness 0.65 mm, length 500 mm) from the factory mechanical polishing the outer surface.

For analysis of the material surface layers was measured roughness (roughness profiler, TR-200), and microhardness (PMT-3, indenter - diamond pyramid with an angle of 136 °) the surface of the samples at various points along the length of the membranes. Several tubes before and after ion treatment were subjected to autoclave tests in the steam-water medium at a temperature of 350 °C and a pressure of 16.5 MPa on a temporary base to 500 hours. The cross structure of the grown oxide are studied by scanning ion microscopy methods (SIM unit Helios-660, a beam of Ga^+, the accelerating voltage of 5 kV).

3. Results and discussion

Figures 4-6 show the SIM- images of samples outer surface in the initial state and after the treatment beam Ar^+ 3×10^{18} cm^{-2} and 9×10^{18} cm^{-2}, respectively. As can be seen, on the surface in the initial state is having a significant amount of both longitudinal and transverse crack with sharp edges (indicated by arrows in Figure 4), the value of the roughness of $R_a = 0.8 \pm 0.1$ μm. With increasing irradiation dose, the number of scratches is reduced, the relief surface becomes smoother character roughness decreases to $R_a = 0.5 \pm 0.1$ μm, while the crystallite boundaries are detected (indicated by arrows in Figure 5).

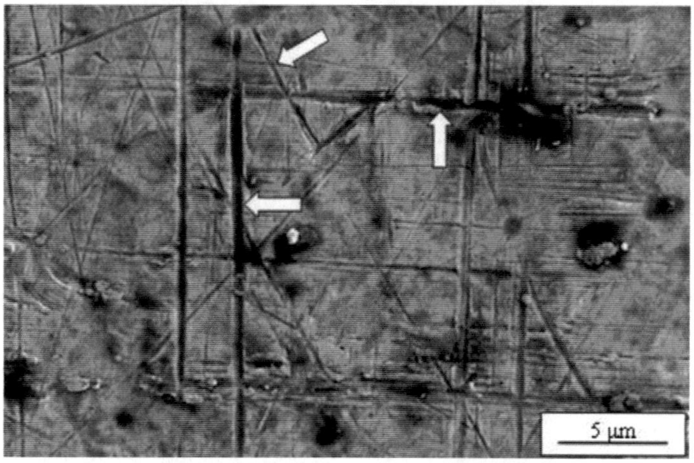

Figure4. SIM-image of the sample from E110 alloy outer surface in initial stage (arrows show the technological scratches).

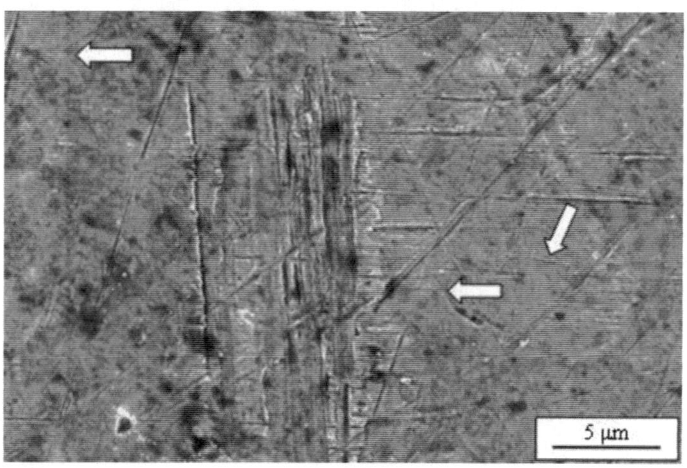

Figure5. SIM-image of the sample from E110 alloy outer surface after treatment by Ar^+ ion beam up to 3×10^{18} ions cm^{-2} (arrows show the crystallite boundaries).

For fluence 9×10^{18} cm^{-2} (Figure 6) vertical drop of neighboring grains, is measured by the photos, amounts to about 0.1 microns with roughness values $R_a = 0.3 \pm 0.1$μm. This value corresponds to the calculated value due to different speed spraying basal and prismatic planes face-centred close-packed lattice of α-Zr. Thus, for the polycrystalline samples surface polishing by ion beam is required to limit the maximum dose, depending of the required final roughness value.

Analysis of the measurement results of the microhardness showed that the matrix treatment of the zirconium alloy by Ar^+ ions beam at modes of cleaning and polishing does not change the average $H\mu$, which is 1.5 GPa, but significantly reduces the scatter in the values of microhardness for sample length, as can be seen from Table 1. The doping of the near-surface layer of the matrix by selected elements (Fe, Al, Ti) at ion mode mixing increases the magnitude of $H\mu$ to 1.6-1.8 GPa while maintaining reduced level of spread of values. The obtained results probably indicate the gradient structure formation on the material surface with high uniformity thickness of about 1 μm (a diamond indenter indentation depth of 4-5 μm).

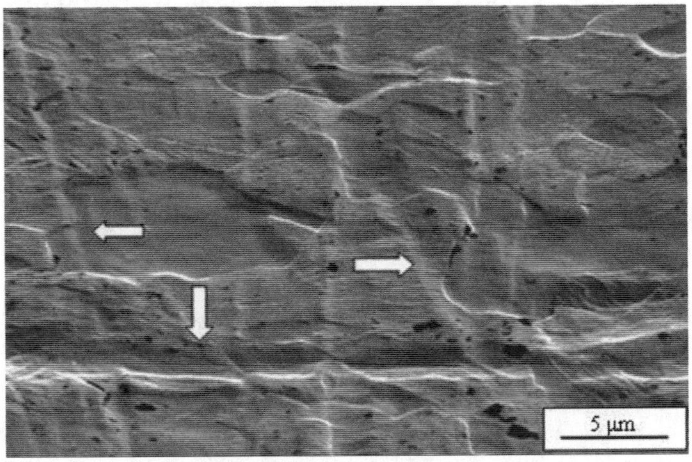

Figure 6. SIM-image of the sample from E110 alloy outer surface after treatment by Ar^+ ion beam up to 9×10^{18} cm^{-2} (shooting at angle 52°, arrows show the smoothed technological scratches edges).

Table 1. Microhardness of tubular samples outer surface.

№	Sample state	H_μ, GPa	Dispersion $\Delta H_\mu/H_\mu$, %
1	initial (mechanical abrasion)	1.5	18
2	Ar^+ treatment up to 5×10^{18} cm^{-2}	1.5	9
3	Ar^+ treatment up to 1×10^{19} cm^{-2}	1.5	15
4	Ar^+ 1×10^{18} cm^{-2} + doping Fe	1.6	11
5	Ar^+ 1×10^{18} cm^{-2} + doping Al	1.8	11
6	Ar^+ 1×10^{18} cm^{-2} + doping Ti	1.6	9

Table2. State of the oxide films on the cladding tubes from E110 alloy outer surface after autoclave tests up to 500 h.

№	Sample state	Oxide film thickness, µm	Pores size and density
1	initial (mechanical abrasion)	0.8±0.1	10±1 % 0.2±0.1 µm
2	Ar^+ treatment up to 5×10^{18} cm^{-2}	0.8±0.1	6±1 % 0.2±0.1 µm
3	Ar^+ treatment up to 1×10^{19} cm^{-2}	0.7±0.2	uniform film
4	Ar^+ 1×10^{18} cm^{-2} + doping Fe	0.4±0.1	uniform film
5	Ar^+ 1×10^{18} cm^{-2} + doping Al	0.5±0.1	uniform film

As can be seen from the results of the autoclave tests, presented in Table 2, the oxide film formed on the ion-modified surface, characterized by increased protective properties (data table obtained by analysing images of the transverse structure of oxide by SIM). On the shell in the initial state the film has a thickness of 0.8 ± 0.1 µm, its body contains pores (gaps) by 0.2 ± 0.1 µm and a bulk density of 10 ± 1%. On the time oxidation basis up to 500 hoursoxide at the samples after ion treatment usually has, as a rule, a smaller thickness and great homogeneity.The best result shows the sample doped with Fe - oxide thickness of 0.4 ± 0.1 µm, pores and discontinuities isn't observed in the body of the film.

4. Conclusion

The influence of the finish ion-beam treatment the external surface the shell pipe length of E110 alloy to 500 mm on the condition of the near-surface layer of material and corrosion productsis studied. Machining was carried out on the installation ILUR-03 in modes of cleaning, polishing and additional alloying of thesample material surface layer by ion mixing radialAr^+ions beam.

It is shown that the selected processing modes reduce the surface roughness of the pipe throughout its length with $R_a = 0.8 \pm 0.1$ µm to $R_a = 0.3 \pm 0.1$ µm.Wherein the homogeneity of the surface and the protective properties of the oxide formed in the autoclave conditions (350 °C, 16.5 MPa, 500 h) increase.

Acknowledgements

This work was performed within the framework of the Center of Nuclear Systems and Materials supported by MEPhI Academic Excellence Project (contract № 02.a03.21.0005, 27.08.2013).

References

[1] Nikulina A V and Malgyn A G 2008 *Atomic Energy* **105** 259
[2] Khusu A P, Vitenberg Yu R and Palmov V A 1975 *Surfaces roughness: theoretical-probability method* (Moscow: Science) p 344
[3] Zaymovsky A S, Nikulina T V and Reshetnikov N G 1994 *Zirconium alloys in nuclear power engineerinig* (Moscow: Energoatomizdat) p 256
[4] Volkov N V, Kalin B A, Valikov R A, Yashin A S and Yalcev V N 2012 *Russian Phys. J.* **55** 186

IOP Conf. Series: Materials Science and Engineering **130** (2016) 012052 doi:10.1088/1757-899X/130/1/012052

Features of relief formation of the polycrystalline samples under the influence of a wide-aperture ion beam

A S Yashin

National Research Nuclear University MEPhI (Moscow Engineering Physics Institute) Kashirskoe Highway 31, 115409 Moscow, Russia

E-mail: yashin_itf@mail.ru

Abstract. The results of investigation of the cylindrical surface state after influence of wide-aperture ion beam are presented in this paper. Experiments were carried out on the parts of cladding tubes from E110 alloy (Zr-1%Nb) up to 500 mm length. The outer surface of the tubes was treated by Ar^+ ions with energy 0.1-1.0 keV up to doses $(1\text{-}10)\times10^{18}$ cm^{-2} on the installation KVK-10. Due to design features of the installation, the angle of incidence of particles on the surface varied from 0 to 90°. It is found that as a result of treatment made, smoothed relief of the surface forms and uniformity of near-surface layer structure improves. The statistical method suggested allowed to reveal features of the polycrystalline samples surface state with random and regular components, based on the results of profilograms analysis.

1. Introduction

As is known, many service properties of the products, such as corrosion resistance, friction, wear, etc. essentially depend on the condition of the surface layer material. In particular, the protective properties of the oxide films formed on the external surface shells ofnuclear reactors fuel elements from zirconium alloys is largely determine by the state of the shells external surface the presence of stressand technological defects and impurities in the near-surface layer of the material etc. [1-2] One of the most effective surface modification methods is ion-beam processing. As a result of charged particle beams the material near-surface layer spraying is taking place, a special relief forms, and new non-equilibrium state can arise in the area of radiation exposure. Because the ion-induced processes are not thermo-activated, the limits of the material surface properties control significantly expanded in comparison with the traditional treatment methods. To increase the treatment homogeneity of polycrystalline materials is the use of a wide-aperture low-energy ion beam of inert gases is interesting.

2. Materials and tools

Formed as samples were used fragments of staff claddings pressurized water reactors of E110 alloy (Zr-1% Nb, an outer diameter of 9.15 mm, length 500 mm), in the state after mechanical factory polishing. The external surface the tubes treated with wide-aperture beam of Ar^+ ions with energies of 0.1-1.0 keV (average 500 eV) to doses $(1\text{-}10)\times10^{18}$ cm^{-2} at the installation KVK-10 [3]. The radiation dose was chosenon the material surface layer spraying calculation the thickness up to 2 μm, which was determined based on the microbalance 1 mg = 0.17 μm. The samples weight was carried out on an analytical balance GH-252 company A&D (sensitivity of 0.01 mg). The outer surface temperature of the samples during the exposure does not exceed 150-200 °C. As can be seen from the circuit in Figure 1, theionsincidence angle at the sample during treatment varies from 0 to 90°, resulting in peak energy release shifts toward the material surface (see. Figure 2) decreases the thermal load on the sample and improves the treatment homogeneity. The using expose geometry allows toavoid the relief development due to the smoothing of the maximum in the angular distribution of the sputtering in this energy range (Figure 3).

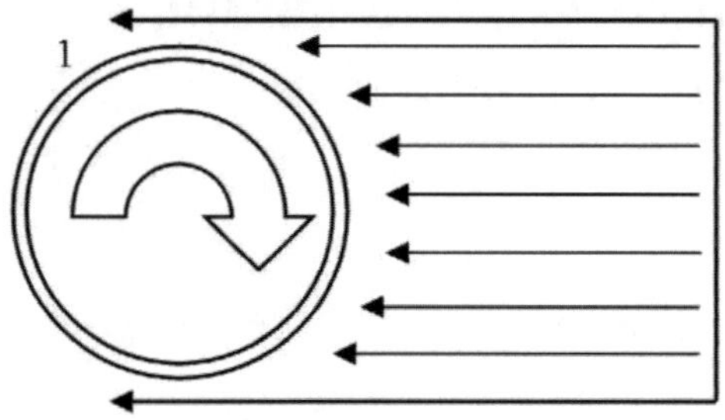

Figure 1. Scheme of treatment of the cylindrical samples on the installation KVK-10 (top view): 1 – sample, 2 – wide-aperture ion beam, arrow shows the direction of the sample rotation.

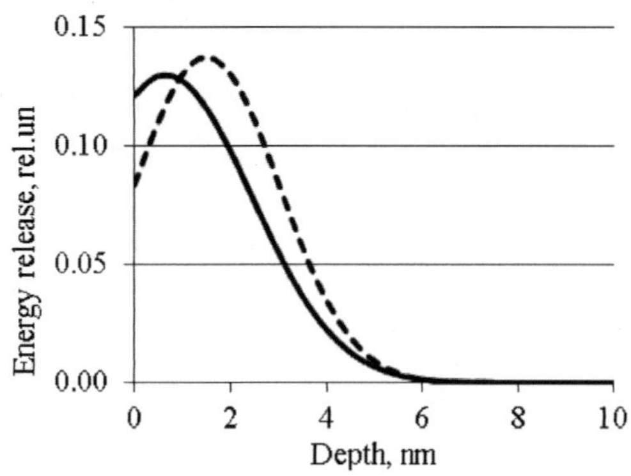

Figure 2. Distribution of energy release by depth of the Zr sample under the influence of focused (dotted line) and wide-aperture (solid line) Ar^+ ion beam (calculation by SRIM-2012).

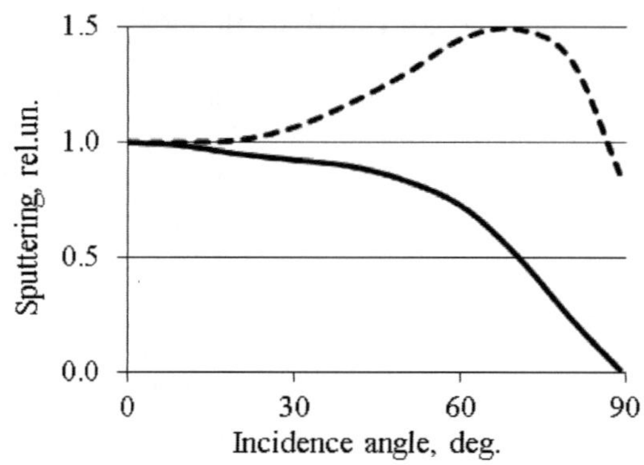

Figure 3. Dependence of the normalized Zr sputtering by wide-aperture Ar^+ 500 eV beam from its angle of incidence on the flat (dotted line) and cylindrical (solid line) surface (calculationbySRIM-2012).

Microhardness measurement of ion-treated surface was carried out on a PMT-3 (indenter - a diamond pyramid with an angle of 136 °, load 70 g, loading time 30 seconds).State of the near-surface layer was studied by optical and scanning ion microscopy (SIM unit Helios-660, a beam of Ga^+, the accelerating voltage of 5 kV). To reveal the material structure on the sample surface using the ion microscope in the milling mode are cut out the hole measuring 15×15 μm, after that obtained cross-section pictures are made. Profilograms ion-treated surface were filmed along a generator tubes at the contact profilograph TR-200 (scanning length of 1.0 mm, 0.5 mm step measurement) with subsequent transfer of data to a PC automatically. As a profile of the center line to a first approximation, the straight line was chosen whose coefficients were determined by the method of least squares (according to GOST 2789-73).

It is known that, for analysis topography convenience real surface profile of a material can be represented as a superposition of systematic and random irregularities [4]:

$$h(x) = h_{syst}(x) + h_{rand}(x), \qquad (1)$$

where $h_{syst}(x)$and $h_{rand}(x)$ – the systematic and random components, respectively.

To systematic component relate regularly arranged elements of the relief, which may occur in the process of permanent technological factors, such as, for example, the shape, size and vibration of the cutting tool. Random component in its turn is formed from a large number of irregular effects on the surface of the material, such as, for example, treatment by the abrasive particles. As a first approximation we can assume that $h_{syst}(x)$- there is some periodic function, and $h_{rand}(x)$- random stationary process with zero mean. As is known, one of the most informative methods of stationary random processes description is the autocorrelation function $K(\tau)$, which characterizes the values dependence at the predetermined point x on the values of the same process at another point $x + \tau$ with displacement τ, and for a limited set of discrete values considering normalization can be evaluated by the formula [4]:

$$K(\tau) = \frac{\sum_{i=1}^{n-1}(h(x_i) - <h>)(h(x_i + \tau) - <h>)}{\sqrt{\sum_{i=1}^{n-1}(h(x_i) - <h>)^2 \sum_{i=1}^{n-1}(h(x_i + \tau) - <h>)^2}}, \qquad (2)$$

where n - is number of measurements h_i, x_i - x coordinate of the measurement point, $<h>$ - arithmetic average of the measured values of $h_i = h(x_i)$. Considering (1), the calculated autocorrelation function of the real profileallows to allocate the systematic and random components and obtain information both about the profile as a whole and its separate components. The solution to this problem seems relevant since identifies mechanisms of relief formation in the treatment process and to correct its parameters according to their stated objectives.

3. Results and discussion

Figure 4 shows distribution histograms values of the microhardness of the samples outer surface in the initial state (after mechanical treatment) and after ion treatment to (3-10)×10^{18} ions cm^{-2}.Performed analysis shows that the ion treatment doesn't change the microhardness mean value of the outer surface, which is about 1.86 GPa. Where in the scatter of values significantly reduces - from 0.18 GPa for the sample in the initial state to 0.08 GPa for the sample after ion-beam processing, and distribution becomes more random. These changes evidence about the near-surface layer homogenization of the material thickness, commensurable with the depth indenter penetration, which is under the chosen conditions of loading of 3-4 μm.

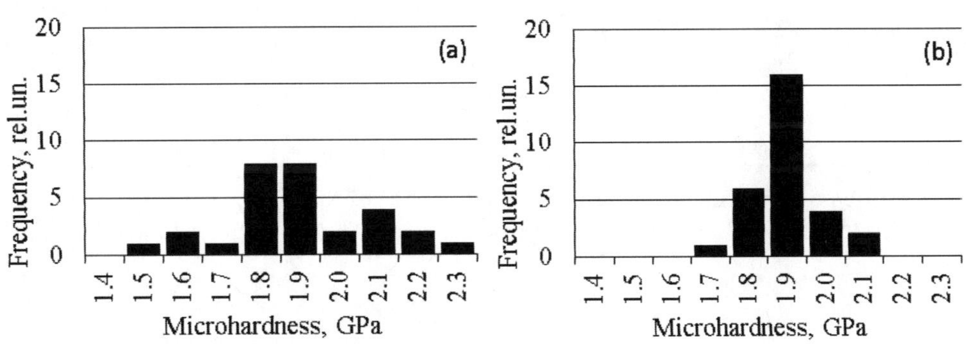

Figure 4. Histograms of microhardness distribution of the samples outer surface after mechanical abrasion (a) and ion treatment to (3-10)×10^{18} cm^{-2} (b).

Figures 5 and 6 show cross-sectional SIM-photos on the samples surface in the initial state and after ion beam treatment to 9×10^{18} cm^{-2}, respectively. Visually, it is seen that in the considered scale of the original surface relief of the sample has a characteristic wavy shape which is almost completely is aligned after ion treatment. Near the untreated sample surface is observed strongly distorted structure (detected layer thickness is about 2 μm - shown by the arrows in Figure 5), which is probably formed as a result of material deformation during the surface mechanical treatment, possibly in combination with saturation of this layer by technological impurities. The structure of the samplenear-surface layer after ion beam treatment consists of well-defined grains with an average size of 3-5 μm (grain boundaries are shown by the arrows in Figure 6) that corresponds to the main state of material volume.

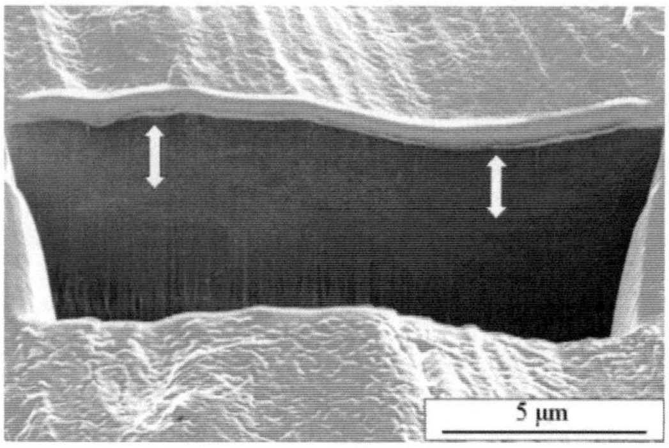

Figure 5. SIM-image of cross section on the outer surface of the sample from E110 alloy after mechanical abrasion (arrows show the boundaries of the defective surface layer).

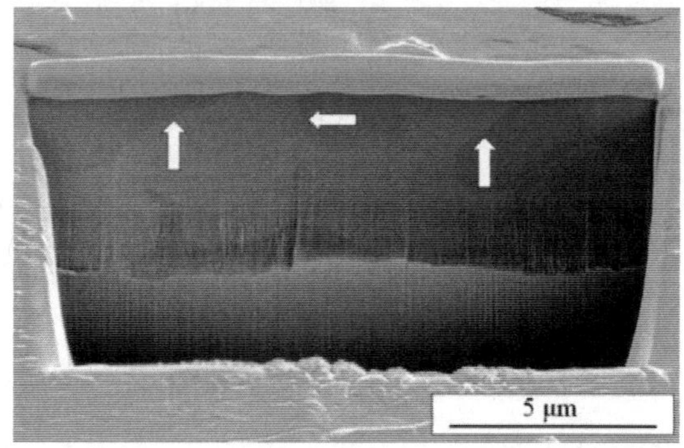

Figure 6. SIM-image of cross section on the outer surface of the sample from E110 alloy after the treatment by Ar^+ beam to 9×10^{18} cm^{-2} (arrows show the grain boundaries near the surface of the material).

Figure 7 shows the profiles (a) and charts of normalized autocorrelation functions (b) the external surface profiles of samples in different states. As can be seen, the autocorrelation function at the coordinate has a characteristic recession and then oscillates around zero. With increasing irradiation dose, the graph $K(\tau)$ becomes smoother, and the periodic component having a wavelength equal to about 0.1-0.2 mm detects at the oscillating portion. These changes point to a reduction in the share of casual and part - systematic profile roughness whose characteristic size amounts less than 100 μm, which confirms the results obtained by the SIM.

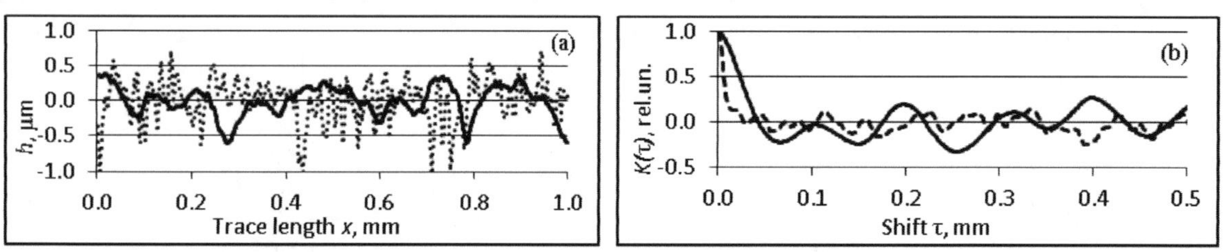

Figure 7. Profilograms (a) and autocorrelation functions (b) of the outer surface of the samples form E110 alloy after mechanical abrasion (dotted line) and ion-beam treatment to 10×10^{18} cm^{-2} (solid line).

4. Conclusion

Investigated the state of the external surface the tubular samples of zirconium alloy E110 (Zr-1%Nb) after treatment with a wide-aperture beam of Ar^+ ions with energies 0.1-1.0 keV.

Analysis of the microhardness measurement results and SIM images showed that by the removal the material with the thickness of 1-2 μm by physical sputtering, the smoothed surface relief structure forms and the near-surface layer material homogenizes.

With the proposed statistical method of processing profilograms it found that the carried out treatmentleads to a reduction in the contribution of the surface topography of the random component that occurs in the process of mechanical material grinding and accompanied by a significant distortion of the structure.

Acknowledgements
This work was performed within the framework of the Center of Nuclear Systems and Materials supported by MEPhI Academic Excellence Project (contract № 02.a03.21.0005, 27.08.2013).

References

[1] Nikulina A V and Malgyn A G 2008 *Atomic Energy* **105** 259
[2] Evdokimov I A, Likhanskii, Aliev T N, Zborovsky V G and Kolesnik M Y 2012 *J. Nucl. Materilas* **424** 190
[3] Kalin B A, Volkov N V, Valikov R A and Yashin A S 2014 *Bulletin ot the Russian Academy of Sciences. Phys.* **78** 553
[4] Khusu A P, Vitenberg Yu R and Palmov V A 1975 *Surfaces roughness: theoretical-probability method* (Moscow: Science) p 344

Determination of mechanical properties from depth-sensing indentation data and results of finite element modeling

M G Isaenkova, Yu A Perlovich, O A Krymskaya and D I Zhuk

National Research Nuclear University MEPhI (Moscow Engineering Physics Institute) Kashirskoe highway 31, Moscow, 115409, Russia

E-mail: dimazhuk@gmail.com

Abstract. 3D finite element model of indentation process with Berkovich tip was created. Using this model with different type of test materials, several series of calculations were made. These calculations lead to determination of material behavior features during indentation. Relations between material properties and its behavior during instrumented indentation were used for construction of dimensionless functions required for development the calculation algorithm, suitable to determine mechanical properties of materials by results of the depth-sensing indentation. Results of mechanical properties determination using elaborated algorithm for AISI 1020 steel grade were compared to properties obtained with standard compression tests. These two results differ by less than 10% for yield stress that evidence of a good accuracy of the proposed technique.

1. Introduction

Indentation is one of the most widely used techniques for determination of mechanical properties of materials. This method has many advantages compared to standard tension/compression tests [1]. It is not required to produce a standard size specimen for testing; any piece of material which can be glued to surface will be enough. In addition, it is possible to get properties of very small (eg. 100 nm) region [2] of the material or in thin films. In classic indentation, the hard indenter is pressed into material with specified force. Then, force is removed and indenter print's size is measured. Usually it is done by using optical microscope [3]. In micro- and nanoindentation this can be a considerable problem. Atomic force microscopy [4] or scanning electron microscopy [5] can be used for this task but it is time-consuming and requires very expensive equipment. To overcome this problem and to obtain as much information about material as possible from single indentation, the depth-sensing indentation was introduced. During such test, the position of indenter is measured while load increases from zero to specified maximum and to zero. The *P-h* diagram is obtained which is the material response to increasing load. This diagram can be processed using various techniques to determine elastic and plastic characteristics of material.

Many methods for processing this diagram exist. The most used among them and standardized in many countries is Oliver-Pharr method that allows obtaining Young's modulus of material [6]. It uses slope of unloading part of *P-h* diagram knowing contact area from geometry relations. This method gives decent accuracy and easy to use but it cannot give any information about material's plastic properties. Being based on Hertz contact problem's solution [7] it cannot be of any use for characterization of plastic deformation during indentation but only for unloading which known to be only elastic. At the same time, the problem of plastic deformation of material during indentation remains unsolved due to presence of many variables such as yield stress, deformation hardness exponent, creep constants. Yet, it is possible to use an empiric method to make a solution of acceptable accuracy. In first approaches, researchers used results of many experiments for different materials, which have been later processed, and then correlation between indentation data and material properties have been established [8]. Practice shows that this relation can be found but this requires much resouces for preparing many materials to undertake many indentations. In addition, mechanical properties can be different in various parts of material and also in different specimens. With increasing of computational power, real experiments are often substituted by finite element (FE) modeling [9]. Advantages of FE method can be of a great use for this field. Material behavior can be easily controlled to represent different materials and calculation can be automated for series of methods.

Purpose of this work was to create such method using results of FE modeling performed and to try it on material with known properties to check results. Work steps as followed:

- Build 3D FE model for material and Berkovich indenter (3-sided pyramid)
- Test model by comparing results of experimental indentation with results of indentation simulation for materials with same properties
- Once model is known to be well representing material's behavior during indentation, perform calculations for materials with different mechanical properties
- Summarize results, find out relations and build dimensionless function to represent them.
- Try method for material with unknown properties and compare its results with same material's compression test's results

2. Modelling & experimental results

2.1. FE Model

In FE representation of deformed material or construction the object is split to a number of elements, each element's behavior is controlled. These elements can be either three, two, one dimensional, or even dimensionless. In this study, where used three-dimensional 20-node (8 nodes in corners of element and 12 nodes on edges) elements to represent volume of indented material and two-dimensional 8-node (4 nodes in corners and 4 nodes on edges).

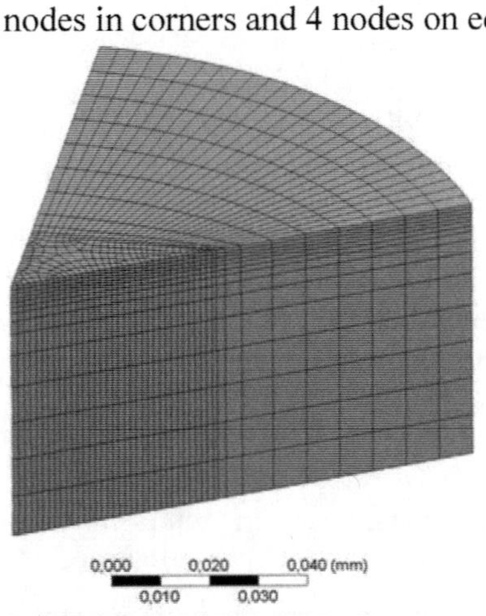

Figure 1. Resulting FE model. Regions with different element density have different color. Indenter is not shown for illustrational purpose.

Elements' response to stresses below yield stress is linear while for stress above yield stress it is multi linear so it can represent exponential hardening law. Element's size is chosen according to gradients of stress and deformation. Higher gradient require smaller elements so under indenter, where element's displacement is maximal it is smaller.

In case of geometrical symmetry, it is common to cut model by a symmetry plane [10]. In our case, we can cut the model to leave only 1/6 and account for remaining parts by involving symmetry boundary conditions on surfaces created by this cutting. Resulting model (Figure 1) can represent full model without loss of accuracy. This allows reducing number of elements to 17414 and nodes to 76851, so time required for process simulation on a computer is much less than for full model. Friction coefficient is chosen 0.2, which accords for dry friction between steel and diamond.

2.2. Model evaluation

Before using model for research purposes, it is required to ensure that it behaves correctly under load. Evaluation can be done by comparing experimental *P-h* diagram to computational for same material. Microindentation experiment requires well prepared surface of specimen. Roughness of surface will lead to great spreading of indentation results [11]. To avoid this, the specimen was polished with sandpaper of different grain size, from bigger to smaller. Later, for removing disturbed metal layer, electrolytic etching was utilized on material surface. Despite of this procedure, experimental microindentation results still have noticeable variation because of different grain orientations and various inclusions in the material. Series of indentation tests performed to minimize this variation. Results of this series were averaged and compared to simulated *P-h* diagram (Figure 2).

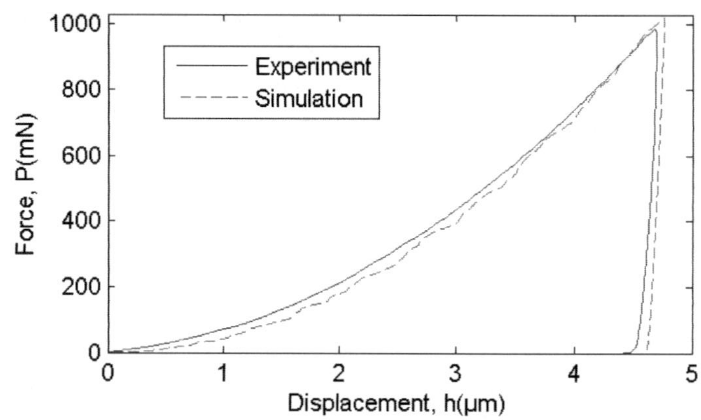

Figure 2. Experimental P-h diagram versus simulated. Experimental line is averaged from series of 12 experiments of microindentation made. Simulated *P-h* diagram made with FE method for material properties obtained from compression test.

In order to compare simulation with experiment, simulation should be made with same material properties as indented material. All needed constants, including yield stress, strain hardening exponent, Young modulus, can be obtained from standard compression test. This type of test, involving all specimen gives macroscopic properties of material, which is in some point of view equivalent to averaging indentation results.

2.3. Simulations

To construct dimensionless functions connecting material's constants and indentation results, series of experiments was made. Using model described in section 2.1 with following parameters: Young modulus 200 GPa, hardening exponent 0.15, yield stress from 150 MPa to 500 MPa with 50 MPa step. Each simulation's *P-h* diagram was processed and graphs for necessary constants were plotted (Figure 3). Usually loading line follows Kick's Law (1), so C is taken as one of the constants to link *P-h* diagram with material properties:

$$P = Ch^2 , \tag{1}$$

where P - force on indenter, h – displacement of indenter, C – material constant.

Other constant S is a slope of unloading curve during starting moment of unloading. Variables h_{max} and h_r are the maximum indentation depth and indentation depth after full unloading consequently. Ratio of elastic to the total energy W_e/W_t is calculated as area ratio under lines of P-h diagram. Plastic energy can be calculated as the difference between area under loading curve and elastic energy, which is area under unloading curve.

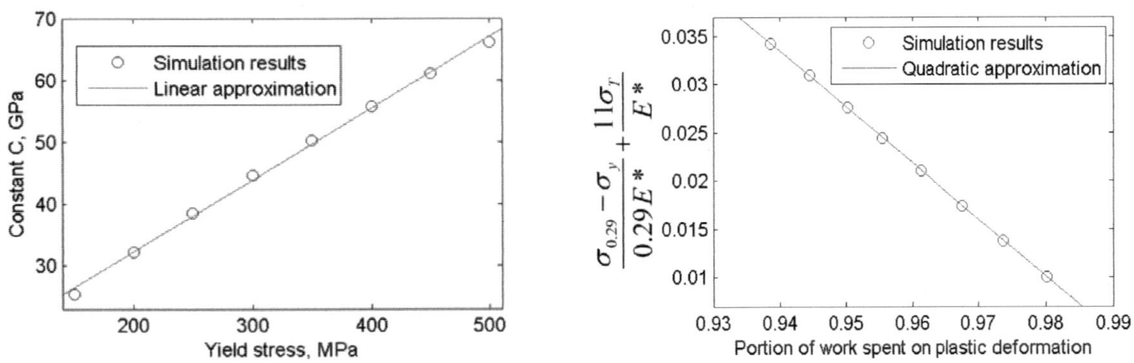

Figure 3. Relation between C constant in Kick's law and yield stress of material (left). Relation between part of dimensionless equation with work ratio (right). Both graphs represent results of FE simulations. Each circle is different material with varied yield stress.

2.4. Method for material properties determination

Using relations given by FE simulation we can construct the method suitable to evaluate material properties with result of FE indentation. It should include following steps.

1. Process experimental P-h diagram to obtain useful values, such as C, S, h_{max}, h_r, W_e/W_t.
2. Using Oliver-Pharr or equivalent method to obtain Young modulus with S constant.
3. Obtain relation between yield stress and characteristic stress (stress at 29% deformation during uniaxial loading) from equation (2)

$$\frac{\sigma_{0.29}-\sigma_y}{0.29E^*}+\frac{11\sigma_T}{E^*}=1.152\left(\frac{W_p}{W_t}\right)^2-2.795\left(\frac{W_p}{W_t}\right)+1.643 \tag{2}$$

   where E^* - Young modulus corrected by elastic deformation of indenter.
4. Substitute this relation to equation (3) and solve for yield stress. This equation is used in many previous works by other authors [12] [13].

$$C=M_1\sigma_{0.29}\left(1+\frac{\sigma_T}{\sigma_{0.29}}\right)\left(M_2+\ln\left(\frac{E^*}{\sigma_T}\right)\right) \tag{3}$$

   where M_1 =6.02 and M_2 = - 0.875 for Berkovich indenter.
5. From exponential hardening law calculate strain hardening exponent

$$n=\frac{ln\left(\dfrac{\sigma_m}{\sigma_{0,29}}\right)}{5{,}01} \tag{4}$$

Method was tested using indentation results for AISI 1020 steel. Set of material properties estimates using method above was compared to compression standard test results. Yield stress value lies within 10% error interval, which is good for indentation methods. For calculation of Young modulus Oliver-Pharr method was utilized. Results of comparison shown in table 1.

Table 1. Results of material's properties estimations using standard compression method and indentation.

Method	Young modulus (GPa)	Yield stress (MPa)	Strain hardening exponent
Compression	215	340	0.13
Indentation	213	375	0.16
Error	0.9%	10%	23%

3. Conclusion

Finite element model for simulation of indentation using Berkovich indenter was build. Model was test by comparing its results to experiment of indentation and good accuracy was observed. Method for estimation of mechanical properties was made using simulations conducted with FE model. Method tested on AISI 1020 steel and 10% error for estimation of yield stress was found comparing to standard compression test.

Acknowledgements
This work was performed within the framework of the Center of Nuclear Systems and Materials supported by MEPhI Academic Excellence Project (contract № 02.a03.21.0005, 27.08.2013).

References

[1] Johnson K L 1970 *J. Mech. and Phys. of Solids* **18.2** 115
[2] Harvey S 1993 *J. of mat. res.* **8** 1291-99
[3] Giannakopoulos A E, Larsson P-L, and Vestergaard R 1994 *J. of solids and str.* **31** 2679-708
[4] Petzold M, Landgraf J, Füting M and Olaf J M 1995 *Thin Solid Films* **264** 153
[5] Byong-Taek L, Nishiyama A and Hiraga K 1993 *JIM* **34** 682
[6] Oliver W and Pharr G 1992 *J. of mat. res.* **7** 1564-83
[7] Spence D A 1975 *J. of elast.* **5** 297
[8] Doerner M F, Gardner D S and Nix W D 1986 *J. of Mater. Res.* **106** 845
[9] Laursen T A and Simo J C 1992 *J. of Mater. Res.* **703** 618
[10] Chandrupatla T R Belegundu A D, Ramesh T and Ray C 1997 *Introduction to finite elements in engineering* (Upper Saddle River: Prentice Hall) pp. 279-300
[11] Gilbert J L, Cumber J and Butterfield A 2002 *J. of biomed. mater. res.* **612** 270
[12] Giannakopoulos A E and Suresh S 1999 Determination of elastoplastic properties by instrumented sharp indentation. *Scripta mater.* **40** 1191-98
[13] Venkatesh T, Van Vliet K J, Giannakopoulos A E and Suresh S 2000 *Scripta mater* **42** 833

Thermo-mechanical finite element modeling of shape memory materials' microindentation

Yu A Perlovich, M G Isaenkova, O A Krymskaya and D I Zhuk

National Research Nuclear University MEPhI (Moscow Engineering Physics Institute), Kashirskoe highway 31, Moscow, 115409, Russia

E-mail: dimazhuk@gmail.com

Abstract. Indentation of shape memory materials and later heating with recovery of indent is studied in this work using finite element modelling. Results of simulations of two types of shape memory materials, with one-way shape memory effect and with superelastic properties compared to experimental indentation with 200μm spherical indenter. Based on results of finite element modeling, several useful quantities plotted for loading, unloading and thermal recovery for various materials with shape memory effect. Recovery of imprint made with Berkovich (three-sided pyramid) compared to recovery of imprint made with spherical indenter.

1. Introduction

Investigation of martensite in 1900s and later the thermally reversible martensitic structure in CuZn and CuAl alloys [1] was a big step for future discovery of "smart" materials and the most significant among them, shape memory materials (SMM). Compared to standard materials they have many extraordinary properties. The most interesting is what they can, as it can be guessed from the name, remember their shape before deformation occurred. The specimen, deformed unrecoverably can restore original form after heating. Also, at certain temperatures they can acquire superelastic properties and deform recoverably to bigger degrees compered to stainless steel. This potential utilized in many fields including aerospace, biomechanical, in nanotechnology as sensors and actuators.

Shape memory alloys (SMAs) can exist in two different phases (austenite and martensite), martensite phase can also have two different structures (twinned martensite, detwinned martensite). This arrangement is possible at different temperatures and stress states depending on principle of minimum free energy and forms six different possible transformations [2]. Such transformations not carried by diffusion but rather by shift of crystal lattice [3]. In summary, we can classify SMA behaviour in three categories.

1. Superelasticity (SE) or Pseudoelasticity (PE) – recoverable deformation up to 10% [4] without change of ambient temperature. This exceptional deformations achieved not by elastic change of interatomic spacing but due to phase transition. Usually a significant hysteresis presents. Stresses of forward transformation start and finish can be much higher than of reverse transformation.

2. One-way shape memory effect – deformation of material can be recovered only after heating. After withdraw of applied force deformation still presents and material can be restored to original shape by heating. Following cooling does not result in change of geometry or volume.

3. Two-way shape memory effect – heating and cooling results in recovery of shape of high and low temperatures. Such materials can be obtained by cyclic deformation training. One cycle includes

deformation at required temperature, unloading, heating and cooling until demanded shape is achieved [5]. As reported, usually only half of one-way shape memory effect strain can be attained [6].

Many applications of SMAs are of very small scale. Such as surgical instruments and stents [7], micro-actuators [8], medical implants [9]. Such small parts require advanced methods for testing of mechanical properties. One of the most widely used techniques for determination of mechanical properties in micro- and nano- scales is instrumented microindentation. During this test, an indenter (usually a pyramid or a sphere) is pressed to material with increasing load. Material response of displacement versus load (P-h diagram) is recorded and analysed. For classical materials without phase transformation a Young modulus can be obtained with high degree of confidence [10]. Several methods exists for determination of yield stress and hardening exponent [11]. Shape memory materials have more complicated behavior including phase transformation, temperature dependence and Young moduli for different phases. As result, the problem of obtaining mechanical properties from indentation results is solved in full. Indentation is a complex loading and it can be very hard to divide one phase from other and from phase transformation process. This paper analyses process of indentation of SMAs, thermal recovery using finite element method and discusses possibilities for retrieving mechanical properties with instrumented indentation.

2. Methods and materials

Finite element modeling (FEM) was utilized to study indentation of shape memory materials with 200 micrometer spherical indenter. Since the testing is axisymmetric, the two-dimensional FE model can be used without loss of accuracy with fitting boundary conditions. Modelled indenter, made from zirconium dioxide and SMA material up to 1000 micrometer from indenter's tip in surface and in depth. Contact's formulation between material and indenter includes friction effect, friction coefficient is taken 0.2. There are 2526 eight-node elements representing material and indenter in model (Figure 1). Nodes in point of initial contact are merged for better convergence and to avoid slipping of indenter to material. As usually, in contact area and in volume of bigger gradients of stresses and deformations the mesh density is higher for more efficient simulation analysis.

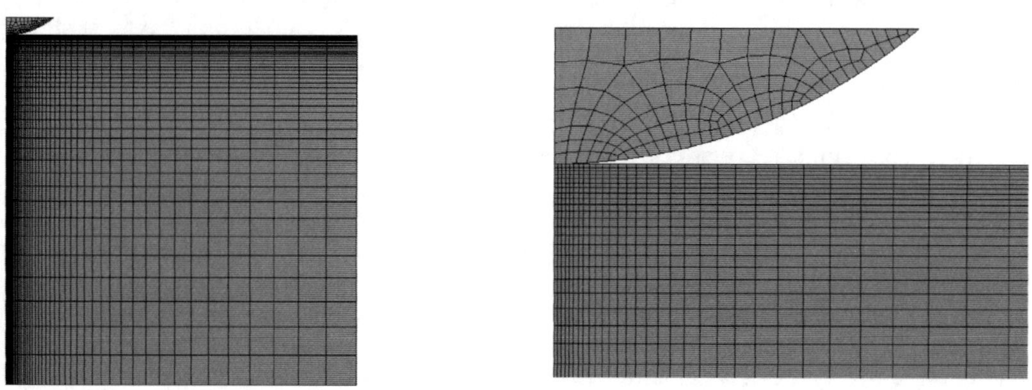

Figure 1. FE model of material (blue) and indenter (grey) produced in specialized software. Image to the right enlarged for illustrational purposes.

SMA has a rather unique behaviour. They can deform elastic to some degree, after that with increasing load the phase transition occur and new phase can deform elastic also. When load it removed, material can ether return to original shape or recover only elastic deformation, depending on temperature. At some conditions, only the change of temperature results in phase transition and change of shape. Evidently, such materials require own model with much higher degree of complexity compared to most elastic and plastic models. The constitutive model for shape memory effect simulations used for this work proposed by Auricchio et al [12-13]. This is a phenomenological constitutive model based on continuum thermodynamics with internal variables. It can represent such thermomechanical behavior as shape memory effect, superelastisity, different response to tension and compression.

$$\Psi(\varepsilon, T, \varepsilon_{tr}) = \frac{1}{2}(\varepsilon - \varepsilon_{tr}) : D : (\varepsilon - \varepsilon_{tr}) + \tau_M(T)\|\varepsilon'_{tr}\| + \frac{1}{2}h\|\varepsilon'_{tr}\|^2 + I_{\varepsilon'_{tr}}(\varepsilon'_{tr}) \quad (1)$$

where D - material elastic stiffness tensor, ε- total strain, ε_{tr} - total transformation strain, $\varepsilon`_{tr}$ - deviatoric transformation strain, $\tau_M(T)$ - a positive and monotonically increasing function of the temperature as $\langle\beta(T - T_0)\rangle+$ in which $\langle\cdot\rangle+$ is the positive part of the argument (also known as Maxwell stress), β - material parameter, T - temperature, T_0 - temperature below which no austenite is observed in a stress-free state, h - material parameter related to the hardening of the material during the phase transformation, ||...|| - indicator function introduced to satisfy the constraint on the transformation norm.

Two NiTi alloys with different compounds were studied for this work. One of them, 55.9 Ni 44.1 Ti is superelastic at room temperature, and the other, 55.1 Ni 44.9 Ti can deform at room temperature and recover original shape when heated above 75 °C. Indentation testing performed at room temperature, while simulation temperatures varied. Physical properties presented in table below (table 1).

Table 1. Transformation temperatures in degrees Celsius and stress at which transformation starts (for alloy at M_s temperature) [14].

Alloy	M_s	M_f	A_s	A_f	σ_0
55.9 Ni 44.1 Ti	-40	-57	-25	-5	85
55.1 Ni 44.9 Ti	18	-22	35	73	100

3. Experiments and simulations

Finite element modelling method allows to change material constants and thus the material behaviour depending on constitutive model. In this particular case for each material it lists elastic moduli for austenite and martensite phases, stress at transformation start at temperature M_s, transformation hardening constant, maximum transformation strain, hysteresis and several thermal constants. Simulation of materials listed in table 1 compared to experiments of indentation to check the consistency of FE model used. Results on picture below (Figure 2).

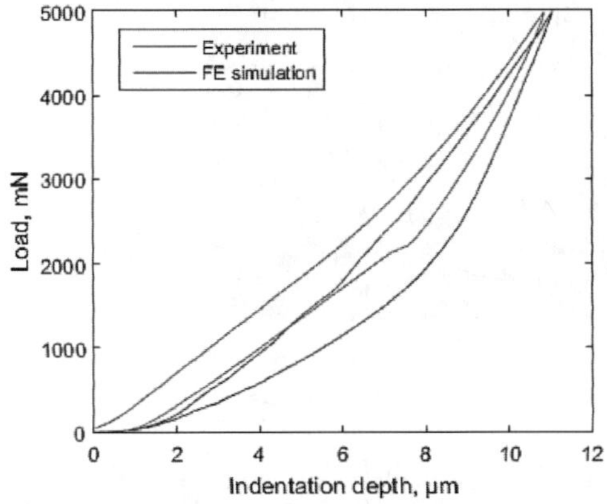

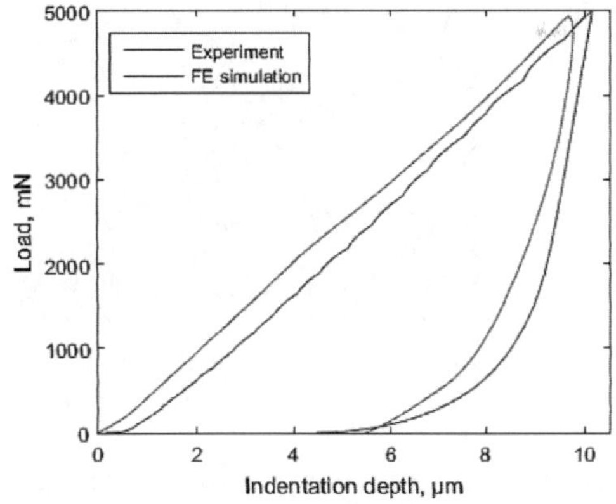

Figure 2. Comparison of experiments and simulations performed. Experimental data taken from averaging of series of indents.

Error can come possibly from inaccuracy in determination of mechanical properties, unevenness of prepared surface, differences of model behaviour and real material deformation and reformations resulted in change of mechanical properties of materials due to discharge of heat in martensite transformation.

Numerical results of simulation can be used to plot areas of material state during loading. At maximum load (Figure 3) this result is similar for both types of materials. Full transformation is observed at 1/10 of indenter's radius below indenter's apex until 1/4 of radius with maximum stress in the middle of this area. Also, we can mark transient area where both phases are present and detwinning/phase transformation process is not fully completed. This material state corresponds with stress plateau during uniaxial extension tests.

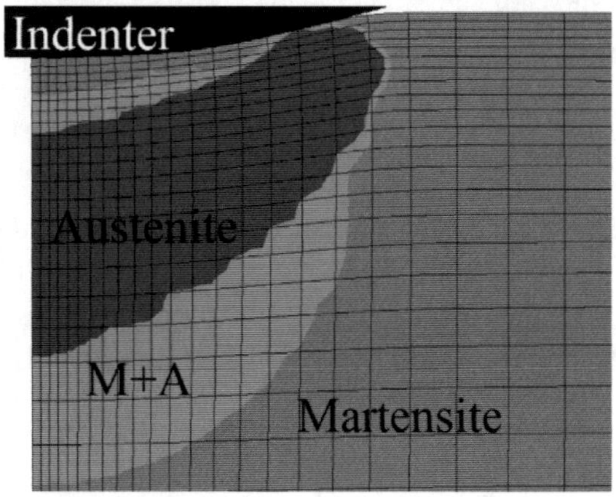

Figure 3. Division of material to different phases during indentation at maximum load (5N). This is axisymmetric cut of model and left line of material shown is axis of force application to indenter.

4. Recovery of microindents

When indent is made below A_f temperature it may not fully recover after the load is removed. Part of material under print still in martensitic phase and thermodynamic condition are so it requires energy to return original shape. With increasing temperature the energy of crystal lattice is also increasing and reverse transformation become possible. Studies show what smaller spherical indents compared to radius of indenter sphere, usually recovers [15]. On other hand, indents made with pyramid indenter (Vickers or Berkovich) do not fully recover due to rotation of structure (Figure 4).

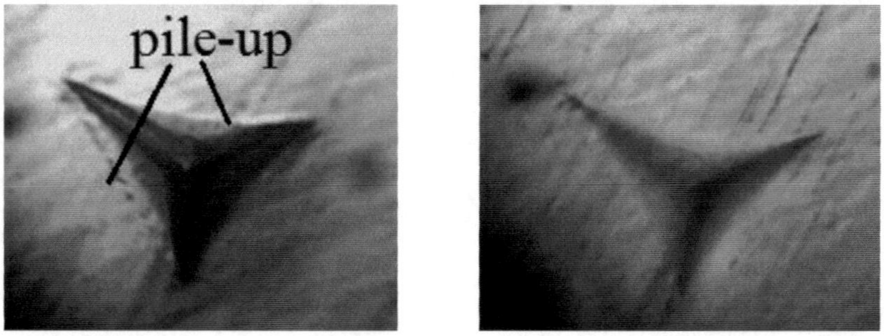

Figure 4. Pictures of one indent taken before heating the specimen (left) and after (right). Indented material - 55.1 Ni 44.9 Ti at room temperature.

Process shown on the pictures above can be modelled with finite element method. This requires coupling of two different solvers with different element formulation and different degrees of freedom. For deformable body we have forces and displacements while in thermal problem it will be temperature. Thermal problem is rather simple: uniform change of temperature is supposed. Because both problems share same model and topology, temperature values over time for every element can be transferred to deformable body solver as a thermal load. Change of temperature lead to change of material behaviour according to constitutive model used and eventually, the recovery of imprint.

Initial loading take place at -10°C temperature at which material shows one-way shape memory effect. As expected at this temperature, material can recover only partially. Later, the temperature is increased and material recovers. Third axis can be added to standard *P-h* diagram to show recovery upon heating (Figure 5).This agrees with obtained experimental data for indentation with spherical indenter and with other works [15]. Coefficient of thermal expansion is set to $8 \cdot 10^{-6}$ °C^{-1} so effect of volume increasing with temperature affect the results less than 1% and can be neglected.

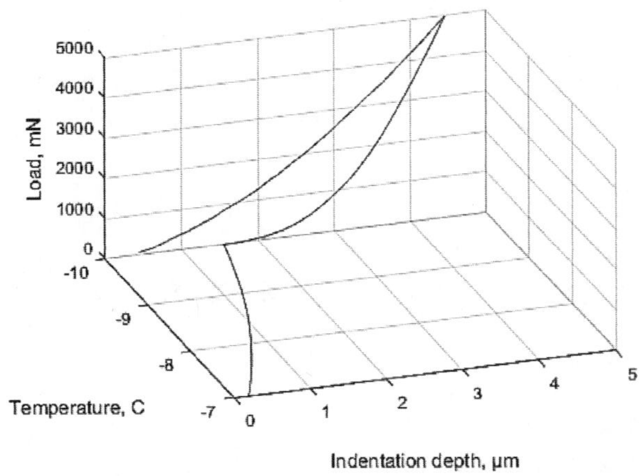

Figure 5. Simulated *P-h* diagram with temperature component. Indentation made with spherical 200 μm indenter. Increasing of temperature follows full unloading.

Furthermore, it is possible research indentation process and thermal recovery not only on changes in *P-h* diagram which can be obtained using modern NI machines but also on changes of volume of different phases during loading, unloading and following change of temperature. This can be of greater interest because such graph cannot be obtained using current technologies.

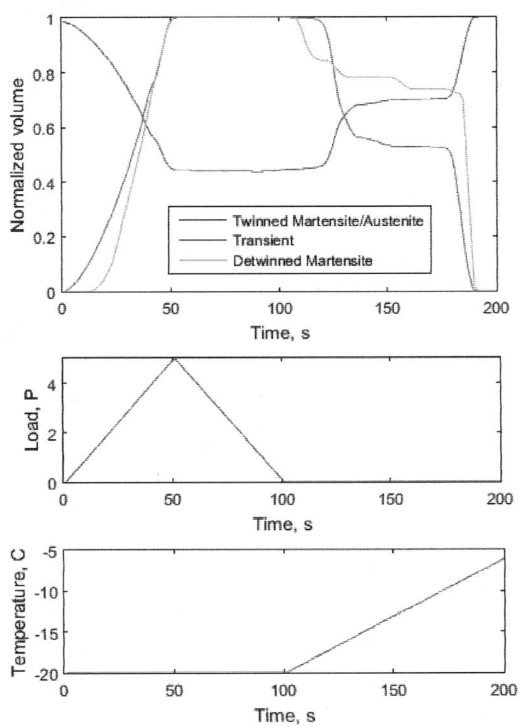

Figure 6. Scaled to maximum volume of each phase over time in indentation process. Volume of initial phase is virtually infinite as it depends on volume of specimen. First 100 secs is loading and unloading and following 100 secs is thermal recovery of imprint.

For simplicity, two dimension plot is used for dependence of each phase volume vs. time (Figure 6). There are also two graphs of temperature and force applied to indenter with respect to time.

5. Conclusion

Indentation of shape memory materials and their thermal recovery is researched using finite element modeling with constitutive thermodynamical model. Results of simulations for OWSME and SE compared to indentation of NiTi alloys with different composition so they have such effects. Using results of simulation, volumetric composition and areas of different phases and structures.

Acknowledgements

This work was performed within the framework of the Center of Nuclear Systems and Materials supported by MEPhI Academic Excellence Project (contract № 02.a03.21.0005, 27.08.2013).

References

[1] Kurdjumov G V and Khandros L G 1949 *Doklady Akademii Nauk SSSR* **66** 211
[2] Mihálcz I 2001 *Mech. Eng.* **451** 75
[3] Lagoudas D C, Entchev P B, Popov P, Patoor E, Brinson L C and Gao X. 2006 *Mech. of Mater.* 430
[4] Leon R T; DesRoches R; Ocel J and Hess G 2001 *Sm.Struct. and Mater.* **227**
[5] Meng X, Chen F, Cai W, Wang and Zhao L 2006 *Mater. transact.* **473** 724
[6] Huang W and W Toh 2000 *J. of mater. sci. lett.* **19** 1549-50
[7] Machado L G and Savi M A 2003 *Braz. J. of Med. and Bio*. Res. **366** 683
[8] Gabriel K J, William S N and Walker A J 1987 *U.S. Pat. No. 4,700,541*
[9] Olsen T W, Loftness P E and Erdman A G *U.S. Pat. App. 11,502,603*
[10] Oliver W C and Pharr G M 1992 *J. of mater. res.* **706** 1564-83
[11] Giannakopoulos A E and Suresh S 1999 *Scripta mater.* **40** 1191-98
[12] Auricchio F, Taylor R L and Lubliner J 1997 *Comp. meth. in app. mech. and eng.* **146** 281
[13] Auricchio F, Fugazza D and DesRoches R 2006 *J. of Eng. Mat. and Tech.* **128** 312
[14] Melton K N and Mercier O 1981 *Acta Metall.* **292** 393
[15] Ni W, Cheng Y-T and Grummon D S 2002 *App. Phys. Lett.* **80** 3310-12

Modern methods of experimental construction of texture complete direct pole figures by using X-ray data

M Isaenkova, Yu Perlovich and V Fesenko

National Research Nuclear University MEPhI (Moscow Engineering Physics Institute), Kashirskoe Shosse 31, Moscow, 115409, Russia

E-mail: MGIsaenkova@mephi.ru

Abstract. Currently used methods for constructing texture complete direct pole figure (CDPF) based on the results of X-ray diffractometric measurements were considered with respect to the products of Zr-based alloys and, in particular, used in a nuclear reactor cladding tubes, for which the accuracy of determination of integral texture parameters is of the especial importance. The main attention was devoted to technical issues which are solved by means of computer processing of large arrays of obtained experimental data. Among considered questions there are amendments of the defocusing, techniques for constructing of complete direct pole figures and determination of integral textural parameters. The methods of reconstruction of complete direct pole figures by partial direct pole figures recorded up to tilt angles of sample ψ=70-80°: the method of extrapolation of data to an uninvestigated region of the stereographic projection, and the method of "sewing" of partial pole figures measured for three mutually perpendicular plane sections of the product. The limits of applicability of these methods, depending on the shape of the test product and the degree of inhomogeneity of the layer-by-layer texture, were revealed. On the basis of a large number of experimental data, the accuracy of the integral parameters used for calculation of the physical and mechanical properties of metals with a hexagonal crystal structure was found to be equal to 0.02, when taking into account the texture heterogeneity of regular products from Zr-based alloys.

1. Introduction

The emergence of new diffractometers of latest generation at X-ray laboratories, which are busy studying of the texture of metallic materials, significantly changed the character of the routine work directly performed by researchers, shifting its centre of gravity toward the trivial manipulation with the enclosed software when calculating the orientation distribution function (ODF) or building 3D pole figures. Automatization of processes of texture diffractometer measuring and computer processing of the measurement results significantly accelerate these procedures and immeasurably reduce their labour input, allowing to operate by the vast arrays of experimental data that had previously been almost impossible. However, while a number of important methodological aspects of the X-ray study of the product texture is out of view of researchers, which consciously or unconsciously rely on the competence of developers of the device and the used software. If you need only information about the general character of texture and about the most remarkable features of its development, the situation is acceptable. But in cases where the texture parameters define the technology and performance of the product or you need a more precise analysis of texture to identify certain processes of structure development in the material, it is necessary to critically estimate the sources of possible errors of the received information and to improve the accuracy of the calculated texture parameters.

Namely such situation is observed at the study of the texture of products from zirconium alloys for nuclear energy and, in particular, the cladding tubes, for which the so-called integral texture Kearns parameters [1] characterize their suitability as an extremely critical elements of reactor design. Meanwhile, during 30 years publications, where the most important for the physical and applied metallurgy questions would be addressed with due methodical thoroughness, do not appear in the scientific literature. Although the fundamental principles of the X-ray method of construction of the direct pole figures (DPFs) for the past decade have not been changed, the current practice of texture analysis is strongly needed in the description of methods of X-ray study of textures, upgraded in accordance with the process automatization and computerization of the treatment process of results. In this work, the gap is filled with reference to the study of the texture of products from zirconium-based alloys. The following methodological developments, which are necessary for the construction and treatment of complete direct pole figures (CDPF) with an accuracy corresponding to the increased possibility of experimental techniques, are described:

1) methods of defocus effect calculation at X-ray measurement of incomplete DPFs (IDPFs);
2) techniques of automatized experimental construction of complete DPF (CDPFs) for products of different types and, in particular, for the cladding tubes;
3) the calculation of integral texture Kearns parameters by using CDPF (0001) for α-Zr.

2. Principles of the construction method of direct pole figures

Crystallographic texture analysis of polycrystalline materials is generally carried out by X-ray diffractometer measurement of direct pole figures {hkl} by tilting the sample (Figure 1) for the corresponding reflections (hkl) [2-6].

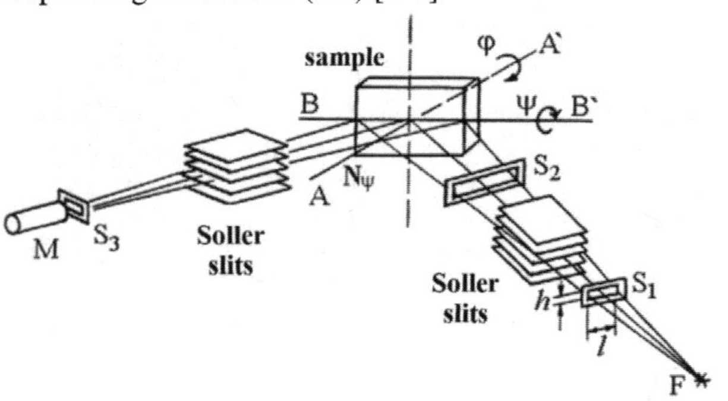

Figure 1. The scheme of DPF measurement by tilt method on reflection of diffracted beam: F – projection of X-ray tube focus, M – detector, S_i – slits; AA′ and BB′ – rotation axes of sample.

DPF {hkl} is the pole density distribution of normals (axes) of the selected type <hkl> in the stereographic projection. Pole density is directly proportional to the intensity of the X-ray beam reflected from the respective planes {hkl}. Therefore, to measure the PPF the intensity of diffracted X-ray radiation from the planes {hkl} is recorded for all possible orientations of the sample relative to the direction of the primary and reflected beams. Planes of the selected type {hkl} reflect at different orientations of the sample if their normals lie in plane containing the primary and reflected beams and divide the angle between them in half. The value of the recorded intensity is proportional to the volume of grain involved in the formation of the diffracted beam.

The procedure of X-ray diffractometer measurement of texture are describe in detail in [2-4] and consist in recording the intensity of X-ray scattering at a certain Bragg angle 2θ and successive positions of the sample, which are determined by the angles ψ and φ, where ψ - the tilt angle of the sample varying between 0 - 80 °, and φ - the rotation angle of the sample around the normal to the surface, varying from 0 - 360 ° (Figure 1). Since these measurements are taken at different angles of the sample, they correspond to the different geometry of the X-ray diffraction - unchanged preserved only angle 2θ, but changing the angle between the surface normal and the incident beam between the normal to the surface and the diffracted beam, changing the area of the irradiated surface of the sample and X-ray penetration depth into the material. The partial violation of the terms of focusing (i.e. the "defocus") caused by sample tilt leads to a broadening of the X-ray line profile and a drop in recorded intensity becoming significant when $\psi > 50°$; moreover, defocus can lead to the angular displacement of the X-ray line. To reduce the effect of defocus it is necessary to limit the vertical divergence of the primary beam and the height of its projection in the sample plane [2]. The system of collimating slit was shown in Figure 1. Soller slits with an angle of divergence 1,5°, installed on the primary beam, help to reduce its vertical divergence, and Soller slit with an angle of divergence 2,5°, set on the detector, prevent it from getting scattered radiation. Even at optimal geometry of measuring it is necessary to consider the maintaining defocus effect by adjusting the obtained data on the basis of a periodic measuring of calibration samples.

Thus, the input data for the construction of PPF are [2]:

1) I_{ij} - the measured intensity values at successive points orientation space (ψ_i, φ_j), where ψ_i varies from 0 to a maximum value ψ_{max}, φ_j - from 0 to 360°; for polycrystalline materials the uniform step for both corners ($\Delta\psi=\Delta\varphi=2.5$ or 5°) is commonly used; for sharp texture or single crystals a detailed measuring near the texture maximum is used;

2) $I_{i\,bg} = I_{\psi}$ – background intensity values of the sample (at $2\theta_{bg}$) for the respective tilt angles ψ_i;

3) K_i – the magnitude of correction coefficient of defocus also varying with the tilt angle ψ_i. Methods of their measurement or calculation are given below.

Then the calculation of the pole density at each point of the stereographic projection carried by the following formula: $P_{ij} = [I_{ij} - I_{ibg}]\cdot K_i/\bar{I}$, where average value of intensity for studied region of IDPF is

$$\bar{I} = \int_0^{2\pi}\int_0^{\psi_{max}} [I(\psi,\varphi) - I_{\psi}]\cdot \sin\psi \cdot d\psi \cdot d\varphi \Big/ \int_0^{2\pi}\int_0^{\psi_{max}} \sin\psi \cdot d\psi \cdot d\varphi\,,$$

and $\int_0^{2\pi}\int_0^{\psi_{max}} \sin\psi \cdot d\psi \cdot d\varphi = s$ - area of studied orientation space of DPF, which equal 2π in case construction of complete DPF.

3. Accounting of defocus of X-ray beam in the construction IDPF by tilting

Construction of DPF is impossible without accurate knowledge of the correction coefficients K_i. The following methods are used for their determination: 1) "sewing" of tilt curves for special composite samples; 2) measuring tilt curves for textureless standard; 3) analytical calculation for sample of small size (the sample size less than area of primary X-ray beam projection on the sample).

The practical difficulty of obtaining powder textureless standards with a density close to the theoretical does not allow to rely on defocus coefficients obtained by the second mentioned methods. In addition, when preparing the textureless powder the crystallographic texture of compression or stacking may be formed in them, what also results in an error in the determination of K_i. Therefore in this paper we consider only the first of these method as more original, since the other two is not difficult for researchers on texture analysis.

As cold-rolled Zr-based alloys are characterized by texture (0001)±30-40°ND-TD<10$\bar{1}$0>, in which the basal normals distributed mainly in the section ND-TD of DPF(0001) (Figure 2), this variant of determining the defocus coefficients comprises preparing a three composite samples, normals to the studied surfaces are oriented over each thirty degrees along section DPF (0001) as shown in figures 2 and 3, where RD - rolling, ND - normal and TD – transverse directions in sheet, correspondingly. Exits of normal to studied sample surfaces shown in Figure 3 are built in DPF (0001) in Figure 2 and are marked as a, b and c, correspondingly.

Assuming that at the optimal geometry of measuring DPFs, defocusing effect does not affect the intensity of the diffracted beam up to the tilt angle of the sample 50°, we will use parts of the tilt to ψ

= 30° as reference (Figure 4). Calculation of the complete cross section of DPF (0001) TD-ND as a result of the construction of reference parts of tilt curves for all samples a, b, c, shown in Figure 3 and the presence of the tilt curve for the sample a recorded up to 80° (Figure 4, curve a) allows us to calculate the defocusing coefficients of the diffracted X-ray beam. The variation of defocusing coefficients with increasing tilt angle is shown in Figure 5. Defocusing coefficients K_i shows how many times is it necessary to increase the intensity of the measured value to obtain its true meaning.

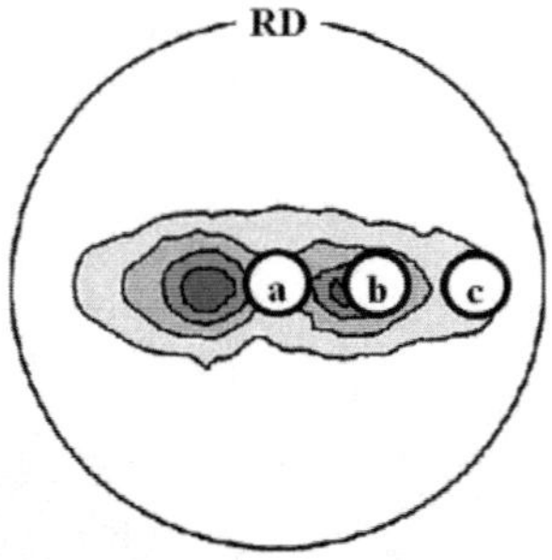

Figure 2. Typical DPF (0001) for rolled sheets from Zr-based alloys.

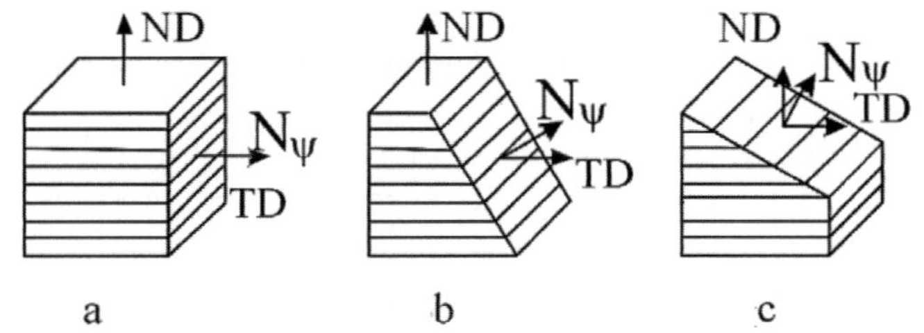

Figure 3. Composite samples used for construction of coefficients of diffracted beam defocusing: a – sample *a*, b – sample *b*, c – sample *c*.

The advantage of this technique is to obtain correction factors for the bulk samples with a density which coincides with the density of the material.

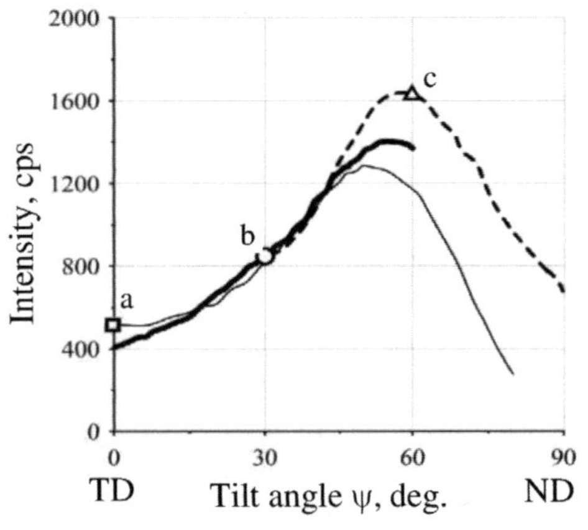

Figure 4. Tilt curves measured for composite samples presented in Figure 3. Exits of normal to studied section are marked by white points.

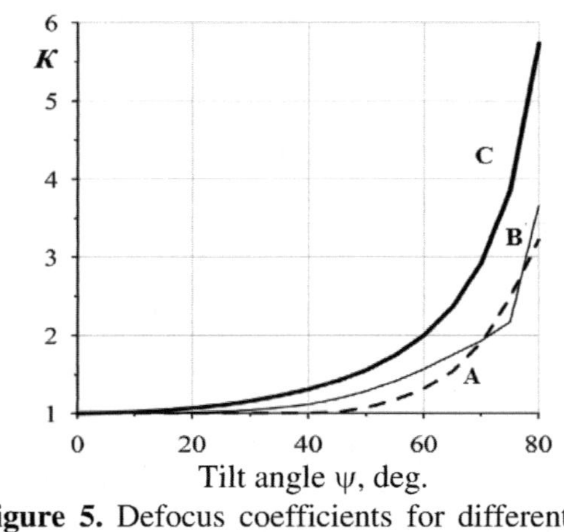

Figure 5. Defocus coefficients for different tilt angles: A - measured for composite samples; B – measured for textureless powder standard; C) calculated for small sample.

Figure 5 also shows the defocus coefficient changing at depending on the tilt angle obtained for the powder sample (B) and sample of small size (C). Comparison of curves A and B shows significant differences in the magnitude of the coefficients of defocusing obtained for textureless standard and composite samples for the same reflection (0002).

4. The methods of construction complete DPFs

For a complete description of the crystallite orientation of hexagonal system it is enough to know the orientation of the basal normal [0001] and rotation around it. That is, for evaluation of the crystallographic texture polycrystal it is necessary to construct two complete DPF (0001) and the

$\{10\bar{1}0\}$ or $\{11\bar{2}0\}$. Having regard to the features of the X-ray spectrum of α-Zr, the texture of products from Zr-based alloys was characterized by DPF (0001) and $\{11\bar{2}0\}$.

In the study of the texture of α-Zr, to quantify integral texture Kearns parameters or the so-called *f*-parameters, it is necessary to construct CDPF. In the case of the measuring of a single section of the sample, only IDPF can be constructed, due to the geometry of a standard method of measuring texture "by reflection". It is filled by following various methods of restore of CDPF from some IDPFs:

1) the completing of peripheral regions of the PF by using a result of the "transmission" measuring;
2) the coincidence, or "sewing" of three IDPF obtained for three mutually perpendicular surfaces;
3) the using of equi-tilted sample;
4) extrapolation of data of IDPF in the uncharted area of stereographic projection;
5) the using of orientation distribution function (ODF) to reconstruct CDPF by several IDPF recorded for the same surface.

Using the method (1) connects with considerable difficulties associated with the preparation of X-ray transparent foil of the same thickness over the entire area of the sample. Method "sewing" (2) is simpler, but requires three times more the initial data and sufficiently complicated software. In the case of "sewing", precision cutting of products and preparation of composite samples are required. In addition, the method of "sewing" cannot be used in the presence of a layerwise sample inhomogeneity or large grains. The complexity of preparation of exactly oriented equi-tilted plane for complex products in the method (3) leads to the fact that it is impossible to distinguish between the methodological error and the existential asymmetry of the sample texture. For the channel and especially for cladding tubes, the method of equi-tilted sample does not apply because of the fundamental impossibility of preserving the external orientation of the axes of the sample across the surface prepared. The quality of the data obtained by the method (4) depends on the accuracy of determining the coefficient of defocusing, the correction of extrapolation algorithm and the location of texture maxima.

It is widely used at present mathematical method reconstruction CDPF using ODF [7, 8] for the products of Zr-based alloys is unacceptable because of significant differences integral Kearns' parameters calculated from the experimentally built CDPF and restored from ODF. In addition, developers ODF indicate the possibility of the formation of additional false peaks in the reduction CDPF [9, 10]. Based on the above as the main methods of CDPF reconstruction are generally used methods (2) and (4) which are effectively applied to the analysis of the crystallographic texture of zirconium products depending on its features and problems solved. The most express method in the study of thin-walled tubes is an extrapolation technique of IDPF on uncharted area of the stereographic projection. In this case the accuracy of the data depends mainly on the accuracy of determining the coefficients of defocusing. The best results are achieved if the orientation of the basal normals is near the normal to the surface under investigation. In the case of thick layers of homogeneous products the method of "sewing" are used.

4.1. "Sewing" of IDPF obtained for three mutually perpendicular cross-sections of the sample.

For three mutually perpendicular planes of the sample (Figures 6, 7a), the normal to which is designated as ND (R), TD (T) and RD (L) are recorded three IDPFs, where R, T, L – radial, tangential, and longitudinal directions of the tube. Scheme of sample preparation from tubes for X-ray study of the texture is shown in Figure 6, for sheets - in Figure 7 a.

Further, in the designation of the surface under study and DPF the name of the appropriate normal will be used. To match the results of the three measuring IDPF, two of them - $IDPF_{TD}$ and $IDPF_{RD}$ - should be brought to the axes of the first $IDPF_{ND}$ by applying to spherical coordinates when they turn on the 90° around one of the axes. Using the formula given in [11], we can obtain the relation between the coordinates of $IDPF_{TD}$ (ψ, φ) and coordinates of $IDPF_{ND}$ (ψ′,φ′):

$$tg\varphi' = (1/tg\psi/\cos\varphi)$$
$$tg\psi' = -1/tg\varphi/\cos\varphi',$$

and also between the coordinates of $IDPF_{RD}$ (ψ, φ) and coordinates of $IDPF_{ND}$ (ψ′,φ′):

$$tg\varphi' = tg\psi * \sin\varphi$$
$$tg\psi' = -tg\varphi / \sin\varphi'.$$

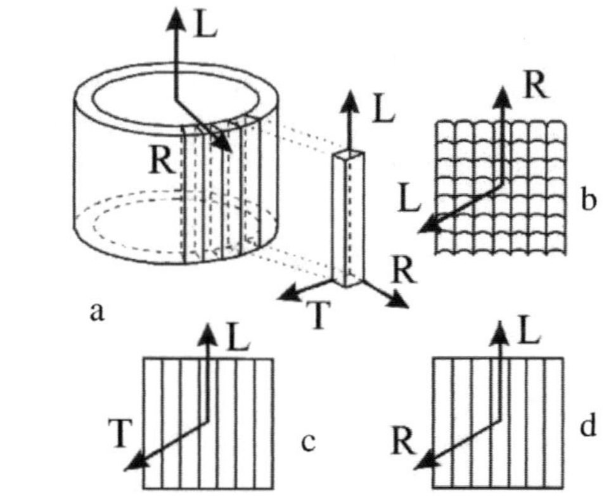

Figure 6. The preparation of the composite samples from the tube for X-ray study of its texture: a – cutting off segments from the tube; b – L-sample; c – T-sample; d - R-sample. Studied surfaces of composite samples are perpendicular to corresponding tube axes.

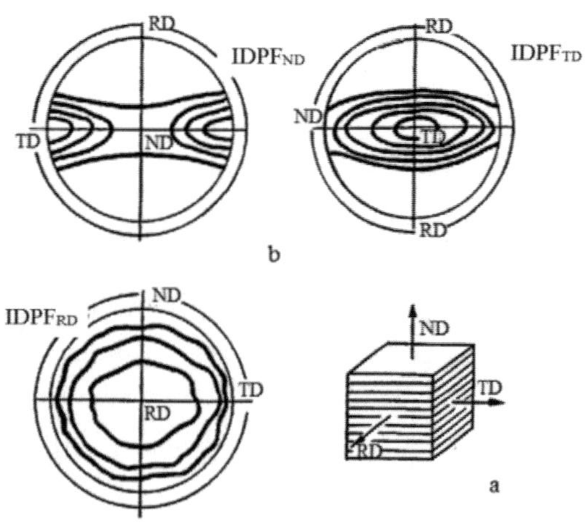

Figure 7. IDPF (0001) (b) built for the three mutually perpendicular planes (a).

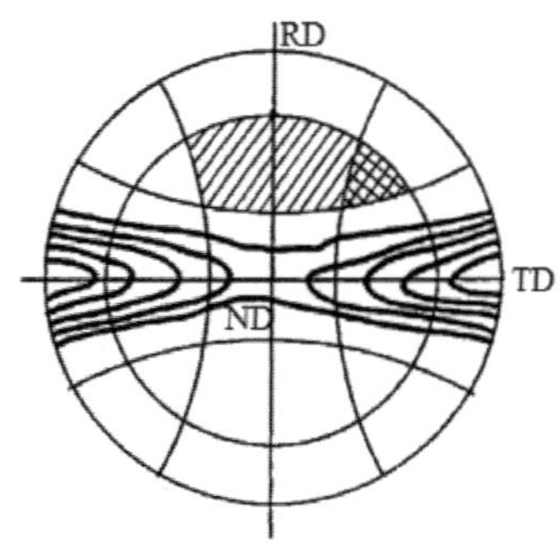

Figure 8. Arrangement of IDPFs Figure 7 on the stereographic projection.

Taken together DPFs cover all areas of the stereographic projection (Figure 8), which allows you to calculate CDPF. If the limit tilt angle of the sample at measuring IDPF is more than 54,7°, besides the double areas of overlapping, triple areas appear, i.e. areas in which the value of the pole density (or intensity proportional to it) at the each point of stereographic projection can be determined from the two or three IDPFs recorded for mutually perpendicular surfaces. "Sewing" of IDPFs is carried out by comparing the intensity of X-ray reflection at equivalent points of different pole figures (Figure 8). Since the intensity measured for different IDPFs at equivalent points belongs to different volumes of the material (as different sections of the sample are explored, even in layerwise uniform sample), and registration of the intensity occurs with an error of pulse count obeying Poisson's distribution [12] you can not expect a perfect coincidence of intensity values at equivalent points. Therefore, to adjust the values of $IDPF_{TD}$ and $IDPF_{RD}$ by using overlapping areas, the average values of the coefficient $\overline{X} = P_k/P_l$, where k and l are indices of one of the directions of ND, RD or TD, are found. If the average value of this ratio is 1, this corresponds to an ideal "sewing" of IDPF.

The measurement error is inversely proportional to the intensity of $I^{1/2}$, then the error of intensity ratio I_1/I_2 is expressed as $E=\sqrt{1/I_1^2+1/I_2^2}$.The higher the recorded X-ray reflection intensity, the greater the accuracy of the measured value (weight). Weight of measured intensity is directly proportional to I_1/I_2, and then the weight of calculated ratio I_1/I_2 is expressed as $W=\sqrt{\frac{I_1 \cdot I_2}{(I_1+I_2}}$. Since the calculation of the coefficient of "sewing" is conducted in the areas of DPF with different pole density, it is necessary to take into account the weight of the calculated values, and hence to carry out the calculation of the average value considering weight. This method of calculating the weighted average coefficients $\bar{X}$ can significantly reduce the error of their determination. The closer the ratio $\frac{\bar{X}_1 \cdot \bar{X}_3}{\bar{X}_2}$ to the unit, the more accurate the coefficients of "sewing". If sample is produced accurate and its installation at the X-ray measuring in the texture holder of attachment is also accurate, and if the layerwise inhomogeneity is absent, then deviation of calculated ratio $\frac{\bar{X}_1 \cdot \bar{X}_3}{\bar{X}_2}$ from unity must be minimum.

4.2. Extrapolation of the data of IDPF in the uncharted areas of stereographic projection.

The measurement of experimental IDPF is carried out to tilt angle 80°, IDPF is completed by the extrapolation of the data in uncharted area of the stereographic projection. Let us consider the distribution of pole density (intensity) in section IDPF specified by some angle φ. Given the equivalence of DPF points with coordinates (ψ = 90°, φ) and (ψ = 90°, φ + 180°), the problem of extrapolation can be reduced to the problem of interpolation. We get a one-dimensional distribution function of the intensity with three uncertain points with coordinates ψ=85, 90, -85°. Using the Lagrange interpolation formula [11], we can calculate the value of the intensity at an intermediate point ψ:

$$I(\psi)=\frac{(\psi-\psi_1)(\psi-\psi_2)\ldots(\psi-\psi_N)}{(\psi_0-\psi_1)(\psi_0-\psi_2)\ldots(\psi_0-\psi_N)}I_0+\frac{(\psi-\psi_0)(\psi-\psi_2)\ldots(\psi-\psi_N)}{(\psi_1-\psi_0)(\psi_1-\psi_2)\ldots(\psi_1-\psi_N)}I_1+$$
$$+\ldots+\frac{(\psi-\psi_0)(\psi-\psi_1)\ldots(\psi-\psi_{N-1})}{(\psi_N-\psi_0)(\psi_N-\psi_1)\ldots(\psi_N-\psi_{N-1})}I_0.$$

Note that the Lagrange interpolation formula gives satisfactory results only in the case of sufficiently "smooth" of the initial data. Therefore, the interpolated function is pre-smoothed using the methods of median and deskewing. CDPF calculation is impossible without accurate knowledge of the correction coefficients K_d, taking into account the deviation of the sample from the focusing circle, which leads to the scattering of the diffracted beam and the intensity loss of X-ray reflection.

In case the location of sharp texture maximum outside the study area restored by extrapolation DPF (0001) does not correspond the texture actually existing in the sample. In case of a deviation of the maxima in distribution of basal normals from the radial direction at an angle more than 60^0 (which can also be assessed by the location of texture maxima on DPF {11$\bar{2}$0} or {10$\bar{1}$0} near the radial direction) the calculation of *f*-parameters must be carried out either by extrapolating DPF (0001), built for the T (TD)-surface, or use the technique of "sewing".

5. The method of calculation of the integral texture Kearns' parameters by CDPF (0001)

Materials with a hexagonal structure characterized by a significant anisotropy of physical and mechanical properties. For example, the Young's modulus for a Zr single crystal changes from 1.25 GPa in the direction of [0001] to 0.99 GPa along <10$\bar{1}$0> [13]. For the polycrystalline materials Kearns [1] introduced the integral textural *f*-parameters which are widely used in the calculation of the properties of materials with a hexagonal lattice and called as *f*-parameters or Kearns' parameters [1, 14].

If polycrystalline is represented as a set of grains with different orientations, i.e. their interaction with each other is ignored, the property value in the selected direction of the polycrystalline material is the sum of the properties of i-th crystals [1, 14-16]:

$$P(\psi) = P_c \sum_i V_i \cos^2\psi_i + P_a \sum_i V_i (1 - \cos^2\psi_i) = f \cdot P_c + (1-f) \cdot P_a,$$

where P_c and P_a are properties along the crystal lattice axes c and a, respectively, and ψ_i - the deviation angle of the selected direction of i-th grain from the normal to the basal plane; from axis c. Then the Kearns' parameters are defined as the sum of effective volume fractions of grains $\sum_i V_i \cos^2\psi_i$, where $V_i = p(\psi,\varphi)*\delta\omega = p(\psi,\varphi)*\sin\psi\ \delta\psi\ \delta\varphi$ - volume fraction of grains calculated from DPF (0001) by pole density value $p(\psi,\varphi)$ registered in solid angle $d\omega$ limited by sections $\psi \div \psi+\Delta\psi$ и $\varphi \div \varphi+\Delta\varphi$ (Figure 9).

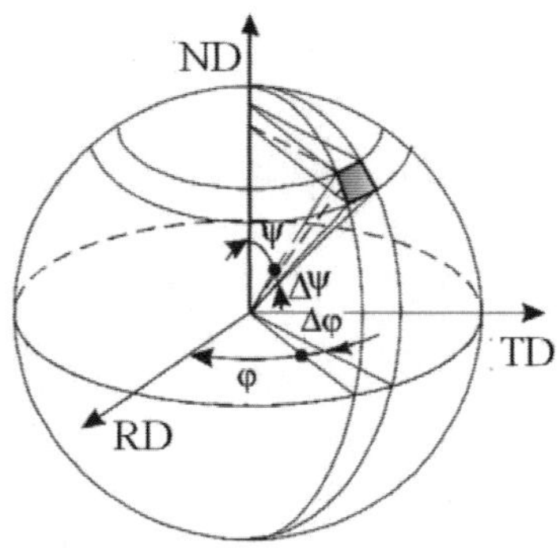

Figure 9. To the question of calculation of f-parameters: angular element on the unit sphere.

The value of f-parameter is represented as the sum (integral) of volume fractions across the all stereographic projection [1, 14-16]:

$$f = \int_0^{2\pi} \int_0^{\frac{\pi}{2}} p(\psi,\varphi) \cdot \sin\psi \cdot \cos^2\psi \cdot d\psi \cdot d\varphi$$

Normally, f-parameters are calculated for the three main directions of products ND, TD and RD in a sheet or R, T and L in the tube [16]:

$$f_R = \int_0^{2\pi} \int_0^{\frac{\pi}{2}} p(\psi,\varphi) \cdot \sin\psi \cdot \cos^2\psi \cdot d\psi \cdot d\varphi$$

$$f_T = \int_0^{2\pi} \int_0^{\frac{\pi}{2}} p(\psi,\varphi) \cdot \sin^3\psi \cdot \sin^3\varphi \cdot d\psi \cdot d\varphi$$

$$f_L = \int_0^{2\pi} \int_0^{\frac{\pi}{2}} p(\psi,\varphi) \cdot \sin^3\psi \cdot \cos^2\varphi \cdot d\psi \cdot d\varphi$$

As seen from the above formulas, the calculation of f-parameters is carried out by means of the complete PF (0001), besides the sum of them should be equal the unit: $f_R + f_T + f_L = 1$.

Since the precision of construction of DPF depends on the intensity recorded in its various areas, so far the texture measurement error is characterized by the orientation distribution of the normals to the reflecting planes in the stereographic projection of the test sample as described in the works [17, 18]. Building such distribution for the instrumental errors by comparing the results of multiple PF recording of the same sample shows that they increase in minima of PF and decrease in texture maxima. Textured heterogeneity is seen as a factor which introduces a mistake in the description of the texture of the material based on the results of the measuring of single sample.

Error of constructing of CDPFs (0001) also affects the values of integral texture parameters. The analysis of the calculated integral *f*-parameters [19] indicates that the error of their calculation, is not more than 0.02 provided that all recommendations by the construction CDPF (0001) are adhered.

6. Conclusions

Construction of a complete direct texture pole figures, even with the modern development of X-ray equipment, and computer processing of the results is very time-consuming task, especially in those cases when it is necessary to define the integral texture parameters with high accuracy as in the analysis of products from Zr-based alloys for nuclear technology. The article discusses ways to correct for defocusing, experimental techniques of construction of complete direct pole figures using the "sewing" texture data for different cross sections of the product and by extrapolation, and method of determining the integral texture parameters. The algorithms of computer processing of the results of the texture measurements used in the study of products from zirconium alloys are described.

Acknowledgements

This work was performed within the framework of the Center of Nuclear Systems and Materials supported by MEPhI Academic Excellence Project (contract № 02.a03.21.0005, 27.08.2013).

References

[1] Kearns J J 1965 *Thermal expansion and preferred orientation in Zircaloy* (WAPD-TM-472, TID-4500. Bettis Atomic Power Lab., Pittsburgh P.A.)
[2] Borodkina M M and Spector E N 1981 *X-ray analysis of texture of metals and alloys* (Moscow: Metallurgiya) p 272
[3] Umanskiy Ya S, Skakov Yu A, Ivanov A N and Rastorguev L N 1982 *Crystallography, X-ray diffractometry and electron microscopy* (Moscow: Metallurgiya) p 632
[4] Gorelik S S, Rastorguev L N and Skakov Yu A 1970 *X-ray diffraction and electron microscopic analysis* (Moscow: Metallurgiya) p 358
[5] Kocks U F, Tome C N, Wenk H-R 1998 *Texture and Anisotropy. Preferred orientation in polycrystals and their effect on materials properties* (Cambridge University Press) p 676
[6] Tailor A 1965 *X-ray metallography* (Moscow: Metallurgiya) p 366
[7] Bunge H-J 1993 *Texture analysis in materials science* (Cuvillier Verlag, Gottingen) p 593
[8] Dahlem-Klein E, Klein H and Park N J 1993 *Program system ODF-analysis for cubic crystal symmetry orthorhombic sample symmetry* Bunge H-J (ed) (Cuvillier Verlag, Gottingen) p 109
[9] Matthies S 1982 *Phys. Stat. Sol.* **B112** K111-14.
[10] Matthies S and Vinel G W 1979 *Phys. Stat. Sol. B* **92** 135
[11] Corn G and Corn T 1974 *Mathematical Handbook* (Moscow: Nauka) p 832
[12] Heiker D M and Zevin L S 1963 *X-ray diffractometry* (Moscow: Phizmatgiz) p 280
[13] Douglas D 1975 *Material science of zirconium* (Moscow: Atomizdat) p 360
[14] Tempest P A 1980 *J. Nucl. Mater.* **92** 191
[15] Nyi J 1967 *Physical properties of crystals* (Moscow: Mir, transl. from English) p 385
[16] Hofer G. 1986 *Experimental Techniques of Texture Analysis* p 331
[17] Kapliy S N, Isaenkova M G and Perlovich Yu A 1992 *Industrial laboratory* **5** 23
[18] Isaenkova M G, Kapliy S N and Perlovich Yu A 1992 *Soviet Atomic Energy* **72** (2) 181
[19] Isaenkova M G, Perlovich Yu A and Fesenko V A 2013 *Industrial laboratory* **79** (7) part 1 25

IOP Conf. Series: Materials Science and Engineering **130** (2016) 012056 doi:10.1088/1757-899X/130/1/012056

Optimization of the procedure for determining integral texture parameters of products from zirconium-based alloys using the orientation distribution function

Yu A Perlovich, M G Isaenkova, O A Krymskaya, V A Fesenko and Y A Babich

National Research Nuclear University MEPhI (Moscow Engineering Physics Institute) Kashirskoe highway 31, Moscow,115409, Russia

E-mail: yuperl@mail.ru, isamarg@mail.ru

Abstract. Integral texture Kearns parameters, or *f*-parameters, are the one of the characteristics that is controlled during the tube production of Zr-based alloys now. These parameters define the anisotropy of physical and mechanical properties of the products. *f*-parameters are the sum of effective fractions of basal normals aligned in a selected sample direction. However, so far there is no standardized procedure for their calculation. The widespread availability of X-ray diffractometers, the high statistical significance of the results and the software development for texture analysis causes the optimality of the *f*-parameters determination by complete pole figures (0001) calculated through orientation distribution function (ODF). Accuracy analysis of the ODF extraction by several incomplete pole figures using LaboTex software was carried out in this paper. The comparison of the calculated data was performed with experimental complete pole figures (0001) obtained by «sewing» method. Various texture types, different number and combinations of incomplete pole figures were considered. For this purpose, 6 incomplete pole figures – (0001), $\{11\bar{2}0\}$, $\{11\bar{2}2\}$, $\{10\bar{1}2\}$, $\{10\bar{1}3\}$, $\{10\bar{1}0\}$ – were registered for each of three mutually perpendicular surfaces of the CANDU-tube sample. By comparison of experimental and calculated by ODF complete pole figures as well as *f*-parameters, procedures of their determination were optimized for products from Zr-based alloys.

1. Introduction

At the present time one of the characteristics, controlled during manufacturing of cladding tubes of Zr-1% Nb alloy, are integral texture Kearns parameters, or f-parameters, which define anisotropy of different mechanical properties [1]. These parameters are the sum of the effective fractions of basis normals (0001), oriented along any selected direction in the sample. They are determined by complete direct pole figures (PF) (0001) [2]. However there is still no standardized procedure for constructing complete PF. Among the most common methods: extrapolation of the incomplete PF data to an unregistered region; "sewing" of incomplete PF registered for three mutually perpendicular surfaces; the calculation through the grain orientation distribution function (ODF), which in turn can be constructed from several incomplete PF or electron backscattering diffraction data (EBSD) [3-4].

Despite widespread occurrence in recent decades of EBSD method for direct measurements of the volume fractions of grains, having specified orientations, and for ODF calculation, it is very local and therefore cannot be used for the products characterizing. Moreover, the reconstruction error of the grain boundaries by Kikuchi lines when analyzing deformed samples do not allow to study the texture evolution of cladding tubes at various stages of their thermomechanical treatment, what is necessary for optimization of the production process.

At the same time the majority of material science laboratories, both in research institutions and in the production, have X-ray diffractometers equipped with texture devices, which allow to perform registration of incomplete PFs. In paper [2], the authors had described in detail the limits of applicability of extrapolation methods and "sewing" for the complete PF reconstruction of products from Zr-based alloys. So the extrapolation method can not be applied in the case when basal normals are oriented outside of the experimentally measured PF area. The results of the "sewing" are affected by the layered inhomogeneity of the sample. One more disadvantage of this method is the necessity to prepare three mutually perpendicular surfaces of the product that associated with certain difficulties in the case of thin-walled cladding tubes of small diameter.

In papers [5-6] the attempts were made for the comparison of *f*-parameters calculated by the complete PFs, which were constructed using mentioned methods. But the authors have not analyzed the features of ODF, reconstructed by different number of incomplete PFs, which could significantly influence on the *f*-parameters calculation. Moreover, in paper [6] during samples preparation of rod only the section perpendicular to the transverse direction (T) had been strictly kept, while two other cross-sections, perpendicular to the longitudinal (L) and radial (R) directions, were not fixed, so *f*-parameters for these sections are inaccurate. Furthermore, the authors didn't take into account during ODF-reconstruction all possible incomplete PFs, which can be obtained for zirconium alloys.

The construction of complete PF through ODF has its limitations and drawbacks, especially as regards the accessibility of required number of incomplete PF for experimental registration. Therefore, in this study in order to optimize the methodology of the *f*-parameters determining the results of their calculation by complete PF, reconstructed through the ODF using a specific set of initial incomplete PF, were analyzed.

2. Experimental data

In the absence of layered inhomogeneity of products, as the more accurate method for the complete PF reconstruction can be considered "sewing" because in this case the construction is implemented entirely on the basic of experimental data. Therefore, the error analysis of the ODF calculation and *f*-parameters determination was conducted by complete PFs obtained by "sewing". For this purpose, for three mutually perpendicular surfaces (longitudinal (L), transverse (T), radial (R)) of the CANDU-tube sample on DRON-3 diffractometer with automatic texture device by using CrK_{α} radiation by incomplete PFs were recorded: (0001), $\{11\bar{2}0\}$, $\{11\bar{2}2\}$, $\{10\bar{1}2\}$, $\{10\bar{1}3\}$, $\{10\bar{1}0\}$ – which available in the spectrum of α-Zr, having a HCP structure. The selection of the sample is determined by its significant difference of the texture obtained for its different surfaces (Fig. 1).This makes it possible to analyze by this example the influence of number of used incomplete PFs, as well as their combinations, on the accuracy of ODF reconstruction. Such an analysis is especially topical for cases of the (0001) PF reconstruction when significant (70-90°) deviation of basis normals from the studied direction occurs (Fig. 1-b), or the presence of additional phase in the material prevents the correct PF (0001) registration.

2. Accuracy analysis of the complete PF reconstruction

The problem of the four-dimensional ODF reconstruction by several incomplete PFs, which are its three-dimensional projections, belongs to the ill-posed class. The uniqueness of its solutions by using any of proposed methods is still questionable. Despite the huge amount of works on accuracy of ODF reconstruction, one of their main drawbacks consists in the determination of error only within the experimentally measured PFs area, which is limited due to geometric features of the X-ray registration with angle 70° when use the "reflection" tilt method.

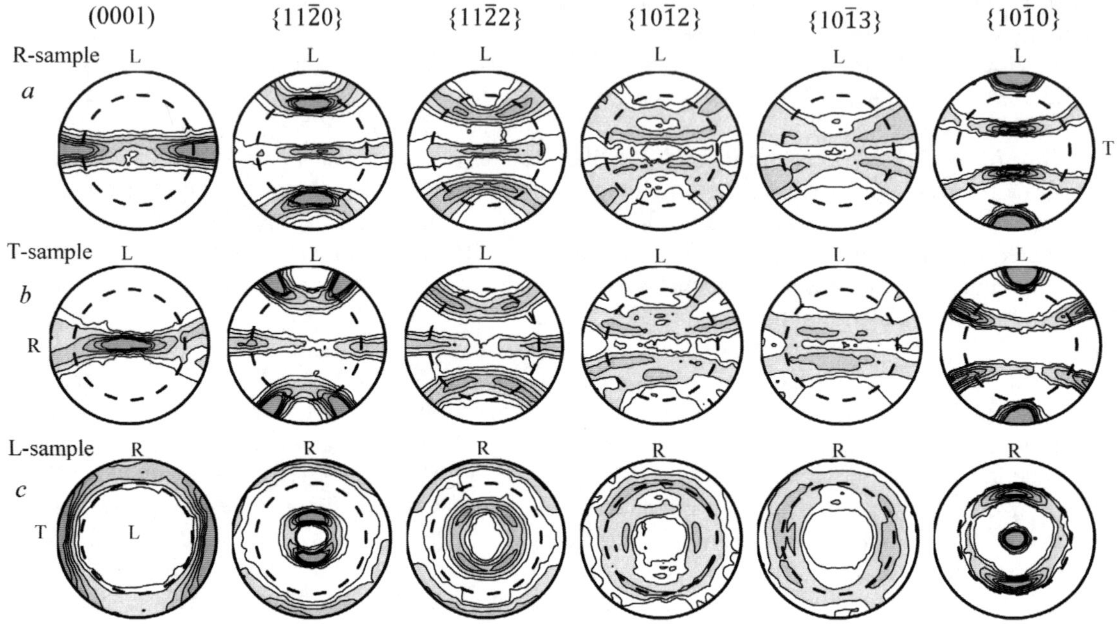

Figure 1. Complete PFs of CANDU-tube sample, constructed by "sewing" method for three mutually perpendicular surfaces: R (*a*), T (*b*) and L (*c*) (dotted line designates the experimentally measured region of stereographic projection with tilt angle 70°)

The quality of the reconstruction is determined by so-called *RP*-factor [7], which is the average deviation of the pole densities *P* in each point *i* of measured region of the stereographic projection by all used PFs:

$$RP_{\{hkl\}} = \frac{1}{I}\sum_{i=1}^{I}\left|\frac{\{P_{exp}\}_i - \{P_{calc}\}_i}{\{P_{exp}\}_i}\right| \cdot 100\% \ ; \quad RP = \frac{1}{N}\sum_{n=1}^{N} RP_{\{hkl\}}, \text{ where}$$

$RP_{\{hkl\}}$ – relative error for each used PF $\{hkl\}$;

$\{P_{exp}\}_i$ – experimentally measured *P* in the *i*-th point of the stereographic projection;

$\{P_{calc}\}_i$ – calculated *P* in the *i*-th point of the stereographic projection;

I – number of experimentally measured points of the stereographic projection;

N – number of incomplete PFs used in the ODF calculation.

The low value of *RP*-factor does not always indicate the high accuracy of the ODF reconstruction, because the main deviations are concerned with the unmeasured region of PF. Thus, analysis of reconstruction errors is an additional problem of quantitative texture analysis, which solution demands the experience and conception of the types of textures.

In the present work for the ODF reconstruction the commercial program LaboTex, based on the discrete method of ADC (Arbitrarily Defined Cells), was used [8, 9]. At that, the analysis of the reconstruction quality was carried out by both *RP*-factor, used in LaboTex, and the average deviation of calculated *f*-parameters from the experimental ones:

$$\Delta f = \frac{\left|f_R^{exp} - f_R^{calc}\right| + \left|f_T^{exp} - f_T^{calc}\right| + \left|f_L^{exp} - f_L^{calc}\right|}{3}.$$

The quality of reconstruction was also evaluated visually by the appearance of complete PF (0001).

Many experimental PFs are not always available for X-ray registration, as in case of the β-phase presence in the sample, whose lines may overlap the reflections of α-Zr. When using the most common X-ray radiation, having wavelength less than CrK_{α}, the reflection (0002) is overlapped by the line (10$\bar{1}$1), and (0004) has a rather low intensity, so there is no possibility to get the basis PF (0001).

Contrariwise for some specific types of textures only PF (0001) has sufficient intensity in the measured region (T-sample), while the others PFs are rather weak within tilt angle of 70°. Therefore, for minimization the number of incomplete PFs, required for the ODF calculation, and optimization of their combination, the influence of total weighting factor *W* [7] of experimental PFs, used in the concrete calculation, was studied. *W* is one of the accuracy characteristics of experimental PFs:

$$W = \sum_{n=1}^{N} \sum_{i=1}^{I} \{P_{exp}\}_i^2 \cdot d\omega_i ,$$

where $d\omega_i$ – is the area of *i*-th element of the unit sphere limited by the tilt angles ($\psi_i - \frac{\Delta\psi}{2}$); ($\psi_i + \frac{\Delta\psi}{2}$), and rotation angles ($\varphi_i - \frac{\Delta\varphi}{2}$); ($\varphi_i + \frac{\Delta\varphi}{2}$), ($\Delta\psi$, $\Delta\varphi$ – variation step of the corresponding angles during PF registration).

3. Results

Different combinations of incomplete PFs used for the ODF calculation, are shown in Table. 1 together with total weighting factors, *RP*-values and *f*-parameters deviations Δf. Combinations of PFs were chosen in a way to maximize the occupation of elementary stereographic triangle for hexagonal lattice. Fig. 2 (*a*, *b*) show reconstructed complete PF (0001) for the cases of minimum and maximum Δf and *RP*, which are marked in bold in the table. 1.

Table 1. The results of the *f*- parameters determination

Calculation variant	R-sample			T- sample			L- sample		
	f_R	f_T	f_L	f_R	f_T	f_L	f_R	f_T	f_L
«sewing»	0.390	0.531	0.079	0.369	0.544	0.087	0.381	0.546	0.074
	W	*RP*	Δf	*W*	*RP*	Δf	*W*	*RP*	Δf
$(0001)\{11\bar{2}0\}\{10\bar{1}0\}\{10\bar{1}3\}\{11\bar{2}2\}$	67.1	45.1	0.013	64.9	44.9	0.022	78.7	53.2	0.024
$(0001)\{11\bar{2}0\}\{10\bar{1}0\}\{10\bar{1}2\}\{11\bar{2}2\}$	67.5	55.9	0.014	64.5	43.0	**0.021**	77.5	59.1	0.021
$(0001)\{11\bar{2}0\}\{10\bar{1}2\}\{10\bar{1}3\}$	44.1	51.7	0.018	44.2	34.5	0.022	46.1	57.7	0.026
$(0001)\{11\bar{2}0\}\{10\bar{1}3\}\{11\bar{2}2\}$	46.8	48.5	0.015	46.6	38.1	0.022	47.1	45.3	**0.009**
$(0001)\{11\bar{2}0\}\{10\bar{1}3\}\{10\bar{1}0\}$	57.8	38.0	0.015	56.2	42.3	0.024	68.5	48.1	0.011
$(0001)\{11\bar{2}0\}\{10\bar{1}0\}\{11.2\}$	60.9	46.5	**0.013**	58.2	45.2	0.025	68.3	51.4	0.027
$\{10\bar{1}0\}\{11\bar{2}0\}\{10\bar{1}3\}\{11\bar{2}2\}$	54.3	45.1	0.023	48.8	45.5	0.085	69.5	63.0	0.028
$\{10\bar{1}0\}\{11\bar{2}0\}\{10\bar{1}3\}\{10\bar{1}2\}$	51.6	48.1	0.014	46.4	41.7	**0.090**	68.5	64.6	0.026
$(0001)\{11\bar{2}0\}\{10\bar{1}3\}$	37.5	39.2	0.019	37.9	33.5	0.027	36.9	55.9	0.013
$(0001)\{11\bar{2}0\}\{10\bar{1}2\}$	37.9	**63.4**	0.021	37.5	28.6	0.032	35.7	45.9	0.028
$(0001)\{11\bar{2}0\}\{11\bar{2}2\}$	40.6	58.4	0.017	39.9	30.7	0.029	36.7	41.7	0.020
$(0001)\{11\bar{2}0\}\{10\bar{1}0\}$	51.6	37.7	0.015	49.5	39.8	0.028	58.1	44.3	0.031
$\{10\bar{1}0\}\{11\bar{2}0\}\{11\bar{2}2\}$	48.1	50.6	0.024	42.1	**51.6**	0.075	59.1	64.8	0.032
$\{10\bar{1}0\}\{11\bar{2}0\}\{10\bar{1}2\}$	45.4	59.0	0.025	39.7	47.2	0.071	58.1	71.9	0.031
$\{11\bar{2}0\}\{11\bar{2}2\}\{10\bar{1}2\}$	34.4	56.4	0.017	30.1	33.4	0.033	36.7	72.9	0.023
$(0001)\{10\bar{1}0\}\{11\bar{2}2\}$	42.4	43.4	0.023	43.1	35.1	0.030	51.0	46.4	0.028
$(0001)\{11\bar{2}0\}$	31.3	47.6	0.023	31.2	**20.0**	0.041	26.5	**24.5**	0.034
$\{10\bar{1}0\}\{11\bar{2}2\}$	29.6	50.2	**0.037**	27.0	46.3	0.079	41.8	66.5	**0.051**
$\{11\bar{2}0\}\{10\bar{1}2\}$	25.1	61.1	0.030	21.4	35.0	0.068	26.5	**76.4**	0.038
$(0001)\{10\bar{1}0\}$	33.1	**28.6**	0.036	34.4	26.1	0.032	40.8	34.8	0.044

As it can be seen from Fig. 2, assessment of the quality at the ODF extraction by widely used *RP*-factor often leads to false results, what can be determined by visual comparison of reconstructed PFs. For example, in the R-sample at minimum *RP* at the overstep border to the unmeasured area of the stereographic projection a discontinuous variation of the pole density is observed, as well as sharp deviation of the maximum towards the L-direction. Such a PF view is mistaken. However, the experimentally measured value of the error is low. In the case of maximum *RP* the complete PF is quite smooth, its view is very similar to obtained by "sewing" and the *f*-parameters deviation is much lower.

For T-sample the presence of a weak maximum at PF {11$\bar{2}$0} in the T-direction (Fig. 1-b) in many calculation variants results in the uncertainty of the orientation of basis normals - the appearance of an additional weak maximum in the L-direction. Since most of the additional PFs is quite diffused, the only ODF-extraction with involved PF {10$\bar{1}$0} gives the result without this maximum. However, PF {10$\bar{1}$0} is located at low Bragg angles and therefore its measurement using radiations with less than the CrK$_\alpha$ wavelength is associated with loss of accuracy due to the increasing of the defocusing coefficients and the overlapping with the reflection (0002). Furthermore, for this type of texture all calculation variants without PF (0001) result in inaccurate ODF reconstruction. This is due to the fact that the location of the basis normals in the center of the stereographic projection is associated with the arrangement of the maxima in other PFs to the periphery of the stereographic projection. So they do not contribute any significant information about the grains orientation.

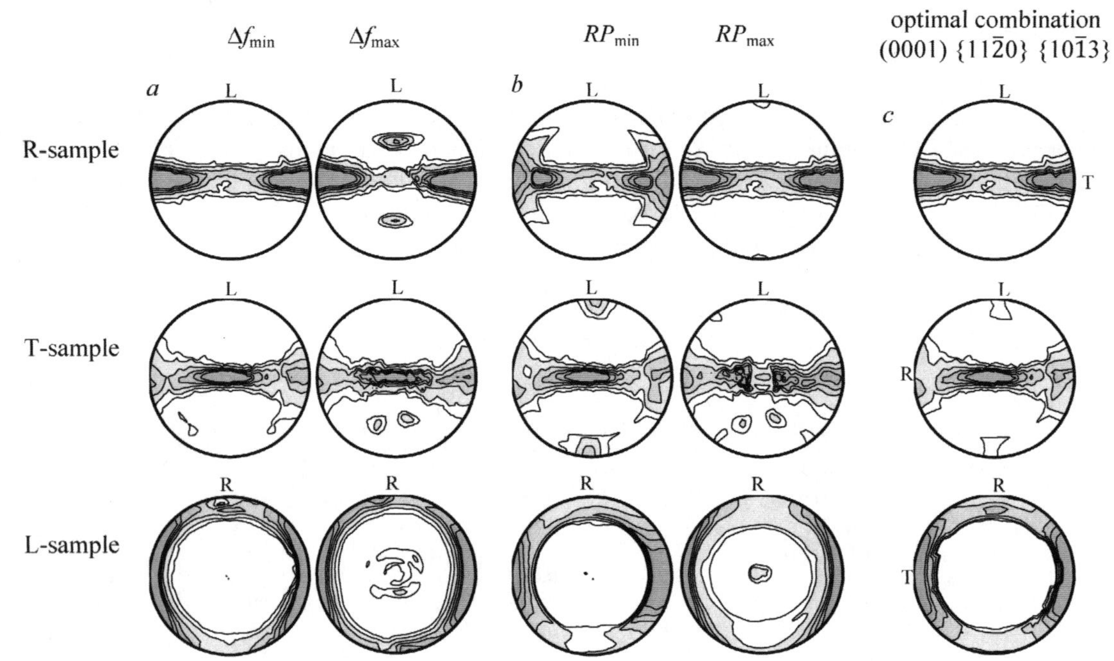

Figure 2. Reconstructed complete PFs (0001) for cases with the minimum and maximum Δf (*a*) and *RP*-factor (*b*), as well as for the optimal combination of incomplete PFs (*c*)

The disagreement to the obtained by "sewing" PF can be seen also for the L-sample at both the minimum and at maximum *RP*. The wider maximum are observed here, and in some cases an additional maximum in the R-direction, which is associated with the ambiguity in case of absence of PF (0001) in the calculation procedure.

The analysis of the PF with the lowest Δf testifies about their maximum similarity with obtained by "sewing" (Fig. 1). Therefore, based on the analysis of table 1 the optimal calculating variant should be selected by the minimum *f*-parameters deviation. Fig. 3 summarizes the results of table 1, which analysis shows that an increasing of total weight of the PFs used for ODF extraction does not result in an reducing of *RP*-factor, and in some cases (T-sample) it even slightly increases. I.e. the increasing of total "precision" of used in calculating PFs has no effect on this characteristic, what once again confirms its failure in the assessment of the reconstruction quality.

The deviation *f*-parameters (Fig. 3-b) tends to decrease with increasing of *W*, but with some plateau in the range *W*= 40-60, which can be considered as optimal values for ODF calculation, because within this range there are calculation variants, having Δf~0.02, what defined by authors of [2] as the minimum error in the *f*-parameters determining. According to Fig. 3-b specified range corresponds to

3-4 used PFs. Due to the necessity of their minimization, one should focus on measuring of 3 incomplete PFs.

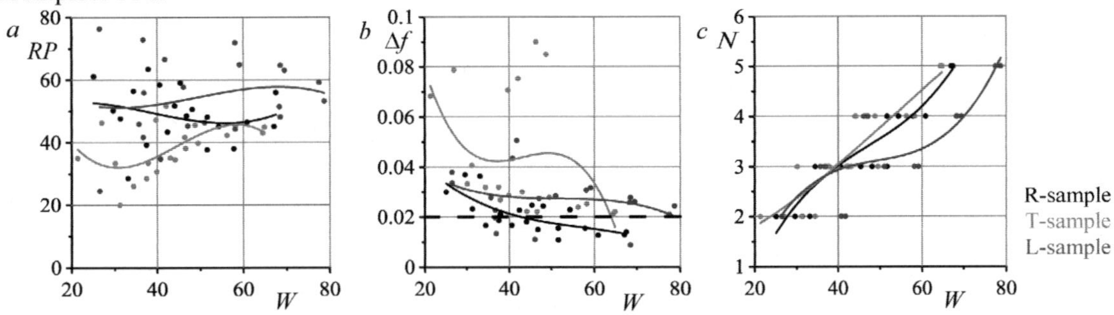

Figure 3. The dependences of *RP*-factor (*a*), Δf (*b*) and number of used incomplete PFs *N* (*c*) on their total total weighting factor *W*

The minimum error in the *f*-parameter determining for all possible combinations observed for R-sample, in spite of the fact that the basis normals is significantly tilted from the normal to the studied surface for this type of texture. The minimum Δf-values for all samples are marked in Table 1 with a red underline, herewith for the T- and L-sample it is the same combination of PFs: (0001) $\{11\bar{2}0\}$ $\{10\bar{1}3\}$. Since in the R-sample for this combination Δf is slightly higher, then it can be considered as optimal for ODF-extraction.

However, in case of unavailability of one of mentioned PF any of combinations with Δf value below 0.03 can be selected. This is particularly actual when there is no opportunity to get basis PF, for example, in case of Zr-alloys with the presence of β-phase. At that, the main problem occurs during ODF extraction for T-sample, for which even the increase of *N* does not lead to an accurate *f*-parameters calculation without PF (0001). The most precise option is a combination of $\{11\bar{2}0\}$ $\{11\bar{2}2\}$ $\{10\bar{1}2\}$, or it may be preferred to calculate the *f*-parameters by extrapolation method in this case [2].

4. Conclusion

Influence of the number of incomplete PFs and their various combinations for different texture types of cladding tubes from Zr-based alloy on quality of ODF extraction was demonstrated. The error analysis of *f*-parameters calculation by complete PF (0001), reconstructed through the ODF, with the experimental data obtained by "sewing" method was implemented. The procedure of *f*-parameters calculation for products from Zr-based alloys was optimized.

Acknowledgements
This work was performed within the framework of the Center of Nuclear Systems and Materials supported by MEPhI Academic Excellence Project (contract № 02.a03.21.0005, 27.08.2013).

References

[1] Kearns J J 1965 *Thermal expansion and preferred orientation in Zircaloy* (Bettice Atomic Power Laboratory Report WAPD-TM-472)
[2] Perlovich Y, Isaenkova M, Fesenko V 2013 *Zavodsckaya laboratoriya* **79** № 7 (1) 25(in Russian)
[3] Randle V, Engler O 2000 *Introduction to texture analysis: macrotexture, microtexture and orientation mapping* (CRC Press, USA) 388 p
[4] Schwartz A J et al 2000*Electron Backscatter Diffraction in Materials Science* (Springer, New York) 403 p
[5] Gruber J A, Brown S A, Lucadamo G A 2011 *J. Nucl. Mater.***408** 176
[6] Mani Krishna K V 2011*J. Nucl. Mater.***414** 492

[7] Bunge H J 1982 *Texture Analysis in Materials Science* (Butterworth, London)
[8] Pawlik K 1986 *Phys. Stat. Sol. (b)***134** 477
[9] LaboTex v. 3.0 by LaboSoft (Krakow, Poland) http://www.labosoft.com.pl

IOP Conf. Series: Materials Science and Engineering **130** (2016) 012057 doi:10.1088/1757-899X/130/1/012057

Express method of construction of accurate inverse pole figures

Yu Perlovich, M Isaenkova and V Fesenko

National Research Nuclear University MEPhI (Moscow Engineering Physics Institute), Kashirskoe highway 31, Moscow, 115409, Russia

E-mail: yuperl@mail.ru

Abstract. With regard to metallic materials with the FCC and BCC crystal lattice a new method for constructing the X-ray texture inverse pole figures (IPF) by using tilt curves of spinning sample, characterized by high accuracy and rapidity (express), was proposed. In contrast to the currently widespread method to construct IPF using orientation distribution function (ODF), synthesized in several partial direct pole figures, the proposed method is based on a simple geometrical interpretation of a measurement procedure, requires a minimal operating time of the X-ray diffractometer.

1. Introduction

The proposed X-ray diffractometer method of constructing the inverse pole figures (IPF) by using tilt curves of the sample has a high accuracy and rapidity. At the present time, following specialists in mathematical methods to describe the texture, it is considered that the most exact and universal method of obtaining IPF is regeneration of the orientation distribution function (ODF). For the construction of ODF it is necessary to have a set of the direct pole figures and a rather complicated computer software for mathematical treatment of data [1-2]. Indeed, this method allows to build IPF for any given section of the test sample without prior diffractometric measuring. However, in most practical cases it is advisable to have more available and express method, although not such universal.

At the simplicity of experimental procedure the proposed method of IPF construction provides a highly accuracy of calculation of weight express texture coefficients P(hkl) for an arbitrarily large number of points within the unit stereographic triangle (for a cubic lattice). This method distinguishes from the traditional ones of construction available IPF often used in practice of X-ray laboratories [3-4].

Proposed procedure for constructing the IPF is considered as applied to materials with a cubic lattice.

2. The basis of express method of constructing the inverse pole figures

Let the slope curves $I_{hkl}(\varphi)$ are registered during the continuous tilt of the test sample and its rapid rotation around the normal to the surface n, where I_{hkl} – the registered intensity of the X-ray reflection (hkl), φ - the tilt angle. Then values $I_{001}(\alpha)$, $I_{011}(\beta)$ and $I_{111}(\gamma)$ accurate within normalization coefficients are proportional to the probability of finding in the sample such grains whose crystallographic axes <001>, <011> and <111> are deviated from the normal to the surface sample n at the angles α, β and γ, respectively, with the accuracy up to normalization coefficients. Judging by the angular sizes of the unit stereographic triangle any external direction n of a cubic lattice of the test material is spaced no more than at 54.74° from the crystallographic axis <001>; at 45° from <011> and 54.74° from <111>; i.e. $\alpha \le 54.74^\circ$, $\beta \le 45^\circ$ и $\gamma \le 54.74^\circ$. Therefore, tilting analyzed polycrystalline sample during recording diffractometric curve at an angle up to 55° for the registration of reflections (001) and (111) and at an angle up to 45° when registering the reflection (011,) we have obtained X-ray reflection at least from one of the families of crystallographic planes {001}, {011} and {111} for each grain (figure 1 *a, b, c*).

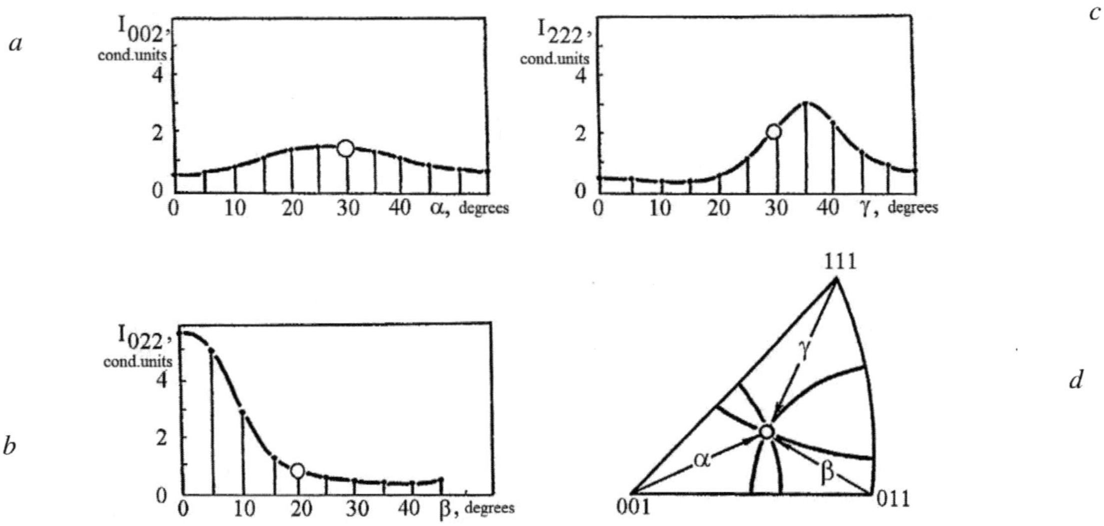

Figure 1. Construction of IPF on sample tilt curves: *a, b, c* – tilt curves $I_{001}(\alpha)$, $I_{011}(\beta)$, $I_{111}(\gamma)$; *d* - a partition of stereographic triangle by circles centered at the poles [[001]], [[011]] и [[111]].

If the axes <001> and <011> of the grains are spaced respectively through angles α and β from the normal n to the sample surface, it may be considered, that these events are as a combination of mutually independent, although fully they are not such. Because at the fixed angle α the angle β cannot take on any value in the range of 0-45°, but only lie within a smaller range, the length and boundary of which depend on the angle α.

Orientation direction of n within the unit stereographic triangle is uniquely defined by two angles: either (α, β), or (α, γ), or (β, γ). At the same time, for the convenience of practical computer build OPF this orientation would be best to set the Cartesian coordinates (x, y) in the plane of stereographic projection.

P(x, y) is denoted the probability that there is in the test polycrystal grain in which the normal n to the sample surface in the stereographic projection plane has coordinates (x, y). Then:

$$P(x,y) = k_1\, I_{001}(\alpha) \times k_2\, I_{011}(\beta)$$
$$P(x,y) = k_1\, I_{001}(\alpha) \times k_3\, I_{111}(\gamma) \qquad (1)$$
$$P(x,y) = k_2\, I_{011}(\beta) \times k_3\, I_{111}(\gamma)\,,$$

where k_1, k_2, k_3 - coefficient of proportionality. Each of the equations (1) with some error defines the same probability of P (x, y). Multiplying together all equations (1), we obtain:

$$P(x,y) = K\,[\Pi(x,y)]^{2/3}, \qquad (2)$$

where

$$\Pi(x,y) = I_{001}(\alpha) \times I_{011}(\beta) \times I_{111}(\gamma)\,. \qquad (3)$$

Thus calculated value of P(x, y) coincides with the probability that in the studied polycrystalline there is a grain whose axes <001>, <011> and <111> have angles α, β and γ with the direction n, respectively (see. figure 1-d).

If the value of P (x, y) is calculated for n points of an elementary stereographic triangle, the normalization coefficient K is found from the following relation:

$$K = \{ \sum^{N} [\Pi(x,y)]^{2/3} \}^{-1} . \qquad (4)$$

When normalizing the sum $\{ \sum^{N}$ P (x, y)$\}$ is not on 1, but on the number of calculation points N, we thereby partially change the meaning of values P (x, y). In this case, P (x, y) is in n times more than the probability of finding the grain with the appropriate orientation, but it already coincides with the meaning of P(hkl), which appears on the IPF constructed by using any of the conventional methods.

In principle, distribution P(x, y) can be calculated by using any of two equations (1). However, in the interest of reducing the possible systematic errors associated with the use of the proposed method of construction of IPF, at the calculations it is advisable to use all three equations (1).

3. The procedure for obtaining experimental IPF

The procedure for obtaining experimental data required to the construction of IPF comprises the following steps:

- diffractometric measurement of tilt curves of sample during spinning around the normal to the its surface at the registration of X-ray reflections from the crystallographic planes {001}, {011} и {111};
- measurement of similar curves by the detector fixed in position of background registration;
- evaluation of the amendments to defocusing when tilting the sample.

When using a computer calculation of IPF by the proposed method, the initial data are ordinates of diffractometric curves of sample tilt $I_{001}(\alpha)$, $I_{011}(\beta)$ и $I_{111}(\gamma)$ measured at regular angular interval $\Delta\varphi$, and the corresponding background values.

The program of the IPF calculation includes:

- determination of the coordinates (x_i, y_i) of all points of ensemble G_o, which are the points of mutual intersection within the unit stereographic triangle of circles with centers at the poles [[001]], [[011]] and [[111]] and with angular radii, multiple $\Delta\alpha$, $\Delta\beta$ and $\Delta\gamma$, respectively (see figure 2-a);

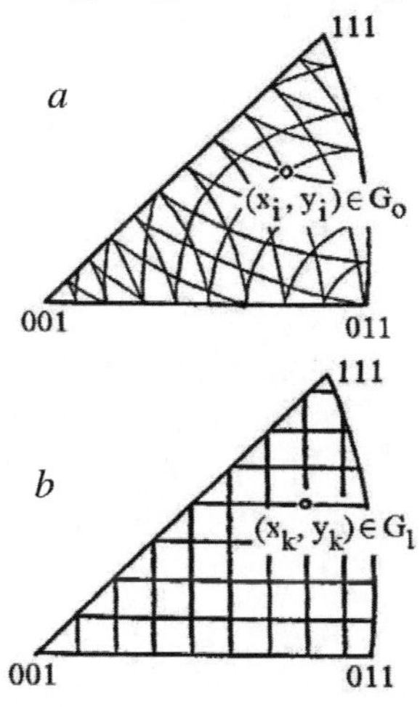

Figure 2. To the calculation of IPF: *a* - G_o set of data points (xi, yi) within the unit stereographic triangle, which are located at the nodes of intersection of the circles with centers at the poles [[001]], [[011]] and [[111]], and with the angular radii, multiple $\Delta\alpha$, $\Delta\beta$ and $\Delta\gamma$, respectively; *b* - G_1 set of calculation points (x_k, y_k), which are located at the nodes of a rectangular grid.

- determination of coordinates (x_k, y_k) of a given set of points G_1, for which will be calculated values *P(hkl)* and which are arranged in sites (nodes) of a rectangular grid covering the entire area of the elementary stereographic triangle (see figure 2-b);
- the calculation of the values of $\Pi(x_i, y_i)$ for all the points of ensemble G_o;
- the calculation of the values of Π (x_k, y_k) for the specified points of the ensemble G_1 by interpolation using as basic units of values P (x_i, y_i) in four (or three) points of set G_o, closest to the given point of the set G_1;

- determination of the coefficients λ_k, which characterize the relative area of a spherical triangle parts corresponding to the different points of ensemble G_1 and differing from each other due to the inhomogeneous distortion of space at a stereographic projection;

- normalization of sum $[\sum^{N} \lambda_k P(x_k, y_k)]$ on the number N, equal to the number of calculation points in the set G_1;

- organization of the array of calculated values $P\,(hkl)$ in the form of IPF and print its image.

The measurement of three tilt curves by means of texture set GP-2 provided with an additional engine requires no more than 0.5 hours.

Figure 3 shows IPFs for the surface layer of the same sample rolled low-alloy chromium constructed by means of the proposed method (a) and method described in [2], based on the procedure for the restore of IPF using ODF (b). (The calculation of the last IPF made by S.F. Kurtasov by the program developed himself.) These IPFs are almost identical in character and only slightly differ by values of pole density at mutual corresponding points.

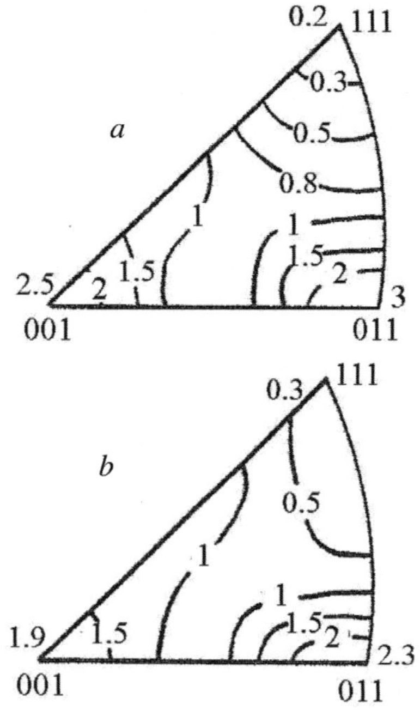

Figure 3. IPFs for the surface layer of the same sample from rolled low-alloy chromium constructed using the proposed method (a) and as described in [2], based on the procedure for the recovery of IPF using ODF (b).

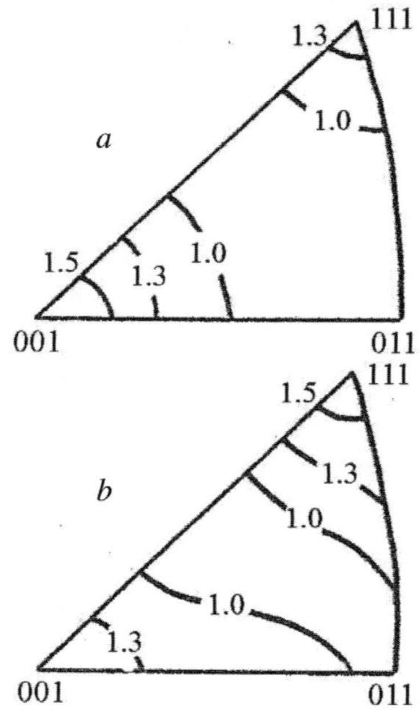

Figure 4. IPF for compound Nb_3Sn in the cross section of plane wire (conductor), calculated by the proposed method (a) and by the more precise method [5] using 14 reflections.

4. Examples of constructing IPFs

The effectiveness of the proposed method is illustrated by the results of the study of the texture of the superconducting compound Nb_3Sn, which has a cubic crystal lattice of type A15. A large number of reflections are observed in the diffraction spectrum of this compound, which allows to obtain relatively accurate information about the texture of the sample, even when using the original method of Harris [5], based on measuring the intensity of the recorded lines. Figure 4 shows IPFs of a sample of compounds Nb_3Sn, calculated by the proposed (a) and improving methods [5] when using 14

IOP Conf. Series: Materials Science and Engineering **130** (2016) 012057 doi:10.1088/1757-899X/130/1/012057

reflections (b). A comparison of obtained IPFs, as in the previous example, confirms that the proposed rapid method is quite accurately conveys features of the studied texture.

It should be noted that the proposed method of construction of IPF is associated with bringing some bias (systematical errors) in calculations, the origin of which is illustrated in figure 5. In the case of increasing the tilt angle of sample above 45° at registering the reflection from the {001} and above 35° for the reflections from the {011} and {111} circle with angular radius φ centered at the corresponding pole oversteps the limits of the stereographic triangles which are mutually symmetric relatively to this pole.

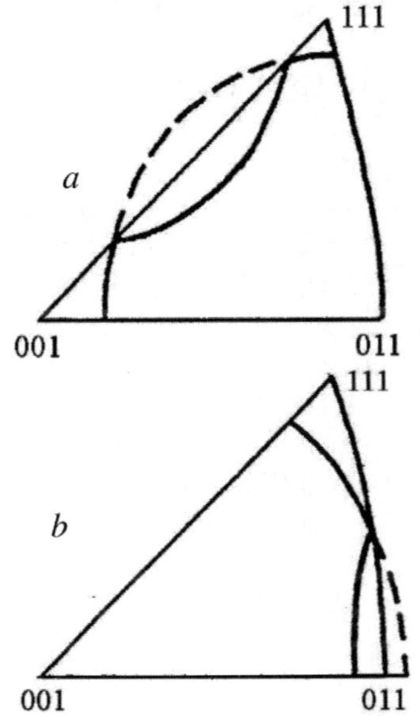

Figure 5. To analysis of the reliability of the proposed method of IPF construction: reflection of outer portion of a circle centered at the poles [[011]] (*a*) and [[001]] (*b*) shown by dotted line inside the stereographic triangle (see explanations in the text).

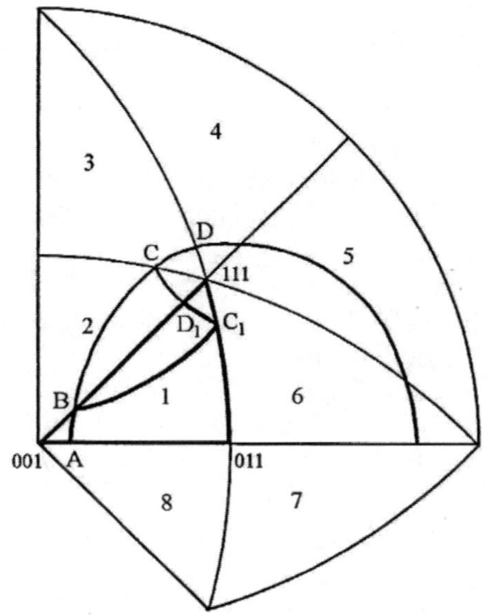

Figure 6. To the question about the formation of the X-ray reflections, recorded at fast rotation of the sample around the normal to its surface. The case of recording the X-ray line (011) (see explanations in the text).

For example, in the case shown in figure 6, the circle with angular radius 40° centered at pole [[011]] oversteps the limits of the stereographic triangles 1, 6, 7 and 8. Therefore, the intensity of X-ray reflection recorded at a rapid rotation of sample about the normal to its surface is determined by averaging the pole density along the polyline ABC_1D_1 constructed within the triangle 1 and consisting from individual segments which correspond to the different sections of the neighboring stereographic triangles. In particular, the segment BC1 corresponds to the section BC of triangle 2 and the segment C_1D_1 - the section CD of triangle 3. Then, the averaged intensity I_{011} (40°), attributed to the points of the arc AB under calculating, actually also depends on the pole density distribution in other areas of the triangle. If the pole density of actual IPF near the poles [[111]] is higher than near the poles [[001]], pole density values attributed to the points near the poles [[001]] under the calculation, are overstated.

However, the error introduced mentioned way in the calculation of IPF is largely reduced due to the simultaneous use of other reflections, at work with which obtained tilt curves characterize the same area IPF without such errors. In this example, these are the reflections from the {001}. The recommended use of three reflections at the sufficiency of IPF construction by any two of them just provides a significant reduction of the pole density in the areas of IPF after the normalization, where recorded average intensity of one of the reflections are too high.

The sharper the investigated texture is, the less the significant errors are due to the procedure of the intensity averaging during measuring of tilt curves of sample. At the same time, when texture is weak, errors of building IPF are insignificant due to the fact that the recorded intensity varies relatively little, depending on the length of the line along which averaging occurs. The analysis shows that the errors of averaging can significantly affect the IPF as a measuring of some specially "designed" texture, almost never occur in real materials.

Routines conversion of polar coordinates to Cartesian and interpolation to calculate the intensity at given points of stereographic projection, which are used in the construction of IPF described by the proposed method are the main components of the program developed by the automated construction of IPF.

Acknowledgements
This work was performed within the framework of the Center of Nuclear Systems and Materials supported by MEPhI Academic Excellence Project (contract № 02.a03.21.0005, 27.08.2013).

References

[1] Bunge H J 1969 *Matematische Metoden der Texturanalyse* (Berlin: Akademie-Verlag) p 330 (in Russian)
[2] Kurtasov S F 1978 *Zavodskaya laboratoriya* **44** (11) 1359-61(in Russian)
[3] Borodkina M and Spector E N 1981 *X-ray analysis of texture of metals and alloys* (Moscow: Metallurgiya) p 272 (in Russian)
[4] Borodkina M M and Kurtasov S F 1979 *Zavodskaya laboratoriya* **45** (9) 830 (in Russian)
[5] Harris G.B. 1952 *Phil. Mag.* **43** 113

IOP Conf. Series: Materials Science and Engineering **130** (2016) 012058 doi:10.1088/1757-899X/130/1/012058

Investigation of radiation-induced transformations in thin NbN films by analytical electron microscopy

K Prikhodko[1, 2], B Gurovich[1], M Dement'eva[1], L Kutuzov[1] and D Komarov[1]

[1]National Research Centre "Kurchatov Institute", Kurchatov sq.1, Moscow, 123182, Russia
[2]National Research Nuclear University MEPhI (Moscow Engineering Physics Institute), Kashirskoe highway, 31, Moscow, 115409, Russia

E-mail: prihodko_ke@nrcki.ru

Abstract. This work demonstrates implementation of low energy electron energy loss technique (EELS) in scanning transmission electron microscopy (STEM) to investigate the changes of free electron density at room temperature in ultra-thin NbN films under composite ion beam irradiation up to the deses of ~3 d.p.a. for nitrogen atoms. It was found the constant value of the free electron density $\sim 1.6 \cdot 10^{29}$ m^{-3} in this dose range while the irradiated material was characterized by metal type of electrical conductivity.

1. Introduction

Last years in the area of cryo-electronics based on superconducting materials comes to the forefront. Nowadays the superconductivity is widely used for radio-frequency electronic devices, high-speed digital switches, the highly sensitive radio receiving equipment. One of the main advantages of the superconductivity are low power consumption and high frequencies of operation.

Niobium nitride has been widely studied because of the high transition temperature and critical current values [1-3]. The main aim of the well-known experiments of irradiation influence on NbN was to understand the nature of high resistance to radiation damage of this material under neutron irradiation [4]; and to study the effects of ion irradiation on its superconducting properties [5]. It have been shown that fast neutron irradiation of NbN films up to the fluence of $1.5*10^{20}$n/cm^2 decreased the transition temperature by ~6% only [4]. This high resistance to radiation damage was attributed to the highly defective nature of the NbN films [4]. At the same time high energy ion irradiation of NbN films by 200 keV Ar^+ [5] and 350 keV Ne^+ [6] ions showed the larger transition temperature reduction due to the damage production. It was found [6] that the radiation-induced resistivity changes saturate at increases after damage value more than 1 d.p.a. per target atom.

Nowadays niobium nitride is one of the most promising materials in the field of low-temperature superconducting electronics to build electronic devices such as bolometers (HEB), superconducting single-photon detectors (SSPD), quantum devices [7].

We are developing the radiation-induced transformation of the thin film niobium nitride to change the electrical properties under low energy ion beam irradiation [8]. We performed the composite ion beam irradiation [9, 10] to get metal type of electrical conductivity of ultra-thin (5nm) NbN films at 4.2 K[11].

Ultra thin NbN superconductive films structure being modified under low energy ion beam irradiation needs to be characterized by the local technique that can provide high spatial resolution together with analytical signal to find out radiation-induced changes in the atomic content of the film. For this purpose, we used analytical STEM techniques on the cross-section samples of the films.

The main task of the present work was to apply low energy loss electron spectroscopy to investigate the free electron density at room temperature changes under composite ion beam irradiation up to the doses of ~3 d.p.a. for nitrogen atoms.

2. Experiment

Ultrathin niobium nitride films with thickness of ~5 nm were deposited by sputtering niobium target by nitrogen ions on the oxidized (~0.15 m amorphous SiO_2) monocrystalline silicon substrate heated to a temperature of about ~800°C [8]. Thus created NbN film characterized by a high superconducting transition temperature of about 12 K and a critical current density of ~$5\cdot10^6$ A/cm^2 at 4.2 K.

Samples were irradiated by composite ion beams extracted from the high-frequency plasma discharge and consisting of OH ions and protons [9,10] with energies (0.1-1) keV in a dose range (1.6-3) d.p.a. for nitrogen. The ratio of OH ions to hydrogen ones in the composite ion beam was monitored by measuring the partial pressures of the residual H_2O component and hydrogen supplied [9,10]. In our case the beam composition was c = $1.2 \cdot 10^{-3}$. The dose rate of nitrogen atoms displacement was performed using SRIM code [12] and in our case was ~0.02 d.p.a./s.

Analytical Transmission Electron Microscopy investigations were performed in STEM mode to get highest spatial resolution together with the detecting of inelastic electron energy loss signal at "Titan 80-300ST" electron microscope at an accelerating voltage of 200 kV. Cross section of the virgin and irradiated samples were made at the Focusing Ion Beam facility "Helios Nanolab 650".

STEM mode was very attractive to perform the chemical elements depth-distribution analysis at cross-section samples, because of the small size of electron probe (~0.14 nm) that the elements distribution profiling of the irradiated ultra-thin film was available.

Electron Energy Loss Spectroscopy (EELS) was used to get analytical information from the thin sample area under electron probe focus. Together with the standard elements relative distribution technique [13] we used low energy loss plasmon peak analysis [14]. The energy of plasmon peak $E_p$was defined [14]:

$$E_p = \frac{h}{2\pi}\omega_p = \frac{h}{2\pi}\sqrt{\frac{ne^2}{\varepsilon_0 m}}, \tag{1}$$

where h – Plank's constant; ε_0 – permittivity of free space; ω_p – plasmon frequency; n– free electron density; e and m – electron charge and mass. Because of the strong influence of the ion irradiation on the electrical properties of the thin NbN film [11], it was important to find out the dependence of free electron density on the dose of composite ion beam irradiation.

3. Results and discussion

Figure 1 shows the plasmon peak energy loss spectra for NbN virgin samples and for samples irradiated by composite ion beam up to doses 1.6 and 3 d.p.a. The intensities of spectra on Figure 1 was normalized by the intensity of the last plasmon peak at ~45 eV. One can use the highest peak on Figure 1 (at ~24 eV, indicated by arrow) to calculate the materials free electron density.

From the shape of the curve on Figure 1 it was obvious the presence of the small first peak at energies ~15 eV, that was lower than the energy of the main plasmon peak. This first peak fitting results indicated on Figure 1 by dotted line. Because the free electron excitations were attributed to the smallest value of electron energy loss, to our opinion, this first peak must be used to calculate the free electron density using equation (1).

Calculations of the value of the free electron density using the first peak at ~15 eV yielded a value of $1.6 \cdot 10^{29}$ m^{-3}, while using a main plasmon peak at ~24 eV yielded a value of $4 \cdot 10^{29}$ m^{-3}. The comparing of these densities with the literature data for NbN showed that the first value is closer to the literature one $n_e \cong 2 \cdot 10^{29}$ m^{-3} [15]. Arguments: 1) that the plasmon excitations correspond to minimum electron energy loss and 2) that the experimental free electron density, calculated from the first peak position, was close to the literature data were the basis for selecting this first peak as the main one in this work.

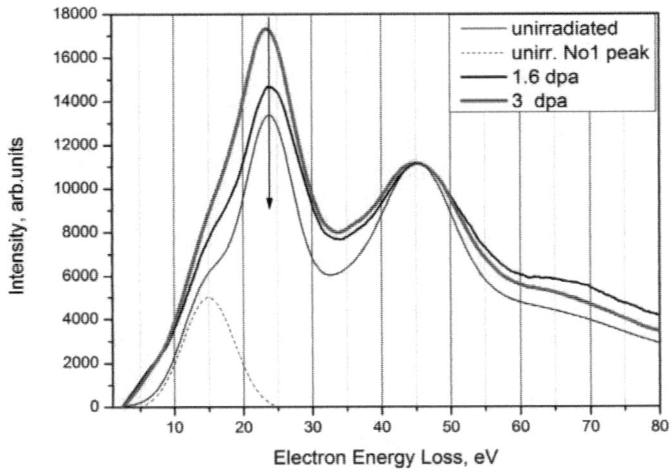

Figure 1. Low energy electron loss spectra for virgin sample and after irradiation up to 1.6 and 3 d.p.a. Dotted line indicates the position of the first peak for virgin sample with lowest energy. Arrow shows the position of the maxima of the curve for virgin sample.

Table 1 shows the experimental results of energy position of the first peak dose dependence. We can see the that up to doses of ~3 d.p.a. the energy position of the first peak did not change significantly and the error was attributed to the energy dispersion of the EELS spectrometer and accuracy of zero-loss peak position determination. The stable value of free electron density suited with the metal type of electrical conductivity for low dose (1.8-9 d.p.a.) irradiated samples [11].

During study of elements concentration evolution under composite ion beam irradiation was shown [16] that the functional conductive part of the film was transformed to the NbNO oxynitride at the range of doses ~(1-4) d.p.a., while the layer close to the irradiated surface of the sample became more oxidized due to the higher degree of nitrogen atoms substitution by oxygen atoms. In spite of this more oxidized upper layer did not take part in electrical transport of the film, we put to Table 1 the value of the first peak position to demonstrate its shift towards zero. This shift was corresponded to the decrease of free electron density on the way of superconductor-metal-insulator transformation under composite ion beam irradiation.

Table 1. Experimental energy position if the first energy loss peak and corresponded free electron density for different doses.

Dose, d.p.a.	E_l (eV)	n_e, $\cdot 10^{29}$ (m^{-3})
0	15.0±0.1	1.62±0.05
1.6	15.1±0.1	1.64±0.05
3	15.2±0.1	1.67±0.05
Top oxidized layer	14.5±0.1	1.51±0.05

In the future investigations we are going to use this low energy electron loss spectra analysis up to high doses of composite ion beam irradiation to characterize the transformation of niobium nitride to niobium oxide from the point of EELS technique in comparing with the electrical properties measurements.

IOP Conf. Series: Materials Science and Engineering **130** (2016) 012058 doi:10.1088/1757-899X/130/1/012058

4. Conclusions

By methods of low electron energy loss (EELS) in scanning transmission electron microscopy (STEM) were studied changes of the free electron concentration on the ultra-thin NbN films under composite ion beam irradiation to different doses. It was found the first peak in EELS spectra at the energy ~15 eV and corresponded free electron density $\sim 1.6 \cdot 10^{29}$ m^{-3} at room temperature. It was shown that composite ion beam irradiation up to dose of ~3 d.p.a. did not significantly change the free electron density of the functional layer.

Acknowledgements

The authors thank Stolyarov V.L. and Olshansky E.D. for the production of the original films of niobium nitride. The work was supported by the Ministry of Education and Science of the Russian Federation - Agreement on Subsidies No 14.607.21.0005 (project code RFMEFI60714X0005) under the Federal Program "Research and development on priority directions of scientific-technological complex for 2014-2020."

References

[1] Aschermann G, Friedrich E, Justi E and Kramer J, *Phys. Z.* 1941, **42** 349.
[2] Rogener H *Z. Phys* 1952 **132.1** 446
[3] Orlando T P, Delin K A, Foundations of Applied Superconductivity, Prentice Hall, 1991, ISBN-13: 978-0201183238
[4] Dew-Hughes D, Jones R *Appl. Phys. Lett.* 1980, **36** 856.
[5] Juang JY, Rudman DA, Talvacchio J, Dover RB, *Physical Review B* 1998 **38**(4) 2354
[6] Siebold Th, Ziemann P, *Physical Review B* 1995 **51**(10) 6328
[7] Guillet B, Arthursson O, Mechin at all. 2008 *Low Temp Phys.* **151** 570
[8] Gurovich B A, Tarkhov M A, Prikhodko K E, Kuleshova E A, Komarov D A, Stolyarov V L, Olshanskii E D, Goncharov B V, Goncharova D A, Kutuzov L V and Domantovskii A G 2014 *Nanotechnologies in Russia* **9** (7-8) 386
[9] Gurovich B A, Prikhodko K E, Kuleshova E A, Maslakov K I and Komarov D A 2013 *Journal of Experimental and Theoretical Physics* **116** 916
[10] Prikhodko K E, Gurovich B A, Komarov D A, Goncharova D A, Kutuzov L V 2014 *Nuclear Instruments and Methods in Physics Research B* **326** 273
[11] Gurovich B A, Prikhodko K E, Tarkhov M A, Domantovsky A G, Komarov D A, Goncharov B V and Kuleshova E A 2015 *Micro and Nanosystems* **7**(3) 172
[12] Ziegler J F, Ziegler M D and Biersack J P 2010 *Nuclear Instruments and Methods in Physics Research B* **268** 1818
[13] Williams D and Carter A Electron microscopy, New York.: Springer 2009 775
[14] Egerton R F Electron Energy-Loss Spectroscopy in the Electron Microscope, New York: Plenum Press, 1986 407
[15] Bartolf H, Fluctuation Mechanisms in Superconductors, Springer Spectrum 2016 136.
[16] Dement'eva M M, Prikhodko K E, Gurovich B A, Kutuzov L V and Komarov D A 2016 *Physics of the Solid State, to be published.*

Relative misorientations of crystals

V I Skrytnyy and V N Yaltsev

National Research Nuclear University MEPhI (Moscow Engineering Physics Institute), Kashirskoe highway 31, Moscow, 115409, Russia

E-mail: VNYaltsev@mephi.ru

Abstract. Orientation relationships between crystals are considered on the base on disorientation distribution function. The distribution of limit disorientation angles is obtained taking into account the symmetry of crystals. Maximum values of limiting disorientation angles for some crystallographic axis are calculated. Distribution density of disorientation angles corresponding to formation of coincidence site lattices is considered. Estimations of interphase disorientations between cubic and hexagonal crystals are carried out.

1. Introduction

Most of the currently used industrial materials are polycrystalline, whose properties are determined not only grains but also the grain boundaries [1]. Grain boundaries connecting by orientation relationship, leading to occurrence coincidence site lattice (CSL) have extreme properties associated with the grain boundaries energy, precipitates at the grain boundaries, grain boundary migration, slip along the grain boundaries, etc. [2].The crystallographic texture describing the direct pole figures, inverse pole figures and orientation distribution function (ODF), characterizes the polycrystalline ensemble applied to the selected sample external coordinate system.

For a description of orientation relationship between the crystals is necessary to consider the misorientation distribution function (MDF).

2. Results and discussions

Since the relative orientation of the two grains may be given by matrix $\boldsymbol{g}$ with a misorientation angle of rotation α and the axis of rotation $\boldsymbol{l}$ or Euler angles φ, θ, ψ, then misorientation can be seen or in the crystallographic space angle α - axis $\boldsymbol{l}$, or in the Euler space. If we express the components of $\boldsymbol{g}$ matrix through the angle α and the direction cosines of the axis of rotation $\boldsymbol{l}$, then matrix $\boldsymbol{g}$ takes the form of

$$\boldsymbol{g}(\boldsymbol{l},\alpha)=\begin{vmatrix} l_1^2(1-\cos\alpha)+\cos\alpha & l_1l_2(1-\cos\alpha)-l_3\sin\alpha & l_1l_3(1-\cos\alpha)+l_2\sin\alpha \\ l_1l_2(1-\cos\alpha)+l_3\sin\alpha & l_2^2(1-\cos\alpha)+\cos\alpha & l_2l_3(1-\cos\alpha)-l_1\sin\alpha \\ l_1l_3(1-\cos\alpha)-l_2\sin\alpha & l_2l_3(1-\cos\alpha)+l_1\sin\alpha & l_3^2(1-\cos\alpha)+\cos\alpha \end{vmatrix}. \quad (1)$$

When considering a grains misorientation as a function $\boldsymbol{g}$ $(\boldsymbol{l}, \alpha)$ performs a misorientation matrix A_0, so that the equivalent misorientation matrices A_i due to the symmetry of the crystal are found as

$$\mathbf{A}_i = \mathbf{A}_0\mathbf{R}_i^{-1}, \quad (2)$$

where $\mathbf{R}_i^{-1}$ - the elements, inverse to elements $\mathbf{R}_i$ group pure rotation of the crystal.

Each element of the group has one and only one inverse element, so if $\mathbf{R}_i$ takes all values elements of group pure rotation, the $\mathbf{R}_i^{-1}$ also takes all the values of the elements of the same group, but in a different order. Given this

$$\mathbf{A}_j = \mathbf{A}_0\mathbf{R}_j, \tag{3}$$

where $\mathbf{R}_j$ - one of the elements of the corresponding group pure rotation.

Thus, for description of a misorientation of the two grains in a cubic material with the symmetry group O there exists 24 equivalent descriptions. The maximum possible value of the minimum rotation angle around the axis **u** with crystallographic indexes [*mnp*] is called the limit rotation angle α_{lim} around [*mnp*]. If the rotation angle $\alpha > \alpha_{lim}$, perhaps equivalent the description of the misorientation angle a $\alpha_i < \alpha$.

Figure 1 shows the distribution of the misorientation limit angles for cubic crystals (symmetry group O). In cubic crystals, limiting rotation angle for the <100> is 45°, for <110> - ≈ 61°, for <111> - 60°. The maximum value of limit rotation angle for cubic crystals is equal to ≈ 62° around <221>.

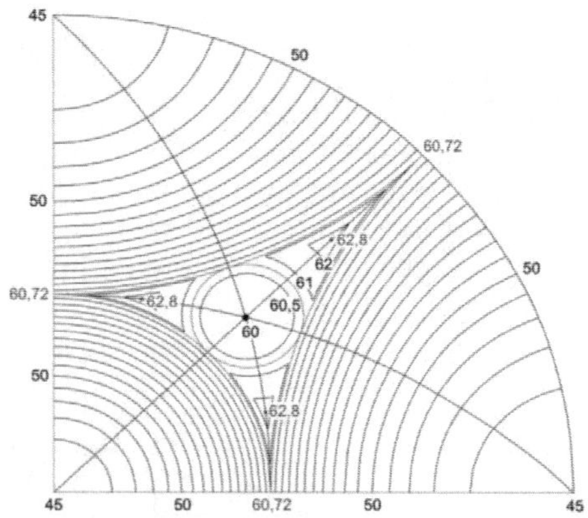

Figure 1. Limit misorientation angles for cubic crystals.

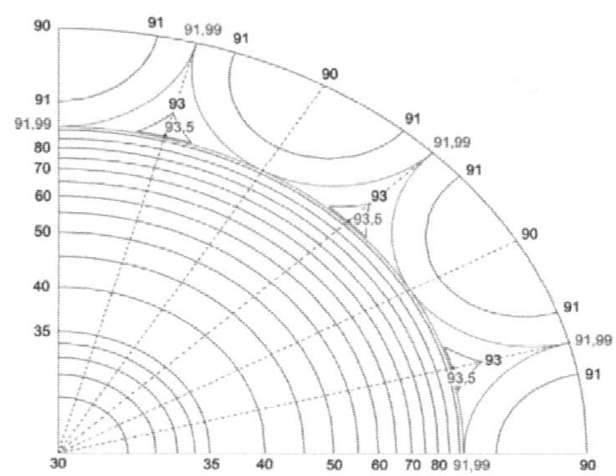

Figure 2. Limit misorientation angles for hexagonal crystals.

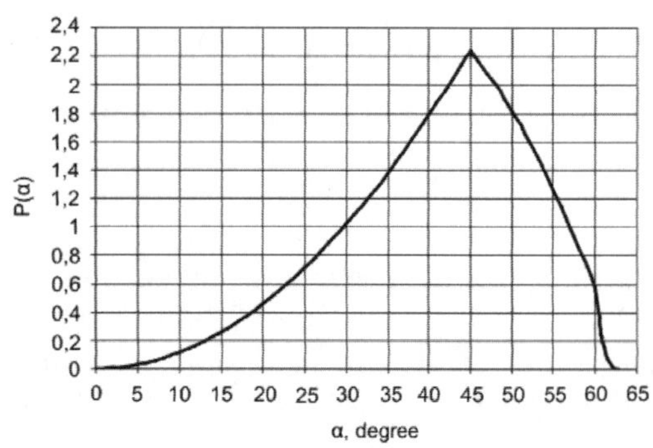

Figure 3. The density distribution of the misorientation angles P (α) for cubic crystals.

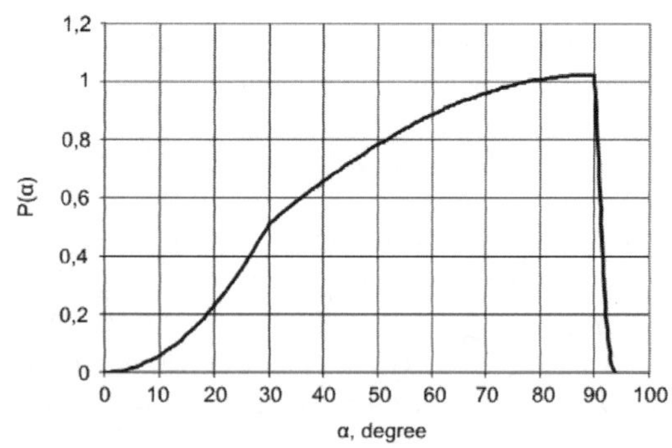

Figure 4. The density distribution of the misorientation angles P (α) for hexagonal crystals.

In the hexagonal crystals for the axis [0001] $\alpha_{lim} = 30^\circ$, for axes $<2\bar{1}\bar{1}0>$ and $<10\bar{1}0>$ $\alpha_{lim} = 90^\circ$, axes $<3\bar{1}\bar{2}0>$ $\alpha_{pr} = 92^\circ$. The maximum value of limit rotation angle for the hexagonal crystals 93°50 for axes $<\bar{1}435>$ (figure 2).For a polycrystalline aggregate of all misorientation angles is a sphere of radius π. Turn in this case is determined by the vector length ***l*** along the axis α, predetermined angles θ and φ. To determine the density distribution of the misorientation angles *P* (α) must perform invariant integration [3] by parameters θ and φ, setting the position of the axes of a reversal, within the scope of the area of the minimum misorientation:

$$P(\alpha) = \frac{(1-\cos\alpha)\int_{Ve} \sin\theta d\theta d\varphi}{4\pi^2}. \quad (4)$$

This integration corresponds to the area of the section calculation field V_E of minimum misorientation angle by sphere of radius α.

Figure 3 and figure 4 show the density relative misorientation angles distribution P (α) for the cubic and hexagonal crystals.

Under certain strictly fixed values of the axis and rotation angle neighboring crystals (ratio Kronberg-Wilson) occurs coincidence site lattice (CSL). To characterize the CSL often use the reciprocal coincidence site lattice density, denoting by Σ - number of lattice points per one coinciding node and the number Σ always simple. So Σ = 1 means a complete coincidence lattices and no border. At Σ = 3 occurs a double in the FCC lattice. Special properties of grain boundaries are preserved under small deviations of the lattices neighboring grains relative orientation from the special. The maximum deflection angle (in radians) of a special orientation, which could possibly lead accommodation using dislocations and grain boundary are saved special properties defined as

$$\Delta\alpha = \frac{10 \div 15}{\pi\sqrt{\Sigma}}. \quad (5)$$

If for description of crystal misorientations using parameters such as the angle α and the axis of a misorientation **u**, then set of all possible misorientations can be represented as a sphere of radius π. The volume of the misorientation space as the orientation space with Euler angles is $8\pi^2$.

Figure 5 within the standard stereographic triangle shows the distribution of misorientations corresponding to emergence of coincidence site lattices to Σ = 99 for cubic crystals.

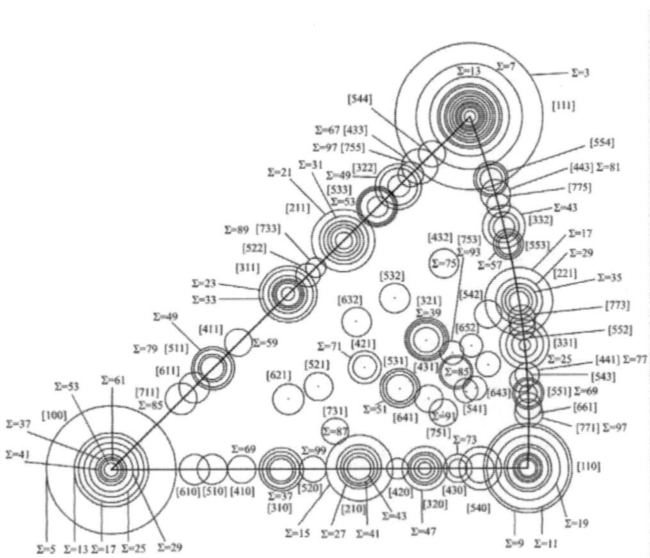

Figure 5. Distribution of misorientations corresponding to emergence of coincidence site lattices to Σ = 99, for cubic crystals.

Figures 6 and 7 show the misorientation distribution corresponding to emergence of coincidence site lattices on the sides of the standard stereographic triangle for cubic and hexagonal crystals.

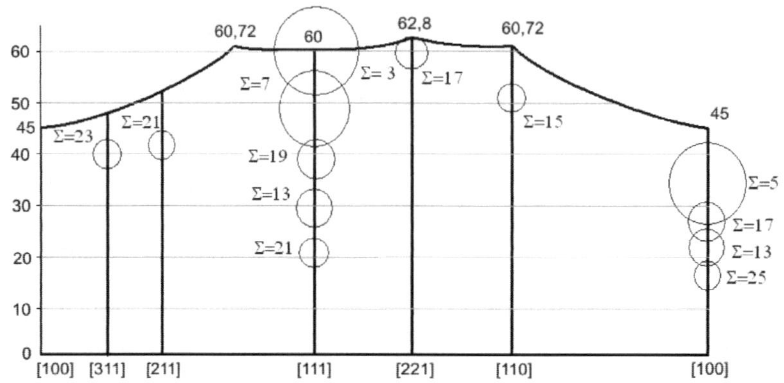

Figure 6. Distribution of misorientations corresponding to emergence of coincidence site lattices to Σ = 25 for cubic crystals on the sides of the standard stereographic triangle.

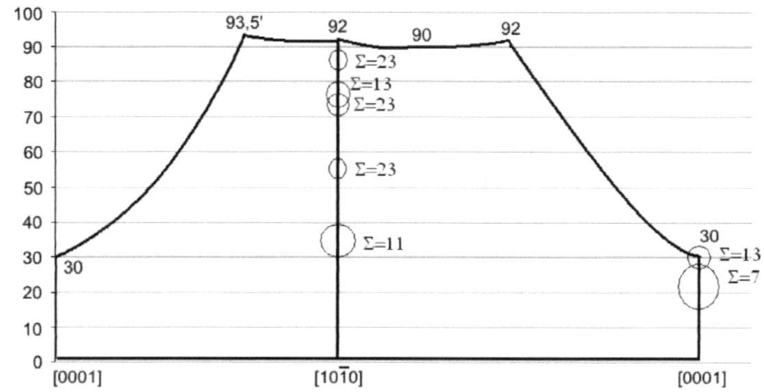

Figure 7. Distribution of misorientations corresponding to emergence of coincidence site lattices to Σ = 23 for hexagonal crystals on the sides of the standard stereographic triangle.

In the case of interphase misorientations the misorientation matrix $\mathbf{A}_k$ has the form $\mathbf{A}_k = \mathbf{R}_i\mathbf{A}_0\mathbf{R}_j$, where $\mathbf{R}_i$ and $\mathbf{R}_j$ are the elements of crystal symmetry the respective phases and the$\mathbf{A}_0$ is the misorientation matrix of the respective phases.

Distribution of the limit angles interphase misorientations for cubic and hexagonal crystals is shown in figure 8. The density of the interfacial misorientation angles P (α) of cubic and hexagonal crystals is shown in figure 9.

IOP Conf. Series: Materials Science and Engineering **130** (2016) 012059 doi:10.1088/1757-899X/130/1/012059

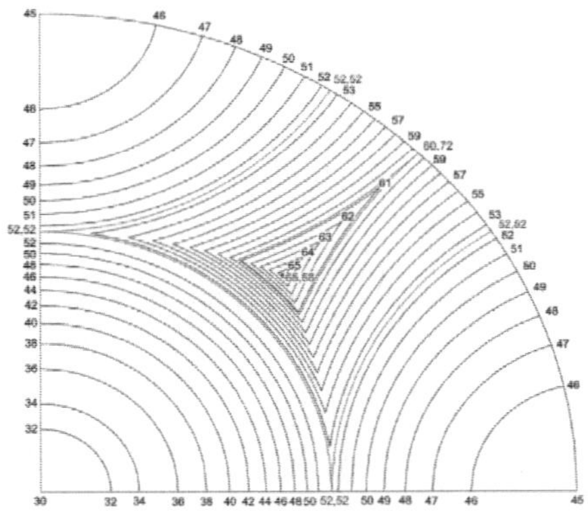

Figure 8. Limit misorientation angles of cubic and hexagonal crystals.

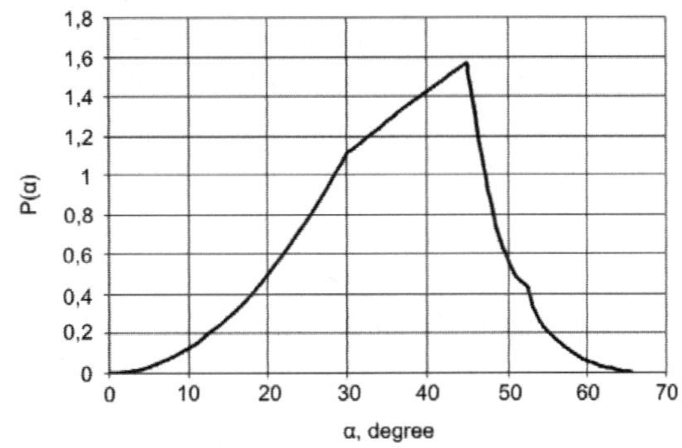

Figure 9. The density of the interfacial misorientation angles P (α) of cubic and hexagonal crystals.

3. Conclusions

The limit misorientation angles and the misorientation angle distributions for the cubic crystals and the hexagonal crystals were determined. The angles corresponding to appearance of the coincidence site lattices for $\Sigma = 99$ cubic crystals and $\Sigma = 25$ hexagonal crystals were determined within the standard stereographic triangle. The limit angles of interfacial misorientation and the distribution of misorientation angles for cubic and hexagonal crystals were calculated.

Acknowledgements
This work was performed within the framework of the Center of Nuclear Systems and Materials supported by MEPhI Academic Excellence Project (contract № 02.a03.21.0005, 27.08.2013).

References

[1] Kaybyshev O A and Valiev R Z 1987 *Grain boundaries and properties of metals* (Moscow: Metallurgiya) p 214
[2] Gleiter H and Chalmers B 1972 *Progress in materials science* 593
[3] Hamermesh M 2002 *Group Theory and its Application to Physical Problems* (Moscow: Editorial URSS) p 588

Double bragg reflections in single crystals and textured polycrystals

V I Skrytnyy and V N Yaltsev

National Research Nuclear University MEPhI (Moscow Engineering Physics Institute), Kashirskoe highway 31, Moscow, 115409, Russia

E-mail: VNYaltsev@mephi.ru

Abstract. Analysis of the detection of the double Bragg reflections (DBR) in single crystals and polycrystals is carried out. Technique of the detection of the double Bragg reflection in single crystals and textured polycrystalline samples using X-ray synchrotron radiation is proposed.

1. Introduction

Sufficient full description of polycrystalline material suggests the knowledge not only the texture but relative crystal misorientation, since many physical and mechanical properties as high-temperature creep, grain boundary diffusion, grain boundary corrosion, grain boundary sliding depends of relative orientation of crystallites [1, 4]. Traditional X-ray texture analysis allows determining the orientation of the crystal relative to the selected external system of coordinates associated with the sample. The possibility of determining the relative orientation of the crystals in a polycrystalline sample is opened by using the double Bragg reflections [2, 3].

2. Double Bragg reflections

The effect of double Bragg reflections originally discovered in single crystals. Thus, the Renninger effect is to observe in diamond forbidden diffractive reflection 222 by sequential reflections from planes (311) and $(1\bar{1}\bar{1})$.

In general, the Laue interference equation for reflection from the plane $(h_1k_1l_1)$ is given by

$$\frac{\mathbf{S_1}-\mathbf{S_0}}{\lambda} = \mathbf{H}_1, \tag{1}$$

and for reflection from a plane $(h_2k_2l_2)$:

$$\frac{\mathbf{S_2}-\mathbf{S_0}}{\lambda} = \mathbf{H}_2\,. \tag{2}$$

After subtracting the second from the first relation we have

$$\frac{\mathbf{S_1}-\mathbf{S_2}}{\lambda} = \mathbf{H_{12}}, \tag{3}$$

where

$$\mathbf{H}_{12} = \mathbf{H}_1 - \mathbf{H}_2. \tag{4}$$

Equation (3) shows that $\mathbf{s_0}$ ray, reflected from the plane $\mathbf{H_2}$, again reflected from the plane $\mathbf{H_{12}}$, goes in the direction $\mathbf{s_1}$. Thus, in the case of single crystals of FCC and BCC metals the double Bragg reflections are superimposed on the basic reflection that does not permit them to watch. Using the X-ray synchrotron radiation polarized in the horizontal plane allows to extinguish the reflection $(h_1k_1l_1)$ in the horizontal plane by choosing a wavelength λ with $2\theta = 90°$ to reflect $(h_1k_1l_1)$. Now, implementation of equations (1) and (2) allows to register the double Bragg reflection $\mathbf{H}_{12}$, which corresponds to the "internal" of double Bragg reflections in single crystals. In the polycrystalline samples possible "external" double Bragg reflections, which carry information about the intergranular misorientations. In this case, relation (1) holds for the first crystal, and the ratio of (2) - for the second. Now $\mathbf{H}_1$ is reciprocal lattice vector of the first crystal, and $\mathbf{H}_2$ - of the second and $\mathbf{H}_{12}$ is:

$$\mathbf{H}_{12} = \mathbf{H}_1 - \mathbf{R}_{12}\mathbf{H}_2 , \qquad (5)$$

where $\mathbf{R}_{12}$ - orientation matrix for the second crystal relative to the first.

3. Results and discussions

Knowing the orientation distribution function (ODF) or the analysis of direct pole figures of the polycrystalline sample allows you to select two peaks with orientation matrices $\mathbf{R}_1$ and $\mathbf{R}_2$, which corresponds to the matrix of crystals misorientation $\mathbf{U}_{12} = \mathbf{R}_2^{-1}\mathbf{R}_1$. Figure 1 a, b shows the geometry of the "external" double Bragg reflection, at first from the plane $\mathbf{n}_1$ (h_1 k_1 l_1) at an angle of Bragg θ_1 for crystal 1, and then the plane $\mathbf{n}_2$ (h_2 k_2 l_2) at an angle of Bragg θ_2 for crystal 2 .On the stereographic projection, the center coinciding with the normal to the sample surface **N** (figure 1 c), showing the outputs of normals $\mathbf{n}_1$ ($h_1k_1l_1$) and $\mathbf{n}_2$ ($h_2k_2l_2$), which are the axes of the diffraction cone with half-angles $\pi / 2 - \theta 1$ and $\pi / 2 - \theta 2$ for the first and second crystals respectively.

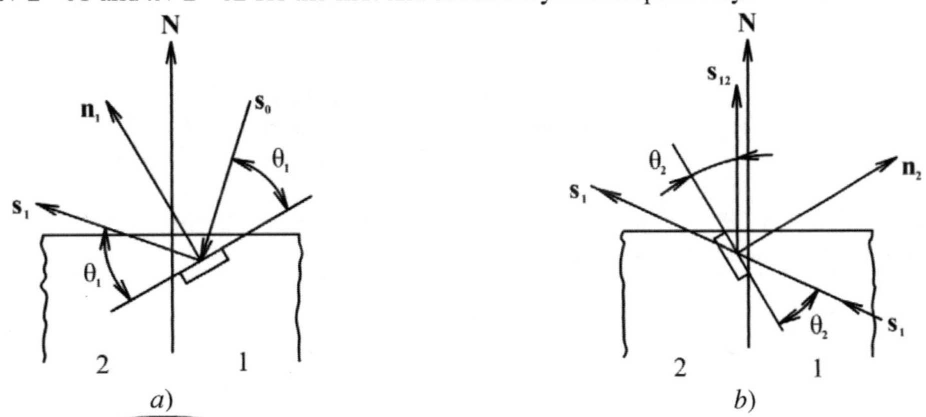

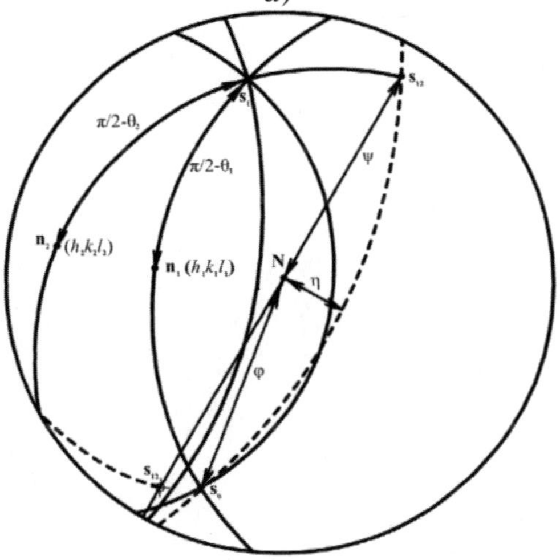

Figure 1. The occurrence scheme of "external" double Bragg reflections:
a) the reflection $\mathbf{s}_1$ from the plane $\mathbf{n}_1$ of crystal 1
b) double "external" Bragg reflection $\mathbf{s}_{12}$ from the plane $\mathbf{n}_2$ of the crystal 2
c) image on the stereographic projection: **N** - normal to the sample surface, $\mathbf{s}_0$ - primary beam $\mathbf{s}_1$ - beam reflected from the plane (h_1 k_1 l_1) of the first die, $\mathbf{s}_{12}$ - twice reflected beam (first from the plane ($h_1k_1l_1$) and then from plane ($h_2k_2l_2$))

The intersection of the circles corresponding to these cones, determines the general direction of $\mathbf{s}_1$, which allows to find the direction of the primary beam $\mathbf{s}_0$ and the direction of double reflected beam $\mathbf{s}_{12}$. Now, the angle φ between **N** and $\mathbf{s}_0$ determines the installation angle of the sample $\pi / 2 - \varphi$ between the primary beam $\mathbf{s}_0$ and a normal to the sample surface, and the angle ψ between **N** and $\mathbf{s}_0$ -

installation angle for the detector on a diffractometer. Position of the plane passing through $\mathbf{s}_0$ and $\mathbf{s}_{12}$, determines the installation tilt angle η of the sample for recording double Bragg reflections in the equatorial plane at the texture attachment for diffractometer. The intensity of the detected reflection proportional to the probability distribution function of the main peak misorientations for a polycrystalline sample.

4. Summary

The use of synchrotron X-ray radiation provides the following advantage:
weak DBR can be easier revealed due to the improved signal/background ratio. Optimum wave length of a primary monochromatized radiation can be selected taking into account an average grain dimension (for polycrystals) and X-ray radiation attenuation coefficients (μ_λ). Double Bragg reflections can find application in investigation of mutual misorientations of interphase boundaries, including the case of considering orientation relationships at phase transformations.

Acknowledgements

This work was performed within the framework of the Center of Nuclear Systems and Materials supported by MEPhI Academic Excellence Project (contract № 02.a03.21.0005, 27.08.2013).

References

[1] *Grain-Boundary Structure and Kinetics* 1980 (Asm Intl; First Edition) p 470
[2] Terminasov Y S and Tuzov L V 1964 *Successes of Physical Sciences* **83** (6) 223
[3] Chang S L 1984 *Multiple Diffraction of X-rays in Crystals* (Springer-Verlag, Berlin-Heidelberg-New York-Tokyo) p 300
[4] Zhilyaev A P 2000 *Effect of grain boundaries statistic on the macro properties of polycrystalline materials* (Ufa, Vestnik UGATU) 9

IOP Conf. Series: Materials Science and Engineering **130** (2016) 012061 doi:10.1088/1757-899X/130/1/012061

Models of thermophysical properties of uranium dioxide

Yu N Devyatko[1], V V Novikov[2], V I Kuznetsov[2], O V Khomyakov[1] and D A Chulkin[2]

[1]National Research Nuclear University MEPhI (Moscow Engineering Physics Institute), 31 Kashirskoe highway, 115409 Moscow, Russia
[2]Join Stock Company "A.A. Bochvar High-Technology Research Institute of Inorganic Materials", 5a Rogova st., 123098 Moscow, Russia

E-mail: homaykov_o_v@rambler.ru

Abstract. Physical models of specific heat at constant volume and thermal conductivity of uranium dioxide are developed based on phonon spectrum data. It is explained an abnormal growth of thermal conductivity coefficient of oxide nuclear fuel in high temperature region. It is shown that the contribution from charge carrier's degrees of freedom into observed thermal conductivity coefficient of uranium dioxide can be neglected up to melting point. Developed models of thermophysical properties do not contain fitting parameters.

1. Introduction

The thermal conductivity of uranium dioxide is measured in the wide range of temperatures [1-4]. An unalloyed (pure) uranium dioxide of stoichiometric composition upon its electrophysical properties represents the typical dielectric because its energy gap width is great and is equal to ε_G=2.7-3 eV [5]. An electron concentration in the conduction band at such energy gap value in the temperature range 1000<T<3100 K is $4\cdot10^{12}<n_e<3\cdot10^{18}$ cm^{-3} that significantly lower than in conductor ($n_e\sim10^{22}$ cм$^{-3}$) or in intrinsic semiconductor ($n_e\sim10^{19}$ cм$^{-3}$). Therefore the contribution of charge carriers to the process of heat transmission in the uranium dioxide is negligible compared with its lattice thermal conductivity in this temperature range.

It's known that the processes of heat transmission are realized in the dielectrics by the quasiparticles - phonons. The thermal conductivity coefficient $\lambda(T)$ falls with increasing temperature according to the power law $\lambda(T>>\theta_D)\sim1/T^x$ (x~1-2) for the most dielectrics at temperatures higher than Debye temperature θ_D [7]. However, on the graph (see Figure 1) of experimental temperature dependence of the uranium dioxide thermal conductivity there is temperature region 1800<T<3100 K in which thermal conductivity coefficient increases with temperature. While the uranium dioxide thermal diffusivity falls monotonically in high-temperature region in according with Debye law (see Figure 2). Such abnormal growth of uranium dioxide thermal conductivity contradicts to the heat transmission theory in the dielectrics and therefore it is connected either with an inaccuracy of measurements or with the calculation error. Therefore for an explanation of high-temperature thermal conductivity is required a physical (uncorrelation) model applicable in temperature range inaccessible for the carrying out of experiment with required degree of accuracy.

The purpose of this work is to develop a model of thermal conductivity of unalloyed, stoichiometric uranium dioxide.

IOP Conf. Series: Materials Science and Engineering **130** (2016) 012061 doi:10.1088/1757-899X/130/1/012061

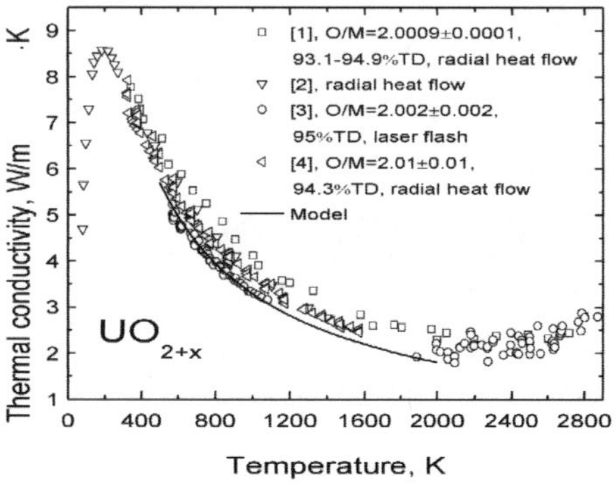

Figure 1. Temperature dependence of uranium dioxide thermal conductivity. The markers are the experimental data [1-4] and the solid line is the calculation within model.

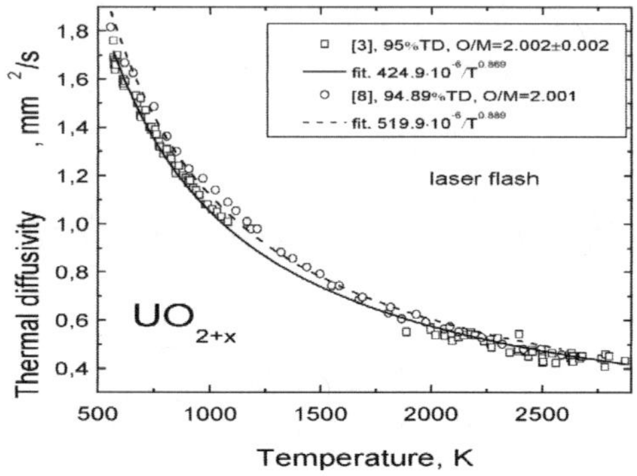

Figure 2. Temperature dependence of uranium dioxide thermal diffusivity. The markers are the experimental data [3,8] and the lines are the approximation by power functions.

2. The abnormal growth of uranium dioxide thermal conductivity in high-temperature range

The thermal conductivity of dielectrics represents the quantity determined at constant volume because according to the Second law of thermodynamics exactly in this case the change of internal energy $(\Delta U)_V$ is equal to heat quantity $(\Delta Q)_V$ received or given by the sample. In the system at constant volume the thermal conductivity equation takes the canonical form:

$$\frac{1}{V}\left(\frac{\partial U}{\partial t}\right)_V = \frac{1}{V}\left(\frac{\partial Q}{\partial t}\right)_V = \nabla\left(\lambda(T)\nabla T\right) \Leftrightarrow C_v\rho\frac{\partial T}{\partial t} = \nabla\left(\lambda(T)\nabla T\right),\ C_v \equiv \frac{1}{\rho V}\left(\frac{\partial Q}{\partial T}\right)_V, \qquad (1)$$

where $C_v(T)$ is specific heat at constant volume, ρ=10.96 g/cm^3 is density of uranium dioxide.

In practice it is impossible to provide the condition of volume constancy due to phenomenon of thermal expansion of solids therefore the measurement of thermophysical properties is carried out at constant pressure. In systems at constant pressure it is accomplished the work on thermal expansion $(\Delta Q)_P = (\Delta U)_P + A$ which necessary to take into consideration in the heat balance equation:

$$C_p\rho\frac{\partial T}{\partial t} = \nabla\left(\lambda(T)\nabla T\right) + \frac{1}{V}\left(\frac{\partial A}{\partial t}\right)_P,\ C_p \equiv \frac{1}{\rho V}\left(\frac{\partial Q}{\partial T}\right)_P, \qquad (2)$$

where $C_p(T)$ is specific heat at constant pressure.

The last term in the thermal conductivity equation (2) is proportional to the heating rate of solid and can be rewritten using directly measured physical magnitudes:

$$\frac{1}{V}\left(\frac{\partial A}{\partial T}\right)_P = (3\alpha)^2 B_T T\frac{\partial T}{\partial t}. \qquad (3)$$

Here $\alpha(T)$ is the coefficient of linear thermal expansion (CLTE), $B_T = \left[-\frac{1}{V}\left(\frac{\partial V}{\partial P}\right)_T\right]^{-1}$ is the compression modulus at constant temperature.

Specific heat at constant volume is connected with specific heat at constant pressure by thermodynamic relation [7]:

$$C_p - C_v = \frac{(3\alpha)^2 B_T T}{\rho}. \qquad (4)$$

Substitution of specific heat at constant pressure from Eq. (4) into thermal conductivity equation at constant pressure (2) leads to its canonical form (1).

Nowadays the laser flash method [9] is used for measurement of thermal conductivity of high-temperature ceramics. Within this method from the non-steady thermal conductivity equation (2) the thermal diffusivity coefficient $\chi(T)$ of studied ceramic sample is being restored. Since thermal conductivity equations in the form (2) and in the form (1) are equal then thermal conductivity coefficient is being calculated as a product of thermal diffusivity and specific heat at constant volume of studied ceramic.

$$\lambda(T) = C_v(T)\rho\chi(T). \quad (5)$$

However the authors of experimental works usually do not take into consideration the anharmonic forces work in the equation (2) formally supposed that $\lambda(T)=C_p(T)\rho\chi(T)$ what leads to systematic error in the thermal conductivity coefficient. The difference between specific heat at constant volume and specific heat at constant pressure increases with temperature faster than thermal diffusivity decreases. The admission of this systematic error leads to abnormal nonphysical growth of the uranium dioxide thermal conductivity coefficient at high temperatures.

For taking into consideration an anharmonic corrections the uranium dioxide thermal conductivity coefficient within laser flash method should be calculated using relation (5). Specific heat at constant volume is design value for which it is necessary to build the corresponding model.

3. A model of the uranium dioxide specific heat at constant volume

The uranium dioxide has the fluorite–type crystal lattice CaF_2 with the lattice constant a=547 pm at room temperature. Since unit cell contains g=3 atom types – uranium atom (U) and 2 oxygen atoms (O) then there is 9 branches in the spectrum of normal modes: 3 phonon branches are acoustic and 6 are optical.

By definition the specific heat of uranium dioxide at constant volume per unit volume [6,7] is given by:

$$C_v(T) = 3gR\frac{\partial}{\partial(k_BT)}\int_0^{\omega_{max}} d\omega D(\omega)\hbar\omega\bar{n}(\omega,T). \quad (6)$$

Here ω_{max} is the maximum frequency in the spectrum of normal modes of uranium dioxide; R=8.31 J/(mol·K) is universal gas constant; $\bar{n}(\omega,T)=\left[\exp(\hbar\omega/k_BT)-1\right]^{-1}$ is Bose-Einstein distribution for phonons, D(ω) is the phonon density of states (fig. 3) [10].

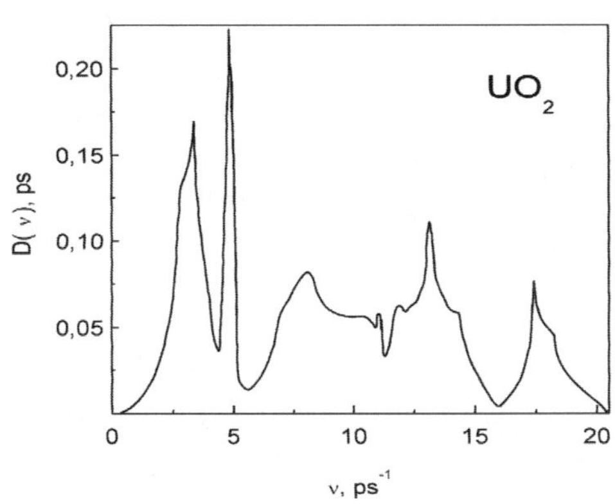

Figure 3. Phonon density of states [10].

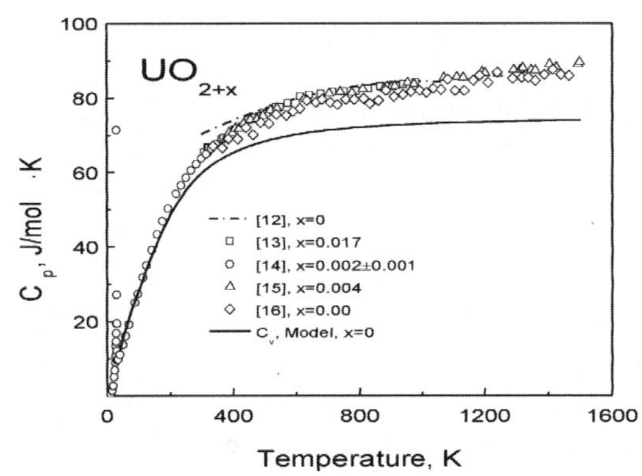

Figure 4. The comparison of the calculated specific heat at constant volume with specific heat at constant pressure experimental data [12-16].

The uranium dioxide specific heat as analytical function of absolute temperature could be obtained introducing the model phonon density of states instead an experimental phonon density of states. This approach is founded because specific heat is an integral value over phonons spectrum. The introduced model phonon density of states should be normalized per unit and must have characteristic frequencies coinciding with characteristic frequencies of experimental phonon spectrum.

We will take into consideration acoustic phonon branches within Debye model and optical within Einstein model. Then the model density of states would be chosen in the following form:

$$D(\omega)=\frac{1}{g}\frac{3\omega^2}{\omega_D^3}\eta(\omega_D-\omega)+\frac{2}{3g}\delta(\omega-\omega_{\mathrm{TO}_1})+\frac{2}{3g}\delta(\omega-\omega_{\mathrm{TO}_2})+\frac{2}{3g}\delta(\omega-\omega_{\mathrm{LO}_2}), \quad (7)$$

where $\eta(x)$ is the Heaviside function; $\delta(x)$ is the Dirac delta function; ω_D=22 THz is the uranium dioxide Debye frequency [11]. The Einstein frequencies for optical branches of normal modes spectrum which determined using fig.3 are equal to $\omega_{\mathrm{TO}_1}=2\pi\nu_{\mathrm{TO}_1}=51.1$ THz, $\omega_{\mathrm{TO}_2}=2\pi\nu_{\mathrm{TO}_2}=82.3$ THz, $\omega_{\mathrm{LO}_2}=2\pi\nu_{\mathrm{LO}_2}=109$ THz, respectively.

Substituting density of phonon states (7) to the Eq. (6) we will get the formula for calculation of specific heat at constant volume within proposed model:

$$C_v(T)=3gR\left[\frac{4}{g}\left(\frac{T}{\theta_D}\right)^3 3I_3\left(\frac{\theta_D}{T}\right)-\frac{3}{g}\frac{\theta_D/T}{\exp(\theta_D/T)-1}+\frac{2}{3g}\sum_{i=1}^{3}\left(\frac{\theta_{Ei}}{T}\right)^2\frac{\exp(\theta_{Ei}/T)}{\left[\exp(\theta_{Ei}/T)-1\right]^2}\right]. \quad (8)$$

Here $I_3(z)\equiv\int_0^z dx\frac{x^3}{\exp(x)-1}$ is the Debye integral of third order. The characteristic temperatures in (8) are equal to $\theta_D=\hbar\omega_D/k_B$=168 K, $\theta_{E1}=\hbar\omega_{\mathrm{TO}_1}/k_B$=391 K; $\theta_{E2}=\hbar\omega_{\mathrm{TO}_2}/k_B$=629 K; $\theta_{E3}=\hbar\omega_{\mathrm{LO}_2}/k_B$=836 K, respectively.

On the figure 4 it is shown the comparison of model specific heat at constant volume of uranium dioxide (8) with experimental temperature dependence of uranium dioxide specific heat at constant pressure [12-16]. It is seen that the calculated specific heat at constant volume C_v in the temperature range $35<T<100$ K coincides with the low-temperature part of specific heat at constant pressure C_p within marker thickness. It indicates a validity of choice of the phonon density of states in (7).

The formula (8) is not applicable for the calculation of specific heat at constant volume in the vicinity of a Neel point and absolute zero, since uranium dioxide in the temperature range $0<T<32$ turns into antiferromagnetic state in which specific heat of this substance determines by degrees of freedom of antiferromagnetic spin waves in general.

4. A model of uranium dioxide thermal conductivity

The operational temperature range of oxide nuclear fuel in the water-cooled nuclear reactor is substantially above Debye temperature of uranium dioxide. At temperature higher than θ_{E3} the thermal diffusivity of uranium dioxide obeys to the Debye law (see. fig. 2):

$$\chi(T)=\chi(\theta_{E3})\left(\frac{\theta_{E3}}{T}\right)^{0.89}, \quad (9)$$

where $\chi(\theta_{E3})$=1.3 mm^2/s is the value of thermal diffusivity at $T=\theta_{E3}$.

To calculate the thermal conductivity UO_2 let us apply developed in the present study a model of UO_2 specific heat at constant volume. Finally, substituting Eq. (8) into relation (5) we obtain:

$$\lambda(T)=\frac{9R}{M}\left[4\left(\frac{T}{\theta_D}\right)^3 I_3\left(\frac{\theta_D}{T}\right)-\frac{\theta_D/T}{e^{\frac{\theta_D}{T}}-1}+\frac{2}{9}\sum_{i=1}^{3}\left(\frac{\theta_{Ei}}{T}\right)^2\frac{e^{\theta_{Ei}/T}}{\left(e^{\theta_{Ei}/T}-1\right)^2}\right]\cdot\chi(\theta_{E3})\cdot\left(\frac{\theta_{E3}}{T}\right)^{0.89}\cdot\rho, \quad (10)$$

where $M=270$ g/mol is molar mass of uranium dioxide.

On the figure 1 it is shown the results of comparison of calculated (equation (10)) and experimental [1-4] thermal conductivity of UO_2 as functions of temperature. It is obvious that in whole temperature range the calculated coefficient of thermal conductivity $\lambda(T)$ is closed to the experimental points. Therefore, for calculation of the thermal conductivity of oxide nuclear fuel in the temperature range $400 < T < 2800$ K is recommended to use semi-empirical formula (10).

5. Conclusion

Unalloyed stoichiometric uranium dioxide shows dielectric properties up to melting temperature since the charge carrier concentration in its conduction band and valence band is negligibly small compared with the typical intrinsic semiconductor. It is shown that the abnormal growth of the thermal conductivity of uranium dioxide at high temperatures ($T > 1800$ K) is the systematic error. This growth is a consequence of neglecting of the anharmonic forces work on the thermal expansion of solids during process of measurement of UO_2 thermal conductivity by laser flash method. Based on the experimental phonon spectrum a model of specific heat at constant volume of uranium dioxide is constructed which describes the experimental data on specific heat of UO_2 at constant pressure C_p in the low temperature range as well as Dulong and Petit limit $C_v = 9R$ in high-temperature range. The calculated thermal conductivity of uranium dioxide obeys the Debye law $\sim 1/T^{0.9}$ at higher temperatures.

Acknowledgements
This work was performed within the framework of the Center of Nuclear Systems and Materials supported by MEPhI Academic Excellence Project (contract № 02.a03.21.0005, 27.08.2013).

References

[1] Scientific report 1976 № 74-3-10A
[2] Moore J P and McElroy D L 1971 *J. Am. Ceram. Soc.* **51** (1) 40
[3] Ronchi C, Sheindlin M, Musella M, Hyland G J 1999 *J. Appl. Phys.* **85** 776
[4] Godfey T G, Fulkerson W, Kollie T G, Moore J P and McElroy D L 1965 *J. Am. Ceram. Soc.* **48** (6) 297
[5] Willardson R K, Moody J W and Goering H L 1958 *J. Inorg. Nucl. Chem.* **6** 19
[6] Ziman J M 2001 *Electrons and Phonons: The Theory of Transport Phenomena in Solids* (Oxford University Press)
[7] Ashcroft N W and Mermin N D 1976 *Solid State Physics* (Harcourt College Publisher)
[8] Hobson I C, Taylor R and Ainscough J B 1974 *J. Phys. D: Appl. Phys.* **7** 1003
[9] Parker W J, Jenkins R J, Butler C P and Abott G L 1961 *J. Appl. Phys.* **32** (9) 1679
[10] Dolling G, Cowley R A and Woods A 1965 *Canadian Journal of Physics* **43** (8) 1397
[11] Devyatko Yu N, Khomyakov O V, Kuznetsov V I, Chulkin D A 2015 *Atomic Energy* **118** (3) 179
[12] Engel T K 1969 *J. Nucl. Mater.* **31** 211
[13] Gronvold F , Kveseth N J, Sven A et al. 1970 *J. Chem. Therm.* **2** 665
[14] Huntzicker J J and Westrum E F 1971 *J. Chem. Thermodyn.* **3** 61
[15] Inaba H, Naito K and Oguma M 1987 *J. Nucl. Mater.* **149** 341
[16] Takahashi Y and Asou M 1993 *J. Nucl. Mater.* **201** 108

IOP Conf. Series: Materials Science and Engineering **130** (2016) 012062 doi:10.1088/1757-899X/130/1/012062

Kinetics of segregation formation in the vicinity of edge dislocation in fcc metals

A V Nazarov[1], A A Mikheev[2], I V Ershova[1] and A G Zaluzhnyi[1]

[1]National Research Nuclear University MEPhI (Moscow Engineering Physics Institute), Moscow, Russia
[2]Moscow State University of Design and Technology, Russia

E-mail: Avn46@mail.ru

Abstract. We use new equations for the interstitial impurity diffusion fluxes under strain to study impurity atom redistribution in the vicinity of dislocations taking into account the strain generated by mentioned defects. Two levels of simulation are applied. First one is evaluation of coefficients that determine the influence of strain tensor components on interstitial diffusion fluxes in fcc structures for different kinds of atom jumps. For this purpose we have developed a model into the framework of molecular static method taking into account an atom environment as near the interstitial site as for the saddle-point configuration. The second level is modeling of interstitial segregation formation based on nonlinear diffusion equations taking strains generated by defects. The results show, that the distributions of the interstitials near the dislocations have quite complicated characters and the vacancy distribution has qualitatively different character as compared with carbon distribution.

1. Introduction

Elastic fields, generated by defects of the structure, influence the diffusion processes. In addition, as a result, it leads to the alteration of the phase transformation kinetic, segregation formation and changes of the system properties [1-4]. Apparently, the first the elastic interaction between a point defect and an edge dislocation has been treated by Cottrell and Bilby [2] under the assumption that the point defect is a center of dilatation in an isotropic continuum. Koehler [3] has treated an additional term in the determination of the vacancy migration energy in a stress field which arises because the vacancy saddle point configuration has lower symmetry than the lattice, and points out that the migration energy is not isotropic in a stress field. Johnson [4] has treated the interactions for a nonspherical defect near a dislocation using isotropic elasticity theory and has determined the first-order shape-effect elastic interaction. In presented work we study redistribution kinetics of defects in the vicinity of dislocations using our theory of diffusion under stress [5-7] and taking into account the strain generated by dislocations.

In the first part of our paper we give a brief consideration to the main features of the theory of diffusion under strain [5-7] and to present the general equations for the fluxes in interstitial alloys under strain. This approach gives the possibility to use the mentioned equations at low temperatures, in conditions where the strain influence on the diffusion fluxes is manifested in maximal degree.

We use evaluation results of coefficients that determine the influence of strain tensor components on interstitial diffusion fluxes in fcc structures and diffusion features [7]. These coefficients are very sensitive to the atomic structure in the nearest vicinity of a defect and still more sensitive to the atomic structure of the saddle-point configuration. Therefore, for these purposes we have used the advanced model developed by us earlier [8-10].

For an illustration of our approach possibilities in the next part we examine a formation of interstitial atom segregation near the dislocations by applying the general equations for the fluxes in interstitial alloys under strain in case, when the strain is generated by dislocation. Wherein the strain in the vicinity of dislocations outside the core is described by isotropic elasticity theory and dislocation line is parallel to <100> direction.

The last part is concerned with a simulation of the vacancy redistribution in the vicinity of dislocations under similar conditions.

2. General Equations

We propose an original microscopic approach to examine the effect of elastic strain of a general type on diffusion. The key moment of this approach is explicit consideration of the fact that positions as well as potential energies of atoms are changed due to the displacement field. This is considered both for the equilibrium and saddle-point configurations. Consequently, strain changes the local activation barriers. Knowing the changes of the activation barriers it is possible to calculate the jump rates. The rates of atom jumps in different directions define the flux density of the defects, explicit form of which can be derived by calculating the balance of different jumps and their contribution to the flux. We obtain the expressions for the vacancy diffusion fluxes and interstitial diffusion fluxes in fcc and bcc structures [6,7].

The carbide atom jumps give the contribution to the flux in the direction of axis X in fcc metal on condition that atoms have the following initial and final positions: (0 ;0; 0)→(± a/2; ± a/2; 0) and (0; 0; 0)→(± a/2; 0; ± a/2), where a is a lattice parameter (Figure 1).

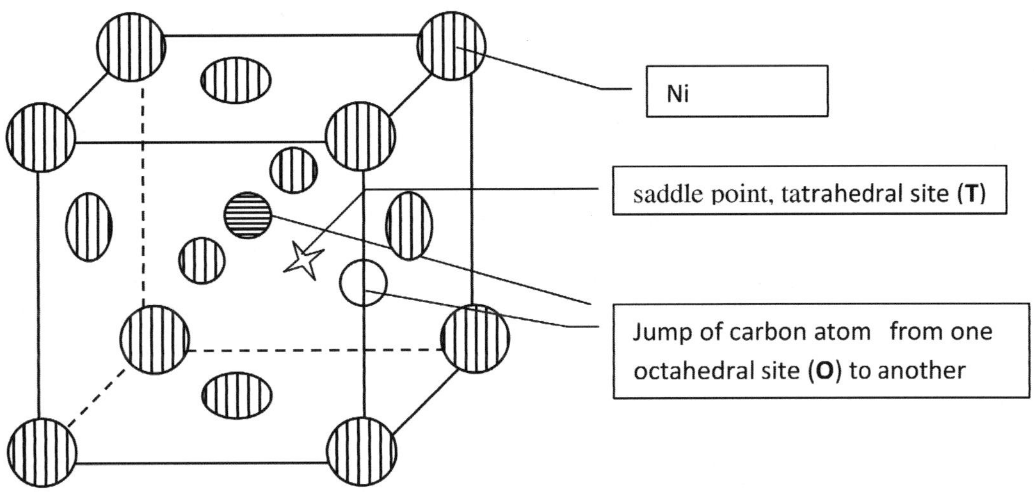

Figure 1. Jumps of the interstitial impurity atom in fcc structures (O – T – O jumps).

For the diffusion of the interstitial atoms in fcc metal under strain twelve (O –T – O) jumps give the contribution to the flux. After calculation of the different jump balance general equations for the X-component of the interstitial atom flux may be written as [7]:

$$J_1 = -\frac{1}{\Omega}\left[\left(D_{11}\frac{\partial c_i}{\partial x} + D_{12}\frac{\partial c_i}{\partial y} + D_{13}\frac{\partial c_i}{\partial z}\right) + c_i\left(\frac{\partial D_{11}}{\partial x} + \frac{\partial D_{12}}{\partial y} + \frac{\partial D_{13}}{\partial z}\right)\right], \quad (1)$$

where c_i is the interstitial atom concentration, Ω is the atomic volume.

For the other axes the equations are similar.

In this case the coefficients D_{ij} depend on the non-diagonal components of the strain:

$$D_{11}=\frac{D}{2}\exp\left(-\frac{K_1\,\mathrm{Sp}\,\varepsilon}{kT}\right)\left[\exp\left(-\frac{K_2\varepsilon_{33}}{kT}\right)\mathrm{ch}\left(\frac{K_3\varepsilon_{12}}{kT}\right)+\exp\left(\frac{K_2\varepsilon_{22}}{kT}\right)\mathrm{ch}\left(-\frac{K_3\varepsilon_{13}}{kT}\right)\right],$$

$$D_{12}=\frac{D}{2}\exp\left(-\frac{K_1 Sp\varepsilon}{kT}\right)\exp\left(-\frac{K_2\varepsilon_{33}}{kT}\right)sh\left(\frac{K_3\varepsilon_{12}}{kT}\right),\tag{2}$$

$$D_{13}=\frac{D}{2}\exp\left(-\frac{K_1 Sp\varepsilon}{kT}\right)\exp\left(-\frac{K_2\varepsilon_{22}}{kT}\right)sh\left(\frac{K_3\varepsilon_{13}}{kT}\right),$$

where ε_{ij} is the strain tensor, D is the diffusion coefficient of interstitial atoms for the perfect crystal. One can readily see that each of these coefficients depends on strain tensor components in a nonlinear way. In corresponding nonlinear equations the functional dependence of strain is determined by coefficients, which are the main characteristics of the strain influence on diffusion (SID coefficients). In the case of the interstitial atom diffusion in fcc metals for (O – T – O jumps) three coefficients K_1, K_2, K_3 determine the influence of strain on diffusion, each of that is a linear combination of the coefficients:

$$K_{xx}^{W}=\frac{1}{2}\sum_{s}\sum_{k\neq s}\frac{\left(x_{ks}^{\mathrm{W}}\right)^2}{R_{ks}^{\mathrm{W}}}\left.\frac{\partial E}{\partial R_{ks}}\right|_{R_{ks}^{\mathrm{W}}},\quad K_{xx}^{0}=\frac{1}{2}\sum_{s}\sum_{k\neq s}\frac{\left(x_{ks}^{0}\right)^2}{R_{ks}^{0}}\left.\frac{\partial E}{\partial R_{ks}}\right|_{R_{ks}^{0}},$$

$$K_{xy}^{W}=\frac{1}{2}\sum_{s}\sum_{k\neq s}\frac{x_{ks}^{\mathrm{W}}y_{ks}^{\mathrm{W}}}{R_{ks}^{\mathrm{W}}}\left.\frac{\partial E}{\partial R_{ks}}\right|_{R_{ks}^{\mathrm{W}}}\quad\text{and analogical them,}\tag{3}$$

where x_k, y_k, z_k are the coordinates of the atom k, $x_{ks} = x_k - x_s$, $y_{ks} = y_k - y_s$, $z_{ks} = z_k - z_s$, $k\neq s$, $R_{ks}=\left|\mathbf{r}_k-\mathbf{r}_s\right|=\sqrt{x_{ks}^2+y_{ks}^2+z_{ks}^2}$ for all atoms k, $E = E(R_{ks})$ is the system energy without strain. These coefficients depend on the atomic interactions and atomic configurations of the defect environments such as: when the atom comes to the saddle point (relative distances between the atoms have index W) and when the atom is at equilibrium position (relative distances between the atoms have index 0). The effect of elastic stress field on the diffusion coefficient matrix is defined by three coefficients K_1, K_2, K_3 for fcc metals:

$$K_1=K_{yy}^{W}-K_{yy}^{0},\quad K_2=K_{xx}^{W}-K_{yy}^{W},\quad K_3=2K_{xy}^{W}.\tag{4}$$

Then we use the solution of the classical theory of elasticity for the strain in the vicinity of dislocations outside the core:

$$\varepsilon_{11}=-\frac{b}{4\pi(1+\nu)}\frac{y}{\left(x^2+y^2\right)^2}\left[\frac{(3+\nu)}{(1-\nu)}x^2+y^2\right]$$

$$\varepsilon_{22}=\frac{b}{4\pi(1+\nu)}\frac{y}{\left(x^2+y^2\right)^2}\left[\frac{(1+3\nu)}{(1-\nu)}x^2-y^2\right]$$

$$\varepsilon_{12}=\frac{b}{4\pi(1-\nu)}\frac{x\left(x^2-y^2\right)}{\left(x^2+y^2\right)^2},\tag{5}$$

where b is the Burgers vector of the dislocation and ν is the Poisson ratio.
Further we substitute the components of the strain tensor equation (5) in equations for D_{ij} (equations (2)). Finally, if we substitute the obtained expressions for fluxes in the continuity equation then we get the diffusion equation, in which the influence of elastic strain on flux is taken into account. In particular, this equation gives the possibility to describe interstitial redistribution under strain.

3. Models and Results

We model redistribution of impurity atoms with the help of a numerical solution of non-linear diffusion equation taking into account the strain generated by dislocations. In this equation the functional dependence on strain is determined by coefficients, which are the main characteristics of the strain influence on diffusion (SID coefficients).

These coefficients are determined with the help of atomic simulation for some interstitial alloys [7] in a similar way like it was done in [8-10] for vacancies. We used N-body potentials developed in [11,12], describing carbon atom interaction with the atom of metals, and our advanced model [8-10], to evaluate these coefficients.

The knowledge of elastic fields let us model the segregation formation kinetics of carbon impurity atoms in the vicinity of dislocations at different temperatures. The uniform initial concentration of impurities was taken and it equals c_i^0.

The distributions of C atom concentration near the dislocation are presented on Figure 2.

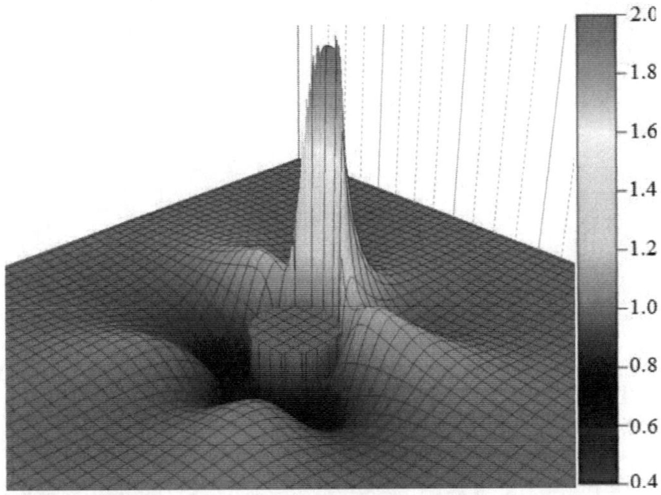

Figure 2. Distribution of carbon impurity atoms in the vicinity of dislocation. $T = 0.4\ T_m$, $t = 5dt$, $dl = 8.7Å$, $c_i^{max}/c_i^0 = 2.1$

Here T is the temperature, T_m is the melting temperature of metal, dt is the conventional time unit, dl is the step along the grid coordinates. Sizes shown in the figure area are (346Å x 346Å).

The results of this simulation show that the distribution of the interstitials near the dislocations has a quite complicated character. A corresponding function $c_i(x,y)$ in the normal plane to the line dislocations has some ridges and valleys between them. The character of the distribution depends on the elastic field generated by the dislocations, on the SID coefficients and their relation to the activation barrier value for a perfect crystal and on the temperature.

Simulation of the vacancy redistribution in the vicinity of dislocations is realized a similar way. But in this case we used the general equations for the vacancy fluxes in metals under strain [6,7] and SID coefficients calculated for Ni. The results are presented on Figures 3 and 4.
The results of this simulation show that the vacancy distribution near the dislocations differs qualitatively from the interstitial one.

It should be noted that in all calculations the concentration near the dislocation core takes a value very close to the maximum in a very short time, and then this value is increased very slowly (Table 1).

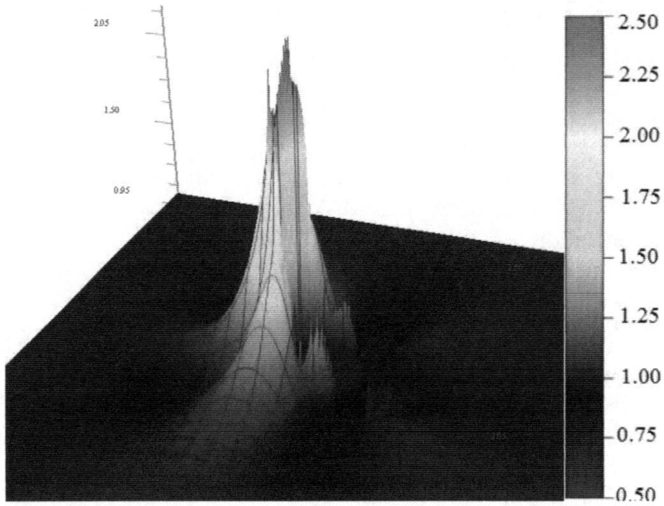

Figure 3. Vacancy distribution in the vicinity of dislocation.
T= 0.4 T_m, t= 5dt, dl= 8.7Å, $c_v^{max}/c_v^0 = 2.58$

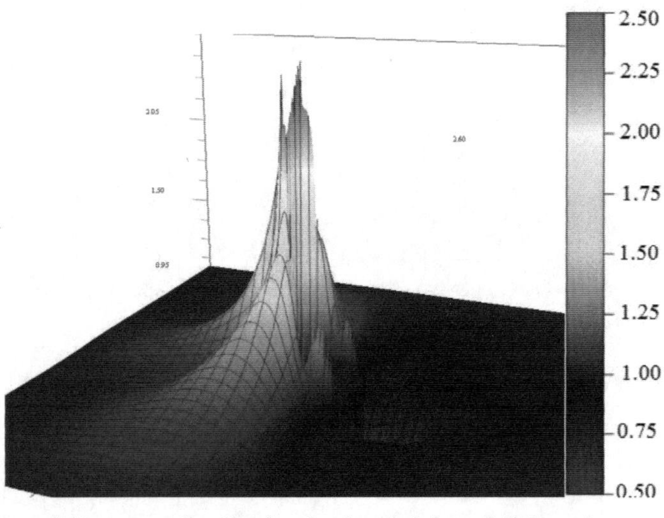

Figure 4. Vacancy distribution in the vicinity of dislocation.
T= 0.4 T_m, t= 6dt, dl= 8.7Å, $c_v^{max}/c_v^0 = 2.6$

Table 1. The dependence of the maximum concentration of vacancies on time from the beginning of the redistribution under the elastic field of the dislocation

	$t3$	$t4$	$t5$	$t6$
$t*10^7$s	3	4	5	6
c_v^{max}/c_v^0	2.35	2.54	2.58	2.6

The concentration increase is much slower at large distances from the core (see figures 3 and 4). This effect is caused by strongly nonlinear equations and can play an important role in the formation of nucleation on dislocations such phases as carbides and hydrides and in the formation of void nucleation in the nearest vicinity of dislocations.

4. Conclusion

Based on the theoretical approach, developed earlier, the nonlinear equation for interstitial diffusion under the strain generated by dislocations in fcc metals has been derived.

Multi-scale model of diffusion, which allows to take into account the influence of the strain on kinetics of interstitial atom redistribution near the dislocation is developed.

Simulations of redistribution kinetics near the dislocation are made for carbon impurity atoms in Ni.

Simulations of vacancy redistribution kinetics near the dislocation are made for different temperatures.

The results of the simulation show that the vacancy distribution near the dislocations differs appreciably from the interstitial one and the distributions have the quite complicated character.

Acknowledgements

This work was performed within the framework of the Center of Nuclear Systems and Materials supported by MEPhI Academic Excellence Project (contract № 02.a03.21.0005, 27.08.2013).

References

[1] Was G S 2007 *Fundamentals of Radiation Materials Science. Metals and Alloys* (New York, Springer)
[2] Cottrell A H and Bilby B A 1949 *Proc. Phys. Soc.* **A 62** 49
[3] Koehler J 1969 *Phys. Rev.* **181** 1015
[4] Johnson R A 1979 *Journal of Applied Physics* **50** 1263
[5] Nazarov A V and Mikheev A A 2004 *Physica Scripta* **T108** 90
[6] Nazarov A V and Mikheev A A 2008 *J. Phys.: Condens. Matter* **20** 485203.1
[7] Nazarov Andrei, Mikheev Alexander, Valikova Irina and Zaluzhnyi Alexander 2011 *Solid State Phenomena* **172-174** 1156
[8] Valikova I V and Nazarov A V 2008 *Def. Diff. Forum* **277** 125
[9] Valikova I V and Nazarov A V 2008 *Phys. of Metals and Metallography* **105** 544
[10] Valikova I V and Nazarov A V 2010 *Phys. of Metals and Metallography* **109** 220
[11] Ruda M, Farkas D and Abriata J 1996 *Phys. Rev.* B **54** 9765
[12] Ruda M, Farkas D and Abriata J 2002 *Scripta Materialia* **46** 349

IOP Conf. Series: Materials Science and Engineering **130** (2016) 012063 doi:10.1088/1757-899X/130/1/012063

Radiation enhanced diffusion processes in UO_2 and $(U,Pu)O_2$

E Smirnov and G Elmanov

National Research Nuclear University MEPhI (Moscow Engineering Physics Institute), Kashirskoe highway 31, 115409 Moscow, Russia

E-mail: GNElmanov@mephi.ru

Abstract. The diffusion mechanisms and migration characteristics of point defects in the fluorite structures of UO_2 and (U, Pu) O_2 were analyzed. The possibility of complex migration mechanisms of anionic and cationic vacancy and interstitial Frenkel defects depending on the stoichiometry of materials was shown.The statistical analysis of published data on the radiation-enhanced diffusion (RED) characteristics of metal atoms in UO_2 and $(U, Pu)O_2$ for temperatures 1303-1673 K was carried out. An approximate temperature dependence of RED coefficient was obtained. Within the framework of understanding of the mechanisms of diffusion processes in fluorite structure, the approximate temperature dependence of the RED coefficient of oxygen for UO_2 and $(U, Pu)O_2$ was proposed.

1. Introduction

Knowledge of characteristics and mechanisms of diffusion processes in UO_2 and $(U, Pu)O_2$, particularly in terms of radiation exposure, is useful for a production methods development, as well as for understanding of the processes of radiation damage during operation (such as radiation creep, grain growth, swelling and so on).

The main results of these studies are presented and discussed in the reviews [1-5]. Unfortunately, most of the works have been published a long time ago. This means that these works do not contain any modern analysis of radiation-accelerated diffusion processes in uranium and plutonium oxides. This publication carries out the forecasting parameters of the radiation-accelerated processes of metal atoms diffusion component and oxygen in UO_2 and (U, Pu) O_2, as well as the analysis of the diffusion processes mechanisms according to the latest published data.

2. Characteristics of diffusion processes in UO_2

The low packaging structure density of UO_2 (0.65) results in the peculiar characteristics of formation and migration of point defects, as well as the diffusion coefficients of oxygen and metal atoms in UO_2 and mixed nuclear fuel $(U, Pu)O_2$.

There were a lot of oxygen diffusion researches in uranium dioxide. It should be noted that the results of these researches were a considerable variation in the measured diffusion parameters of oxygen. It relates to significant differences of techniques applied, as well as to purity and to method of preparing materials, and mostly with abrupt dependence of diffusion characteristics from the stoichiometry of the materials being under research.

Statistical analysis of the experimental data presented in the reviews [1–5] allows us to get for the nominally stoichiometric materials following temperature dependence of the oxygen diffusion coefficients in UO_2:

$$D_{O\to UO_2} = (0.335 \pm^{0.65}_{0.22}) \exp\left(-\frac{(2.61 \pm 0.11)\,\mathrm{eV}}{kT}\right), \frac{\mathrm{cm}^2}{\mathrm{s}}. \quad (1)$$

This is almost with the same as the proposed earlier in work [5] the temperature dependence

$$D_{O\to UO_2} = 0.26 \exp\left(-\frac{2.6\,\mathrm{eV}}{kT}\right), \frac{\mathrm{cm}^2}{\mathrm{s}}.$$

However, obtaining reliable data for stoichiometric compositions is complicated by a low possibility of obtaining such materials. Single-phase region of UO_2 compositions is very wide, and an excess of oxygen in the dioxide is difficult to avoid. In this case, reasonable criteria of reliability characteristics of oxygen diffusion processes in nominally stoichiometric compositions UO_2 could be developed through an analysis of diffusion mechanisms.

Assuming that the activation energy for oxygen diffusion is equal to

$$Q = \frac{1}{2} E^f_{FO} + E^m_O,$$

where E^f_{FO} – a formation energy of oxygen Frenkel defect in the fluorite structure of UO_2, and E^m_O – anion energy migration. To estimate Q we need data on the characteristics of defects in UO_2. The formation energy value of anion Frenkel defect equals to (3.0–4.0) eV [2, 4]. The values of anion migration energy E^m_O are different for interstitial and vacancy mechanisms of oxygen diffusion in the UO_2 anion sublattice, and depends on the stoichiometry of oxide. Thus, the activation energy for oxygen diffusion is determined by E^m_O. This is confirmed by a data scattering in activation energies of oxygen diffusion in UO_2 [2–6] and, accordingly, by an experimental and calculated data on the migration energy of vacancy and interstitial anion Frenkel defects.

The value E^m_O for interstitial anion defect equals to (0.8–1.1) eV for the stoichiometric oxide UO_2, and this in the best way is consistent with the above mentioned activation energy of an oxygen diffusion. Indeed, in this case, the activation energy for oxygen diffusion is $Q \approx \frac{1}{2}(3{,}5 \pm 0{,}5) + (0{,}9 \pm 0{,}1) \approx (2{,}65 \pm 0{,}35)$ eV.

The results of experimental researches of the uranium and actinides diffusion in UO_2 and UO_{2+x} also demonstrate a very significant impact of stoichiometry. The authors of papers and reviews [2, 4, 5] have proposed the following parameters of the temperature dependence of the uranium diffusion coefficients in quasi-stoichiometry uranium dioxide:

$$D_{U\to UO_2} = 0.65 \exp\left(-\frac{5.6\,\mathrm{eV}}{kT}\right), \frac{\mathrm{cm}^2}{\mathrm{s}}. \quad (2)$$

In accordance with this analysis for the diffusion of uranium in a "quasi stoichiometric" dioxide, it is also proposed the diffusion mechanism with the participation of Frenkel defects, i.e., cation interstitials and vacancies in the cation sublattice.

3. Characteristics of radiation-enhanced diffusion in UO_2 and $(U,Pu)O_2$

The fundamental characteristics of radiation-enhanced diffusion (RED) metal atoms of U and Pu in UO_2 and (U, Pu) O_2 are of an interest for estimation of the of diffusion processes mechanisms that determine the intensity of radiation damage in the process operation of these materials under irradiation. Unfortunately, in a global experimental practice, the only study of this problem was carried out [7, 8].

Investigations were carried out with the UO_2, (U, 2.5%Pu)O_2 and (U, 1.5%Pu) samples, using ^{233}U and ^{238}Pu in the temperature range (1300–1400) °C in neutron fluxes $(7\cdot10^{12}-1.2\cdot10^{14})\ \mathrm{n\cdot cm^{-2}\cdot s^{-1}}$

during (56–334) hours at fission speeds $\dot{F} = (7\cdot10^{11} f - 6.4\cdot10^{13} f)\ \mathrm{n\cdot cm^{-3}\cdot s^{-1}}$ and doses of $F = (4.2\cdot10^{17} f - 3.1\cdot10^{19} f)\ \mathrm{n\cdot cm^{-3}}$.

The results for temperatures below ~1300 K show that the RED coefficients do not depend on the temperature (athermal region) and their values are

$$D^{*}_{RED} \approx (5 \pm ^{9.5}_{3.2})\cdot 10^{-17},\ \frac{\mathrm{cm^2}}{\mathrm{s}}.$$

Statistical analysis of the data for the temperature range 1303–1673 K allows to obtain the approximate temperature dependence of RED for the specified interval, which is of the following form:

$$D^{U,Pu\to UO_2}_{RED} \approx (3.62 \pm ^{29.6}_{3.2})\cdot 10^{-13} \exp\left(-\frac{(1.0\pm0.21)\ \mathrm{eV}}{kT}\right),\ \frac{\mathrm{cm^2}}{\mathrm{s}}. \quad (3)$$

The results presented by now are unique. Attempts to estimate the coefficients of RED in UO_2 using computer simulation techniques for low temperatures in the ballistic mixing and high temperatures cannot be considered being satisfactory, because there is no observed correlation of these results with experimental data. It is noted [10] that athermal RED in UO_2 at temperatures below 1300 K depends essentially on the behavior of short-lived gaseous fission fragments, and the coefficients of RED are linearly dependent on the speed of fission. The instant effects of fission are the displaced atoms and recombining Frenkel pairs [10]. In the case of the dominant recombination process, it can be assumed that this process is thermally activated. Then the RED coefficient is $D^{*}_{RED} \sim \sqrt{K_o D_v}$ (K_o – the generation rate of defects, D_v – the mobility of vacancies); the migration energy of recombining defects $E^d_m \approx 2Q^{*}_{RED}$ (Q^{*}_{RED} – activation energy in (2)) [11, 12]. In this case, for the specified temperature range $E^d_m \approx 2{,}0 \pm 0{,}2\,eV$, and this is consistent with formula for the temperature dependences of migration of uranium vacancies ($D^{UO_2}_{U_v}$) and interstitials ($D^{UO_2}_{U_i}$) in UO_2 proposed in [2]:

$$D^{UO_2}_{U_v} = 2\cdot10^{-3} \exp\left(-\frac{2.4\ \mathrm{eV}}{kT}\right), \frac{\mathrm{cm^2}}{\mathrm{s}}; \qquad D^{UO_2}_{U_i} = 1\cdot10^{-1} \exp\left(-\frac{2.0\ \mathrm{eV}}{kT}\right), \frac{\mathrm{cm^2}}{\mathrm{s}}. \quad (4)$$

Using temperature dependences $D^{U\to UO_2}_{TAD}$ и $D^{U\to UO_2}_{RED}$ presented above, where D_{TAD} – thermally activated diffusion coefficient, it is possible to estimate the coefficient of radiation diffusion enhancement for metal atoms in UO_2 $R^{U\to UO_2}_{RED} = \frac{D_{RED}}{D_{TAD}}$, which haves the following form:

$$R^{U\to UO_2}_{RED} \approx (5.6 \pm ^{160}_{5{,}2})\cdot 10^{-13} \exp\left(\frac{(4.6\pm1.0)\ \mathrm{eV}}{kT}\right). \quad (5)$$

Temperature dependence $R^{U\to UO_2}_{RED}$ in logarithmic representation of reduced temperature is shown in Figure 1 (line 1):

$$\lg R^{U\to UO_2}_{RED} = -(12.25\pm1.32) + (7.46\pm1.57)\frac{T_m}{T}, \quad (6)$$

where the melting temperature of UO_2 is $T_m \sim (3100 \pm 30)$ K.

The upper limit of applicability of given expression (6) is determined by a condition of $R = 1$, for which $T \approx (0.61\pm0.04)T_m$, that corresponds to the value obtained by statistical processing of all available data in the publications devoted to RED [13]. The lower limit is determined by the transition temperature to athermal region. For comparison, figure 1 shows the results of statistical processing of the coefficients R for metal elements [13] (line 2).

As follows from the results shown in figure 1, the radiation enhancement of diffusion processes in UO_2 is comparable with those for pure metals. The values T_m/T for $R = 1$ corresponds to each other within experimental accuracy.

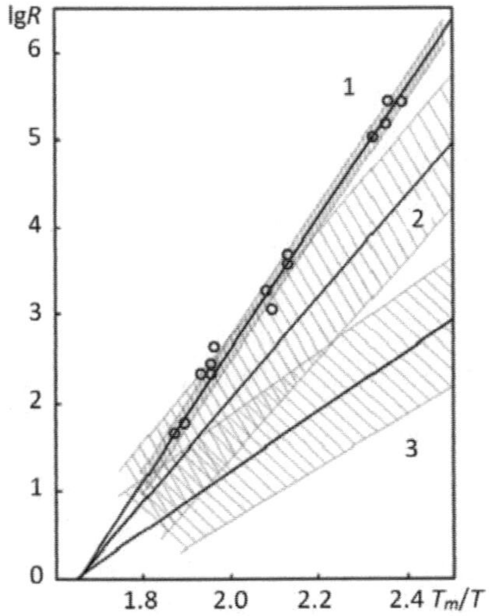

Figure 1. Dependence of coefficient R_{RED} for U и O in UO_2 from temperature:
1 – $U \to UO_2$, $(U, Pu)O_2$ – calculation by the data [7, 8] using the equations (2) and (3):

$$\lg R_{RED}^{U\to UO_2} = -(12.25 \pm 1.32) + (7.46 \pm 1.57)\frac{T_m}{T};$$

2 – metals for comparison – according to the data [13]:

$$\lg R_{RED}^{met} = -(9.70 \pm 0.38) + (5.88 \pm 0.16)\frac{T_m}{T};$$

3 – $O \to UO_2$ – according to the data of the present work:

$$\lg R_{RED}^{O\to UO_2} = -(5.73 \pm 0.28) + (3.48 \pm 0.18)\frac{T_m}{T}.$$

Taking into account the diffusion mechanisms, which were mentioned above, and experimental errors presented in this paper, the characteristics of RED anionic defects in UO_2 could be predicted.

If we consider that Frenkel pairs are the main defects in the cation and anion sublattices of UO_2, and assume consistent ratio of the migration energies of the these defects to the activation energy of thermally activated diffusion of uranium and oxygen in a close-packed structure of UO_2, the estimated migration energy of oxygen defects is

$$E_m^{def\,(O)} \approx E_m^{def\,U)} \cdot \frac{Q_{TAD}^{O\to UO_2}}{Q_{TAD}^{U\to UO_2}}, \tag{7}$$

where $E_m^{def\,(O)}$ and $E_m^{def\,U)}$ – migration energies of the oxygen and uranium defects.

Hence, the RED value of oxygen activation energy at the dominant recombination process during the annealing of radiation defects is

$$Q_{RED}^{O\to UO_2} = \frac{1}{2}E_m^{def\,(O)} \approx \frac{(2.0 \pm 0.2)\cdot(2.61 \pm 0.11)}{2\cdot 5{,}6} \approx (0.47 \pm 0.13)\ \text{eV}.$$

The value $E_O^m \sim (1.0 \pm 0.2)$ eV that results from the formula above is consistent with the estimated value of (0.8–1.1) eV, obtained in the papers [2–6].

Taking into account the total value of the ratio T_m/T for $R = 1$, we obtain the expression for the predicted temperature dependence of coefficient of radiation enhancement of oxygen diffusion in UO_2

$$R_{RED}^{O\to UO_2} = (1.885 \pm^{1.74}_{0.91})\cdot 10^{-6} \exp\left(\frac{(2.14 \pm 0.72)\ \text{eV}}{kT}\right). \tag{8}$$

This dependence corresponds to line 3 at figure 1. The real value of this enhancement factor can be significantly lower due to the capture of radiation defects by the fragments of fission in the process of formation of complexes gas atom - defect [5]. Thus, the radiation-enhanced diffusion coefficient of oxygen in UO_2 according to (1) and (8) is

$$D_{RED}^{O\to UO_2} = R_{RED}^{O\to UO_2}\cdot D_{TAD}^{U\to UO_2} \approx (6.31 \pm^{3.28}_{2.14})\cdot 10^{-7} \exp\left(-\frac{(0.47 \pm 0.13)\ \text{eV}}{kT}\right), \frac{\text{cm}^2}{\text{s}}. \tag{9}$$

3. Conclusion

1. The analysis of diffusion mechanisms and point defects characteristics of migration in the fluorite structure UO_2 and (U, Pu)O_2 was carried out. A possibility of migration mechanisms of anionic and cationic vacancy and interstitial Frenkel defects, depending on the stoichiometry of materials, was demonstrated.

2 Statistical processing of published data on the characteristics of radiation-enhanced diffusion (RED) of metal atoms in UO_2 and (U, Pu)O_2 was carried out for the temperature range of 1303-1673 K. An approximate temperature dependence of RED coefficient was obtained.

3. Within the knowledge about diffusion processes mechanisms in fluorite structure, an approximate temperature dependence of the RED oxygen in UO_2 and (U, Pu)O_2 was proposed.

Acknowledgements
This work was performed within the framework of the Center of Nuclear Systems and Materials supported by MEPhI Academic Excellence Project (contract № 02.a03.21.0005, 27.08.2013).

References

[1] Fedorov G and Smirnov E 1984 *Diffusion in Reactor Materials* (New Delhi: Oxonian Press)
[2] Matzke Hj 1987 Atomic Transport Properties in UO_2 and Mixed Oxides (U,Pu)O_2 *J.Chem. Soc. Faraday Trans.* **83** 1121–42
[3] Murch G, Richard C, Catlow A Oxygen Diffusion in UO_2, ThO_2 and PuO_2 1987 *J.Chem. Soc. Faraday Trans.* **83** 1157–69
[4] Matzke Hj 1992 Diffusion Processes in Nuclear Fuels *Diffusion Processes in Nuclear Materials* Agarwala R (North Holland) 9–69
[5] Matzke Hj 1990 Atomic mechanisms of mass transport in ceramic nuclear fuel materials *J. Chem. Soc. Faraday Trans.* **86** 1243–56
[6] Nagels P, Yan Lierde W and al 1966 *Thermodynamics* **2** (Vienna) 311–26
[7] Hoh A and Matzke Hj 1973 Fission-enhanced self-diffusion of uranium in UO_2 and UC *J. Nucl. Materials* **48** 157–64
[8] Matzke Hj 1983 Radiation enhanced diffusion in UO_2 and (U, Pu)O_2 *Radiation Effects* **75** 317–25
[9] Martin G, Maillard S, Van Brutzel L and al 2009 A molecular dynamics study of radiation induced diffusion in uranium dioxide *J. Nucl. Materials* **385** 351–57
[10] Matzke Hj, Lucuta P G and Wiss T 2000 Swift heavy ion and fission damage effects in UO_2. *Nucl. Instr. Methods in Phys. Research B* **166-167** 920–26
[11] Was Gary S 2007 *Fundamentals of Radiation Materials Science* (Berlin-Heidelberg: Springer) p 815
[12] Ferry C, Lovera P, Poinssot C and Garcia P 2005 Enhanced diffusion under alpha self-irradiation in spent nuclear fuel *J. Nucl. Materals* **346** 48–55
[13] Smirnov E and Shmakov A 2001 Radiation enhancement of diffusion in metals and alloys *Defect and diffusion forum* **194-199** (Paris) Diffusion in materials Part 1 pp 1451–56

The comparison of different approaches to the modeling of the structural properties σ–phase of Fe-Cr system

A L Udovsky[1,2] and M V Kupavtsev[1]

[1]Federal state budgetary organization of the science
Baikov Institute of Metallurgy and Materials Science, Russian Academy of Sciences
[2]National Research Nuclear University MEPhI (Moscow Engineering Institute)

E-mail: udovsky@imet.ac.ru; kmv@imet.ac.ru

Abstract: The three- sub-lattice model (3SLM) for description of atom's distribution of two components with different coordination numbers (12, 14 and 15), into σ–phase structure depended on composition and temperature is depictured in this paper. Energetic parameters of 3SLM were calculated by fitting procedure fixed to results obtained by ab-initio calculations conducted for paramagnetic states of differently ordered complexes stayed at the sigma-phase's crystal structure for Fe-Cr system at 0 K. Respective algorithm and computer program have allowed to calculate an atom distribution of components upon the sub-lattices of σ–phase at 300 – 1100 K. The temperature dependences of filling atoms on the model three sub-lattices for alloys compositions 40, 50 and 60 at. % Fe was calculated. There is satisfactory agreement between calculated results and the experimental data obtained by neutron and structural research methods. The equilibrium between BCC solutions and σ– phase of Fe-Cr system was calculated. The satisfactory consent of results of calculation with experimental data for education temperature σ– phases from BCC– solution and some divergences with experiments is received at 800 K.

1. Introduction

The crystal lattice of sigma-phase contains 5 sub-lattice in which are placed 30 atoms possessing various coordination of the next neighbors (12, 14 and 15. The last ten years of the sigma-phase became object of modeling of numerous quantum and mechanical calculations [1-6] which are applied to calculations of energies of communication and a mixture enthalpy at 0K for the ordered complexes. The analysis of results of numerous calculations on modeling of structural, magnetic and thermodynamic properties of alloys with structure of sigma-phase showed that there are more many not solved problems. In particular, yet it was not possible to carry out calculations of total energies of communication, equilibrium parameters of a lattice, the volume module of elasticity, the partial magnetic moments of atoms sigma-phase taking into account equilibrium static mixtures of atoms from different sub-lattice from sigma-phase. Especially there is open a question of establishment of influence of static mixtures on magnetic properties of the atoms located in five sublattice of sigma - phase, and the average magnetic moment depending on structure.

2. The model formulation for σ– phase

In this work, in order to describe the atom component distributions depending on alloy composition and temperature into the sigma-phase structure the real structure of sigma phase have been modeled by

3 sub-lattices structure. This simplification is caused by to provide possibility of the decision of system of the equations state, received as a result of minimization of a functional of free energy of mixture on internal (configuration) degrees of freedom. According to 3SLM the real structure of σ–phase $A_2^{12}B_4^{15}C_8^{14}D_8^{12}E_8^{14}$ (including 5 sub-lattices) was replaced by 3 sub-lattices. All of the model's sub-lattices are being filled by atoms of two components with coordination numbers (12, 14 and 15).

$$a^{(12)} = a^{(A)} + a^{(D)} = 10$$

$$a^{(14)} = a^{(C)} + a^{(E)} = 16$$

$$a^{(15)} = a^{(B)} = 4.$$

The Gibbs free energy of mixing of σ– phase of alloys with regard to σ– phases of both components according to 3SLM is equal

$$\begin{aligned}
{}^{\sigma}\Delta^{\sigma}G^{\sigma}(x,T) &= y_1^{12}\cdot y_2^{12}\cdot\delta E_{12}^{12} + y_1^{12}\cdot y_2^{14}\cdot\delta E_{12}^{14} + y_1^{12}\cdot y_2^{15}\cdot\delta E_{12}^{15} \\
&+ y_1^{14}\cdot y_2^{12}\cdot\delta E_{14}^{12} + y_1^{14}\cdot y_2^{14}\cdot\delta E_{14}^{14} + y_1^{14}\cdot y_2^{15}\cdot\delta E_{14}^{15} + \\
&y_1^{15}\cdot y_2^{12}\cdot\delta E_{12}^{15} + y_1^{15}\cdot y_2^{14}\cdot\delta E_{14}^{15} + y_1^{15}\cdot y_2^{15}\cdot\delta E_{15}^{15} + \\
&R\cdot T\cdot\begin{bmatrix} \alpha^{12}\left(y_1^{12}\cdot\ln(y_1^{12}) + y_2^{12}\cdot\ln(y_2^{12})\right) + \\ \alpha^{14}\left(y_1^{14}\cdot\ln(y_1^{14}) + y_2^{14}\cdot\ln(y_2^{14})\right) + \\ \alpha^{15}\left(y_1^{15}\cdot\ln(y_1^{15}) + y_2^{15}\cdot\ln(y_2^{15})\right) \end{bmatrix},
\end{aligned} \tag{1}$$

3. The method of calculations

According to the equation (1) functional of Helmholtz energy of mixture of sigma-phase the equation depends on external arguments (concentration of one of components and temperature) and from internal parameters (y_1^{12}, y_2^{15}) – configuration degrees of freedom.

The system of the equations of state was received by a way of minimization of a functional of free energy of Helmholtz (1) (an alloy depending on structure, temperature and two independent internal parameters (y_2^{12}, y_2^{15}) - the part of filling of atoms two sorts of components for three sub-lattices of inter-metallic compound) on independent internal degrees of freedom at the fixed values of temperature and alloy structure. Equilibrium values of internal parameters of model (at constant values of structure of an alloy and temperatures) are as the decision of system of equations of state (2a) at observance of conditions of local stability an inequality (2b).

$$\begin{cases} \dfrac{d^{\sigma}\Delta^{\sigma}G^{\sigma}(x,T)}{dy_2^{12}} = 0 \\ \dfrac{d^{\sigma}\Delta^{\sigma}G^{\sigma}(x,T)}{dy_2^{15}} = 0 \end{cases} \tag{2a}$$

$$\det\begin{Vmatrix} \dfrac{d^2\Delta^{\sigma}G^{\sigma}(x,T)}{dy_2^{(12)^2}} & \dfrac{d^2\Delta^{\sigma}G^{\sigma}(x,T)}{dy_2^{(15)}dy_2^{(12)}} \\ \dfrac{d^2\Delta^{\sigma}G^{\sigma}(x,T)}{dy_2^{(12)}dy_2^{(15)}} & \dfrac{d^2\Delta^{\sigma}G^{\sigma}(x,T)}{dy_2^{(15)^2}} \end{Vmatrix} > 0. \tag{2b}$$

Using the set equation of material balance transition to oblique-angled system of coordinates (y_2^{12}, y_2^{15}) was carried out at the fixed value of composition of an alloy, then carrying out transition since the beginning to Cartesian, and then to polar system of coordinates with a pole choice in a point $(y_2^{12} = 0, y_2^{15} = 0)$.

It allowed to present, first, system of the equations of state (2a) in orthogonal system of coordinates, and, secondly, to carry out matrix diagonalization (2b). Transition to polar system of coordinates at the fixed structure and temperature allowed to parametrizate of system (2b) and to carry out dividing of roots of system (2a). Execution performance (2b) allowed finding areas of local stability in space of internal parameters of model. For system (2a) at each fixed value of composition (x) one of components and temperatures (T) received the decision of system of the equations (2a) which allowed describing concentration dependences of distribution of atoms of components on modeling sub-lattice of sigma-phase. Authors was developed original algorithm and the program for the numerical decision of system of the equations of state (2a).

4. The calculation of energetic parameters of model

A set of equations described the Gibbs free energies of σ–phase (in relation to meta-stable σ–phases of pure components) for compositions of σ–phase being a base solution which equal to compositions of ordered complexes of $A_2{}^{12}B_4{}^{15}C_8{}^{14}D_8{}^{12}E_8{}^{14}$ – type for σ– phase [7] for which in turn quantum-mechanical calculations of formation energies [1, 4] were carried out, was constituted in order to calculate the 3SLM energetic parameters. The linear system of the equations of relationships [7] between the 3SLM energetic parameters and results of quantum-mechanical calculations of formation energy of the ordered complexes of type $A_2{}^{12}B_4{}^{15}C_8{}^{14}D_8{}^{12}E_8{}^{14}$ was obtained as a result. Thus the restrictions were lifted as the inherent quantum mechanical calculations (calculations energies of formation of complexes completely ordered only when 0K) and phenomenological models (a large number of energy parameters, the inability to describe the distribution of the component atoms over the sub-lattices σ– phase) thermodynamic properties of alloys with σ– phase structure.

The solution of linear system concerning the 3SLM energetic parameters allowed receiving the following values of energetic parameters of model presented in table 1.

Table 1. The data of energetic parameters of 3SLM model for alloys with the σ–phase structure.

$\Delta E_{12}^{12} = -6500$ (J/mol.)	$\Delta E_{12}^{14} = 520$ (J/mol.)	$\Delta E_{12}^{15} = -7020$ (J/mol.)
$\Delta E_{14}^{12} = -19500$ (J/mol.)	$\Delta E_{14}^{14} = -13520$ (J/mol.)	$\Delta E_{14}^{15} = 7540$ (J/mol.)
$\Delta E_{15}^{12} = 5460$ (J/mol.)	$\Delta E_{15}^{14} = -10660$ (J/mol.)	$\Delta E_{15}^{15} = -26000$ (J/mol.)

5. The calculation results

The solution of system of the state equations allows to describe structural (distributions of atoms of components on three model sub-lattices) and thermodynamic properties of sigma-phase depending on alloy composition and temperature. Developed 3SLM algorithm and the computer program were applied to modeling of distribution of Fe and Cr atoms on model sub-lattices of sigma-phase depending on composition and temperature.

Figures 1 -2 shows a comparison of the calculated distribution of the atoms over the sub-lattices depending on the composition obtained in this work, both with Joubert data [6] on, so their comparison with the experimental data at T = 500 and 1000 K. We also calculated the temperature dependence of filling atoms on the model three sub-lattices of σ-phase for alloy compositions 40, 50 and 60 at. % Cr (figures 3-5).

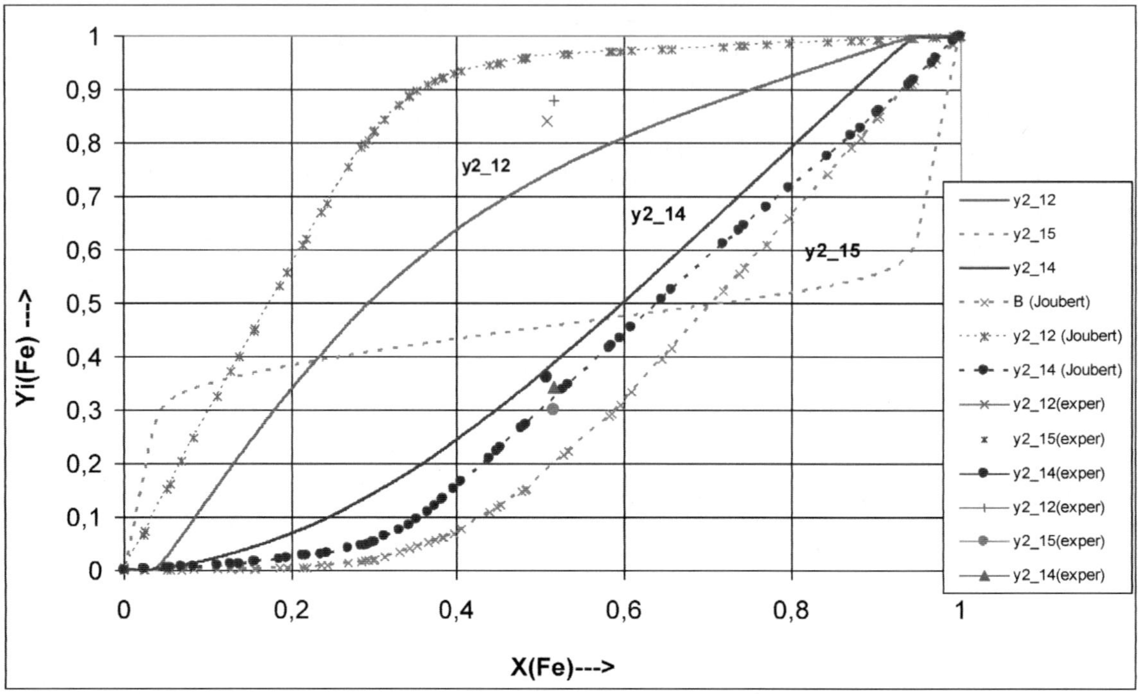

Figure 1. Comparison of the distribution of Fe atoms over the sub-lattices obtained in this study with data Joubert [6] and their collation with experimental data at 1000 K [9-11].

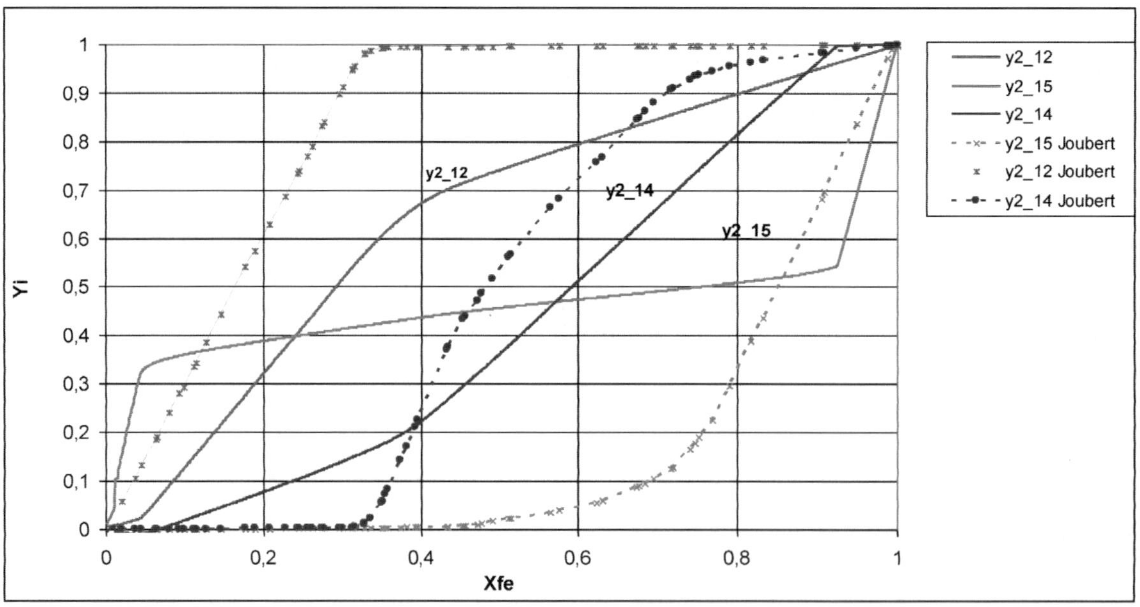

Figure 2. Comparison of the distribution of atoms over the sub-lattices obtained in this study with data Joubert [6] at 500 K.

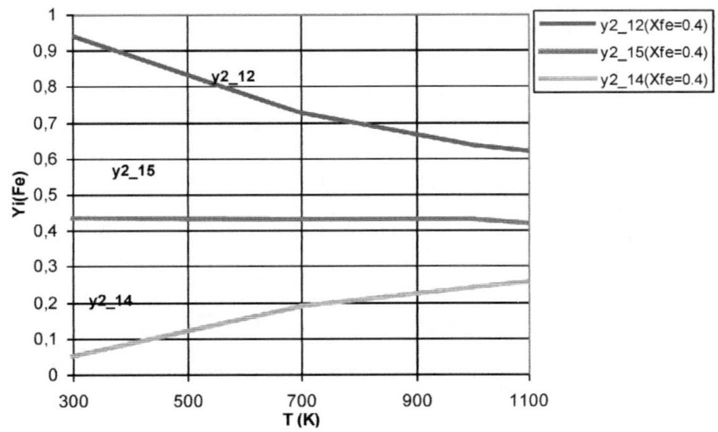

Figure 3. Temperature dependence of filling atoms on the three sub-lattices model for the alloy composition 40 at. % Fe.

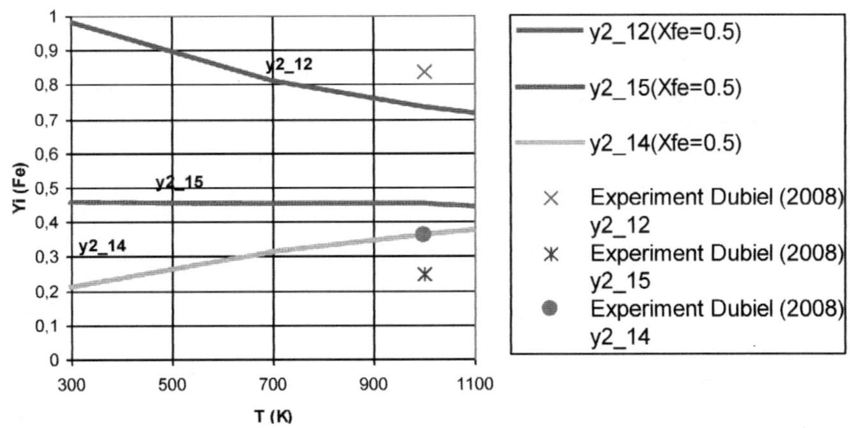

Figure 4. Temperature dependence of filling atoms on the three sub-lattices model for the alloy composition of 50 at.% Fe, the experiment at 1000 K [9-11].

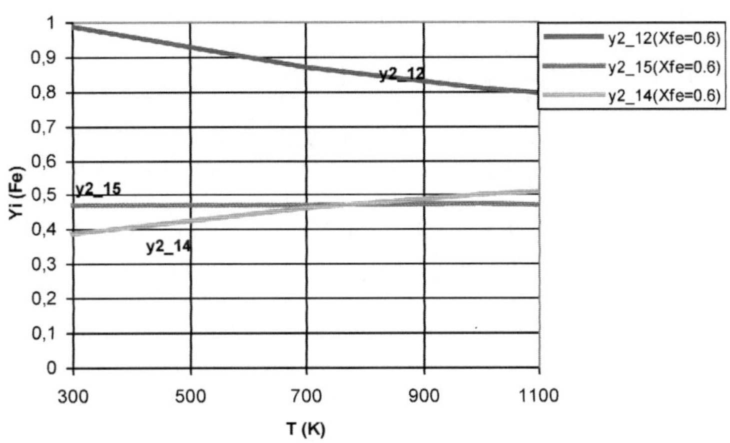

Figure 5. Temperature dependence of filling atoms on the three sub-lattices model for the alloy composition 60 at. % Fe.

IOP Conf. Series: Materials Science and Engineering **130** (2016) 012064 doi:10.1088/1757-899X/130/1/012064

The results obtained allowed to calculate the equilibrium between the bcc - and σ– phases (figure 6).

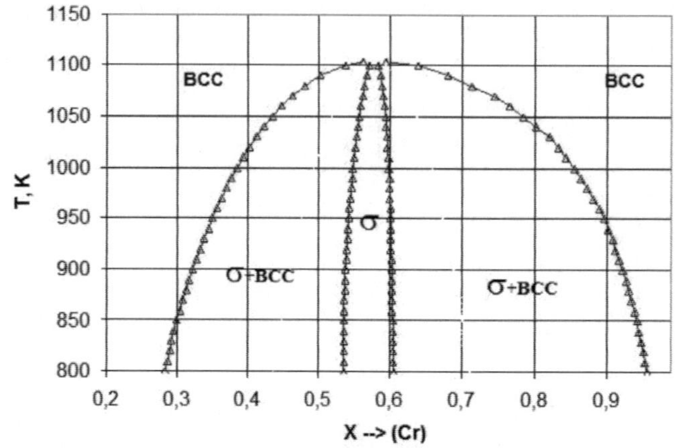

Figure 6. The fragment of BCC-Sigma equilibrium of Fe-Cr system.

6. Conclusion

For 3SLM were calculated concentration and temperature dependences of the structural (filling model sub-lattice atoms of components) and thermodynamic properties of the sigma phase and evaluated equilibria between the sigma - and bcc - phases in the system Fe-Cr.

Acknowledgments

The research was supported by grants for RFBR projects 09-03-00983-a, 13-03-00482-a and by Chemistry and Materials Science Branch of the Presidium of Russian Academy of Sciences, project OKhNM-02.

Reference

[1] Havrankova J, Vrest'al J, Wang L J, Sob M 2001 *Phys. Rev. B* **63** 174104
[2] Houserova J, Friak M, Sob M, Vrestal J B 2002 *Comp. Mater. Science* **562** 25
[3] Fries S G, Sundman B. 2002 *Phys. Rev. B* **66** 012203
[4] Kabliman E A, Mirzoev A A, Udovskii A L 2009 *The Physics of Metals and Metallography* **108** (5) 435–440
[5] Kabliman E, Blaha P, Schwarz P K, Peil O E, Ruban A V and Johansson B 2011 *Phys. Rev. B* **84** 184206
[6] Joubert J-M, Crivello J-C 2012 *Appl. Sci.* **2** 669-681
[7] Udovsky A, Kupavtsev M 2012 *J. Min. Metall. Sect. B-Metall.* **48** (3) 483
[8] Watson A, Hayes F H 2001 *Journal of Alloys and Compounds* **320** 199
[9] Cieslak J, Reissner M, Steiner W, Dubiel S M 2007 *Journal of Magnetism and Magnetic Materials* **613–615** 310
[10] Cieslak J, Tobola J, Dubiel S M and Sikora W 2010 *Phys. Rev. B 82* 224407
[11] Dubiel S M, Cieślak J 2011 *Critical Reviews in Solid State and Material Sciences* **36** (4) 191

IOP Conf. Series: Materials Science and Engineering **130** (2016) 012065 doi:10.1088/1757-899X/130/1/012065

Verification of ab-initio mixing enthalpy using thermodynamic simulation of phase equilibrium and the temperature dependences of the heat capacity of the bcc Fe-Cr alloys

A L Udovsky[1,2] **and D A Vasilyev**[1]

[1]Baikov Institute of Metallurgy and Material Sciences of RAS, Leninskiy pr. 49, Moscow, 119991, Russia
[2]National Research Nuclear University MEPhI (Moscow Engineering Physics Institute), Moscow, Kashirskoe highway 31, Moscow, 115409, Russia

e-mail: udovsky@imet.ac.ru; vasilyev-d@yandex.ru

Abstract. The paper deals with application of physical-empirical models for the thermodynamic description of the bcc Fe-Cr alloys and phase equilibrium, as well as prediction of behavior of the temperature dependences of the specific heat of alloys. This approach allowed performing verification of ab-initio calculations results obtained by different authors for the mixing enthalpy at 0K which were used to assess the chemical part of the mixing enthalpy. Analysis of calculated phase diagram fragments and the temperature dependences of heat capacities for two alloy compositions and their comparison with experimental data, has allowed us to estimate the degree of reliability of various approximations used in ab-initio calculations, and thereby realize their verification for further practical use.

1. Introduction

Ferritic steels are considered as promising for use as reactor fuel element cladding of the new generation IV, as well as the first wall of future fusion reactors. This is due to their low creep and good corrosion resistance at elevated temperatures [1]. Long duration of the material exploitation under irradiation and elevated temperatures is responsible for kinetic processes, due to diffusion, which lead to the changes of its structural and physical properties that in turn can lead to cracking and fracture of the material. For understanding the diffusion processes taking place in those materials is expedient to conduct the computer simulations. In its turn the suitable interatomic potentials (IP) which are used, for example, in the embedded atom model (EAM) are needed to simulate the kinetic processes. IP parameters are frequently calculated by fitting to the ab-initio calculations. The Fe-Cr system is a base for development of ferritic steels, therefore great attempts of obtaining reliable IP for ferromagnetic bcc Fe-Cr alloys are carried out for a long time. In recent 10 years ab-initio calculations of the mixing enthalpy (ΔH) of ferromagnetic bcc alloys at 0K were carried out in many works, see for example [3-5] and Figures 1 – 3. In particular, in the work [6] "a fitting" to values of ab-initio calculations of ΔH calculated in [4] was made for obtaining the IP.

Calculation of ΔH differs between ΔH obtained in [4, 6] and ΔH obtained in [3]. In this work, a verification of ab-initio calculations of ΔH conducted in [3-5] is carried out by the thermodynamical simulation.

2. Description of the model

According to [7], the Gibbs energy of mixing of bcc Fe-Cr alloys depending on composition and temperature without taking into account a short range order was formulated as a partition function of non-interacting subsystems:

$$^{\alpha\alpha}\Delta G^{\alpha}(x,T) = \Delta G^{\alpha}_{chem}(x,0) + \sum_{j}^{4} {}^{\alpha\alpha}\Delta G^{\alpha}_{j}(x,T) + RT\sum_{i}^{2} x_i \ln(x_i) \ , \qquad (1)$$

where the molar mixing energy of bcc alloys is described with respect to the bcc phase of the pure components by the formula:

$$^{\alpha\alpha}\Delta G^{\alpha}_{j}(x,T) = G^{\alpha}_{j}(x,T) - \left((1-x)G^{\alpha}_{j,Fe}(T) + x\cdot G^{\alpha}_{j,Cr}(T)\right). \qquad (2)$$

In the formulas (1) and (2) x is the mole fraction of Cr, T – temperature, R – the gas constant, j – index runs on all those non-interacting statistical subsystems: 1 - elastic, 2 - vibration, 3 - electronic and 4 - magnetic parts of the Gibbs energy of mixing.

$\Delta G^{\alpha}_{chem}(x,0)$ – the chemical part of the mixing energy at 0K was described as a difference between the free mixing energy of alloys calculated by ab-initio calculation at 0K and the energy sum caused by the static displacements of atoms from their ideal lattice nodes formulated in a frame of the elastic theory, and the magnetic part of the free mixing energy, see the following formula.

$$\Delta G^{\alpha}_{chem}(x,0) = \Delta H(x,0) - \Delta G_{elast}(x,0) - \Delta G_{mag}(x,0) \ . \qquad (3)$$

Details of used physical - empirical models are described in [7].

3. Calculated results

The concentration dependencies of physical properties of bcc Fe-Cr alloys which were approximated according to experimental data of the alloys taken from literature, namely, heat capacity electronic coefficients, thermally expansion coefficients and the mean magnetic moments, Debye and Curie temperatures, were used for calculating the energy parts (2).

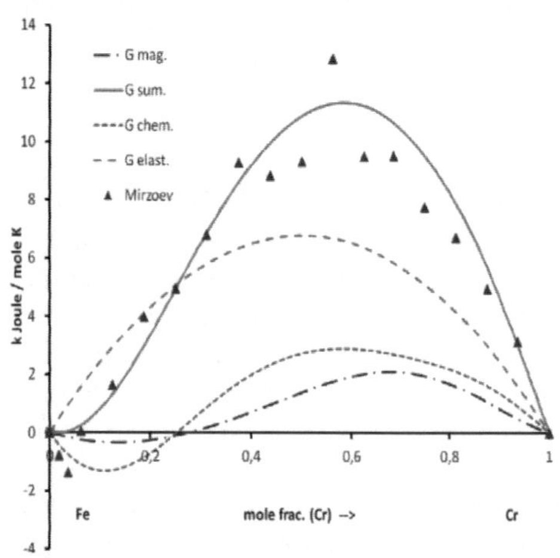

Figure 1. The calculated concentration dependences: ▲ - the mixing enthalpy ΔH of ferromagnetic bcc Fe-Cr alloys at 0K, according to [3]; G_{elast} - the energy of elastic distortions of the crystal lattice; G_{mag} , G_{chem} – the magnetic and the chemical energy parts.

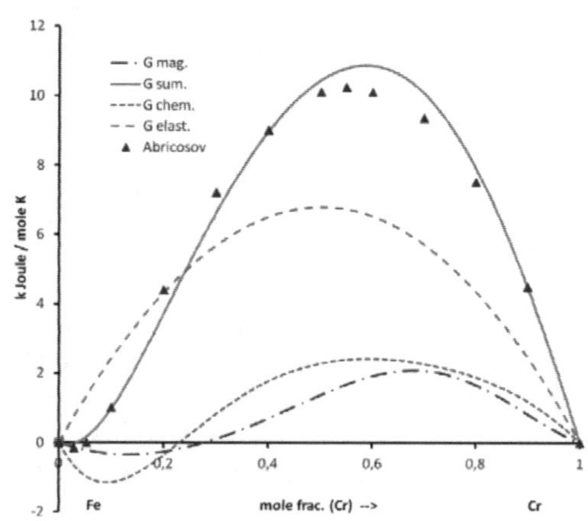

Figure 2. The calculated concentration dependences: ▲ - ΔH of ferromagnetic bcc Fe-Cr alloys at 0K, according to [4]; G_{elast} - the energy of elastic distortions of the crystal lattice; G_{mag}, G_{chem} – the magnetic and the chemical energy parts.

Calculated Gibbs mixing energy was used for calculation boundaries of bcc Fe-Cr alloys (a more detailed description is given in [7]) and the temperature dependence of the isobaric heat capacity. In contrast to the work [7], in this paper, various options to describe the enthalpy of mixing solutions, ΔH, obtained using ab-initio calculations by different authors: Mirzoev, Abrikosov and Korzhavyi [3-5] have been discussed.

The graphs of enthalpy of mixing alloys obtained by those authors and corresponding calculated mixing energy of chemical interaction, $\Delta G^{\alpha}_{chem}(x,0)$, are shown in figures 1-3. The bcc Fe-Cr alloy boundaries and spinodal line calculated in this work using different input data of enthalpy of mixing alloys ΔH at 0 K obtained in [3-5], are shown in Figure 4 in comparison with experimental data taken from [9-11].

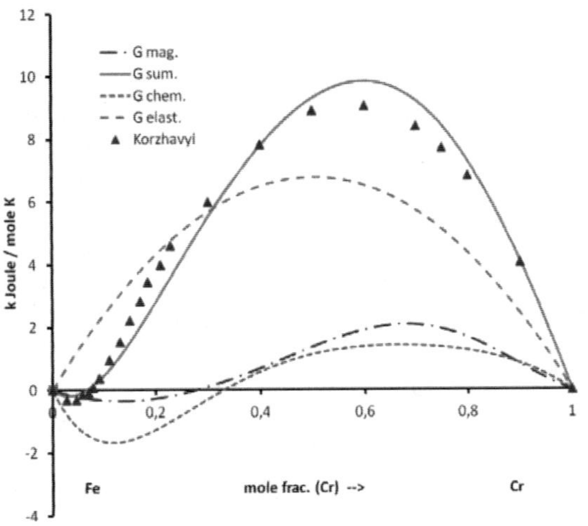

Figure 3. The calculated concentration dependences: ▲ - ΔH of ferromagnetic bcc Fe-Cr alloys at 0K, according to [5]; G_{elast} - the energy of elastic distortions of the crystal lattice; G_{mag}, G_{chem} – the magnetic and the chemical energy parts.

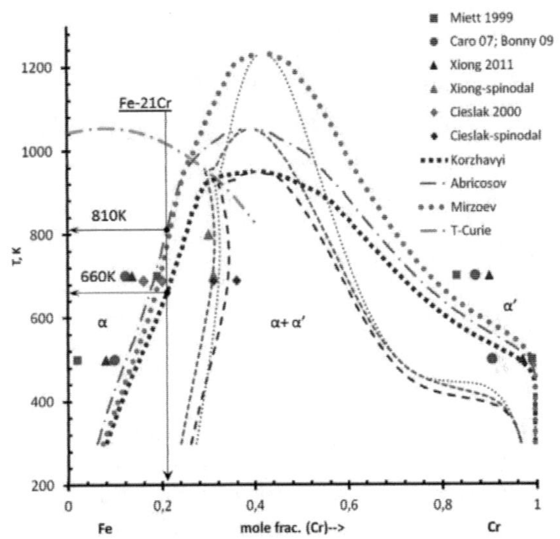

Figure 4. The calculated solubility and spinodal curve of bcc Fe-Cr alloys obtained by ab-initio calculations, shown in Figures 1-3, in comparison with the experimental data taken from [9-11]; the arrows show coordinates of configurational points of the Fe-21 at.% Cr alloy.

4. Discussion

The boundaries of bcc Fe-Cr alloys are shown in Figure 4 without the σ - phase which exist in the experimental phase diagram [8]. Namely, the upper part of the calculated phase boundaries, being above 748K is in a metastable state. If the $\Delta H(x,0\ K)$ dependence obtained by Mirzoev will be used for bcc Fe-Cr alloy boundary calculations then the critical point becomes above 1200 K and therefore the results of work [3] can not be considered as reliable evidence.

The critical point of the curve of bcc alloys boundary obtained by the data calculated in work [4] is about coincided with the critical point of the experimental diagram [8]. But because of the present the σ- phase in the equilibrium diagram such critical point location can not be trusted as well. Therefore, in our opinion, the data obtained in [5] should be preferred.

The calculated dependencies of isobaric heat capacity of Fe-6.7 at.% Cr and Fe-21 at.% Cr alloys are shown in Figure 5 in comparison with experimental data taken from [9]. The two-phase area (α+α') of the phase diagram is calculated by the formula (4) [12]:

$$C_p^{\alpha+\beta}(x,T) = \alpha(x,T)\left(C_p^{\alpha}(x^{\alpha}) + T\left(\frac{\partial x^{\alpha}}{\partial T}\right)^2\left(\frac{\partial^2 G^{\alpha}}{\partial x^2}\right)_{x^{\alpha}}\right) + \beta(x,T)\left(C_p^{\beta}(x^{\beta}) + T\left(\frac{\partial x^{\beta}}{\partial T}\right)^2\left(\frac{\partial^2 G^{\beta}}{\partial x^2}\right)_{x^{\beta}}\right), \quad (4)$$

where β - the mole fraction of bcc - phase of the Cr-rich alloys (or α'- phase).

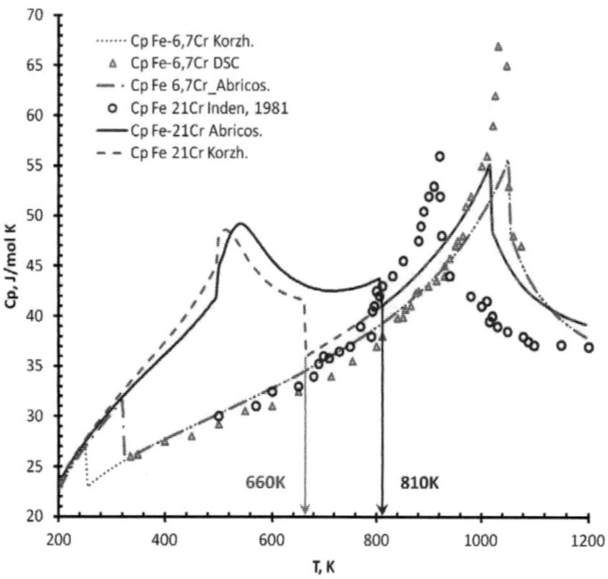

Figure 5. The temperature dependences of isobaric heat capacities for Fe-6.7 at.% Cr and Fe-21 at.% Cr alloys calculated using the results of ab-initio calculations of the mixing enthalpy at 0K [4, 5] in comparison with the experimental data (Δ -Xiong, ○ -Inden) taken from [9]. The temperatures of the heat capacity jumps of Fe-21 at.% Cr alloy are shown by arrows.

One can learn from those graphs that when "moving" down by temperature from the high temperature area to the lower area the first heat capacity jumps could be seen when the alloys transit from paramagnetic to ferromagnetic state. The difference in heat capacity jumps between experimental and calculated data of the Fe-21at.% Cr alloy is due to insufficiently accurate approximation of Curie temperature, it can be expected that a more accurate description would eliminate this misleading.

By the further lowering temperature, the second jumps could be seen (at 660K or 810K- figure 5 using data of work [5] or [4] accordingly) when the configuration point for composition of the Fe-21 at.% Cr alloy transits from one phase area (α) to the two phase area (α+α') of the phase diagram (shown by arrows in figure 4).

This obtained result can be used for verification of models employed in ab-initio calculations of the mixing enthalpies at 0K. For example, if one will take the Fe-6.7 at.% Cr or Fe-21 at.% Cr alloy and carry out the heat capacity measurements at 280K or 780K correspondingly then according to the obtained results in the present study the heat capacity jumps don't be expected at these temperatures.

5. Conclusions

The calculated boundary of bcc Fe-Cr alloys and temperature dependencies of heat capacity of two alloys were accomplished by using ab-initio calculations of the mixing enthalpy [3-5] and presented in Figures 4 and 5. The calculated results of alloy heat capacities can be recommended the experimenters for accurate definition of the phase boundaries of bcc Fe-Cr alloys, in particular for the Fe- rich alloys. The obtained result allows to conduct verification of those ab-initio data by caring out the experimental researches and thus to help researchers involved in the kinetic simulations in the choice of the inter-atomic potentials.

Acknowledgments

The research was supported by grants for RFBR projects 09-03-00983-a, 13-03-00482-a and by Chemistry and Materials Science Branch of the Presidium of Russian Academy of Sciences, project OKhNM-02.

References

[1] Bonny G, Pasianot R C, Malebra L, Caro A, Olsson P and Lavrentiev MYu 2009 *J. Nuc. Mater.* **385**(2) 268

[2] Daw M and Baskes M 1984 *Phys. Rev. B* **29** 6443.

[3] Mirzoev A A, Yalalov M M and Mirzaev D A 2003 *Phys. Met. Metallogr.* **97** (2) 310

[4] Olsson P, Abrikosov I A, Vitos L et all. *J.Nucl. Mater*. 2003 **321** 84

[5] Korzhavyi P A et all. *Phys.Rev. B* 2009 **79** 054202

[6] Wallenius J et all. *Phys. Rev. B* 2004 **69** 094103

[7] Udovskii A L and Vasil'ev D A 2015 *Russian Metallurgy (Metally)* **3** 237

[8] Massalski T, Okamoto H and Subramanian P R 1990 Binary Alloy Phase Diagrams, 2nd edition (3 volumes) (SAM International (OH))

[9] Xiong W et al. 2011 *CALPHAD* **35** 355

[10] Cieslak J, Dobiel S M and Sepoil B 2000 *J. Phys.: Condens. Matter* **12** 6709-17

[11] Miettinen J 1999 *CALPHAD* **23** (2) 231

[12] Udovsky A L 2011 *Russian Metallurgy (Metally)* **9** 920